AF684828

Batteries for Electric Vehicles

ELECTRONIC & ELECTRICAL ENGINEERING RESEARCH STUDIES

POWER SOURCES TECHNOLOGY SERIES
Series Editor: **N E Bagshaw, MA CEng FIM FBIM**

2. Batteries for Automotive Use
   P Reasbeck *and* **J G Smith**

3. Maintenance-Free Batteries
   Lead-Acid, Nickel/Cadmium, Nickel/Metal Hydride
   A HANDBOOK OF BATTERY TECHNOLOGY, SECOND EDITION
   D Berndt

4. Batteries for Electric Vehicles
   D A J Rand, R Woods, R M Dell

Batteries for Electric Vehicles

D A J Rand
Manager, Novel Battery Technologies,
CSIRO Division of Minerals, Port Melbourne, Australia

R Woods
Adjunct Professor of Metallurgical Engineering,
University of Utah, Salt Lake City, USA

R M Dell
Formerly Head of Applied Electrochemistry
at the Atomic Energy Research Establishment, Harwell, England

RESEARCH STUDIES PRESS LTD.
Taunton, Somerset, England

JOHN WILEY & SONS INC.
New York · Chichester · Toronto · Brisbane · Singapore

RESEARCH STUDIES PRESS LTD.
24 Belvedere Road, Taunton, Somerset, England TA1 1HD

Copyright © 1998, by Research Studies Press Ltd.

All rights reserved.
No part of this book may be reproduced by any means, nor transmitted, nor translated into a machine language without the written permission of the publisher.

Marketing and Distribution:

Australia and New Zealand:
Jacaranda Wiley Ltd.
Sydney Office, Suite 4A, 113 Wicks Road, North Ryde, NSW 2113, Australia

Canada:
JOHN WILEY & SONS CANADA LIMITED
22 Worcester Road, Rexdale, Ontario, Canada

Europe, Africa, Middle East and Japan:
JOHN WILEY & SONS LIMITED
Baffins Lane, Chichester, West Sussex, UK, PO19 1UD

North and South America:
JOHN WILEY & SONS INC.
605 Third Avenue, New York, NY 10158, USA

South East Asia:
JOHN WILEY & SONS (ASIA) PTE. LIMITED
2 Clementi Loop #02-01
Jin Xing Distripark, Singapore 129809

Library of Congress Cataloging-in-Publication Data

Rand, D. A. J. (David Anthony James), 1942-
Batteries for electric vehicles / D.A.J. Rand, R. Woods, R.M. Dell.
p. cm. -- (Electronic & electrical engineering research studies. Power sources technology series ; 4)
Includes bibliographical references and index.
ISBN 0-86380-205-2 (Research Studies Press). -- ISBN 0-471-96799-8 (Wiley)
1. Electric vehicles--Batteries. I. Woods, R. II. Dell, Ronald. III. Title. IV. Series.
TL220.R34 1997
629.25'02--dc21 97-33106
CIP

British Library Cataloguing in Publication Data
A catalogue record for this book is available from the British Library.

ISBN 0 86380 205 2 (Research Studies Press Ltd.) *[Identifies the book for orders except in America.]*
ISBN 0 471 96799 8 (John Wiley & Sons Inc.) *[Identifies the book for orders in USA.]*

Printed in Great Britain by SRP Ltd., Exeter

Editorial Foreword

At the beginning of the twentieth century, electric vehicles powered by batteries were in regular use on the roads of many countries in the Western World. The major advances in the internal-combustion engine in the early 1900s, that resulted in improved performance and greater reliability, caused an increase in popularity of vehicles using such engines. Consequently, by the middle of the century, only a relatively few vehicles using batteries for propulsion, such as milk-delivery floats, were left on the roads.

The demise of the road electric vehicle was largely due to the limitations of its power source. The batteries then available were capable of delivering only low levels of energy per unit weight and, accordingly, the range of the vehicle was severely restricted; the recharge time was several hours before the vehicle was again ready for use; and the capital cost of the battery was comparatively high. The advantages of the internal-combustion-engined vehicle in terms of long driving range, short refuelling time and lower cost, together with the greater flexibility in operation, left battery-powered vehicles (literally) standing.

Although interest in battery propulsion was renewed for short periods during the various oil crises, it has been environmental concerns — in particular, the pressing need to alleviate air pollution in city areas — that have provided the impetus to a major revival. This started with the initiative of the California Air Resources Board in 1989 by which rules were laid down for future vehicle sales in California, with an increasing proportion of zero-emission types in the vehicles sold per year into the next century. The mandate resulted in the formation of the US Advanced Battery Consortium which is providing funds for the development of new power sources, the formation of the Advanced Lead-Acid Battery Consortium which is supporting research into advanced designs of lead/acid battery systems, and the onset of other moves in various parts of the world to progress both batteries and fuel cells.

It is now eight years since the Californian initiative. A great deal of research and development has already been carried out, and is still under way, on all aspects of electric vehicles — but especially on improving the power source. It is an opportune time to review batteries for electric vehicles and this book gives the up-to-date status of the different technologies.

The authors come from opposite ends of the world — two from Australia, and one from the UK. They all have well-established, international reputations. Dr David Rand, the lead author, is Manager of Novel Battery Technologies at CSIRO in Melbourne, Australia, and is particularly well known for his contributions to the advances in lead/acid battery technology. He is often invited to present his latest findings to electric vehicle and battery conferences in the Pacific Rim, Europe and America. Dr Ron Woods, who also carried out his research into power sources at CSIRO, is currently Adjunct Professor of Metallurgical Engineering at the University of Utah in the USA. He too has presented results at symposia and colloquia around the world, and particularly in America. Dr Ron Dell was formerly Head of Applied Electrochemistry at the Atomic Energy Research Establishment in Harwell, UK. He is a recognized authority on the newer battery systems which are being developed, as is demonstrated by his many contributions in the literature. By combining their knowledge and experience, these three authors have produced a well-balanced and authoritative book on batteries for electric vehicles. It is a very welcome addition to the *Power Sources Technology Series*.

N.E. Bagshaw
Editor
January 1998

Preface

Battery-powered electric vehicles (EVs) first became commercially available in the 1890s and their popularity grew rapidly throughout the first decade of the twentieth century. At that time, approximately equal numbers of vehicles propelled by steam engines, petrol engines and batteries were operating on the roads of major cities such as London, Paris and New York. The early popularity of the EV was based upon its cleanliness, simplicity of operation and reliability — attributes which were not offered by the contemporaneous internal-combustion-engined vehicle (ICEV). Gradually, as both the performance and the reliability of the ICEV improved, and the inherent limitations of the EV became obvious, the pendulum swung in the direction of the ICEV and EVs became almost entirely restricted to off-road duties. Ironically, it was the invention of the battery-operated self-starter in 1911 that ultimately sounded the death knell of the early battery-propelled road vehicles.

Throughout the subsequent years, EVs have continued to find niche markets, particularly in duties where the pollution and/or noise of ICEVs is unacceptable. The favoured situations include many enclosed locations (industrial warehouses, hospitals, airports, underground mines), and environmentally sensitive outdoor places (city centres, parks, golf courses). Such applications have been sufficient to sustain a viable EV industry which manufactures fork-lift trucks, tractors, tugs, invalid carriages, and similar 'vehicles'. The provision of traction batteries (predominantly lead/acid types) for these EVs is also an established industry.

A renaissance of interest in road EVs began in the 1970s as a result of the world 'oil crises' of 1973 and 1979. People began to recognize that supplies of readily extractable petroleum are finite and that, with a growing world population and with increasing demands for personal mobility, an ever-expanding world fleet of ICEVs is not feasible in the long-term. This view was reinforced in the 1980s by a deepening world-wide concern over atmospheric pollution in many large cities. Legislation was passed to improve urban air-quality by the fitting of catalytic converters, but realization gradually dawned that, ultimately, it would be necessary to restrict the number of ICEVs in cities, either by legislation or by price barriers. The EV is, first and foremost, a non-polluting vehicle and it is in this context that a new era is emerging for the urban electric car, delivery van and transit bus — with major automotive manufacturers throughout Europe, Japan and the USA finally taking the EV option seriously and investing heavily in its development. The EV programmes are progressing hybrid drive-trains in addition to pure-battery power.

It has long been understood that a limitation to the acceptance of EVs lies in the moderate performance of the traditional lead/acid battery; in particular, the range limitation which this system imposes, the lengthy time taken for recharge, and the modest service life. Many new battery chemistries with potentially superior properties have been identified in the laboratory, as well as various types of fuel cell. All of these might be developed as power sources for EVs. Throughout the world, major industrial teams of scientists and engineers have been established to seek better batteries, often with financial support and encouragement from governments, who have recognized the industrial potential inherent in EVs and batteries. Only gradually, however, has the intrinsic difficulty of developing advanced batteries for motive power become appreciated, with the associated high costs and long timescales to practical reality. For example, some individual battery development projects have lasted for more than twenty years and have incurred expenditures of US$ 10 M – 100 M without a commercial product yet to command a market. Nevertheless, the search continues unabated. The lead/acid battery industry, too, has recognized the opening up of a new opportunity, as well as the threat which is posed by other battery chemistries, and has responded with a vigorous collaborative research programme to adapt and improve the performance of its present technology. Given these initiatives, many different systems, both old and new, are in contention as the traction battery of the future.

This book provides an overview of the progress which has been made to date in the development of rechargeable traction batteries. In addition, a brief résumé of fuel cells is presented in the concluding discussion on the future for pure-battery and hybrid EVs. The story begins with an account of different types of vehicles which have been built over the years, with photographs of some of the more interesting or important. In the second Chapter, the fundamentals of storing energy in batteries are described. Chapter 3 then sets out the performance criteria which are sought in an EV traction battery; these include the targets prescribed by the United States Advanced Battery Consortium. From a consideration of the criteria, it begins to become apparent why the development of a high-performance traction battery is a major task. This leads to a discussion, in Chapter 4, of the laboratory testing of EV batteries and the simulated urban driving cycles which have been proposed for realistic evaluation under conditions which are typically encountered in city driving.

The 'core' of the book resides in Chapters 5–9. These provide an account of the advances which have been made in lead/acid batteries, alkaline batteries, flow batteries, sodium batteries, and lithium batteries. Scientific and technical aspects of each battery type are described in detail, together with a summary of the present performance of the battery and a review of its strengths and weaknesses in comparison with the other competing systems. Mention is made also of the industrial involvement in each battery, and in cases where the technology has reached an advanced state, a brief account is given of the development history.

In the final Chapter, a broader look is taken at the factors which are likely to influence the rate and extent to which EVs are taken up by society at large. First, the various batteries are compared and their principal attractions and limitations are listed — it is still very much an open contest between the candidates. The final choice of battery for mass manufacture will depend not only on technical performance, but also on commercial and economic considerations, public acceptability, safety, and suitability for recycling. Fuel cells and hybrid drive systems are summarized, before leading into a discussion of the many infrastructure considerations which are relevant to the development of a mass market for EVs, especially the availability of battery-charging facilities. The likely views of public authorities (governments and local administrations) on the desirability of EVs are contrasted to the free-market views of vehicle manufacturers, as well as those of consumers who may prefer, individually, the freedom afforded by the traditional ICEV. Ways in which divergent expectations might be reconciled, through a combination of taxes, subsidies and national or local legislation, are discussed briefly. It is concluded that the day of the EV is dawning, but much remains to be done before a pure-battery vehicle, or perhaps a battery-hybrid vehicle, becomes the major mode of personal transportation. Finally, in a glimpse over the horizon, consideration is given to those factors which would facilitate the mass introduction of EVs.

The authors wish to thank all those holders of copyright who have kindly consented to the use of their illustrations. While every effort has been made, it has not always been easy to identify the correct copyright holders, especially where an illustration has appeared in more than one publication or where the holder is no longer in business. The authors apologize should any omissions have inadvertently occurred.

The authors acknowledge all the many fruitful discussions they have had with friends and colleagues, as well as the considerable satisfaction which they have derived from working in this stimulating field. They wish also to express their appreciation of the dedication and expert skills of Ms Rita Spiteri (CSIRO) in producing the complete text for publication. In addition, they are grateful to Ms Lee McRae (CSIRO) for draughting many of the illustrations. Final thanks must go to the authors' respective wives who have provided sustained encouragement during the researching and writing of this book.

D.A.J. Rand
R. Woods
R.M. Dell

January 1998

Contents

Chapter 10

CHAPTER 1

Battery-Powered Electric Vehicles

1.1. Historical Perspective

Ever since 1837, when electric traction was first demonstrated by Robert Davidson in Aberdeen, Scotland [1], it has been the ambition of enthusiasts, entrepreneurs, engineers and scientists, as well as government agencies and industrial organizations, to develop an effective electric road vehicle. Initially, interest in the battery-powered electric vehicle (henceforth, designated: EV) stemmed from a natural human desire to outpace and outdistance the horse-drawn carriage. In more recent times, well after the internal-combustion-engined vehicle (henceforth, ICEV) had established its supremacy in road transportation, concern over both increasing atmospheric pollution and diminishing petroleum supplies has led to renewed activity in EV development.

The batteries which were available for the storage of electricity at the time of Davidson's pioneering EV trials were low in both sustained power and delivered energy. The practicality of electric transportation awaited the discovery of batteries which gave higher electrical output and could be recharged effectively. The first of these — the lead/acid system — was demonstrated by Gaston Planté to the French Academy of Sciences on 26 March 1860 [2]. A further twenty or so years were to pass, however, before steady improvements in both system design and manufacturing method enabled lead/acid batteries to be produced on a commercial scale. A short time later, the rechargeable alkaline systems of nickel/cadmium [3] and nickel/iron [4] also entered the market-place. At the end of the nineteenth century, with the mass-production of these rechargeable batteries, EVs became a common sight on the roads of industrialized nations [5–8]. By 1912, several hundred thousand electric cars and trucks were in service throughout the world as city cars, taxicabs, delivery vans, trams, and buses.

Nevertheless, the popularity of the personal EV proved to be short lived. A ten-fold increase in world oil-production between 1890 and 1920 [9], the invention of the ICEV self-starter by Charles F. Kettering in 1911 (ironically using a lead/acid battery), and the implementation in 1913 of the first assembly-line production of gasoline cars by Henry Ford with the Model-T vehicle, all enabled ICEVs to satisfy more completely the public's rapidly evolving demand for greater range, speed and mobility — all at cheaper cost. Despite the fact that, between 1910 and 1925, the storage capacity of batteries was increased by 35%, their life by 300%, and the driving range of electric trucks by 230%, while maintenance costs dropped by 63% [10], the performance of the EV power source still could not match that of heat engines using combustible fuels. The same is still true today. For example, one litre of a modern lead/acid battery weighs 2.4 kg, contains 0.07 kWh, and will propel a passenger car only a few hundred metres; by comparison, one litre of petrol weighs 0.85 kg, contains 11 kWh, and will drive the vehicle about 10 km. In other words, petrol has a specific energy (Wh kg^{-1}) which is far greater than that of a battery — in traction applications, it is 100 times lighter and takes up 40 times less space. Furthermore, petrol pumps refuel ICEVs at the rate of 45 l min^{-1} (500 kWh min^{-1} in energy terms), whereas domestic electricity sockets can only deliver energy at a rate of about 50 Wh min^{-1}, i.e., 10 000 times more slowly. Thus, replenishing the fuel tank of an ICEV is accomplished in a matter of a few minutes, but recharging the battery of an EV in the home takes hours. In summary, the two overriding factors which reduced the competitiveness of the first personal EVs (and, indeed, still restrict their usage today) were short daily range and long refuelling time.

By the end of the First World War, the emphasis in electric traction had shifted to the direct use of mains power in public transportation networks, i.e., in trains, trams, and trolley buses. In turn, the latter two forms of transport became largely superseded by the diesel-engined bus — not always because of inferior speed or acceleration, but because of limited manoeuvrability and the ever-increasing costs of both maintaining and extending the current-supply networks. One notable exception to the demise of the road EV has persisted in the UK where, to this day, a large fleet of battery-powered vehicles (over 24 000) has been used for the doorstep delivery of milk. The duties of the 'milk float' are ideally suited to the characteristics of EV operation (see Section 1.8).

Over the years, many off-the-road specialist applications have emerged for EVs in situations where short range is not a serious handicap and freedom from exhaust and/or noise pollution is of marked benefit. Battery-driven vehicles have found use in: (i) plant operations, e.g., fork-lift trucks for materials handling; (ii) track services, e.g., locomotives for minerals haulage; (iii) recreational pursuits, e.g., golf carts; (iv) transportation for disabled persons and medical supplies, e.g., wheelchairs; (v) airport ground-support operations, e.g., aircraft push-out tractors; (vi) personnel and baggage carrying services, e.g., in hotel complexes, railway stations.

But still there continued the quest for a ***road*** EV — one that would prove traffic-compatible with its ICEV counterpart. A revival of interest in the electric passenger car occurred both during the Second World War and during the immediate post-war years of petrol rationing in Europe and Japan. Vehicles were marketed widely and included the registration of more than 3000 EVs during 1943 in Japan alone. The EVs of the 1940s did not, however, have high performance: they had a range of 50–60 km and a maximum speed of 30–35 km h^{-1}. Since that time, improvements in electric motors, power-control systems, regenerative-braking procedures and battery-recharging equipment, together with advances in vehicle-construction techniques, have resulted in EVs becoming lighter, more efficient, and easier to use and maintain. At the same time, scientists and technologists have worked to reduce the limitations of range, acceleration, refuelling time and cost that are imposed on EVs by the battery.

Following the 'energy crisis' of 1973, and the recognition of the need to reduce the amount of petroleum consumed by the road transportation sector, three approaches were adopted to overcome the battery problem: (i) improvement of the performance of state-of-the-art lead/acid batteries; (ii) development of other batteries (the so-called 'advanced' or 'alternative' systems) with potentially superior energy-storage characteristics to the established lead/acid battery; (iii) investigation of hybrid vehicle designs as a possible means of circumventing the range limitations of EVs. All three approaches were studied in the USA following the enactment of Public Law 94-413, 'The Electric and Hybrid Vehicle Research, Development, and Demonstration Act of 1976'. About the same time, other notable research programmes were commenced in Australia, France, Germany, Italy, Japan, Sweden, Switzerland, the UK, and the USSR [11].

In addition to efforts to raise the performance of the lead/acid battery for electric traction, more than thirty alternative systems, using a wide range of chemistries, have received attention. The operating principles and status of these candidates is discussed in Chapters 5 to 9 of this book.

A number of different hybrid electric vehicle (HEV) concepts have been investigated. When one of the power sources is a battery, there is a choice between three basic configurations [12]: (i) an all-electric hybrid, i.e., mains–battery (catenary), high-energy battery–high-power battery, or fuel cell–high-power battery; (ii) a series heat-engine–battery hybrid, in which a small heat engine (e.g., an ICE) generates electricity which is fed into the battery–controller system so that the vehicle is driven exclusively by an electric motor; (iii) a parallel heat-engine–battery hybrid, in which either, or both, power sources can propel the vehicle. These hybrid concepts together with flywheel assistance, are shown schematically in Fig. 1.1. The use of a flywheel to store braking energy and to augment peak-power output of the battery introduces difficult mechanical engineering problems associated with continuously-variable transmissions (CVTs). A major advantage of heat-engine–battery hybrid technology is that the resulting vehicles would be compatible with existing ICEVs and their associated refuelling infrastructure, although battery charging operations do require additional infrastructure if full use is to be made of substituting liquid fuel energy with electricity. In the final analysis, the choice of hybrid system

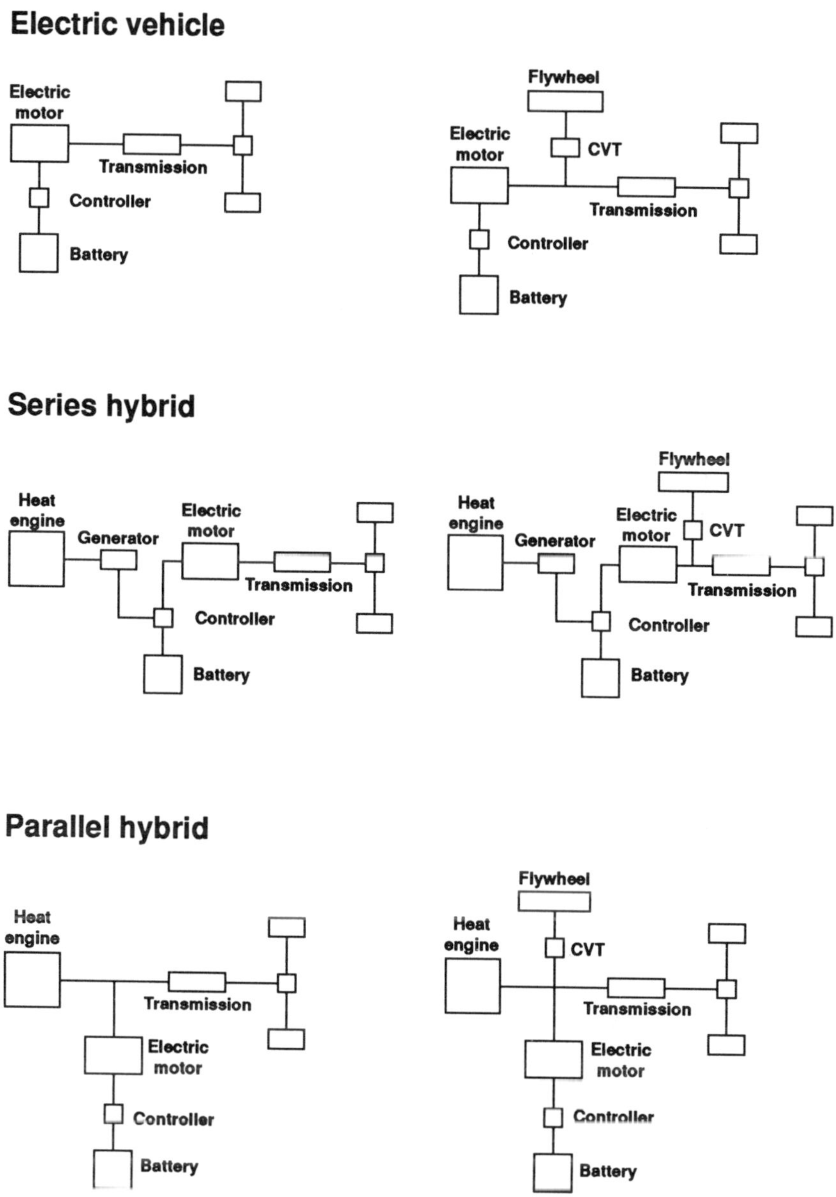

Fig. 1.1. Power-trains in electric and hybrid vehicles, with and without flywheel assistance.

depends upon the required duty of the vehicle, the degree of engineering complexity, and the cost. For further discussion of the prospects for HEVs, see Section 10.4, Chapter 10.

Despite such activity, road applications of EVs have been restricted to small fleets of experimental urban delivery and service vehicles (e.g., in Japan, the UK, and the USA), as well as to a few city-centre buses (e.g., in Germany, the UK, and the USA). Many electric cars have been designed and constructed in low numbers, but no production line has operated successfully for more than a short while. The major impediment to progress remains the restricted performance of the battery.

In summary, it is generally considered that further penetration of EVs into the various transportation sectors will depend upon factors such as: (i) improvements in EV technology, especially better battery performance; (ii) competing developments and advances in ICEV technology; (iii) relative economics of vehicle purchase, operation and maintenance; (iv) infrastructure requirements of EV fleets, e.g., electricity-supply networks, battery-charging facilities; (v) the availability and price of petrol and substitute liquid or gaseous fuels; (vi) social, legislative and political dictates. With regard to the last factor, EVs have received special attention since the mid-1980s as a possible means for reducing the urban pollution which is caused by the harmful emissions of carbon monoxide, nitrogen oxides and hydrocarbons from automobiles [13–16]. An example of such concern was the decision, made in 1988, of the City Council of Los Angeles, USA to sponsor an international competition for the design, construction and marketing of 10 000 EVs — 5000 vans and 5000 two-seat passenger vehicles [17,18]. Following on from the 'Los Angeles Initiative', the California Air Resources Board (CARB) introduced, on 13 December 1989, new regulations for vehicle emissions that were to take effect in the late 1990s. These regulations involved the introduction of new categories of vehicles that included transitional low-emission vehicles (TLEVs), low-emission vehicles (LEVs), ultra-low-emission vehicles (ULEVs), and zero-emission vehicles (ZEVs). The last mentioned are to produce no 'tailpipe' emissions at all. The legislation mandated that 2% of all new vehicles offered for sale in California by the major automobile manufacturers should be ZEVs in model years 1998–2000, 5% in 2001–2002, and 10% in 2003 and beyond. The ZEVs must not emit directly any of the regulated pollutants, i.e., carbon monoxide, nitrogen oxides and non-methane organic gases. In effect, ZEVs have to be powered by batteries and/or hydrogen-based fuel cells. Following the initial CARB mandate for ZEVs, the States of both New York and Massachusetts enacted similar legislation. Under pressure from the oil and automotive industries, CARB modified the California ZEV legislation in March 1996. This eased the short-term mandates, but maintained the target for year 2003. Between 1998 and 2000, the seven major automakers (viz., Chrysler, Ford, General Motors, Honda, Toyota, Mazda and Nissan) agreed to supply 3750 advanced-technology ZEVs with ranges of more than 160 km per charge. Each manufacturer's pro rata share of the ZEVs is based on the manufacturer's share of the California market for passenger cars and light-duty trucks up to 1.7-t curb weight. The new agreements include provisions for the seven companies to collaborate in a development partnership with CARB, so that the commercialization of advanced batteries and further work on ZEV technology are both accelerated. As many as 800 000 EVs are expected to be on the roads in California by the year 2010. Whereas Massachusetts subsequently followed California's decision to back away from the initial 1998 requirement for ZEVs, the US District Court ruled on 5 August 1997 that New York State could retain the original mandate.

1.2. In-Plant Vehicles

Electric vehicles have a long history of service in the handling of materials and products. First introduced for horizontal transport around 1910, and then as the fork-lift truck in 1922, such vehicles rapidly found increasing favour for the lifting, moving and stacking of goods in industrial operations. In-plant EVs operate almost exclusively on lead/acid batteries and account for the great majority of the sales of the traction version of this battery.

Today, six main categories of battery-powered industrial truck are used in materials handling:

- **Straddle/stacker:** The centre of gravity of the load falls within the truck's wheelbase. Outrigger arms on the front of the truck run on small wheels to 'straddle' the load or pass under the racking. The truck can be either rider- or pedestrian-controlled.

- **Platform:** The fixed type of truck is normally rider-controlled. The elevated version runs under stillages and raises the load for transport; it can be either rider- or pedestrian-controlled. Loads of up to 20 t can be carried. Most models are adapted for special customer requirements.

- **Pallet:** Long flat arms run through the pallet on small front wheels. Although usually pedestrian-controlled, it is becoming increasingly designed for rider operation. Models are offered with load capacities of up to 3 t.

- **Fork-lift:** In the counterbalanced type, the load (typically up to 5 t) is carried forward of the front wheels and is balanced by the truck and battery weight. In the reach type, the fork carriage and/or mast extends beyond the front wheels to pick up the load and then retracts within the wheelbase for travel. This design results in a compact and highly manoeuvrable vehicle, well able to do high-lift work in narrow aisles. The side loader is a variety of reach truck in which the forks are mounted at right-angles to the

Counterbalanced

Reach

direction of travel and thereby allow handling of long loads in confined spaces. All three vehicles are usually rider-controlled.

Side loader

- **Order picker:**

The operator controls the lift and travel from a cage that can be elevated. This allows fast, safe, efficient, mixed-order selection along narrow passages in high-density warehouses.

- **Turret truck:**

This is a compact, versatile truck with 180°-rotating, side-shift forks for high stacking of loads on either side of a narrow aisle. The width of the latter is dictated by the size of the load. Lift heights of up to 13 m are possible. The vehicle is always rider-controlled and is available with load capacities of up to 2 t.

The size of the lead/acid batteries used in each of these categories of truck naturally depends on the size of the load to be shifted or lifted. Battery packs range from 12 to 72 V with capacities (C_5/5 rate) up to 680 Ah for a 48-V system, and up to 1700 Ah for a 72-V system. The trucks are usually fitted with a control facility which ensures that no more than 80% of the battery's capacity is utilized during a work period. Such an arrangement extends battery life, which is expected to be between four and six years, i.e., 1200 to 1800 charge–discharge cycles.

Other in-plant and associated applications of battery-driven equipment include: personnel carriers; tugs and tractors for hauling a wide variety of loads; floor scrubbers and sweepers; vacuum cleaners; elevating platforms; lawn mowers; and snow blowers. Compared with ICEVs, overall owning and operating costs are claimed to favour electrics, despite the higher initial price and the purchase or rental of a battery charger. In an analysis conducted by the Electric Vehicle Association of Great Britain [19], it was estimated that an electric fork-lift truck can show up to 31% reduction in both short- and long-term operating costs through lower fuel bills, lower maintenance/labour/spares requirements, and longer vehicle service-life. Other attributes of electric traction include fume-free operation in confined areas, reduction in noise levels, and comparative simplicity in operation.

To increase profitability, manufacturers are continually seeking ways to improve the efficiency of their materials handling and equipment selection. For example, an early study showed [20] that in the manufacture of some products, direct labour was as low as 9 to 11% of the cost of sales, whilst material handling could be 30 to 40%. This realization, together with the appearance of the micro-processor and accompanying advances in information technology, has given rise to a marked interest in battery-powered, automatic-guided vehicles (AGVs). The AGV is seen as an essential link in many high-technology manufacturing processes and plant services (e.g., Fig. 1.2), and is becoming increasingly associated with operations which employ robots. The vehicle is driverless, follows some type of guidepath (typically, a wire embedded in concrete), and operates automatically when given a command. An on-board micro-computer handles all the AGV functions such as start–stop, speed, load–unload, etc. A decentralized system of micro-computers manages the traffic flow through communication with each AGV so that missions proceed smoothly, i.e., the traffic control stops collisions, forms queues when necessary, and guides vehicles to the best route in order to reach swiftly their respective destinations. A central computer provides the mission control, i.e., it instructs AGVs where to collect and deliver loads, selects AGVs for the required operational tasks, and directs empty AGVs to either parking or charging areas. AGVs typically operate on 24-V, 36-V or 48-V packs of lead/acid batteries of the industrial golf-cart (see Section 1.4) or automotive (i.e., ICEV starter) design, but the use of valve-regulated, 'maintenance-free', lead/acid technology (see Section 5.2.7, Chapter 5) is rapidly growing.

Fig. 1.2. Battery-powered, automatic-guided vehicle. (By courtesy of Eaton-Kenway, USA.)

1.3. Locomotives and Trains

For a great many years, battery-powered locomotives (Fig. 1.3) have been widely applied in mining operations throughout the world to haul personnel, minerals, and other materials. For example, a fleet of about 60 locomotives are used underground to transport loads of up to 300 t of ore in the copper and the lead/zinc operations of Mount Isa Mines Limited in Queensland, Australia; whilst 349 locomotives have been employed in UK coal mines [21]. Typically, the units have 84-V, 112-V or 350 V packs of lead/acid traction batteries that are assembled from individual cells of 704-Ah, 528-Ah and 340-Ah capacity (C_5/5 rate), respectively. Such locomotives are considered to provide a clean, quiet and economic means of transportation. Because of the potentially hazardous conditions of mining, strict legislation (usually specific to each country) has been implemented to ensure that the battery and all auxiliary electrical equipment (e.g., chargers) are explosion-proof and well ventilated, and that the external inter-cell connectors are electrically insulated and of sufficient strength to withstand the mechanical stresses which result from braking and/or sudden stops of the locomotive.

Fig. 1.3. Battery-powered locomotive used in mining operations.

Battery-powered vehicles which run on pneumatic tyres, rather than on tracks, are also used in some mining operations, e.g., coal haulage, in both the UK [22] and the USA (Fig. 1.4(a)). These vehicles have low profiles and, therefore, are similar in design to aircraft push-out tractors (see Section 1.6). Trolley–battery hybrid vehicles are also employed, for example, the Swedish ABB truck (Fig. 1.4(b)). For most of the time, this truck works from a 1000-V a.c. overhead pick-up wire, but at tunnel intersections in the mine and when dumping, it runs on power supplied by a pack of nickel/cadmium batteries (250 V; 70 Ah). Trolley–battery locomotives were also used during the construction of the Eurotunnel that now links France and the UK.

Battery locomotives are employed in the London Underground network to haul trains which carry maintenance workers along with servicing equipment and heavy track-refurbishing materials (e.g., replacement rails,

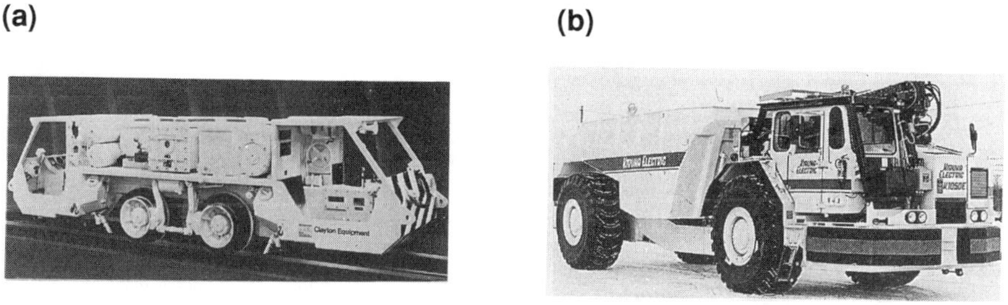

Fig. 1.4. (a) Low-height locomotive; (b) trolley–battery mining truck.

sleepers, concrete and ballast). A fleet of 43 such locomotives are brought into operation every night after passenger services have ceased and the electricity power-supply has been switched off. The traction batteries consist of a pack of 160 lead/acid cells of 1026-Ah capacity, and have a working life of up to eight years. The batteries are recharged when the locomotives return to the depot in the early morning, simply by connection to the conductor rails of the passenger-train track.

Battery-powered passenger trains, or more specifically railcars, have been in operation in Germany since the early 1900s (Fig. 1.5). In 1979, approximately 20% of the German network of railways (i.e., a total distance of 6000 km) was served by electric railcars [23]. Over 200 units were in use each day and covered 19×10^6 km in that year, for which the batteries supplied 42×10^6 kWh. The duties included express traffic, short-distance journeys, and feeder service to city railways. The railcar operations were supported by an infrastructure of 100 charging stations. The traction batteries were lead/acid types which were manufactured and maintained by VARTA AG. The railcar battery pack consisted of 220 cells which weighed 21 t (i.e., 20% of total wagon weight), stored 650 kWh, and provided a daily range of 250–400 km. On average, the service-life of the batteries was four years, or 4×10^5 km. In 1977, only one battery problem was experienced in 4.5×10^6 km of service. The total frictional losses (e.g., air-drag, rolling resistance) of battery trains are only one-third those of road vehicles. Hence, the limited specific energy of the lead/acid battery is a less critical parameter in rail, as opposed to road, transportation.

Fig. 1.5. Battery-powered railcar (Model 515) operated by the German railways. (Taken from [23] with permission of JEC Press, Inc.)

From the German experience, it is claimed [23] that the advantages of battery-powered train/railcars include: great reliability, long service-life, low noise-level, simple operation and maintenance, no pollution, and low-cost operation.

In the early 1970s, British Rail examined the viability of battery traction as an alternative to diesel on non-electrified railway lines. It was calculated [24] that over 50% of the then existing diesel-train services (i.e., ~50×10^6 km per year) could be covered with a battery energy which was equivalent to 425 kWh per railcar. Further, it was shown that in services which run for a substantial part of their journey over electrified track, a mains–battery hybrid railcar with a battery of 250 kWh would be adequate for about 85% of possible passenger journeys (i.e., ~5×10^6 km per year), and a mains–battery hybrid locomotive with a battery of 675 kWh would complete a distance of 4×10^6 km per year in freight-haulage operations. These calculations for hybrid units were based on the assumption that the batteries could be recharged at high rates whilst the train was in electrified sections of the track. Encouraged by the fact that pure-battery or mains–battery hybrid trains, compared with mains-only electric trains, do not require the installation of new and expensive lineside equipment, and, compared with diesel trains, have lower fuel costs (this was the time of the oil crisis), British Rail embarked upon an active research programme aimed at the development of the sodium/sulfur battery for railway applications. This battery, which has a higher specific energy than lead/acid, has been investigated by many groups around the world, but has yet to reach commercialization. For the present status of the sodium/sulfur system, the reader is referred to Chapter 8.

1.4. Recreational Vehicles

The major use of EVs in this category is in the conveyance of equipment and players on golf courses. These vehicles are intended to speed up the play, to protect the ground surface, and to assist those golfers who are physically disabled and/or elderly. On some courses, the use of EVs is mandatory — to the chagrin of those who look upon the sport more as a form of exercise! The vehicles (Fig. 1.6) are either pedestrian-controlled (i.e., golf carts) or rider-controlled (i.e., golf buggies). Specifically designed lead/acid batteries, now known generically as 'golf-cart batteries', have been developed for such applications and their cell construction is quite distinct from that of automotive or industrial-truck types. Golf-cart batteries are produced as either 6-V or 12-V units and have capacities typically ~200 Ah (C_{20}/20 rate).

Because of their fume-free operation, EVs are finding increasing application in other sporting arenas, e.g., to allow journalists, cameramen and officials to cover footraces (Fig. 1.6). Mention should also be made of the dedicated group of enthusiasts who regularly take part in EV endurance runs, or 'electrathons' as these events have now become known. Modern EV racing competitions began in 1979 in England and rapidly became popular with Electric Vehicle Associations in many countries. Unlike events at the turn of the century, where speed determined the winner, the aim has been to drive EVs with a given electrical energy storage (usually set in terms of the total weight allowed for the batteries) as far as possible in a set time. Nevertheless, the basic human desire of devising means to cover maximum distance in minimum time has not diminished. In 1983, two Australians, Hans Tholstrup and Larry Perkins, drove a solar-powered EV (Fig. 1.6) across their continent from Perth to

Golf cart

Golf buggy

Olympic Games EV, 1984

Solar EV, 1983

EV racing, 1890s

EV racing, 1990s

Sightseeing EV

Electric narrowboat

Fig. 1.6. EVs for recreational pursuits.

Sydney; they completed the 4084 km trip in 172 h. The solar collector consisted of 720 cells arranged in 20 modules and the electricity produced was stored in two 12-V automotive batteries used singly or in tandem. Feats such as this have given rise to solar-powered EV races ('Whispering Grand Prix'). The first such race was held in Switzerland in 1985 over a 380-km course from Lake Constance to Lake Geneva [25], and the most demanding to date has been the World Solar Challenge from Darwin to Adelaide (3010 km) in Australia. At the inaugural World Solar Challenge, in 1987, the winning car — the General Motors 'Sunraycer' — covered the distance in just over five days at an average speed of 67 km h^{-1}. The finishers in the first three places are shown in Fig. 1.7(a)–(c). In 1990, the Ingenieurschule Biel's 'Spirit of Biel/Bienne II' (Fig. 1.7(d)) took the chequered flag with an average speed of 65 km h^{-1} and, despite the much poorer overall weather conditions, the average speed of the second to ninth cars exceeded that of the runner-up in 1987. In 1993, all the five leading cars eclipsed the race record and the winner, the Honda 'Dream' (Fig. 1.7(e)), travelled at a remarkable average speed of 85 km h^{-1}. Honda also won the 1996 race, with a two-seater car (Fig. 1.7(f)), at a record-breaking average speed of 90 km h^{-1} [26] — strong testimony to the significant technological advances and improved strategy that have been achieved during the intervening years.

Despite the present enthusiasm for solar-powered transport, it should be pointed out that whilst such travel challenges the ingenuity of the engineer, solar flux is too dilute for on-board collectors to serve as the main energy source for EVs in general. Nevertheless, there are valuable spin-offs. The most notable of these was the introduction, in 1991, of the 'Impact' electric car by General Motors (see Fig. 1.22). The manufacturers asserted that the vehicle capitalized on engineering and design principles first put into practice with the Sunraycer. In particular, the Sunraycer project led to important new advances in electronics, motor design, structural materials and tyres, as well as to improved aerodynamics and reduced rolling resistance.

(a)

First: General Motors (Sunraycer)

(b)

Second: Ford Motor Company (Model S)

(c)

Third: Ingenieurschule Biel (Spirit of Biel)

(d)

Spirit of Biel/Bienne II

(e)

Honda Dream, 1993

(f)

Honda Dream, 1996

Fig. 1.7. (a)–(c) Vehicles finishing in the first three positions of the 1987 World Solar Challenge. Winner in: (d) 1990; (e) 1993; (f) 1996.

Another useful application of EVs is to provide sightseeing services in tourist attractions such as amusement and holiday parks, nature reserves, places of historic interest, resorts, and exposition centres. A wide variety of multi-passenger type EVs have been offered, for example, by Japanese manufacturers (Fig. 1.6).

Finally, there has been a resurgence of interest in battery-driven boats for surface operation on canals, rivers and lakes [27,28]. The first recorded experiment to propel a boat with battery power was made in St. Petersburg in 1838 with the use of a Grove battery [1]. In 1880, M. Gustav Trouvé of France invented the first outboard motor. This was an electric motor which was driven by two batteries (each of six bichromate cells) and used to propel a boat along the River Seine at a speed of 9 km h^{-1} [29]. Remarkably, a lead/acid-battery-driven boat crossed the English Channel in 1886. By 1888, a fleet of 22 electric pleasure boats was operating on the River Thames in London, and was supported by an infrastructure of five floating and four land-based charging stations. The largest of these launches, the 75-passenger 'Viscountess Bury', was powered by a pack of Fauré-type lead/acid batteries that delivered 28 A at 85 V for a full speed of 25 km h^{-1} [30]. The electric boats were readily accepted and their numbers had grown to 100 by 1893 [31]. Electric propulsion has been used continuously since 1909 for sightseeing trips on the Königsee in Bavaria, Germany [32], while other electric pleasure craft were operated on the rivers of France, on the Danube in Austria, on the canals of Venice in Italy, and as ferry boats in Norway. The popularity of electric boats in the late 1890s was due to their significant advantages compared with vessels powered by manual labour, sails, and even steam and naphtha engines. They were quiet, clean, and easy to operate. Ultimately, as with the automobile, the electric boat was defeated by the superior energy and power, as well as the long range, of the internal-combustion engine.

Nowadays, with a standard voltage available everywhere for easy recharging of batteries, with sophisticated electronic controls for added safety, and with the above-mentioned environmental benefits of reduced noise, less vibration, low wash, no smelly fumes and no oily pollution, electric boating is again becoming attractive [33]. Depending on the size of the craft, the lead/acid batteries range from 12 to 120 V and from 50 to 800 Ah (C_{20}/20 rate). Commercial use of electric boats in the UK centres on relatively large (18–21 m long) self-drive canal and river boats for the holiday trade (Fig. 1.6), although a growing number of smaller craft are being built for private use. A consortium of private and commercial UK interests formed the Electric Boat Association in 1982. In the USA, small battery-powered outboard-motor boats are gaining increasing popularity with fishing enthusiasts and recreational boaters. Electric waterbuses are even returning to the canals of Venice [34]. In October 1989, the first of a planned fleet of between 10 and 20 passenger-carrying 'vaporetto elettrico' was commissioned. The vessel has a special low-wash hull which will cause less disturbance to the canal-side palaces, and is propelled by 12 t of tubular-plate lead/acid batteries (240 V, 1450 Ah). On a sporting note, a world speed record for electric boats of 81.8 km h^{-1} was set by Lady Arran in the UK on 22 November 1989. The craft was of a hydroplane design and was powered by lead/acid aircraft-starter batteries.

The development of battery-powered vessels for underwater operations commenced around the time that surface electric boats were making their appearance in large numbers. Military applications provided the initial impetus, and the vessels became known as 'submarines'. The alternative term — 'submersible' — is used to denote underwater craft which are used for peaceful purposes, e.g., for surveillance, exploration, rescue operations, fishing. Prototype pure-electric submarines were built in the late 1880s in France, Spain and the UK; by 1913, 16 of the world's navies had a total of over 400 submarines [35]. Lead/acid cells were used exclusively and their performance was steadily improved over the years: cells of capacity 1920, 5350 and 8800 Ah were employed in the years 1901, 1943 and 1978, respectively, and these had corresponding specific energies of 8.7, 23.0 and 32.6 Wh kg^{-1} [32]. The battery of a present-day diesel–electric submarine is truly massive. For example, the battery of an Oberon-class submarine of the Royal Australian Navy weighs 230 t and contains 145 t of lead. The pack consists of 448 $\times$ 2-V / 8400-Ah cells which produce either 448 or 996 V by parallel- or series-connection — a flexibility which gives either 'sprinting' or 'staying' power, according to the submarine's tactical needs. In recent times, nuclear energy has been increasingly adopted for underwater propulsion by the world's major navies. Nevertheless, nuclear submarines still require an emergency standby battery. For a comprehensive account of both surface and underwater battery-powered craft, the reader is referred to *Batteries on Ships* by N.E. Bagshaw [32].

1.5. Vehicles for Disabled Persons / Medical Services

There are thousands of disabled people who have achieved greater personal freedom through battery power. In addition to the well-known wheelchair, there are battery cars available for a wider range of activities, e.g., for travelling to and from work, shopping, and social outings (Fig. 1.8). Increased mobility for the negotiation of steps, stairs, kerbs, etc., has been obtained through advanced vehicle technology. For example, one such conveyance, built in Australia and called the 'Transcender', operates over level surfaces or ramps on four small wheels, but when the rider approaches stairs or wants to get in or out of motor vehicles, a switch lowers composite plastic tracks which enable the ascent or descent to be made (Fig. 1.8). Invalid wheelchairs and cars are usually powered by conventional automotive or golf-cart lead/acid batteries. There is scope, however, for designing a lead/acid battery specifically for invalid assistance and, in particular, for taking advantage of the maintenance-free operation (i.e., water additions not required) of the developing valve-regulated technology (see Section 5.2.7, Chapter 5).

Battery-driven vehicles, by virtue of their moderate speed and quietness of operation, also make positive contributions to the comfort and safety of hospital patients, in the form of site ambulances, trolleys, and mobile intensive-care units (e.g., battery-operated baby incubators), Fig. 1.8. By virtue of their cleanliness, many EVs (including AGVs) are also employed in large hospital complexes to transport medical supplies, laundry, foodstuffs, etc.

Conventional wheelchair

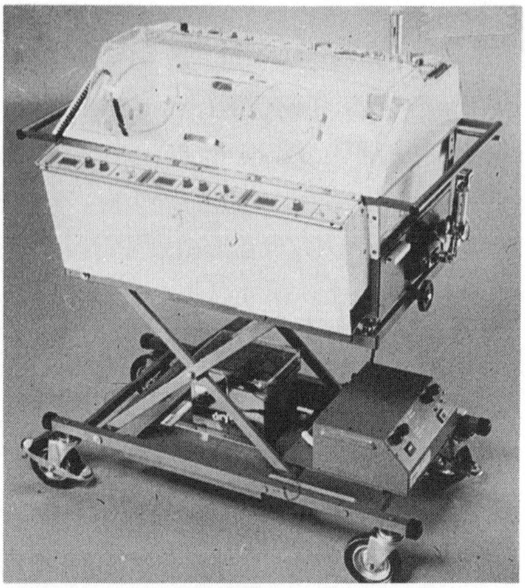

Baby incubator

Climbing wheelchair (Transcender)

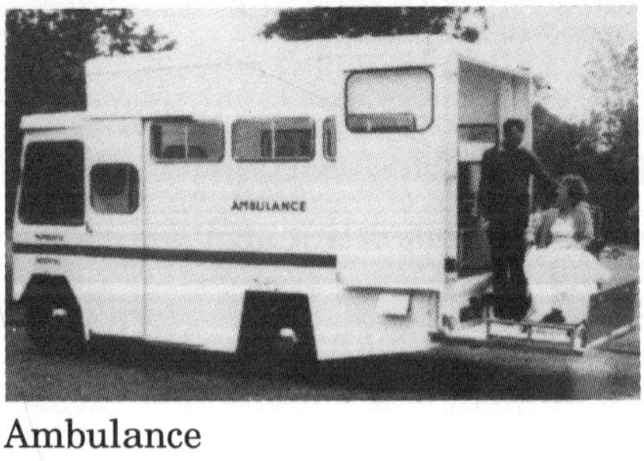

Ambulance

Invalid car

Fig. 1.8. EVs for disabled persons and medical services.

1.6. Airport Ground-Support Vehicles

In the early 1980s, airline operators and airport authorities started to change to EVs for a wide range of services [36]. Electrics were found to fit smoothly into the flow of aircraft loading and unloading operations. In 1984, Harvey [37] estimated that about 50 000 mobile pieces of equipment were in operation at the approximately 100 major airports of the world, and that about 10% of these units were battery powered.

Types of EV now in widespread use include: aircraft push-out tractors; baggage trucks; belt loaders (i.e., cargo and baggage conveyors); pallet/container loaders and transporters; passenger steps; cabin lifts for disabled passengers; telescopic access/work platforms; passenger and personnel carriers and vans; lavatory, water, and catering trucks; fork-lift trucks and warehouse tugs; floor and runway sweepers; de-icers; and even fire tenders (Figs. 1.9(a),(b)). The claimed advantages are: significant cost savings (i.e., low fuel price, no 'idling' consumption of fuel); reduced maintenance; reduction of noise and exhaust/fuel odours (a key objective in securing the comfort of passengers, airport workers and local residents); smoother acceleration during aircraft push-out; instant readiness for use in extremes of climate; greater manoeuvrability; lower fire risks when aircraft are being refuelled.

The lead/acid batteries used in airport ground-support equipment vary in size from the 'jumbo' packs of aircraft push-out tractors to the more modest units of baggage trucks. Push-out tractors operated by some airlines in the USA are powered by two 600-Ah batteries which each have 64 cells and together weigh 5500 kg. In Germany, the F 246 tractor is driven by a lead/acid battery which uses copper-stretch-metal (CSM) technology to improve power performance [38]; the battery consists of 120 cells of 888-Ah capacity ($C_5/5$ rate) assembled in two steel boxes. The number of pushes possible per unit to 80% of maximum discharge (the limit usually applied) varies with aircraft weight and engine idle thrust, distance per push, and tractor design. The expected life of the F 246 battery is 1600 cycles at 80% depth-of-discharge (DoD), or 4–6 years [37].

Push-out tractor

Baggage truck

Belt loader

Container loader

Passenger steps

Work/service platform

Fig. 1.9(a). Airport ground-support EVs.

Fork-lift truck

Water bowser

Passenger transporter

Floor sweeper

Fig. 1.9(b). Airport ground-support EVs.

1.7. Scooters, Motorcycles and Bicycles

Electric tricycles were introduced in the early 1880s by Ayrton and Perry in the UK [39], and by Trouvé in France [29]. These were the first applications of lead/acid batteries to personal land transportation. Since those early days, a number of battery-driven scooters, motorcycles, and pedal-assisted bicycles have been marketed. In the early 1970s, Solo-Kleinmotoren GmbH produced and sold around 10 000 electric scooters (Fig. 1.10). These were powered by two automotive lead/acid batteries of 50-Ah capacity ($C_{20}/20$ rate) [40]. The scooter had a top speed of 25 km h^{-1} and a range of about 40 km [7]. In the 1980s, Australia Post conducted field trials into the suitability of electric scooters for mail delivery. The machines were manufactured both locally and overseas, and although the driving characteristics met the approval of postmen, the vehicles had

Solo scooter (Germany)

Daihatsu scooter (Japan)

ZES 2000 (Taiwan)

Corbin-Gentry motorcycle (USA)

Pedelec Stella (UK)

Citibike (UK)

Fig. 1.10. Battery-powered scooters, motorcycles and bicycles.

inadequate ranges. Various designs of electric scooter have been offered by Japanese manufacturers (Fig. 1.10). In the USA, Corbin-Gentry produced a lead/acid-powered motorcycle for commuter use with a top speed of 48 km h^{-1} and a range of 48–64 km [7]. On 19 August 1974, this Company set a land speed record of 266 km h^{-1} with a specially built vehicle which was propelled by a 120-V pack of silver/zinc cells. (Note, just five days later, a home-built car called the 'Battery Box' and powered by lead/acid batteries was driven at 282 km h^{-1}; on 26 September 1996 in the USA, Lloyd Healey set the world speed record of 335 km h^{-1}.)

In 1983, the UK Government passed legislation which permitted the use of 'electrically-assisted cycles' by anyone aged 14 or over — without a driving licence, road tax, or compulsory insurance. This incentive gave rise to a number of machines of which the Pedelec Stella (Fig. 1.10) was the first purpose-designed, electrically-assisted bicycle to have been independently certified to meet all the British standards. The bicycle was powered by a 12-V lead/acid battery which weighed 10.9 kg and had a capacity of 28 Ah (C_5/5 rate). Two later products were the Sinclair Zike and the Citibike (the latter is shown in Fig. 1.10). The former uses nickel/cadmium batteries to drive a small motor and provide a maximum speed of ~19 km h^{-1} over a short range. By contrast, the Citibike is powered by lead/acid batteries and has a range of ~30 km at 24 km h^{-1}. An optional solar panel can extend the range by a further 8 km after a day in the sun. The Citibike folds into the boot of a car and weighs 35 kg, nearly three times that of the Zike. A market is also developing for devices which bring electrical power to conventional bicycles. In China, the Shanghai Xinlian Electrical Vehicle Corporation is offering a wide range of such machines (the CRANES TF 100–Z series) and these have a top speed of 20 km h^{-1} and a range of 20–30 km.

The increasing pressures which are being placed on governments to improve air quality in traffic-congested cities have rekindled interest in the development and introduction of battery-powered motorcycles. In Taiwan, for example, the Environmental Protection Agency has introduced a mandate which requires 2% of the scooters (or motorcycles) sold by every manufacturer in the year 2000 to be electrically powered. At present, more than 10 million scooters crowd the urban and rural regions of Taiwan. Following the mandate, the major manufacturers and related industries joined together to develop both electric scooters and electric motorcycles. The prototype scooter — the ZES 2000 (Fig. 1.10) — is powered by four 12-V, 28-Ah, valve-regulated lead/acid batteries which provide a normal urban driving range of 30 km [41]. Similar development projects are being undertaken by Peugeot in France, and by both Sciex and Honda in Japan.

1.8. Delivery and Service Vehicles

An obvious application for electric traction in light commercial markets lies in those duties which involve multiple stop–starts, modest daily distances, and long/regular idle periods suitable for battery recharging. A classic example of matching EV performance characteristics to the job in hand is the milk float used in the UK for the door-to-door distribution of milk, together with dairy and food products (Fig. 1.11). The milk float is required to stop at almost every house in every street on its delivery round. The wear and tear on an ICEV in such an intensely stop–start operation would be considerable, given this vehicle's more numerous moving parts. Equally important, the distance to be covered is short (up to 45 km) and high speeds are not necessary. Refuelling with cheap off-peak electricity, and silent delivery whilst customers are asleep, are added advantages. A cost analysis undertaken by the Electric Vehicle Association of Great Britain [19] demonstrated that a 1.53-t payload vehicle delivering milk to 250 houses over 40 km every day of the year was 40% cheaper to operate than a diesel-driven counterpart. Milk-float voltages range from 30 to 72 V; the capacity of the lead/acid cells varies considerably and depends on the payload, number of stop–starts, and distance and terrain to be covered. Generally, the cells are of the tubular-plate type (see Section 5.3.1, Chapter 5).

A similar type of service for which EVs have been considered is mailbox clearance and doorstep delivery of postal items. Road trials have been conducted by post offices in several countries, e.g., Australia, Austria, the UK, and the USA. The US Postal Service employed several hundred EVs

Milk-float

Meter reader

Refuse-collection trucks

Postal vans

Fig. 1.11. Road EVs for door-to-door services.

which included 350 trucks (Fig. 1.11) that were produced by the American Motors General Corporation [42]. The battery pack consisted of 27 lead/acid cells (300-Ah capacity, $C_6/6$ rate) and weighed 577 kg. By the late 1970s, the Austrian Postal Service had used EVs for over 70 years to deliver packages in Vienna and other cities and large towns; some vehicles still in operation were over 30 years old [43]. Recently, Deutsche Post AG has conducted field trials with Mercedes-Benz 180E electric vans powered by zinc/air batteries (see Section 6.9, Chapter 6).

Other house-to-house services which have come under the scrutiny of EV developers include utility-meter reading (electricity, gas, water) and refuse collection (Fig. 1.11). In the latter application, both pedestrian-controlled and driver-controlled vehicles were field-tested in the UK. Meanwhile, in France, 135 refuse-collection trucks were operated in Paris alone, with another 245 in service throughout the remainder of the country. The reduced noise level made these trucks very popular. A comprehensive range of electric refuse-collection vehicles in a wide choice of body styles was made available in the USA by U.S. Electricar.

Next in terms of prospective markets for on-the-road EVs are urban and suburban services where the stops are more widely separated, but still frequent. Typical users are: deliverers of newspapers and magazines in bulk; couriers; mobile shops and libraries; catering, laundry, meat-product, confectionary and florist businesses; ice-cream vendors; service engineers; and local tradesmen (Fig. 1.12). Even the concept of an electric taxi has been evaluated. In the UK, pioneering work resulted in the development of two classes of vehicle for delivery duties: light vans (about 1-t payload) and medium vans (about 2-t payload). Lucas Industries concentrated on the former type of vehicle, whereas the Chloride Group investigated the latter. In 1981, the EV activities of these two companies were merged by the formation of Lucas Chloride EV Systems Limited (LCEVS). Four major vehicle programmes were conducted by LCEVS. The light vans were the Bedford CF and the Freight Rover Sherpa (Fig. 1.13). Power to the d.c. motor in each vehicle was provided by 36 tubular-plate lead/acid batteries constructed in 6-V modules (type EV 5T) which each had a capacity of

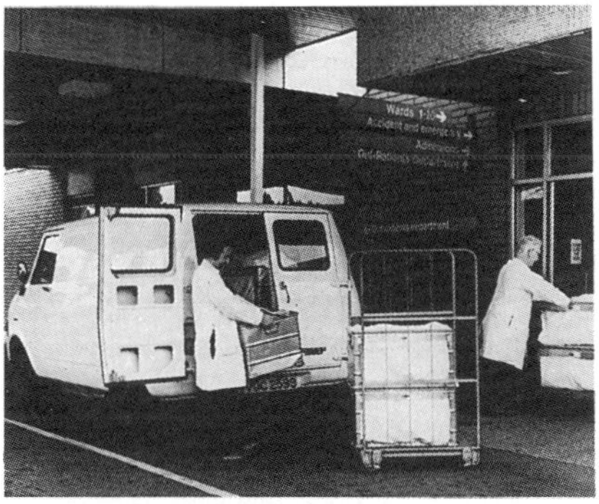

Laundry delivery

Mobile library

Newspaper delivery

Service engineers

Courier service

Electric taxi

Fig. 1.12. Road EVs for urban and suburban duties.

Bedford CF

Freight Rover Sherpa

Dodge 50 Series Electric

Leyland Electric Roadrunner

GM Griffon

Ford Ecostar

G-Van

TEVan

Fig. 1.13. Electric vans developed in the UK and the USA.

184 Ah and an a specific energy of 34 Wh kg^{-1}, both at the $C_5/5$ rate. The medium vans were the Dodge 50 'Series Electric' and the Leyland 'Electric Roadrunner' (Fig. 1.13); again propelled by tubular-plate batteries, but now in 2-V, 420-Ah ($C_5/5$ rate) units. With the exception of the Electric Roadrunner, the electric models were built on conventional assembly lines alongside their petrol or diesel counterparts, and the drive systems were fully compatible with normal build procedures. The vehicles were marketed by the manufacturers as an integral part of their product range and distributed through the normal dealer networks. Ownership costs of the vehicles were claimed to be directly competitive with ICEVs [44]: the higher initial capital cost of the EV was offset by significant savings in fuel, maintenance and licence costs.

Bedford CF electric delivery vans, manufactured by LCEVS in the UK, were incorporated into the service fleets of 11 electric utilities and 11 other companies in North America. The vehicle was changed to left-hand drive and re-named the GM 'Griffon'. It had a maximum speed of 85 km h^{-1} and a range of 104 km when operating under the SAE J227a/C driving schedule [45] (see Section 4.3.1, Chapter 4). Griffon vans were used for customer service calls, carpools, shuttles at public-relations events, mail transfers, and shop deliveries of items which ranged from flowers to auto parts. In July 1986, Bedford suspended production of the electric version of the CF van. Nevertheless, GM decided to continue the project in the USA. The planned successor was the GM 'G-Van'. Before the appearance of this vehicle, however, Lucas Industries decided, in March 1987, to pull out of the joint venture with the Chloride Group. The latter took over the Lucas interest and, as Chloride EV Systems Limited (CEVS), continued to develop electric drive/control systems and batteries. At that stage, over 500 vans and trucks had been fitted with Chloride drive-systems and had been produced by five major European vehicle manufacturers. Over 16×10^6 km had been accumulated by those vehicles in service with 90 users in 10 countries.

The G-Van was a prototype conversion of a US-produced van and was appreciably larger than the Griffon van, with some 25% more cargo space.

An improved tubular-plate lead/acid battery — known as the 3ET 205 and again manufactured by the Chloride Group — enabled the G-Van to deliver a level of performance similar to that achieved by its predecessor, i.e., a maximum speed of 83 km h^{-1} and an SAE J227a/C driving range of 96 km [45]. Three initial proof-of-concept vehicles were built by a Detroit-based company called Cars & Concepts, Inc., working closely with CEVS. Following successful field trials, 33 prototype vehicles were produced in late 1989 by Vehma International (a subsidiary of the Canadian-based firm, Magna International) with funding assistance from the Electric Power Research Institute. It appeared that the prospects for the G-Van were boosted in November 1989 by Vehma being named the winner of phase 1 of the initiative of the Los Angeles City Council to reduce air pollution through the introduction of 10 000 EVs [17,18]. Unfortunately, the initiative failed to eventuate. It has, however, been superceded by the CARB mandate for ZEVs, as discussed above.

In parallel with the above GM project, Chrysler developed an electrically-powered van, known as the 'TEVan'. The vehicle was based on the Chrysler T-115 minivan body and, like the GM vans mentioned above, had a d.c. motor. Traction power was supplied by a 180-V nickel/iron battery manufactured by Eagle-Picher Industries, Inc. This battery offered better performance (105 km h^{-1}) and range (177 km) than the lead/acid batteries used in the G-Van. It should be pointed out, however, that both the payload capacity and the cargo space of the TEVan were considerably smaller. The TEVan had an automatic battery water system, an on-board charger, and an electronic instrument cluster which displayed battery state-of-charge and the remaining driving range (predicted from the previous 10 km travelled). In competition, Ford announced its 'Ranger' electric van for delivery to dealerships in 1998. This vehicle is powered by a pack of 39 valve-regulated lead/acid battery modules, each of 8 V.

Battery-powered vans with advanced propulsion systems were included in the US Department of Energy (DOE) Electric and Hybrid Vehicles Program. The key outcomes of this work are discussed in Section 1.11. In Japan, the Suzuki Carry and Daihatsu Jet commercial vehicles have been available for lease. Numerous European companies are conducting road trials with electric vans, e.g., Daimler-Benz, Hill Electric (UK), PSA, Renault, and Volkswagen. The trial being conducted in Finland by the Post Office is particularly interesting on account of the low winter temperatures which are experienced; 51 Elcat vans are being operated, sometimes in temperatures as low as –30 °C.

1.9. Buses

Surveys conducted in several countries have shown that a large fraction of urban bus fleets on well-defined routes cover limited distances, or are only used at morning or afternoon travel peaks. That is, the buses operate under conditions which are eminently suitable for electric conversion. There are three basic classes of electric traction that can be applied to public transportation systems:

(i) **mains only:** these vehicles have no on-board energy-storage facility and are totally dependent for their power supply on fixed overhead catenary, sub-surface conduit, or surface contact systems that extend the full length of any route, e.g., the classic trolley bus;

(ii) **hybrid:** such vehicles depend on power from a limited on-board energy-storage unit (e.g., battery, ICE or flywheel) together with an external supply from overhead wires, a third rail, or other forms of pick-up; operation with the external supply would be limited to the most heavily used transit corridors, a switch to battery power would enable negotiation of complicated junctions and provide suburban road-loop services;

(iii) **battery only:** power to this class of vehicle is provided by a pack of batteries which, when exhausted, can either be recharged or replaced by a fresh (charged) set; pure-battery buses eliminate the need for the costly and aesthetically displeasing electricity-supply infrastructure which is required by mains vehicles and, at the same time, increase vehicle manoeuvrability.

The development of pure-battery (and mains–battery) buses has been pursued in many countries since the early 1980s (Fig. 1.14). The most significant initial effort was made in Germany [46] where field trials were conducted on pure-battery buses, operated both with battery changeover (Fig. 1.15(a)) and with boost charging during stops en-route, as well as on battery–trolley (i.e., the 'DUO-BUS', Fig. 1.15(b)) and battery–diesel hybrid systems. Results of laboratory testing and field operations showed [47,48] that interim charging improved battery service-life; the batteries employed consisted of 180 tubular-plate lead/acid cells of 455-Ah capacity ($C_5/5$ rate). By the end of 1985, 20 pure-battery buses (operating in Mönchengladbach and Düsseldorf) had travelled a total distance of over 8.5×10^6 km, and single buses up to 0.55×10^6 km. The best battery life was equivalent to 2399 cycles (100% DoD), or 0.15×10^6 km. The battery–diesel hybrid buses had half the number of cells of the pure electric and could travel for 26% of the route (i.e., 50% of the time) powered by the batteries alone.

The production and utilization of electric buses, powered by batteries or other non-stationary means of generation and storage of electric power, are attracting considerable interest in the USA. It is considered that electric buses are ideally suited to public transit routes, i.e., the buses typically operate at relatively low speeds and drivers can work on shifts which match the range capabilities of current technologies. Moreover, buses are sufficiently large vehicles to carry the number of lead/acid batteries that are required to provide a workable range. Finally, US government agencies are coming to realize the significant public relations advantages of electric buses — it is not often that a transit system can boast of lowering operating costs while reducing both air pollution and dependence on imported oil! In 1996, 81 electric buses were operating on public transit systems in 28 cities throughout the USA. A notable success has been the fleet of eight shuttles employed by the Santa Barbara Metropolitan Transit District. Each shuttle completes a service of up to 130 km during 8 h each day on a single battery charge. The public has responded very favourably — ridership increased ten-fold during the first year of all electric-bus operation, i.e., from 0.1 to 1 million passengers per year. During four years, the shuttles have logged more than 0.48×10^6 km and 60 000 h of

Australia

Japan

UK

UK

USA

USA

USA

USA

Fig. 1.14. Battery-powered buses.

(a)

Battery-powered bus

(b)

Mains–battery bus

(c)

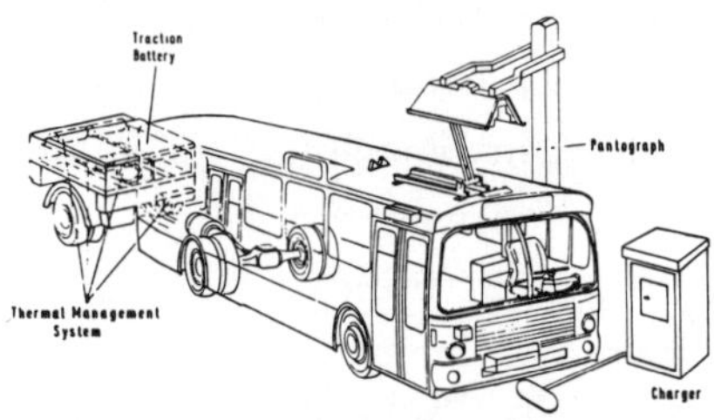

(d)

En-route charging: concept and practice

Fig. 1.15. Electric bus development in Germany.

service during the course of some 8000 driving cycles. Following this success, 15 more shuttles were ordered in April 1996. Other potential applications for electric buses include school services, airport shuttles, and hotel courtesy transport.

1.10. Passenger Cars

In terms of size of market, the ultimate prize for EV developers is the private battery-driven car. Activity in this sector has been considerable (especially in the years immediately following the 1973 energy crisis and the 1989 California ZEV mandate) and has generally been the province of individual engineers and inventors, small new companies, university departments, and, occasionally, a major ICEV manufacturer. Readers of specialist EV journals will know that announcements of promising new electric cars often go hand-in-hand with reports of the demise of earlier ventures. It would therefore be unprofitable to discuss here in detail the design and performance of any one particular model. Nevertheless, some of the EVs which have attracted attention are shown in Figs. 1.16–1.22. There has been a marked trend towards small cars which are designed specifically for the city-centre environment. This market is ideally suited to EVs because acceleration and efficient stop–start driving are more important than range and top speed. Examples of such 'microcars' are the BMW E1, Citroën Citela, Fiat Zic, Mercedes Vision, PSA Peugeot-Citroën Tulip, and Volkswagen Chico.

Major research and demonstration programmes into electric cars have also been conducted by the governments of several countries, notably by the Ministry of International Trade and Industry in Japan [49], and by the DOE in the USA [50]. The first DOE battery-driven test vehicle, the 'ETV-1', was introduced in June 1979 [51], Fig. 1.23. Built by the General Electric Company (with Chrysler Corporation as a major sub-contractor), this was a lead/acid-powered, four-passenger, urban/suburban car with a 120 km range at 56 km h^{-1}, a top speed of 112 km h^{-1}, and an acceleration of zero to 48 km h^{-1} in 8.8 s. About the same time, a second vehicle, designated the 'ETV-2' (Fig. 1.23), was produced by the Garrett Corporation and employed a flywheel-assisted power sub-system [52]. Since hybrid technology was considered by some to satisfy more adequately the private motorist's criteria for an alternative to conventional ICEVs, the ETV-1 technology was combined with a fuel-injected petrol engine in a parallel ICE–battery hybrid version of the vehicle, the 'HTV-1'. The electric drive was intended for use on short journeys and in traffic congestion (range 56 km), and the gasoline engine for use on highways to give improved range. The lead/acid pack (108-V) for these three vehicles was developed by Johnson Controls, Inc. Two sizes of 12-V module were designed [53] for the ETV vehicles, namely, the EV-2300 (183 Ah, 39.5 Wh kg^{-1}; both at the $C_3/3$ rate) and the EV-3000 (240 Ah, 41.5 Wh kg^{-1}; both at the $C_3/3$ rate). One size of 12-V module was developed [54] for the HTV-1 vehicle, namely, the EV-1300 (105 Ah, 37.5 Wh kg^{-1}; both at the $C_3/3$ rate). The ETV-1 and HTV-1 projects terminated in 1981.

Partridge Wilson (UK)

Lunar Rover (USA)

Enfield (UK)

Sinclair C5 (UK)

Hope Whisper (Denmark)

Volkswagen CitySTROMer (Germany)

Fig. 1.16. Passenger cars: 1930s–1980s.

Fiat X1/23 (Italy)

3P (Italy)

Zagato (Italy)

Phase II Electric (Australia)

L38-V (Japan)

Commuta-car (USA)

Fig. 1.17. Passenger cars: 1970s–1980s.

El car (USA)

Copper Electric Runabout (USA)

Globe Union Endura (USA)

Bradley GTE Electric (USA)

Sears Roebuck XDH-1 (USA)

Briggs & Stratton hybrid (USA)

Fig. 1.18. Passenger cars: 1970s–1980s.

Citroën Citela (France)

Citroën AX (France)

Peugeot 106 (France)

Renault Clio (France)

PSA Tulip (France)

BMW E2 (Germany)

Fig. 1.19. Passenger cars: 1990s.

Mercedes Vision (Germany)

Volkswagen Chico (Germany)

Fiat Zic (Italy)

Microcar Electric (Italy)

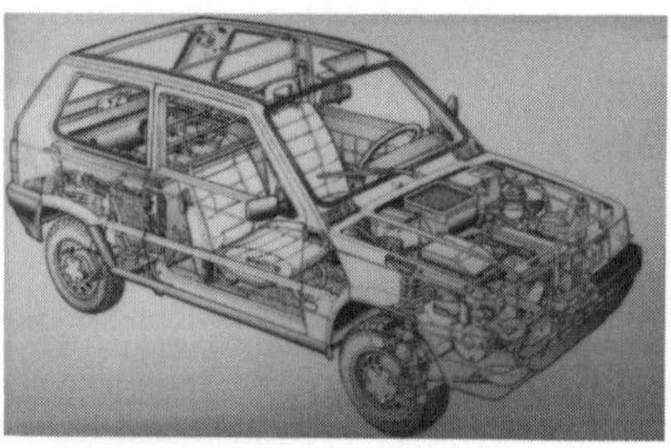

Panda Elettra (Italy)

Pininfarina: Ethos 3EV (Italy/USA)

Fig. 1.20. Passenger cars: 1990s.

Toyota EV-50 (Japan)

Honda EV Plus (Japan)

Nissan FEV-1 (Japan)

Nissan FEV-2 (Japan)

Toyota RAV4-EV

Kia KEV-4 hybrid (South Korea)

Fig. 1.21. Passenger cars: 1990s.

Chrysler: Intrepid ESX hybrid (USA)

Ford Ghia Connecta (USA)

General Motors Impact (USA)

General Motors EV-1 (USA)

Solectria Force (USA)

Solectria Sunrise (USA)

U.S. Electricar Sedan (USA)

Fig. 1.22. Passenger cars: 1990s.

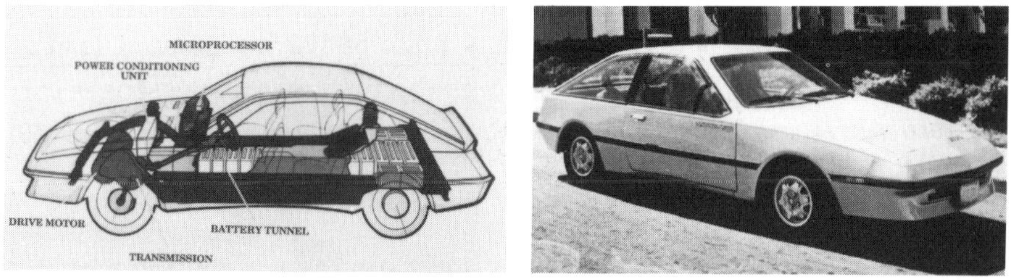

ETV-1

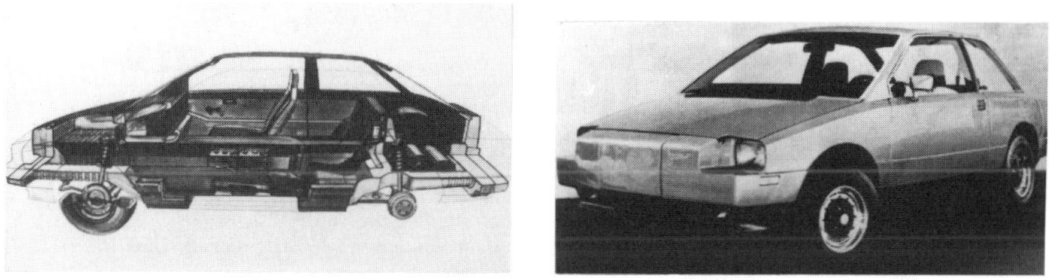

ETV-2

Fig. 1.23. DOE electric and hybrid test vehicles.

In more recent times, General Motors demonstrated its electric Impact car. This vehicle had two a.c. induction-type motors (each of which drove one of the front wheels) and was powered by 32 × 10-V, valve-regulated, lead/acid batteries (see Section 5.2.7, Chapter 5). The car accelerated from zero to 100 km h^{-1} in 8 s and had a projected top speed of over 160 km h^{-1}. Such performance rivals ICEVs and the range capability is adequate for the daily needs of most urban drivers. In one programme — the Electric Vehicle PREView Drive — General Motors teamed up with utilities in 12 cities to allow over 1000 consumers to test drive the Impact for up to a month and then provide feedback on vehicle performance, capabilities, and infrastructure needs. On-board computers also gathered data on a wide range of technical issues, from charging and driving patterns to power and durability. On 4 January 1996, General Motors unveiled the 'EV-1', a two-seater sports car which has been derived from the Impact (see Fig. 1.22). This is one of the first models to be designed exclusively as an EV and not adapted from an existing ICEV. It incorporates many novel features which include inductive charging of the batteries (see Section 10.5.1, Chapter 10) and a lightweight, aluminium frame. Power is provided by a three-phase,

102-kW, a.c. motor and this gives the car a top speed of 125 km h^{-1} and brisk acceleration. The range with a pack of valve-regulated lead/acid batteries is 110–140 km. The EV-1 was the first electric passenger car to go into series-production in the USA and was made available in 1997 through dealers in both California and Arizona.

Other major automobile companies have been providing competition on a global level, e.g., Honda with the 'EVX', Toyota with the 'EV-50' and the 'RAV4-EV', and Nissan with the 'FEV-1' (see Fig. 1.21). France has become most aggressive in bringing vehicles to the market-place. For example, PSA (Peugeot-Citroën) has a dedicated EV assembly line with a capacity of 30 vehicles per day (Peugeot 106 and Citroën AX, see Fig. 1.19). As of mid-1997, there were about 800 EVs operating in Paris. As part of its national strategy, the French government is working with the country's car manufacturers and electric utilities. In Germany, BMW, Volkswagen and Mercedes are all demonstrating vehicles which are equipped with batteries of diverse chemistry. Meanwhile, in Italy, Fiat is producing several EV models in limited production runs.

Despite all these efforts, as far as the private passenger vehicle market is concerned, the battery-powered car has failed to attract other than EV enthusiasts as buyers because of its high cost, limited range, and long refuelling time. It has been widely argued that the EV would make an ideal second (or 'commuter') car, but in most families the second car often doubles as the first car when the latter is unavailable and, thereby, could be required occasionally to undertake much longer journeys. In addition, economic considerations often dictate that the second car is a pre-owned, cheap model. Obviously, this is a market to which the new EV is not ideally suited.

Notwithstanding this negative outlook, under the stimulus of rising urban air pollution and the resultant political pressures in Europe, Japan, and the USA, at least six major automotive companies have committed themselves to the series-production of electric cars. These are listed in Table 1.1.

Table 1.1. Major electric passenger car programmes.

Manufacturer	Country	Car model	Battery
Peugeot	France	106E1	Nickel/cadmium
Citröen	France	AX	Nickel/cadmium
Renault	France	Clio	Nickel/cadmium
General Motors	USA	EV-1	Lead/acid
Toyota	Japan	RAV4-EV	Nickel/metal hydride
Honda	Japan	CUV-4	Nickel/metal hydride

Attempts have also been directed towards making use of EVs as public rental cars; the two most celebrated examples were the Witkar project in The Netherlands and the PREET system in Japan (Fig. 1.24). The Witkars were small, two-seat, three-wheel vehicles, and were designed for maximum trips of about 5 km between recharges and top speeds of about 35 km h^{-1}. The power was supplied by nickel/cadmium batteries which were recharged at strategically located stations. The PREET system was developed by the Daihatsu Kogyo Company, and used small two-passenger cars. There were zone stations servicing 2 or 3 cars and area stations accommodating 10 to 20 cars. The latter stations permitted charging of the batteries (lead/acid) and routine inspection of the cars. Users received a membership card and when this card was inserted into the terminal equipment of the station computer system, the number of the car to be used was indicated. On returning the car, re-insertion of the card calculated the rental charge in accordance with the period of vehicle use, and the charge was automatically paid from the driver's bank account. A similar EV-rental system was proposed in 1987 by City Wheels Limited for service in Monaco, Italy, and the USA. It was based on the use of French-designed two-seater 'Ligier' cars which were capable of 70 km h^{-1} and a range of 20 to 30 km (Fig. 1.24).

A new city-car rental scheme has been announced by PSA Peugeot-Citroën in France. Known as TULIP (Transport Urbain, Libre, Individuel et Public), it envisages vehicles (Fig. 1.19) being strategically placed around cities and operated by PIN entry devices [55]. There will be a central control station and a network of parking and automatic inductive

PREET system (Japan)

Witkar system (The Netherlands)

Ligier vehicles

Fig. 1.24. Public rent-an-EV concepts.

recharging stations. TULIP is seen to combine the advantages of the private car (flexibility, ease of use, door-to-door transportation) with the benefits of public transport (no vehicle purchase, no maintenance costs, fee-for-use payment system). More than 20 French cities have expressed an interest in the TULIP concept, as have cities in several other countries.

1.11. Power-Train Development

Following development of the ETV-1 and ETV-2 test vehicles, the DOE Electric and Hybrid Vehicles Program was directed towards the synergistic integration of the power-train (i.e., electric motor, control electronics, transmission, axle and wheels) and battery technologies which are unique to EV propulsion. Two approaches were adopted: the Single-Shaft Electric

Fig. 1.25. ETX-I vehicle undergoing tests on dynamometer. (Taken from [56] with permission of the US Department of Energy.)

Propulsion System Technology Development Program and the Dual-Shaft Electric Propulsion System Technology Development Program (DSEP). The former project was initiated jointly by Ford and General Electric in 1981 and the work examined the concept of single-shaft, a.c., power-train technology. In this, the electric motor and transmission are mounted on a common axis in an integral unit which operates within the same housing and with common lubricating oil. The power-train of the first-generation system was installed in a Mercury LN7 test vehicle, re-named the 'ETX-I' (Electric TransaXle), see Fig. 1.25. Extensive dynamometer and track testing [56] verified that the power-train met, or exceeded, the technical goals which were set for energy consumption, acceleration, gradeability, and driveability. The power-train programme was completed in August 1985 [57]. Three batteries were evaluated in the ETX-I project, namely, a full-scale advanced lead/acid battery designed and built by Lucas Chloride

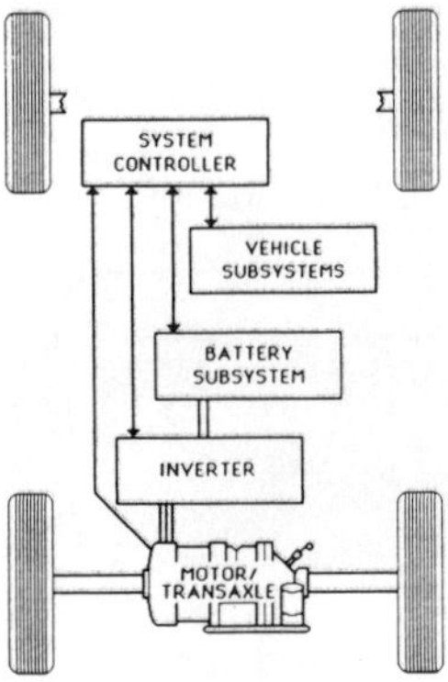

System block diagram [61]

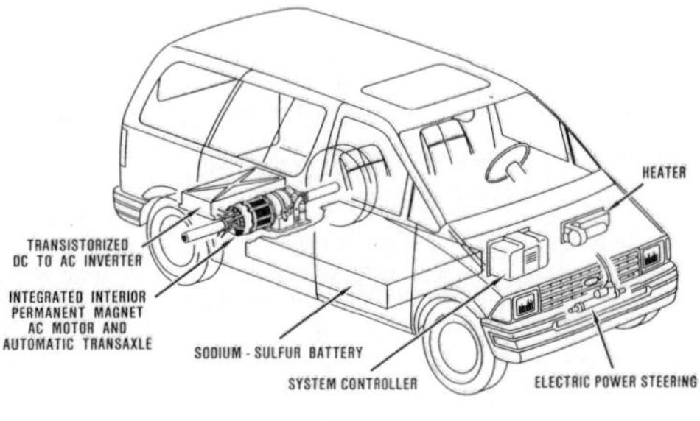

System in test vehicle [61]

Transaxle [63]

Test vehicle

Fig. 1.26. ETX-II: integrated single-shaft a.c. propulsion system. (Taken from [61,63] with permission of the US Department of Energy.)

EV Systems Limited [58], a one-third scale sodium/sulfur battery produced by Ford Aerospace and Communications [59], and a full-scale zinc/bromine battery developed by Exxon Research and Engineering Company [60].

During the ETX-I programme, it was established that EVs would most likely be used first in urban van applications. Consequently, a new project was established, in 1985, to develop a second-generation, single-shaft, a.c. propulsion system for vans. The test vehicle — the 'ETX-II' — was a modified Ford Aerostar van (Fig. 1.26) [61] and was used to evaluate

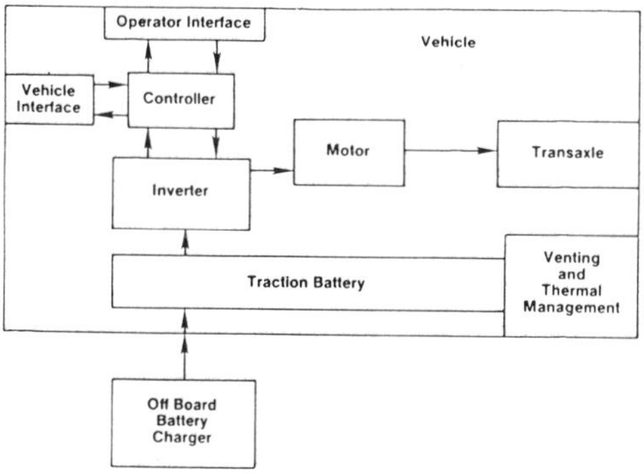

System component block diagram [65]

Motor/transaxle assembly [64]

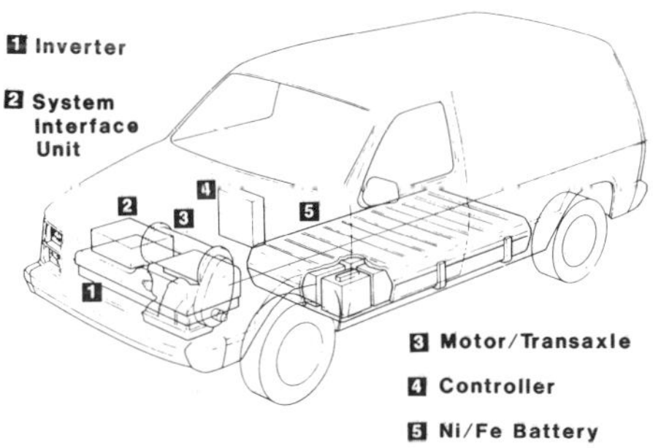

Subsystem placement in T-115 van [65]

Test vehicle

Fig. 1.27. DSEP: dual-shaft electric propulsion system. (Taken from [64,65] with permission of the US Department of Energy.)

sodium/sulfur batteries from Powerplex Technologies, Inc. (a joint-venture company formed between Magna International and Asea Brown Boveri) [62] and Chloride Silent Power Limited. The projected maximum speed and urban range of the vehicle were 97 km h^{-1} and 182 km, respectively (cf., performances of the G-Van and the TEVan, above). Improvements over the ETX-I included: a slightly lighter, more compact d.c.-to-a.c. inverter; integrated digital control of the motor and the inverter, rather than separate analog/digital control; a permanent magnet a.c. motor, rather than an induction a.c. motor; an unsprung transaxle integrated with the rear axle, which eliminated the need for constant-velocity joints.

In the dual-shaft (DSEP) concept [63], the electric motor and the transmission operate on separate parallel axes (Fig. 1.27). The goal of the DSEP project was to develop a lightweight van for urban/suburban services [64,65]. The power-train was developed by the Eaton Corporation and the battery — a nickel/iron type — by Eagle-Picher Industries [50]. The test-bed vehicle met, or exceeded, nearly all its performance goals, namely, maximum speed 112 km h^{-1} (goal: 90.6 km h^{-1}), range 93.3 km, (goal: 104.7 km).

1.12. Outlook

In general, prospects for the wider application of EVs in off-the-road service appear to be good. It is also not unreasonable to conclude that increasing numbers of electric delivery vehicles will be seen on the roads themselves, and that urban public transit operators may move more towards battery-powered buses. The horizon for the widespread use of electric cars is more distant (see Chapter 10); market growth depends heavily on legislation such as that enacted in California. Despite all other influences, however, the key to bringing the electric car nearer to reality, and indeed to increasing the acceptance of all types of EV, is a better battery — one that weighs less, is more compact, can store–deliver more energy, lasts longer, can be recharged more rapidly, and costs less — than present ones. This is a formidable scientific and technological challenge.

References

1. MACKIE, D. *The Practical Mechanic and Engineer's Magazine*, November 1842, pp. 48–52.

2. PLANTÉ, G. *C.R. Acad. Sci.*, 1860, **50**, pp. 640–642.

3. JUNGNER, W. *Swedish Patent 10, 177* (issued 1899).

4. EDISON, T.A. *British Patent 20, 960* (1900).

5. ERICKSON, C.R. *Electric Vehicle News*, 1972, **1**, pp. 26–29.

6. ADLER, C.A., and REIBSAMEN, G.G. *Electric Vehicles at a Glance* (McGraw-Hill, New York, USA, 1978).

7. SHACKET, S.R. *The Complete Book of Electric Vehicles* (Australia & New Zealand Book Company, Sydney, Australia, 1979).

8. WAKEFIELD, E.H. *History of the Electric Automobile* (Society of Automotive Engineers, Inc., Warrendale, Pennsylvania, USA, 1994).

9. TIRATSOO, E.N. *Oilfields of the World* (Scientific Press, Palo Alto, California, USA, 1973).

10. *Electrical World*, 1925, **25**, pp. 665–668.

11. COLLIE, M.J. (Ed.). *Electric & Hybrid Vehicles, Energy Technology Review No. 44* (Noyes Data Corp., New Jersey, USA, 1979).

12. BURKE, A.F., and SOMUAH, C.B. *Computer-Aided Design of Electric and Hybrid Vehicles, General Electric Report No. 81CRD212* (General Electric Company, Schenectady, New York, USA, 1981).

13. ALDOUS, A.F. *Proc. 9th International Electric Vehicle Symposium (EVS-9)*, 13–16 November 1988, Toronto, Canada (Electric Vehicle Association of Canada, 1988), Paper No. EVS88-P03.

14. ADAMS, W.A., and HARVEY, L.D.D. *Proc. 9th International Electric Vehicle Symposium (EVS-9)*, 13–16 November 1988, Toronto, Canada (Electric Vehicle Association of Canada, 1988), Paper No. EVS88-P04.

15. DeLUCHI, M.A., WANG, Q., and SPERLING, D. *Transpn. Res.-A*, 1989, **22A**, pp. 1–27.

16. LANDGREBE, A.R., ADAMS, S.D., and HURWITCH, J.W. *The Potential Impact of Electrochemical Technologies on the Nation's Air Quality* (U.S. Department of Energy, Office of Utility Technologies, Washington, DC, USA, May 1990).

17. BRAUDE, M. *Proc. 9th Electric Vehicle Symposium (EVS-9)*, 13–16 November 1988, Toronto, Canada (Electric Vehicle Association of Canada, 1988), Paper EVS88-037.

18. BRAUDE, M. *Proc. 10th International Electric Vehicle Symposium (EVS-10)*, 3–5 December 1990, Hong Kong (International Research Centre for Electric Vehicles, Department of Electrical and Electronic Engineering, University of Hong Kong, 1990), pp. 1067–1075.

19. *Are You Forking out Too Much on Transport?* (Electric Vehicle Association of Great Britain Limited, 1985).

20. DELANEY, M.J. *Proc. Battery Council International 1985 Convention*, Toronto, Canada (Battery Council International, Chicago, Illinois, USA, 1985), pp. 31–33.

21. KIBBLE, J.D. *Electric Vehicle Developments*, 1980, **5**, pp. 17–19.

22. KIBBLE, J.D. *Batteries International*, 1989, **2**, pp. 22–25.

23. GLOCK, W. *Prog. Batteries Sol. Cells*, 1979, **2**, pp. 200–203.

24. CAREY, W.C.M. *A Cost Comparison between Sodium/Sulphur Battery Railcars and Diesel Electric Rail Cars as Replacement for Existing Diesel Railcars*, British Rail Research and Development Division Technical Note TNEC 6, 1974.

25. SCHMIDT, T. *Electric Vehicle Developments*, 1989, **8**, pp. 133–136.

26. MAYER, M.G., and RAND, D.A.J. *The Battery Man,* 1997, **39,** pp. 16, 18–22, 24, 29–31.

27. *Battery Electric Boats for Inland Waterways* (Lead Development Association, London, UK, 1981).

28. *Recent Advancements in Battery Electric Boats for Inland Waterways* (Lead Development Association, London, UK, 1985).

29. TROUVÉ, G. *The Telegraphic Review & Electrical Review,* 1 September 1881, Vol. IX, No. 206.

30. MARTIN, T.C., and SACHS, J. *Electrical Boats and Navigation* (C.C. Shelley, New York, USA, 1894), p. 25.

31. AYRTON, W.E. *In: The Royal Commission for the Chicago Exhibition 1893, Official Catalogue of the British Section.* (William Clowes & Son Limited, London, UK), p. 245.

32. BAGSHAW, N.E. *Batteries on Ships*. Battery Applications Book Series No. 1, BARAK, M. (Ed.) (Research Studies Press, John Wiley & Sons, Inc., Chichester, UK, 1982), p. 9.

33. PORTER, D.F. *Electric Propulsion*, 1986, **1**, pp. 1, 5.

34. DESMOND, K. *Batteries International*, 1990, **3**, p. 68.

35. HEZLET, A. *Electronics and Sea Power* (Stein and Day, New York, USA, 1975).

36. *Airport Electric '84. Report on a Seminar at Gatwick Airport* (Lead Development Association, London, UK, 1984).

37. HARVEY, A. *Airport Electric '84. Report on a Seminar at Gatwick Airport* (Lead Development Association, London, UK, 1984), pp. 29–30.

38. KIESSLING, R. *J. Power Sources*, 1987, **19**, pp. 147–150.

39. *Scientific American,* 25 November 1882, Vol. XLVII, No. 22, p. 343.

40. HEUMANN, G.W. *Machine Design*, 1973, pp. 20, 23.

41. LIU, C.T., LIN, B.M., and PAN, J.S. *J. Power Sources,* 1996, **59,** pp. 185–187.

42. CRANE, D.P., and BOWMAN, J.R. *Proc. 4th International Electric Vehicle Symposium,* 1976, **1**, pp. 1–28.

43. BRUCKER, F. *Prog. Batteries Sol. Cells*, 1979, **2**, pp. 197–199.

44. *Are You Driving a Good Deal Lately?* (Electric Vehicle Association of Great Britain Limited, 1984).

45. DRIGGANS, R.L., and WHITEHEAD, G.D. *Proc. 9th International Electric Vehicle Symposium (EVS-9)*, 13–16 November 1988, Toronto, Canada (Electric Vehicle Association of Canada, 1988), Paper No. EVS88-039.

46. *Electric Road Vehicles. Urban Transport Research 28* (Federal Minister of Transport, Transport Policy Department, Bonn-Bad Godesberg, Federal Republic of Germany, 1982).

47. KLEIN, F.H. *J. Power Sources*, 1986, **17**, pp. 257–266.

48. KLEIN, F.H. *Proc. Electrafleet '86 Conference*, Birmingham, England.

49. AKIKAWA, Y., TAKESHI, M., and SAKAMOTO, T. *Proc. 9th International Electric Vehicle Symposium (EVS-9)*, 13–16 November 1988, Toronto, Canada (Electric Vehicle Association of Canada, 1988), Paper No. EVS88-P23.

50. *Electric and Hybrid Vehicles Program, 12th Annual Report to Congress for Fiscal Year 1988*, U.S. Department of Energy, Washington, DC, USA, February 1989.

51. KURTZ, D.W. *The DOE ETV-1 Electric Test Vehicle, Phase III Final Report: Performance Testing and System Evaluation.* NASA-CR-168817 (Jet Propulsion Laboratory, Pasadena, California, USA), prepared for the U.S. Department of Energy, Washington, DC, USA, 1981.

52. AiResearch Manufacturing Company of California. *Near Term Electric Test Vehicle ETV-2, Phase II — Final Report.* DOE/CE/51213-01, prepared for the U.S. Department of Energy, Washington, DC, USA, April 1981.

53. BULLOCK, K.R., MAHATO, B.K., BRILMYER, G.H., and WIERSCHEM, G.L. *In* BULLOCK, K.R., and PAVLOV, D. (Eds.) *Proc. Symposium on Advances in Lead–Acid Batteries* (The Electrochemical Society, Pennington, USA, 1984), Proc. Vol. 84–14, pp. 451–465.

54. *Development of Aqueous Batteries for Electric Vehicles.* Summary Report October 1981 – December 1983, ANL/OEPM-83-5 (Argonne National Laboratory, Illinois, USA, 1984).

55. BUREAU, N. *Electric and Hybrid Vehicle Technology '95* (UK & International Press, Dorking, UK, 1995), pp. 20–22.

56. *Electric and Hybrid Vehicles Program, 8th Annual Report to Congress for Fiscal Year 1984*, U.S. Department of Energy, Washington, DC, USA, February 1985.

57. Ford Motor Company and General Electric Company, *ETX-1: First Generation Single-Shaft Electric Power Propulsion System Program, Vol. I, Final Report.* DOE/NV/10308-H1, prepared for the U.S. Department of Energy, Washington, DC, USA, February 1987.

58. McWHINNIE, J., KELLEY, B.P., SIMS, R.I., and CLEGG, G.A. *In* KEILY, T., and BAXTER, B.W. (Eds.) *Power Sources 12, Research and Development in Non-Mechanical Electrical Power Sources* (International Power Sources Symposium Committee, Leatherhead, England, 1988), pp. 93–111.

59. HASKINS, H.J., and REITZ, G.A. *Proc. 8th International Electric Symposium (EVS-8)*, 20–22 October 1986, Washington, DC, USA, pp. 328–337.

60. BELLOWS, R.J., EINSTEIN, H.E., KANTNER, E., GRIMES, P.G., and MALACHESKY, P. *Proc. 20th Intersoc. Energy Conv. Eng. Conf.*, 18–23 August 1985, Miami Beach, Florida, USA (Society of Automotive Engineers, Inc., Warrendale, Pennsylvania, USA, 1985), pp. 2.70–2.78.

61. BATES, B., PATIL, P.B., and CICCARELLI, M.F. *Proc. 8th International Electric Vehicle Symposium (EVS-8)*, 20–22 October 1986, Washington, DC, USA, pp. 396–400.

62. ALTMEJD, M., and DZIECIUCH, M. *Proc. 9th International Electric Vehicle Symposium (EVS-9)*, 13–16 November 1988, Toronto, Canada (Electric Vehicle Association of Canada, 1988), Paper No. EVS88-024.

63. *Electric and Hybrid Vehicles Program, 11th Annual Report to Congress for Fiscal Year 1987,* U.S. Department of Energy, Washington, DC, USA, March 1988.

64. KALNS, I. *Proc. 8th International Electric Vehicle Symposium (EVS-8)*, 20–22 October 1986, Washington, DC, USA, pp. 179–182.

65. KALNS, I. *Proc. 9th International Electric Vehicle Symposium (EVS-9)*, 13–16 November 1988, Toronto, Canada (Electric Vehicle Association of Canada, 1988), Paper No. EVS88–047.

CHAPTER 2

Storing Energy in Batteries

The battery is the fuel tank of the EV. It stores chemical energy in the form of reactive species (i.e., during 'charging') and is designed to release this energy as electrical power (i.e., during 'discharging'). In order to appreciate the extent to which the various candidate battery systems can fulfil the propulsion requirements of EVs, it is first necessary to understand how batteries work.

In this Chapter, a brief review is presented of the basic concepts of battery operation for the benefit of those readers who are not familiar with this aspect of electrochemical science.

2.1. Cells and Batteries

The basic components of an electrolytic **cell** are a **positive electrode**, a **negative electrode**, and an **electrolyte**. During cell operation, ions are created and consumed at the two electrode/electrolyte interfaces by oxidation/reduction reactions. The positive electrode can be an oxide, a sulfide or some other compound that is capable of being reduced during discharge of the cell, while the negative electrode is generally a metal or an alloy that is capable of being oxidized. Both electrodes must be electronically conducting in order to convey electrons to and from the reaction sites; where the conductivity of the positive electrode is inadequate, an electronic conductor (such as carbon) is intimately mixed with the active material. The electrolyte provides the internal circuit between the electrodes, and is an ionizing liquid or solid which has a high and selective ionic conductivity for the ions that take part in the electrode reactions. The electrolyte must be a non-conductor for electrons in order to avoid **self-discharge** of the cell. Other components of a cell include metallic **current-collectors** which conduct current to and from the electrodes, and **separators** which ensure that the electrodes do not touch each other and create a short-circuit. Where a solid electrolyte is employed, this serves also as the separator. In some cell designs, each electrode polarity may consist of a number of **electrode plates** joined in parallel.

A **battery** consists of one or more electrolytic cells that are electrically connected in series. The **voltage** of the battery is the individual cell voltage multiplied by the number of series-connected cells. For example, the familiar 12-V ICEV starter battery consists of six, 2-V lead/acid cells series-connected in a single enclosure. A **battery pack** is a collection of these battery modules that can be joined together in series and/or in parallel configurations to yield the desired voltage and current capability. (Note, generally, the cells in a battery module are series-connected, but in some large batteries — notably, sodium/sulfur, see Chapter 8 — a battery module may be a low-voltage unit with three or four series-connected cells and many parallel chains.)

Primary batteries are designed for a single discharge, after which they are discarded. Examples of this type are the manganese dioxide batteries used to power torches, radios, etc., and the lithium batteries employed in watches, cameras, etc. Batteries which can be recharged by passing a current in the opposite direction to that flowing on discharge are termed **secondary**. Batteries for EV propulsion are required to be rechargeable, and hence are of the secondary type.

2.2. Conversion of Chemical Energy to Electrical Energy

The **energy** stored in a battery is the difference in free energy, ΔG, between the chemical components in the charged and discharged states. In order to illustrate how chemical energy is converted into electrical energy on demand, consider the reaction in which copper is displaced from solution by zinc:

$$Cu^{2+} + Zn \rightarrow Cu + Zn^{2+} \tag{2.1}$$

This reaction is the basis of the Daniell cell which was developed just before the time of Robert Davidson's first EV experiments (see Section 1.1, Chapter 1) and is now of historical interest only. The standard free energy, ΔG^o, of reaction (2.1), that is the free energy for unit activity of Cu^{2+} and Zn^{2+} ions, is -212 kJ mol^{-1} at 25 °C. Since this value is negative, the conventions of thermodynamics demand that the reaction will proceed spontaneously in the direction shown. Thus, if zinc metal is added to a solution containing Cu^{2+} ions, copper metal will be precipitated. In fact, reaction (2.1) is employed in the extraction of zinc from its ores for the removal of copper impurities from the solution which enters zinc-electrowinning cells. In this metallurgical application, the contained chemical energy is not utilized but is simply dissipated as heat. Similarly, metallic corrosion is an electrochemical reaction with a negative ΔG^o and takes place spontaneously.

Reaction (2.1) proceeds by individual electrochemical steps, namely:

$$Cu^{2+} + 2e^- \rightarrow Cu \tag{2.2}$$

$$Zn \rightarrow Zn^{2+} + 2e^- \tag{2.3}$$

In the process discussed above for the removal of copper from electrowinning solutions, both reactions occur simultaneously on the surface of the zinc metal. If, however, the copper and zinc components are separated so that reactions (2.2) and (2.3) must take place at different sites (electrodes), then the overall process can continue only if the electrons generated in reaction (2.3) are able to flow through an external circuit that connects the two electrodes. Furthermore, the flow of electrons can now perform useful work. Thus, in the Daniell cell (Fig. 2.1), reaction (2.1) is harnessed so that its chemical energy is released only on demand and is utilized as electrical power.

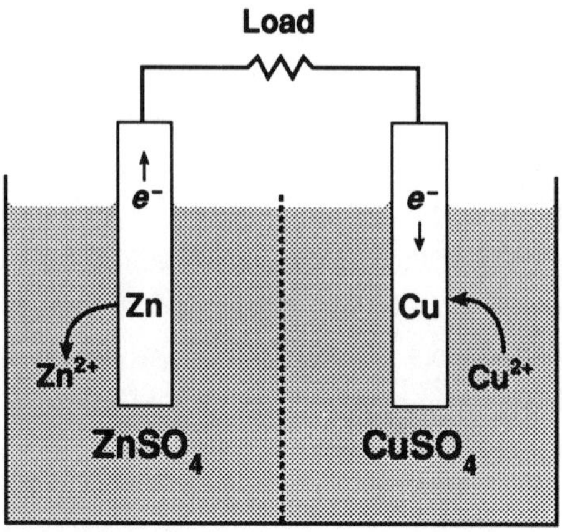

Fig. 2.1. Schematic representation of the Daniell cell.

2.3. Electrical Parameters

2.3.1. Reversible potentials and voltages

The driving force for the external current derived from a cell is the difference in the **reversible potentials** of the two electrode polarities. This difference is the **reversible voltage** of the cell, V_r , and is related to the Gibbs free energy which is liberated in the chemical reaction, ΔG, through:

$$\Delta G = -nFV_r \tag{2.4}$$

where n is the number of electrons transferred per ion and F is the faraday (96 485 coulombs per mole). Under standard conditions (ΔG^o), this relationship gives the **standard cell voltage**, V^o. The value of V^o for the Daniell cell, as calculated from the free energy of reaction (2.1), is 1.1 V at 25 °C.

An electrolytic cell must have two electrodes and thus it is not possible to measure the potential of a single electrode. Nevertheless, it is convenient to consider the potentials of individual electrodes separately.

This is achieved by arbitrarily setting the value of one electrode — the **standard hydrogen electrode** (SHE) — at zero volts. The SHE consists of hydrogen gas at unit fugacity in equilibrium with hydrogen ions at unit activity:

$$2H^+ + 2e^- \rightleftarrows H_2 \qquad (2.5)$$

The single-electrode potential of reaction (2.2) can be calculated by coupling the electrode with the SHE, i.e.,

$$Cu^{2+} + H_2 \rightleftarrows Cu + 2H^+ \qquad (2.6)$$

The ΔG^o of reaction (2.6) is –64.9 kJ mol^{-1} at 25 °C, and thus the V^o of the cell is 0.337 V. By convention, the value of V^o for a cell consisting of an electrode in tandem with the SHE is taken to be the **standard electrode potential** (E^o) of that reaction. Thus, the E^o for reaction (2.2) is 0.337 V. The International Union of Pure and Applied Chemistry (IUPAC) has adopted the system in which single-electrode potentials are derived on the basis of reduction processes. To calculate the zinc electrode potential, the reaction must be written as the reverse of reaction (2.3), i.e.,

$$Zn^{2+} + 2e^- \rightleftarrows Zn \qquad (2.7)$$

Coupling this reaction with reaction (2.5) gives:

$$Zn^{2+} + H_2 \rightleftarrows Zn + 2H^+ \qquad (2.8)$$

The ΔG^o of reaction (2.8) is +147 kJ mol^{-1} at 25 °C and, therefore, E^o = –0.763 V. The voltage of any cell is the difference in the potentials of the two component electrodes. For example, the V^o of the Daniell cell is the difference in the E^o values for the copper and zinc electrodes, i.e., $V^o = 0.337 - (-0.763) = 1.1$ V.

In practice, experimental measurements of single-electrode potentials are made against a reference electrode (e.g., calomel, silver/silver chloride, mercury/mercurous sulfate, mercury/mercuric oxide), the potential of which is well established on the SHE scale [1]. It should be noted that the SHE itself is a conceptual electrode [2]; an actual SHE cannot be constructed because this would require preparation of a solution with a known single-ion activity.

The change of free energy in a chemical process is a function of the activities of the solution species and, therefore, the reversible cell voltage must vary in an analogous manner. From eqn. (2.4) and the dependence of ΔG on the reactant activities, the **Nernst relationship** is derived:

$$V_r = V^o - \frac{RT}{nF}\ln\left[\frac{\Pi\text{ (activities of products)}}{\Pi\text{ (activities of reductants)}}\right] \qquad (2.9)$$

For the Daniell cell at 25 °C:

$$V_r = 1.1 - 0.0296 \log \{[Zn^{2+}] / [Cu^{2+}]\} \qquad (2.10)$$

Note, the Nernst relationship is equally applicable to individual electrode reactions, with E_r and E^o replacing V_r and V^o, respectively, in eqn. (2.9).

In a cell, the electrode that exhibits the more positive potential is the **positive**; the other electrode is the **negative**. The terms **cathode** and **anode** are also used, and denote electrodes at which cathodic and anodic processes occur, respectively. A cathodic process involves the flow of electrons from the electrode to reduce solution species, e.g., reaction (2.2). An anodic process is the reverse situation in which solution species are oxidized and electrons are transferred to the electrode, e.g., reaction (2.3). A cell that spontaneously produces a voltage is a **galvanic cell** and, in this case, the positive electrode is the cathode (Fig. 2.2(a)). A cell that is driven by the application of an external voltage is an **electrolysis cell** and the positive electrode is now the anode (Fig. 2.2(b)). Naturally, this leads to some confusion with secondary batteries because the electrodes that serve as cathodes on discharge become anodes on charge, and *vice versa* (Fig. 2.2(c)).

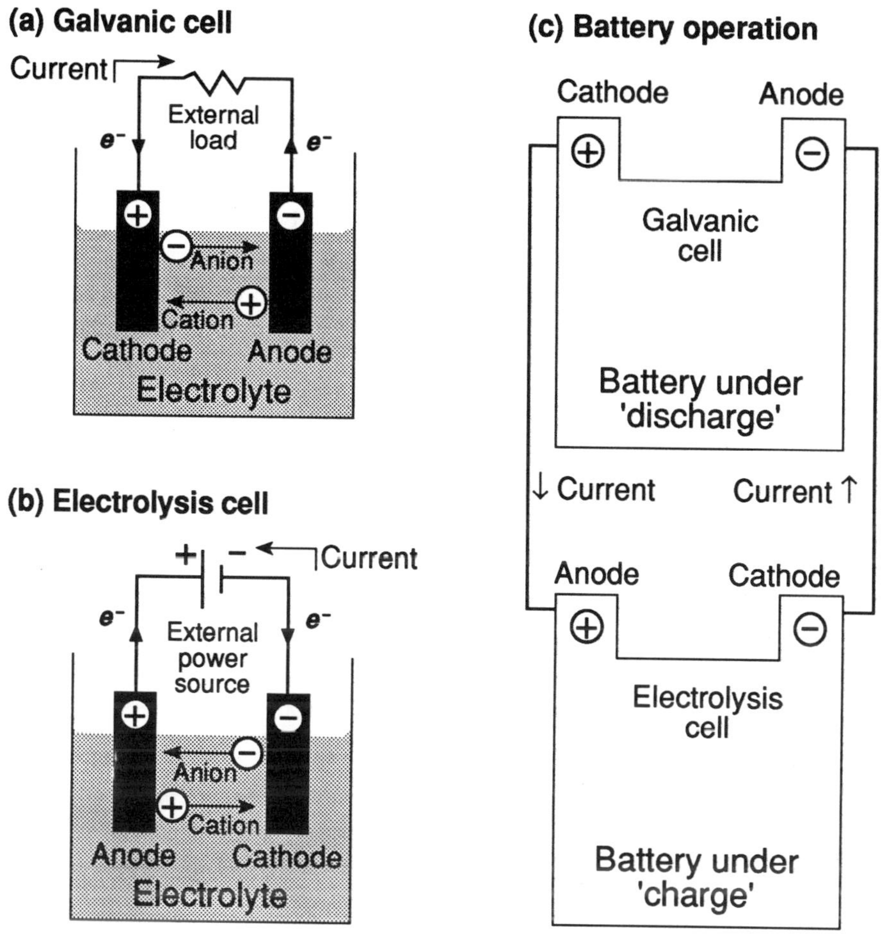

Fig. 2.2. Terminology used in the operation of electrolytic cells and batteries.

Where the terms anode and cathode are used for batteries, it is normal to employ the terminology of the galvanic cell, i.e., the positive electrode in each cell is the cathode and the negative is the anode. It is best, however, to confine the terms cathodic and anodic to currents and use 'positive' and 'negative' to characterize electrodes. Not only does the latter terminology remain the same on charge and discharge, but it is also particularly appropriate because the terminals of commercial cells and batteries are clearly marked ⊕ and ⊖ .

2.3.2. Overpotential losses

The free energy of a battery reaction is the maximum amount of useful work that is available. To obtain maximum work, there must be no losses in potential within the system. This condition is fulfilled exactly only when there is no current flow, but is approximately true when the current is small. Under the current loads required from traction batteries, there is a departure from equilibrium conditions and the useful work performed by the battery is less than the maximum value.

When current is flowing, the shift in potential of an electrode away from the reversible (equilibrium) value is termed the **overpotential**, η:

$$\eta = E_i - E_r \qquad (2.11)$$

where: E_i is the potential under the current load during charge or discharge; E_r is derived from the Nernst relationship. The term **polarization** describes the state that leads to the appearance of an overpotential. Under such a condition, an electrode (or cell) is said to be **polarized**.

Kinetic limitations to the charge-transfer process at an electrode surface give rise to an **activation overpotential**, η_A. This is also referred to as a **charge-transfer overpotential**. The relationship between this overpotential and the rate of the electrochemical reaction (i.e., the current flowing) has been derived from Absolute Rate Theory (see, for example, [3]). Briefly, the Theory is as follows.

Consider the charge-transfer process in which an oxidized species (O) is converted to a reduced species (R):

$$\mathrm{O} + ne^- \rightleftarrows \mathrm{R} \qquad (2.12)$$

The rates of the forward (cathodic) reaction and the reverse (anodic) reaction are each expressed as a **current density** *(i)*, which is the total current (I) divided by the electrode area, i.e.,

$$i_c = nFk_c a_O \exp(-\Delta G_c^\ddagger / RT) \quad (2.13)$$

$$i_a = nFk_a a_R \exp(-\Delta G_a^\ddagger / RT) \quad (2.14)$$

where: k_c, k_a are rate constants; a_O, a_R are the activities of the respective species; $\Delta G_c^\ddagger$, $\Delta G_a^\ddagger$ are the free energies of activation of the respective reactions; R is the gas constant (8.3145 J mol^{-1} K^{-1}).

The net current density, i, is given by:

$$i = i_a - i_c \quad (2.15)$$

When $i = 0$, the reaction will be at equilibrium and the electrode potential will be at its reversible value. Under these conditions, the component current densities, i_a and i_c, will still be finite, but will be equal and opposite to each other. The current that flows equally in each direction is called the **exchange-current density**, i_0.

If, for example, a net anodic current is flowing, and the potential is shifted from the reversible value by an amount η_A, then the activation energy for the anodic process will be decreased by $\alpha nF\eta_A$, where α is the fraction of the electrical energy effective in assisting the process. Correspondingly, the activation energy of the reverse (cathodic) process will be increased by $(1 - \alpha)nF\eta_A$. Thus:

$$i_a = nFk_a a_R \exp[-(\Delta G_a^\ddagger - \alpha nF\eta_A)/RT] \quad (2.16)$$

$$i_c = nFk_c a_O \exp[-\{\Delta G_c^\ddagger + (1-\alpha)nF\eta_A\}/RT] \quad (2.17)$$

Since i_0 is the value of i_a and i_c when $\eta_A = 0$, then:

$$i_a = i_0 \exp(\alpha nF\eta_A / RT) \quad (2.18)$$

$$i_c = i_0 \exp\{-(1-\alpha)nF\eta_A / RT\} \quad (2.19)$$

Thus, the net current flowing is given by:

$$i = i_0 \; [\exp\{\alpha n F \eta_A / RT\} - \exp\{-(1-\alpha) n F \eta_A / RT\}] \qquad (2.20)$$

This is the **Butler–Volmer equation**. It should be noted that the expression is equally applicable in the case where a net cathodic current is flowing.

Figure 2.3(a) shows the potential dependence of i_a, i_c, and the net current density i. It can be seen that there is an approximately linear relationship between overpotential and current at low values of η_A, namely:

$$\eta_A = (RT / nF)\,(i / i_0) \qquad (2.21)$$

At high overpotentials, the rate of the back reaction becomes negligible compared with that of the forward process, and the relationship becomes logarithmic (Fig. 2.3(b)), i.e.,

$$\eta_A = (RT / \alpha nF)\,\ln\,(i / i_0) \qquad (2.22)$$

This is the **Tafel equation** and is generally presented in the form:

$$\eta_A = a + b \log i \qquad (2.23)$$

The expression for η_A (eqns. (2.21) and (2.22)) shows that the activation overpotential losses are much lower in those batteries based on electrode processes that have high exchange-current densities. Another way in which η_A can be reduced is to employ electrodes with large surface areas. This will decrease the current density for a given load and, in turn, will decrease i / i_0. In addition to the simple approach of purely increasing the geometrical size, the electrode area can be enhanced by using active materials with a high intrinsic surface area, e.g., porous matrices.

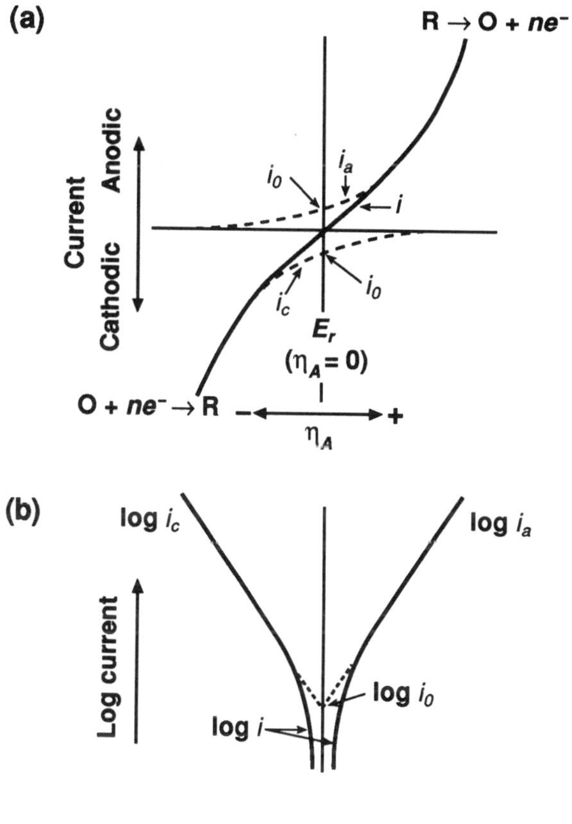

Fig. 2.3. Schematic representation of current density (i) as a function of the activation overpotential (η_A) of an electrode reaction: (a) Butler–Volmer exponential plots; (b) Tafel logarithmic plots.

In the theory developed so far, it has been assumed that the activities of the reacting species at the electrode/solution interface remain constant. These species will become depleted, however, if the current passed is greater than the rate at which the species are replenished from the bulk solution by the mass-transport processes of diffusion, migration and convection. This depletion gives rise to a **concentration overpotential**, η_C (sometimes called a **mass-transfer overpotential**). The form of the relationship between current and η_C is shown in Fig. 2.4. It can be seen that, initially, the current increases rapidly with η_C and then reaches a

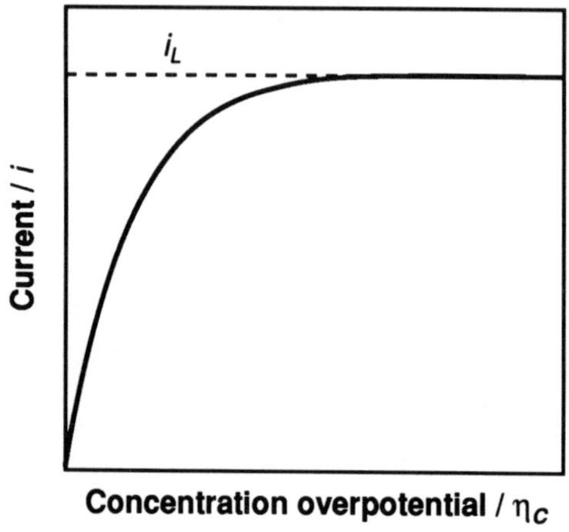

Fig. 2.4. Schematic representation of the dependence of current density (i) on the concentration overpotential (η_C) of an electrode reaction.

limiting value, i_L. This behaviour is given by the relationship:

$$\eta_C = -(RT/nF) \ln (1 - i / i_L) \qquad (2.24)$$

At low overpotentials, eqn. (2.24) becomes:

$$\eta_C = (RT / nF) (i / i_L) \qquad (2.25)$$

These relationships indicate that concentration overpotential losses are minimized under conditions where i_L has its highest value. The limiting current can be increased through the use of high concentrations of reactant species (i_L is proportional to reactant activity), or through enhancing mass-transport rates by, for example, forced electrolyte flow.

The total overpotential is the sum of the activation and concentration components, i.e.,

$$\eta = \eta_A + \eta_C \qquad (2.26)$$

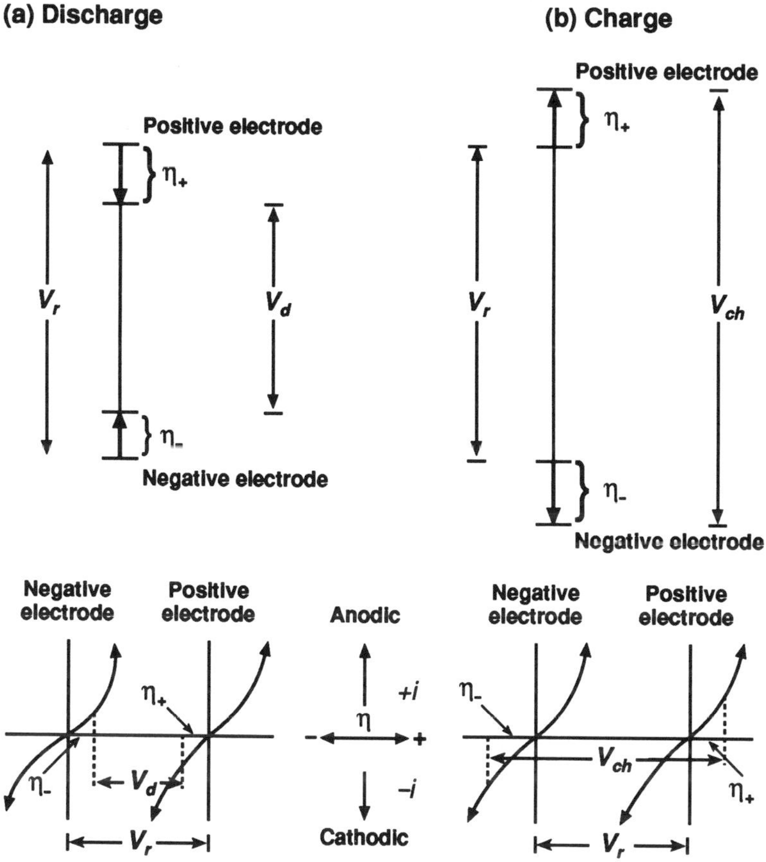

Fig. 2.5. Schematic representation of the variation of cell voltage (V) with current density (i) for a galvanic cell during: (a) discharge; (b) charge. Subscript r denotes equilibrium conditions ($i = 0$); subscripts ch and d denote charge and discharge, respectively; subscripts + and – denote the positive and negative electrodes, respectively. Also included are schematics of the variation in electrode overpotential (η_-, η_+) with anodic (i_a) and cathodic (i_c) current density.

Overpotentials can exist at both electrodes; η_+ is the total overpotential at the positive and η_- is that at the negative. The way in which these overpotentials affect the cell voltage on both discharge and charge is shown in Fig. 2.5. It can be seen that the influence of overpotential is to cause the cell voltage on discharge (V_d) to be lower than the reversible value (V_r), and to be greater than this value on charge.

In addition to the overpotentials at the electrode interfaces, there will be a voltage drop, $I\mathcal{R}$, across the internal resistance ($\mathcal{R}$) of each cell in the battery. This resistance is the sum of the contributions from the various cell components. The $I\mathcal{R}$ drop acts in the same way as the electrode overpotentials and is often referred to as the **resistance overpotential**; it decreases the cell voltage on discharge and increases the voltage on charge. To minimize this source of energy loss, electrolytes with high conductivities are required. In batteries based on aqueous systems, the most effective electrolytes are strong acids and alkalis because hydrogen and hydroxyl ions display the highest molar conductivities. The solution component of the $I\mathcal{R}$ drop can also be minimized by placing the electrodes as close together as practicable.

The cell voltage on discharge is given by:

$$V_d = V_r - \eta_+ - \eta_- - I\mathcal{R} \tag{2.27}$$

It can be seen from eqns. (2.21) and (2.25) that, at low overpotentials, the relationship between i and η takes the same form as Ohm's Law. Thus, η_A and η_C for each electrode polarity can be expressed in terms of an activation overpotential resistance, $\mathcal{R}_A$, and a concentration overpotential resistance, $\mathcal{R}_C$, respectively:

$$\eta = i\mathcal{R}_A + i\mathcal{R}_C \tag{2.28}$$

In other words, eqn. (2.27) can be considered in the form:

$$V_d = V_r - I\mathcal{R}' \tag{2.29}$$

where $\mathcal{R}'$ is the sum of the internal resistance of the cell and the equivalent resistances of the activation and concentration overpotentials at both electrodes. Similarly, the cell voltage on charge, V_{ch}, will be:

$$V_{ch} = V_r + I\mathcal{R}' \tag{2.30}$$

In summary, overpotentials result in a decrease in the useful work obtained during discharge of a battery, and an increase in the work required to charge a battery.

2.3.3. Energy efficiency

The efficiency of storing and recovering electrical energy from a battery is determined by two factors. First, there is a loss in voltage due to: (i) the overpotentials at the positive and negative electrodes on both charge and discharge; (ii) the internal cell resistance, as discussed in the previous Section. In terms of voltage performance, the instantaneous efficiency during discharge is given by V_d / V_r where V_d is the instantaneous value of the battery voltage. From eqn. (2.29), this efficiency is equal to $(1 - I_d \mathcal{R}' / V_r)$, where I_d is the instantaneous discharge current. Thus, the efficiency decreases with increase in discharge current and is favoured by a low resistance, as well as by a high value of V_r. There is a similar efficiency on charge, viz., $V_r / V_{ch} = (1 + I_{ch} \mathcal{R}' / V_r)^{-1}$, where V_{ch} and I_{ch} are the instantaneous voltage and current values, respectively. It is more practical to consider battery efficiency over a combined charge/discharge operation. The resulting **voltaic efficiency** is the product of the average efficiencies on discharge ($\overline{V}_d / V_r$) and on charge ($V_r / \overline{V}_{ch}$) or, in other words, the ratio of the average voltage during discharge ($\overline{V}_d$) to the average voltage during charge ($\overline{V}_{ch}$).

The presence of parasitic processes is the second factor by which the energy efficiency can be impaired. For example, during the charging of batteries using aqueous electrolytes, energy can be wasted on the electrolysis of water. The proportion of the charge that is involved in the useful cell reaction is known as the **coulombic** (or **ampere-hour (Ah)**) **efficiency**.

The **energy** (or **watt-hour (Wh)**) **efficiency** is the proportion of the energy used in charging the battery that is recovered as useful work during the discharge process. The energy efficiency can be measured directly by integrating the power into and out of the battery, i.e.,

$$\text{energy efficiency} = \frac{\int_0^{t_d} I_d V_d \, \mathrm{d}t}{\int_0^{t_{ch}} I_{ch} V_{ch} \, \mathrm{d}t} \tag{2.31}$$

where t_{ch} and t_d are the charge and discharge times, respectively. The energy efficiency is equal to the product of the voltaic and coulombic efficiencies and is sometimes referred to as the **round-trip efficiency**.

2.3.4. Capacity

The **capacity** of a battery is the electrical charge that it will deliver under specified conditions, and is expressed in terms of ampere hours (Ah). The capacity is determined by the mass of active material contained in the battery, and how much of this material undergoes reaction before the battery can no longer deliver the required current at a useful voltage. Individual cells range in capacity from fractions of an ampere hour to many thousands of ampere hours. In this respect, cell capacity is usually related directly to cell size, i.e., larger cells usually mean higher capacities. In practice, the voltage decays gradually with discharge time and then falls rapidly as one of the component active materials becomes depleted. Taking the schematic curve of Fig. 2.6 as a typical example of discharge, the capacity is $I_d t_d^*$ where t_d^* is the time taken for the voltage to fall to a value at which the battery can no longer accomplish effectively the required task. At this stage, the battery is in a fully-discharged state. The **depth-of-discharge** (DoD) is the amount of charge withdrawn compared with the total amount that is available at the same discharge rate. For example, a battery that has been taken to 80% DoD will have been discharged for a time, t_d, such that $t_d = 0.8\ t_d^*$.

It should be noted that the capacity is dependent upon the rate at which charge is withdrawn, i.e., the value of the discharge current. The capacity decreases with increase in discharge current at a rate that

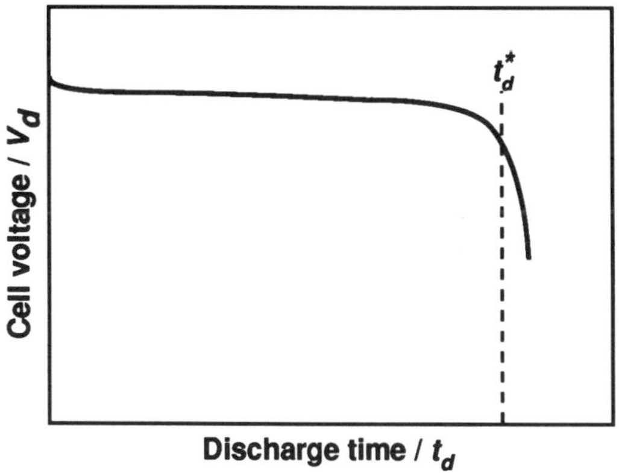

Fig. 2.6. Variation of cell voltage (V_d) under constant-current load.

depends upon the specific system and its cell design. Battery manufacturers quote a **rated**, or **nominal**, capacity for their products. This capacity is the amount of charge that is expected to be delivered from a fully-charged cell or battery for a specified discharge rate, temperature and discharge cut-off voltage. Lead/acid batteries, in particular, exhibit a marked rate dependency (see Section 5.2.4, Chapter 5). Thus, it has become accepted practice to normalize the current (amperes) during discharge to the rated capacity, and to express the discharge rate as C_X / t, where X is the hour rate for the rated capacity, and t is a time specification, in hours. For example, a battery with a nominal capacity of 175 Ah at the $C_5/5$ rate at 25 °C would supply 35 A for 5 h at that temperature. To ensure that batteries will fully meet the requirements of their applications, and to account for losses during use, actual capacities are often engineered to be greater than the nominal values.

2.3.5. Specific energy

The **theoretical specific energy** of a battery is the maximum energy that can be generated per unit total mass of cell reactants. Since the maximum available energy is $-\Delta G$, then:

$$\text{theoretical specific energy} = -\Delta G / \Sigma M_i \qquad (2.32)$$

where ΣM_i is the sum of the molecular weights of the individual species involved in the battery reaction.

In practice, the electrode reactants are supported on inactive current-collectors that carry the current to and from the battery terminals. Other non-reactant materials include the separators (used for preventing internal shorting between electrodes), the container, and electrolyte ingredients added to enhance conductivity. The presence of these materials, together with the fact that the electrode reactants themselves are not fully utilized, results in the **practical specific energy** being much less than the theoretical value given by eqn. (2.32). Furthermore, as discussed above, the energy released is less than the maximum value due to overpotential losses. In the general case:

$$\text{delivered energy} = \int_0^{t_d^*} I_d V_d \, dt \qquad (2.33)$$

Under constant-current discharge (Fig. 2.6):

$$\text{delivered energy} = I_d \, t_d^* \, V_d \qquad (2.34)$$

The term **energy density** is taken to represent the ratio of energy output from a cell, battery, or battery/pack system to its volume (i.e., Wh l^{-1}). It is not the same as specific energy (i.e., Wh kg^{-1}). The energy of a battery depends on the discharge rate and, hence, in reporting specific energy or energy density values, this rate should also be given. Because of restrictions on the mass of batteries that may be carried on an electric road vehicle and also the volume available to accommodate them, both specific energy and energy density are important criteria. These two battery parameters are referred to also as the **gravimetric energy density** and the **volumetric energy density**, respectively.

2.3.6. Specific power

The **specific power** of a battery is the capability to deliver power, I_dV_d, per unit weight (likewise, **power density** is defined in terms of unit volume). The output during peak (short-duration) demand is greater than the power provided during sustained periods. The power available from a battery falls off progressively as the discharge proceeds. Further-more, the power capability is dependent upon the state-of-charge of the battery [4]. Thus, as in the energy case, power densities should be given in conjunction with the particular discharge to which they relate. A frequently used specification is the peak power output of the battery at 80% DoD.

2.4. Heat Generation

The chemical energy that is converted to electrical energy when a cell functions reversibly under isothermal conditions is the free energy of the cell reaction, ΔG. The heat energy component is the enthalpy, ΔH, and this is related to ΔG by the relationship:

$$\Delta G = \Delta H - T\Delta S \tag{2.35}$$

where ΔS is the change in entropy of the cell reaction. Since there must be a heat balance between the initial and final states, the difference between ΔG and ΔH (i.e., $-T\Delta S$) must be due to energy given to, or taken from, the surroundings. If ΔS is negative, then heat will be liberated during the cell reaction, while if the reverse situation prevails, heat will be absorbed. Since the discharge and charge reactions of a cell are the reverse of each other, the thermodynamic parameters for discharge will be the same as those for charge, except that the signs will be reversed. In other words, the entropy term will result either in heating on discharge and cooling on charge, or *vice versa*. Moreover, under practical conditions, a battery does not operate (during either charge or discharge) at its reversible voltage but at a value that is shifted by $\eta_+ + \eta_- + I\mathcal{R}$, see eqn. (2.27). These overpotential and resistive losses result in the generation of heat. Thus, the total quantity of heat generated per second, Q, when a reaction takes place in an electrolytic cell is:

$$Q = -I(T\Delta S / nF) + I(\eta_+ + \eta_-) + I^2\mathcal{R} \tag{2.36}$$

For low overpotentials, eqn. (2.28) holds and hence:

$$Q = -I(T\Delta S / nF) + I^2\mathcal{R}' \tag{2.37}$$

The effect of the above thermodynamic concepts can be illustrated by an analysis of the lead/acid battery system, the discharge reaction of which is:

$$Pb + PbO_2 + 2H^+ + 2HSO_4^- \rightarrow 2PbSO_4 + 2H_2O \tag{2.38}$$

From the values reported [5] for ΔG and ΔH, the value of $T\Delta S$ at 25 °C and 4.6 mol dm^{-3} H_2SO_4 is 11.23 kJ mol^{-1}, and $T\Delta S / nF = 0.058$ V. Thus, the entropic term in eqn. (2.37) will be negative during discharge of a lead/acid battery and will result in the absorption of heat. The second (overpotential) term in eqn. (2.37) represents heat generation. Overall, the battery will cool at currents at which $I^2\mathcal{R}' < I(T\Delta S / nF)$, i.e., when $I < 0.058 / \mathcal{R}'$, and will heat up under the opposite condition. For a typical 100-Ah lead/acid battery, the value of $\mathcal{R}'$ is ~1.5×10^{-3} Ω and, hence, the battery is expected to cool when the discharge current is less than ~40 A. By contrast, the discharge reactions of the nickel/iron and nickel/zinc batteries have negative entropies and for these alkaline systems, therefore, both terms in eqn. (2.37) result in heating.

When batteries are subjected to consecutive charge–discharge cycles, the entropic term in eqn. (2.37) cancels out, irrespective of the sign of ΔS, whereas $I^2\mathcal{R}'$ is positive for both the charging and discharging operations. Thus, all batteries generate heat on cycling and this must be dissipated to the environment to avoid the temperature rising continuously. In traction applications, the $I^2\mathcal{R}'$ heating is greater on discharging than on charging because the latter is conducted over a longer period and in a more controlled manner. Clearly then, it is beneficial to have discharge reactions that give rise to positive entropy changes since the associated cooling will help to offset the greater $I^2\mathcal{R}'$ heating. Thus, from the above considerations, thermal management is a simpler task with lead/acid than with alkaline batteries.

2.5. Battery Design

It is usual to construct batteries from cells in a **monopolar** configuration. Figure 2.7(a) shows the basic concepts of this design. Each cell consists of a number of parallel-connected positive plates together with a similarly connected group of negative plates to build up the required capacity. The cells are joined in series to form a battery. In this configuration, the capacity of the battery is the same as that of the cell since current flows equally through all cells.

Batteries employing **parallel** connections of cells are used when a higher capacity than that provided by the individual cell is needed. It is generally less costly and more reliable to employ a cell with a higher capacity than it is to connect multiple cells in parallel. Situations do arise, however, where larger cells are either unavailable or unsuitable, so a battery that consists of parallel strings of cells is used. In such cases, the battery voltage is that of the individual cell, while the battery capacity is the individual cell capacity multiplied by the number of cells in the battery.

Batteries should never be constructed from other than identical cells. Mixing of cell types, cell sizes, cells from different manufacturers or, even, from different manufacturing lots may result in catastrophic failure of the battery.

Cells can also be combined in a **bipolar** configuration, as illustrated in Fig. 2.7(b). The cells are connected through plates that each, in turn, act as the current collector for the positive electrode in one cell and for the negative in the adjacent cell. With this arrangement, the current pathway through the intercell connections is considerably shorter than that of the conventional monopolar design. Thus, resistance losses are less in bipolar batteries and materials of lower conductivity can be employed as current

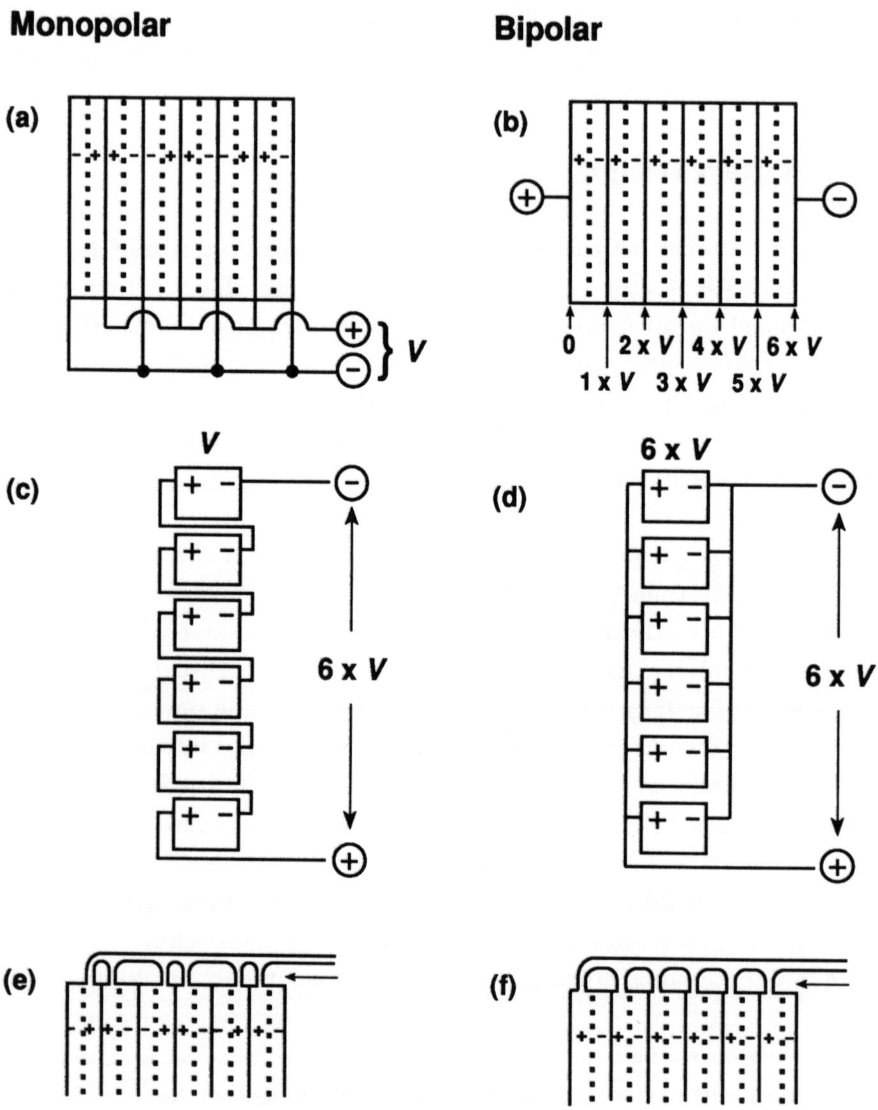

Fig. 2.7. Schematic representations of monopolar and bipolar battery designs. (a),(b) cell geometry (dotted lines = separators). (c),(d) cell connections: series for monopolar; parallel for bipolar. (e),(f) electrolyte manifold (only positive inlet shown): monopolar — no shunt current (all electrodes at same potential); bipolar — shunt current flows through manifold (all electrodes at different potentials).

collectors. The capacity of a bipolar battery, like that of the monopolar type, is the same as that of the individual cell.

For a given size and number of plates, a bipolar module will have less capacity, but a greater voltage, than its monopolar counterpart. By judicious selection of series–parallel connections, however, battery packs can be assembled from either type to give systems with identical capacity and voltage. For example, the series-connected monopolar cells shown in Fig. 2.7(c) have the same total capacity and voltage as the parallel-connected bipolar cells in Fig. 2.7(d).

Bipolar batteries are usually designed to allow electrolyte flow through each cell. This arrangement results in a current path between the individual cells via the electrolyte manifold. Since all electrodes in a bipolar battery are at different potentials (Fig. 2.7(f)), a **shunt current** will flow through the manifold and will result in self-discharge and, hence, an energy loss. This current can be minimized by using narrow flow-tubes and long pathways. Shunt-current protection is particularly important in flow batteries (see Section 7.3.5, Chapter 7). Note, there is no shunt current in the monopolar configuration (Fig. 2.7(e)) since all electrodes of a given polarity are at the same potential.

In the ideal case, it is desirable to maximize the specific energy, the specific power, and the overall electrical efficiency. This would involve the incorporation, as far as possible, of the following features.

- A cell reaction with a large theoretical specific energy that corresponds to a high reversible voltage and reactants of low equivalent weight.
- Fast electrode reactions to minimize activation overpotential.
- High concentrations of reactants in solution and an optimum physical structure of the electrodes to extend the range of currents that can be drawn and to increase the specific power.
- Smallest practical electrode separation and maximum electrolyte conductance to minimize energy-wasting ohmic losses which occur upon current flow, and to enhance further the specific power.

- Electrodes and electrolytes that are compatible and stable over long periods of time so that a long shelf-life is ensured.

Other important parameters are the life of the battery (usually measured as the number of charge–discharge cycles that are sustainable), the range of temperature over which the battery will operate effectively, the existence of overcharge and overdischarge mechanisms, the safety of the battery under abuse or accident conditions, the toxicity of the chemicals involved, and the cost of manufacture. In practice, it is usually impossible to meet all of these conditions simultaneously and a compromise has to be accepted according to the intended usage. Even within one field, such as EV traction, the relative priorities differ according to the application. For example, the detailed battery specification will depend on the particular EV, the duty cycle envisaged, the ambient temperature encountered and so on, as discussed in the following Chapter.

References

1. IVES, D.J.G., and JANZ, G.J. *Reference Electrodes: Theory and Practice* (Academic Press, New York, USA, 1961).

2. BIEGLER, T., and WOODS, R. *J. Chem. Ed.*, 1973, **50**, pp. 604–605.

3. BOCKRIS, J. O'M., and REDDY, A.K.N. *Modern Electrochemistry*, Vol. 2 (Plenum Press, New York, USA, 1970).

4. SPINDLER, W.C. *In* TRAN, T., and SKYLLAS-KAZACOS, M. (Eds.) *Electrochemistry: Current and Potential Applications. Proc. Seventh Australian Electrochemistry Conference,* 15–19 February 1988, The University of New South Wales, Sydney, NSW, Australia (The Electrochemistry Division, The Royal Australian Chemical Institute, 1988), pp. 28–31.

5. BERNDT, D. *Maintenance-Free Batteries* (Research Studies Press, Taunton, England, 1993), p. 37.

CHAPTER 3

Performance Criteria for Electric-Vehicle Batteries

3.1. Vehicle Propulsion Factors

In all the diverse applications of EVs, the battery provides both the power needed to perform the specific tasks and the energy necessary to sustain vehicle operation over a period of time.

The power, P_V, required to drive a vehicle is given by:

$$P_V = v\,(F_r + F_w + F_a + F_h) \tag{3.1}$$

where v is the speed of the vehicle and the 'F' terms are the forces applied in overcoming rolling resistance (F_r), counteracting wind resistance (F_w), accelerating the vehicle (F_a), and climbing hills (F_h).

The force required to overcome rolling resistance is a quadratic function of the velocity [1], i.e.,

$$F_r = kMG\,(1 + 4.7 \times 10^{-3}\,v + 1.3 \times 10^{-4}\,v^2) \tag{3.2}$$

where: k is a constant; M is the mass of the vehicle; G is the gravitational constant; and v is expressed in m s^{-1}. At low vehicle speeds, eqn. (3.2) approximates to:

$$F_r = kMG \tag{3.3}$$

The wind resistance can be expressed as:

$$F_w = \rho C_d A_f v^2 / 2 \tag{3.4}$$

where: ρ is the air density; C_d is the air-drag coefficient; A_f is the frontal area of the vehicle. Since F_w is proportional to the square of the velocity, its impact increases markedly with speed. At speeds below ~30 km h^{-1}, F_w becomes negligible compared with the other resistive forces.

The force required for acceleration is:

$$F_a = M \, \mathrm{d}v/\mathrm{d}t \tag{3.5}$$

Finally, the force necessary to move a vehicle uphill is:

$$F_h = MG \sin\theta = MG \sin(\tan^{-1} b) \tag{3.6}$$

where θ is the angle of incline, and b is the gradient of the hill.

3.2. Power and Energy Requirements of Batteries

The power demand, P_B, on the battery of an EV is given by:

$$P_B = P_V / \varepsilon_M \varepsilon_E \tag{3.7}$$

where ε_M and ε_E are the respective efficiencies of the mechanical and electrical components of the drive-train in the EV. The energy which the battery must deliver is the integral of the power demands of the vehicle between recharges. Naturally, the power and energy requirements of an EV depend upon the duty which the vehicle must perform in its given application. From a battery point of view, the use of EVs can be considered in terms of two main categories, namely, in-plant operation and on-the-road service.

3.2.1. In-plant vehicles

The duty profiles of in-plant EVs are usually regular and well-defined. For example, electric trucks employed in materials handling carry out specific routines of lifting and moving loads within working shifts. In such an application, it is possible to calculate, with reasonable accuracy, the energy which is required for the given task.

Vehicle manufacturers provide customers with data which quantify the energy demanded from the battery in performing the individual vehicle manoeuvres typically associated with materials handling. To utilize this information, operators analyze their duty profiles in terms of the kind and amount of work that has to be accomplished. This involves an assessment of: (i) the extent of the journey on the flat; (ii) the elevation of any ramps to be negotiated; (iii) the weights of the loads and the heights to which they need to be lifted. In terms of power demand (eqn. (3.1)), item (i) covers the forces F_r and F_a, while item (ii) concerns F_h; the remaining force, F_w, is negligible because of the low speeds encountered in this type of service. Note, additional power is needed in lifting operations; this is proportional to the total weight that has to be raised. In order to arrive at the total energy demand placed on the battery, the sum of the energy requirements of each run is multiplied by the number of journeys to be accomplished within a given working period. A battery pack is then selected and is designed to have an excess capacity over the calculated value. This ensures that overdischarging does not occur and also accommodates the decline in battery performance with service. The lead/acid battery readily meets the power and energy demands of electric trucks which are working under normal conditions. In lifting operations, the weight of the battery pack is an advantage since it counterbalances the payload.

Similar data are available for calculating the battery performance required by track vehicles. The only complication is on complex routes which have varied sets of track conditions, but even in these situations, with a systematic approach and by taking each set of conditions in turn, an accurate assessment of the battery criteria can be made.

3.2.2. Road vehicles

Lead/acid batteries have proved to be quite adequate for powering on-the-road EVs in situations where the distances to be covered in the duty profiles are relatively short. The most longstanding and successful of these applications has been the doorstep delivery of milk in the UK. The performance parameters of the milk float are presented in Table 3.1. As discussed in Section 1.8, Chapter 1, battery-powered delivery vans have also found useful niche markets. The characteristics of a typical example of one of these vans are also given in Table 3.1.

Table 3.1. Typical parameters of delivery vehicles.

	Milk float	**Delivery van**
Vehicle weight (t)	2.3	2.5
Payload (t)	1.5	1.0
Maximum speed (km h^{-1})	30	80
Range on typical duty cycle (km)	40	80
Stored energy (kWh)	30	40
Energy consumption (Wh t^{-1} km^{-1})	200	140

An effective measure of EV performance is the energy consumed in moving the vehicle over unit distance. Since three of the four forces which resist propulsion are linearly dependent on the mass of the vehicle (see eqns. (3.1) to (3.6)), the energy demand on the battery is approximately proportional to this mass. Therefore, the energy per tonne per kilometre (Wh t^{-1} km^{-1}) is an appropriate measure of the performance of different vehicles. This parameter depends on the driving profile since the resistive forces are a function of velocity and acceleration (eqns. (3.2)–(3.6)). For example, the greater energy consumption of a milk float compared with a van (Table 3.1) is due to the multiple stop–start nature of the former's service.

In order to explore the ability of batteries to power EVs in a wide spectrum of applications, it is necessary to define a duty profile which is representative of general driving behaviour. Each vehicle owner has different transportation needs, although most of the trips would be expected to follow a regular pattern, e.g., daily commuting by the private motorist. Further, most driving will be undertaken in urban/suburban locations and therefore each driver will, in general, face similar road conditions and carry out similar manoeuvres, e.g., acceleration, cruising, braking, stopping at traffic lights. Standard driving schedules have been developed that represent typical vehicle service; these are discussed in detail in Section 4.3.1, Chapter 4.

An early example of the usefulness of standard driving schedules can be found in the work of Cairns and Hietbrink [1]. With the use of eqns. (3.1) to (3.6), these researchers calculated the power and energy needed to drive an EV over the SAE J227a/C and J227a/D driving schedules that have been developed in the USA by the Society of Automotive Engineers. The findings are presented in rows 1 and 2 of Table 3.2, together with a number of parameters that were derived when a commercial lead/acid battery was used to power the vehicle. At the time, this battery was considered to be state-of-the-art. Rows 3 and 4 give the power and energy demands on the battery; it was assumed that the efficiencies of the mechanical and electrical power-trains of the vehicle were 95 and 70%, respectively. From the point of view of vehicle design and handling, a maximum value of 0.3 was recommended for the fraction of the vehicle weight that could be safely allocated to the battery pack. Taking this value, the power and energy demands on the battery are those presented in rows 5 and 6. Present-day commercial lead/acid batteries are well able to meet the power requirements. On the other hand, the stored energy is limited; at best, state-of-the-art batteries deliver ~40 Wh kg^{-1} (C_3/3 rate). A typical average speed for the J227a/C schedule is ~33 km h^{-1} [1] and, therefore, a 40 Wh kg^{-1} battery would provide a vehicle range of 100 km (row 7) over a discharge time of 3 h. The more severe duty of the J227a/D schedule (typical average speed ~39 km h^{-1} [1]) will involve a discharge time that is appreciably less than that at which the aforementioned specific energy is rated. The capacity of a lead/acid battery decreases markedly with increase in the discharge rate (see Section 5.2.4, Chapter 5)

Table 3.2. Calculated performance parameters for a typical EV powered by lead/acid batteries.

	SAE driving schedule	
	J227a/C	J227a/D
Performance requirements per unit gross vehicle weight		
1. Peak power (W kg^{-1})	13.3	17.5
2. Energy demand (Wh t^{-1} km^{-1})	80	100
3. Power from battery (W kg^{-1})	20	26
4. Energy from battery (Wh t^{-1} km^{-1})	120	150
Performance requirements per unit weight of battery pack		
5. Power demand (W kg^{-1})	67	87
6. Energy demand (Wh kg^{-1} km^{-1})	0.4	0.5
Derived vehicle range per battery discharge		
7. Initial (km)	100	70
8. End-of-life (km)	75	53

and, in the present example, calculations show that a 40 Wh kg^{-1} battery (C_3/3 rate) will supply only 35 Wh kg^{-1} over the J227a/D schedule to give an initial vehicle range of 70 km (row 7). The service life of a battery can be considered to be at an end when the capacity reaches 75% of its initial value (see Section 4.3.2, Chapter 4). At this stage, vehicle range has decreased by about 25%; end-of-life values are given in row 8, Table 3.2.

To power a vehicle for 3 h over the SAE J227a/D schedule requires a battery with a specific energy of 58.5 Wh kg^{-1} at the C_3/3 rate. This 45% increase in specific energy (from 40 to 58.5 Wh kg^{-1}) would provide a 67% increase in range (from 70 to 117 km). In other words, any factor that leads to an increase in battery specific energy will increase the available discharge time and will have a greater than proportional effect on the vehicle range.

The driving ranges derived from the above calculations are too limited for EVs to make a major impact on the road transportation market. One obvious way of extending the range is to include more batteries, but this would be done at the expense of road-handling characteristics. Some EVs have been designed with a battery weight fraction of 0.4 and this will produce a driving range that is one-third greater than the values listed in Table 3.2, provided that the total vehicle mass is kept constant, i.e., the increased mass of the battery is offset by a reduction in the structural mass. In practice, however, a heavier battery will normally require more structural steel and thus the increase in driving range will be much less than proportionate. Another approach to improving range is through better drive-train efficiency. There is not a great deal of scope here, however, since the efficiencies of present EVs are already quite high.

The above calculations have not considered the influence of regenerative braking on EV performance. The incorporation of such a facility has become commonplace in modern EV design; it uses the motor as a generator during deceleration and hill descent, and thus converts the vehicle's potential energy into electrical energy and puts charge back into the battery. Range extensions of up to 30% have been reported for simulated travel over SAE J227a/D duty [2–4]; by contrast, only a 15% increase in range is obtained [5] for operation over the Australian Electric Vehicle Association (AEVA) schedule. On the other hand, because of their more frequent stop–start duty, a gain in range of up to 50% is possible with buses [6]. It is clear that the benefits of regenerative charging are dependent upon the service required from the vehicle.

The above methods for extending the driving range of road EVs are limited in scope. To achieve major improvements, batteries with greater specific energies are required. Simultaneous attainment of specific energy and life goals is difficult, however, because engineering design features which enable the former goal to be reached tend to distance the latter through causing an acceleration in battery degradation processes.

3.2.3. Hybrid vehicles

The development of a battery–ICE hybrid vehicle is aimed at alleviating air pollution and conserving petrol, while maintaining a driving range which is commensurate with that obtained from a conventional ICEV. This is achieved by: (i) running the ICE under its most efficient conditions, and switching in battery power to assist during acceleration; (ii) using the battery alone on short trips. The battery is recharged from the mains supply and/or from an on-board generator. For more discussion of hybrid electric vehicles, see Section 10.4, Chapter 10.

The hybrid vehicle of the DOE programme (HTV-1, see Section 1.10, Chapter 1) was specified to have an all-electric range of ~50 km. Compared with an ICEV equivalent, the hybrid was ~20% heavier (because of the weight of the added lead/acid batteries) and had a greater overall energy consumption [7]. To conserve energy, and not just petrol, a hybrid with an appreciable all-electric range requires a battery with a specific energy which is substantially greater than that available from conventional lead/acid technology. Thus, the performance criteria for batteries used to power hybrid or pure-battery EVs are essentially the same.

3.3. Meeting Battery Performance Criteria

To provide affordable road EVs with acceptable performance in mixed traffic, the motive-power batteries must have the following characteristics (shown schematically in Fig. 3.1):

- high energy per unit mass/volume → for long driving range
- high peak power → for good acceleration and hill-climbing
- high energy efficiency → for economic use of input electricity
- low self-discharge → to minimize energy loss on standing
- fast recharge → for rapid refuelling
- long service (cycle) life → for low depreciation cost
- low cost → to gain customer acceptance
- zero (or low) maintenance → so that little routine attention is required

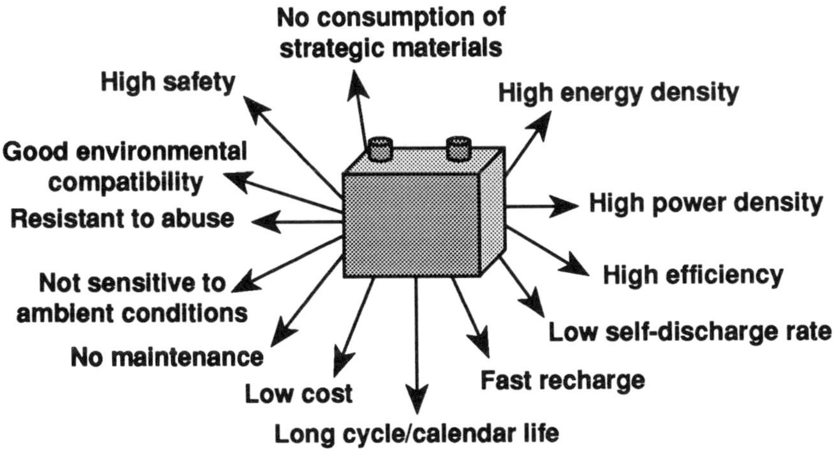

Fig. 3.1. Characteristics required for an ideal EV battery.

- independent of ambient conditions → resistant to both extremes and variations in local climate
- robust design and operation → tolerant of abuse (both electrical and mechanical)
- environmentally benign → non-toxic materials
- proven safety → safe to use under both normal driving and crash conditions
- readily available constituents → made from non-strategic materials.

Among these requirements, the key feature of an EV battery is its specific energy, closely followed by specific power, cycle life, and cost.

3.3.1. Greater specific energy

The specific energy of a battery depends on both the free energy of the cell reaction and the molecular weights of the reacting species (eqn. (2.32), Chapter 2). Thus, the 'ideal' couple would be derived from a highly electronegative element and a highly electropositive element, both of low atomic weight. Hydrogen, lithium or sodium would be the best choice for the negative reactant, and the lighter halogens, oxygen, or sulfur for the positive. To put such couples together in a battery requires electrode designs for effective utilization of the contained active materials, as well as electrolytes of high conductivity that are compatible with the materials in both electrodes. These constraints result in oxygen and sulfur being used in some systems as oxides and sulfides, rather than as the elements themselves. For operation at ambient temperature, aqueous electrolytes are advantageous because of their high conductivities. Here, alkali-group metals cannot be employed as electrodes since these elements react with water; it is necessary to choose other metals which have a reasonable degree of electropositivity, such as zinc, iron, or aluminium. When considering electrode couples, it is preferable to exclude those elements which have a low abundance in the earth's crust, are expensive to produce, or are unacceptable from a health or environmental point of view.

Examination of possible electrode couples has resulted in the study of more than thirty different battery systems with a view to developing a reliable, high-performance, inexpensive power-source for electric traction. The theoretical specific energies of the systems championed for EVs are presented in Table 3.3. Practical specific energies, however, are well below the theoretical maxima. Apart from electrode kinetic and other restrictions that serve to reduce the cell voltage and prevent full utilization of the reactants, there is a need for construction materials which add to the battery weight but are not involved in the energy-producing reaction.

In order to appreciate the extent to which the practical value of the specific energy is likely to differ from the theoretical value, it is instructive to consider the situation of the well-established lead/acid battery. A breakdown of the various components of a lead/acid battery designed to give a practical specific energy of 45 Wh kg^{-1} is shown in Fig. 3.2. The shaded area is the proportion of the battery weight which is directly involved in producing electrical energy. This constitutes only 26% of the total weight. The remainder is made up of: (i) potential cell reactants which are not discharged at the rates required for EV operation; (ii) water used as the solvent for the electrolyte (sulfuric acid alone is not suitable); (iii) lead grids for current collection; (iv) 'top lead', i.e., terminals, straps and inter-cell connectors; (v) cover, container and separators.

Table 3.3. Theoretical specific energies of candidate batteries for EVs.

Battery ⊕	⊖	Cell reaction: Charge ⇐		Discharge ⇒	Specific energy (Wh kg^{-1})
Acidic aqueous solution					
PbO_2	Pb	$PbO_2 + 2H_2SO_4 + Pb$	⇔	$2PbSO_4 + 2H_2O$	171
Alkaline aqueous solution					
NiOOH	Cd	$2NiOOH + 2H_2O + Cd$	⇔	$2Ni(OH)_2 + Cd(OH)_2$	217
NiOOH	Fe	$2NiOOH + 2H_2O + Fe$	⇔	$2Ni(OH)_2 + Fe(OH)_2$	267
NiOOH	Zn	$2NiOOH + H_2O + Zn$	⇔	$2Ni(OH)_2 + ZnO$	341
NiOOH	H_2	$2NiOOH + H_2$	⇔	$2Ni(OH)_2$	387
MnO_2	Zn	$2MnO_2 + H_2O + Zn$	⇔	$2MnOOH + ZnO$	317
O_2	Al	$4Al + 6H_2O + 3O_2$	⇔	$4Al(OH)_3$	2815
O_2	Fe	$2Fe + 2H_2O + O_2$	⇔	$2Fe(OH)_2$	764
O_2	Zn	$2Zn + 2H_2O + O_2$	⇔	$2Zn(OH)_2$	888
Flow					
Br_2	Zn	$Zn + Br_2$	⇔	$ZnBr_2$	436
Cl_2	Zn	$Zn + Cl_2$	⇔	$ZnCl_2$	833
$(VO_2)_2SO_4$	VSO_4	$(VO_2)_2SO_4 + 2VSO_4 + 2H_2SO_4$	⇔	$2VOSO_4 + V_2(SO_4)_3 + 2H_2O$	114
Molten salt					
S	Na	$2Na + 3S$	⇔	Na_2S_3	760
$NiCl_2$	Na	$2Na + NiCl_2$	⇔	$2NaCl + Ni$	790
FeS_2	LiAl	$4LiAl + FeS_2$	⇔	$2Li_2S + 4Al + Fe$	650
Organic lithium					
$LiCoO_2$	Li–C	$Li_{(y+x)}C_6 + Li_{(1-(y+x))}CoO_2$	⇔	$Li_yC_6 + Li_{(1-y)}CoO_2$	320*

* For maximum value of x of 0.5 and $y = 0$

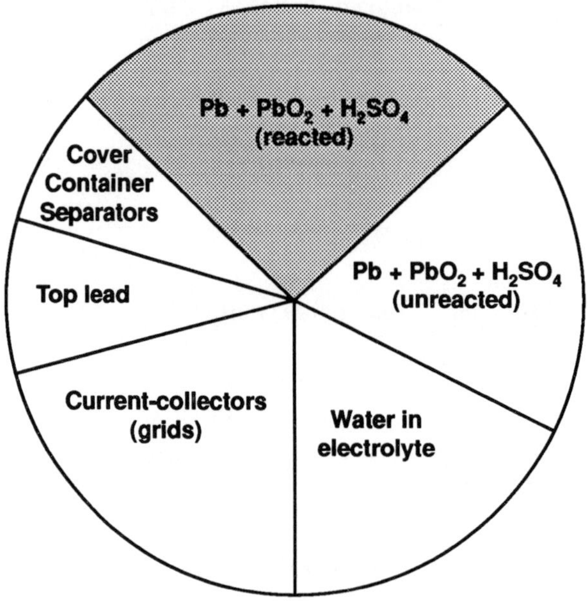

Fig. 3.2. Weight distribution of the components of a lead/acid EV battery with a specific energy of 45 Wh kg^{-1} at the $C_5/5$ rate.

A similar ratio of practical-to-theoretical specific energy must be expected for each of the candidate systems listed in Table 3.3. Present values realized by experimental cells and prototype batteries are listed in Table 3.6 (see later) and will be discussed in Chapters 4–9, together with the characteristics and status of the various systems.

3.3.2. Higher specific power

The specific energy problem is compounded by the fact that an increase in this parameter is generally achieved at the expense of specific power. Three factors dictate the trade-off between specific energy and specific power. First, both overpotentials and the inherent internal resistance of the battery cause a current-dependent drop in output voltage, and thus reduce the total available energy. Second, the operating characteristics of some battery systems change at high currents and, thereby, lower the theoretical value of the specific energy. Third, designing a battery to handle high currents requires its current collectors to be enlarged and this displaces active material, and again lowers the specific energy.

The difficulty of simultaneously maximizing a battery's energy and power capabilities has given rise to the hybrid, or dual-battery, concept. In this, the main unit would be optimized for range (specific energy), and a smaller one would be used for acceleration and hill-climbing (specific power). During less-demanding driving, the vehicle's energy-management system would charge the power battery from the range unit.

3.3.3. Longer cycle life

The cycle life is the number of times that the battery can be recharged. Ideally, the battery should last as long as the vehicle itself. Ironically, this goal is particularly hard to reach with EVs because they enjoy a very long life by virtue of their greater mechanical simplicity.

Measuring battery life purely in terms of charge–discharge cycles is complicated by the fact that the battery is likely to experience differing degrees of discharge between charges, in accordance with the driver's travel demands. Furthermore, some battery technologies yield a constant energy-throughput, while others benefit from shallower cycles and last longer under such a service regime. Still others are more appropriately described in terms of calendar life rather than cycles, particularly those systems which employ highly corrosive reactants.

3.3.4. Lower cost

The eventual factor which determines the commercial success of any new product is, of course, cost. This is especially true of motive-power batteries which may have to be replaced one or more times during the life of a vehicle.

3.4. Setting New Targets for Battery Performance

Following the regulations for zero-emission vehicles in California, major research and demonstration programmes for EVs have been implemented in the USA, Europe and Japan. In particular, the United States Advanced Battery Consortium (USABC) has been formed to stimulate and oversee work by battery developers so that appropriate power-source technology can be made available within the timeframe of the California plan. The USABC formulated two sets of criteria for the required battery packs (Table 3.4). The purpose of the mid-term criteria was to develop realistic battery packs for the 1995–1998 time period. By contrast, the target of the long-term criteria is to gain a vehicle performance which will be directly competitive with present petrol-powered vehicles. Within each set of criteria, primary and secondary objectives have also been defined. More recently, the USABC developed [8] the intermediate commercialization criteria to detail the level of battery performance needed to support successfully an EV market launch and to enable EVs to begin competing with ICEVs. These criteria are also included in Table 3.4.

Not all of the USABC goals are necessarily seen to be reasonable by everyone. For instance, a cycle life of 600 cycles over a calendar life of five years corresponds to 120 recharges per year, or one every three days, whereas most EV drivers will wish to recharge their battery every night. Over five years, it is more likely that there will be 1500 partial recharges rather than 600 full recharges. Again, the long-term goals, taken as a whole, seem more idealistic than realistic. Each of the individual targets will be attained by one or more battery types, and several may accomplish most of them; for a single battery to attain all the targets will be a major achievement. Thus, it will probably be necessary to select the battery type to suit the vehicle application, having due regard to the battery deficiency which is the least critical. It should also be noted that for hybrid EVs (see Section 10.4, Chapter 10) the battery targets may be quite different from those for pure EVs.

Table 3.4. USABC objectives for EV battery packs.

USABC objective:	Mid-term	Commercialization	Long-term
Primary objectives:			
• Specific energy, Wh kg^{-1} (C_3/3 discharge rate)	80 (100 desired)	150	200
• Energy density, Wh l^{-1} (C_3/3 discharge rate)	135	230	300
• Specific power, W kg^{-1} (80% DoD* per 30 s)	150 (200 desired)	300	400
• Specific power (regen.), 20% DoD / 10s, W kg^{-1}	75	150	200
• Power density, W l^{-1}	250	460	600
• Recharge time (20 → 100% SoC†), h	<6	6 (4 desired)	3 to 6
• Fast recharge time, min	<15 (40 → 80% SoC)	<30 @ 150 W kg^{-1} (20 → 70% SoC) (<20 @ 270 W kg^{-1} desired)	<15 (40 → 80% SoC)
• End-of-life	20% degradation of power & capacity specification	20% degradation of power & capacity specification	20% degradation of power & capacity specification
• Calendar life, years	5	10	10
• Life, cycles	600 @ 80% DoD	1000 @ 80% DoD 1600 @ 50% DoD 2670 @ 30% DoD	1000 @ 80% DoD
• Life, urban miles	100 000	100 000	100 000
• Ultimate cost, US$ / kWh	<150	<150 (75 desired)	<100
• Operating environment, °C	–30 to +65	20% loss at extremes of –40 & +50 (10% desired)	–40 to +85
• Continuous discharge in 1 h (no failure), % rated energy capacity	75	—	75
Secondary objectives:			
• Efficiency, % (C_3/3 discharge rate, 6-h charge, at end-of-life)	75	80	80
• Off-tether pack energy loss			
– self-discharge, %	<15 (in 48 h)	12 days: cumulative loss <25% with some performance loss @ extreme temperature limits	<15 (in 1 month)
– thermal loss	<3.2 W / kWh (<15% in 48 h)	3 days: <15% with no performance loss	<3.2 W / kWh (<15% in 48 h)
• Maintenance	zero	—	zero

* DoD = depth of discharge; † SoC = state of charge.

Table 3.5. ALABC plan for advanced lead/acid batteries for EVs.

Improved energy-to-weight ratio

- 50 Wh kg^{-1} at $C_3/3$ discharge rate

Improved power-to-weight ratio

- 150 W kg^{-1} at 80% DoD

Improved rapid charging

- 100% in 4 h; 80% in 15 min; 50% in 5 min

Improved battery life

- 3 years over 500 cycles of the simplified federal urban driving schedule (SFUDS) with less than 20% capacity loss

Improved cost per battery pack

- US$ 150 per kWh

In the opinion of the USABC, neither of the two most developed technologies — lead/acid and nickel/cadmium — will be suitable for EVs. It is argued that the first stores too little energy, while the second suffers from a debilitating combination of high cost and toxicity problems.

Given that further development of the lead/acid battery is essential for the establishment of a legitimate EV industry by the turn of the century, an international group of battery manufacturers and associated industries has set up a consortium — the Advanced Lead-Acid Battery Consortium (ALABC) — to provide a design which will satisfy a significant share of the EV market in the short- to mid-term. The ALABC has five critical research and development objectives (Table 3.5) that concentrate on both material and system design aspects with an overall goal of improved material utilization, that is, greater specific energy (and, thus, vehicle driving range).

The status of the most promising battery systems for EV propulsion is given in Table 3.6. The information has been obtained from a survey of recent literature, e.g., [9]. Where possible, data relevant to the criteria set down by the USABC have been listed. This is particularly difficult for

Table 3.6. Status of battery systems in competition with lead/acid for EVs.

System	Specific energy (Wh kg^{-1})	Peak power (W kg^{-1})	Energy efficiency (%)	Cycle life	Self-discharge (% per 48 h)	Cost (US$ / kWh)
Acidic aqueous solution						
Lead/acid	35–50	150–400	>80	500–1000	0.6	120–150
Alkaline aqueous solution						
Nickel/cadmium	40–60	80–150	75	800	1	250–350
Nickel/iron	50–60	80–150	65	1500–2000	3	200–400
Nickel/zinc	55–75	170–260	70	300	1.6	100–300
Nickel/metal hydride	70–95	200–300	70	750–1200+	6	200–350
Aluminium/air	200–300	160	<50	?	?	?
Iron/air	80–120	90	60	500+	?	50
Zinc/air	100–220	30–80	60	600+	?	90–120
Flow						
Zinc/bromine	70–85	90–110	65–75	500–2000	?	200–250
Vanadium redox	20–30	110	75–85	—	—	400–450
Molten salt						
Sodium/sulfur	150–240	230	85	800+	0*	250–450
Sodium/nickel chloride	90–120	130–160	80	1200+	0*	230–345
Lithium/iron sulfide (FeS)	100–130	150–250	80	1000+	?	110
Organic/lithium						
Lithium-ion	80–130	200–300	>95	1000+	0.7	200

* No self-discharge, but some energy loss by cooling.

prototype batteries because exact details of the system design (e.g., cell, module or battery ± auxiliaries) and/or the operating conditions (e.g., discharge rate, depth-of-discharge, temperature) are not always provided. In general, battery performance is strongly dependent upon such factors. For these battery types, therefore, the quoted energy, power and cycle life should be taken as only approximate values.

It should also be noted that the data, particularly for alternatives to lead/acid and nickel/cadmium (the so-called 'advanced' batteries), usually relate to the results of laboratory studies and that battery packs in vehicles rarely achieve the performance levels predicted by bench tests on cells or monoblocs. In any event, the wide ranges of values that are often quoted for each performance parameter, as well as for the manufacturing cost, emphasize the severity of the problems facing scientists and engineers in the development of reliable and cost-effective batteries to meet the USABC specifications.

References

1. CAIRNS, E.J., and HIETBRINK, E.H. *In* BOCKRIS, J. O'M., CONWAY, B.E., YEAGER, E., and WHITE, R.E. (Eds.) *Comprehensive Treatise of Electrochemistry Vol. 3: Electrochemical Energy Conversion and Storage* (Plenum Press, New York, USA, 1981), pp. 421–504.

2. YAO, N.P. *J. Power Sources*, 1982, **7**, pp. 317–329.

3. BURKE, A.F. *In* McNICOL, B.D., and RAND, D.A.J. (Eds.) *Power Sources for Electric Vehicles* (Elsevier, Amsterdam, 1984), pp. 975–1005.

4. HORNSTRA, F. *J. Power Sources*, 1986, **17**, pp. 81–87.

5. HILL, R.J., RAND, D.A.J., and WOODS, R. *In* PEARCE, L.J. (Ed.) *Power Sources 10, Research and Development in Non-Mechanical Electrical Power Sources* (The Paul Press Limited, London, UK, 1985), pp. 459–480.

6. KORDESCH, K.V. *In* KORDESCH, K.V. (Ed.) *Batteries Vol. 2: Lead-Acid Batteries and Electric Vehicles* (Marcel Dekker, Inc., New York, USA, 1977), pp. 201–413.

7. BURKE, A.F. *Proc. 17th Intersoc. Energy Conv. Eng. Conf.*, 8–12 August 1982, Los Angeles, California, USA (Institute of Electrical and Electronics Engineers, Inc., New York, USA, 1982), pp. 284–289.

8. *USABC: News For and About The United States Advanced Battery Consortium,* Winter 1996.

9. KALHAMMER, F.R., KOZAWA, A., MOYER, C.B., and OWENS, B.B. *Performance and Availability of Batteries for Electric Vehicles: A Report to the Battery Technical Advisory Panel, 11 December 1995,* prepared for the California Air Resources Board, El Monte, California, USA.

CHAPTER 4

Laboratory Testing of Electric-Vehicle Batteries

4.1. Testing Philosophy

Battery testing plays a vital role in EV development programmes; it serves to monitor progress in the light of programme goals. The most realistic way to test a traction battery is in the EV itself. Indeed, this must be the ultimate method since no other test reproduces fully the conditions which correspond to actual driving of a vehicle. Evaluating batteries under simulated duty profiles in the laboratory does, however, provide much important information on the performance which may be expected in EV operation. The advantage of the laboratory approach is that the characteristics of both single cells and battery modules can be obtained; there is no need to have a complete battery pack. Also, the effects of changes in individual vehicle parameters can be readily explored. Such information is difficult to obtain from field trials with vehicles as many operational parameters may vary simultaneously in such trials.

The performance of an EV is determined primarily by the power available from the battery. Therefore, accurate evaluation of EV batteries should be based on schedules in which power is the control variable. Traditionally, however, bench-testing of traction batteries (primarily of the lead/acid type) has involved applying discharge regimes under constant-current conditions. Whilst this duty can be a useful approach for battery comparison purposes, it does not lead to confident predictions of the performance experienced in actual EV service where the load on the battery is far from constant. For example, the range cannot be obtained simply by determining the discharge time for a constant current which represents the average value during driving. Further, discharge under variable current can have an appreciably different effect on battery life from that of using a constant current with the equivalent average value. One possible cause of the change in the battery degradation rate could be differences in the generation of heat within the battery. In the relationship derived between heat output and current flow (eqn. (2.37), Chapter 2), the term that arises from overpotentials and internal resistance ($I^2\mathcal{R}'$) contains the square of the current. For the same average value of I, a patterned discharge must have a higher average I^2 than a constant-current discharge and, hence, give rise to greater heating.

The best method of simulating EV service involves the application of discharge regimes which reproduce the variable load encountered by the vehicle in its intended application. Charging is also a major function in test programmes designed to assess battery service-life. The charging regime is independent of the EV characteristics and, therefore, is usually selected in terms of what is considered by the manufacturer/developer to be the most appropriate for the battery system under investigation. Nowadays, increasing attention is being directed towards the development of fast-charging techniques in order to increase customer acceptability of EVs (see, for example, Section 5.5.4, Chapter 5).

4.2. Simulating the Service of In-Plant and Track Vehicles

In-plant and track vehicles usually operate over a routine profile which is specific to the task in hand. To simulate the duty of such vehicles, the power profile is measured as the vehicle is driven over its required service. It is normal practice to simplify the duty to a repetitive schedule which incorporates the salient operational features of the vehicle, namely: acceleration, cruising, coasting, braking, stopping, and standing.

An example of a power profile used in testing a track vehicle is shown in Fig. 4.1. This profile was devised by the Australian Sugar Research Institute for assessing the applicability of lead/acid-powered locomotives

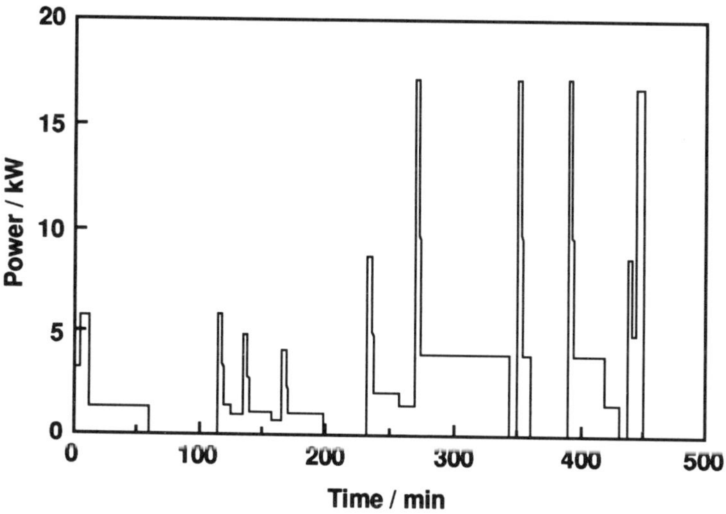

Fig. 4.1. Scaled-down power–time profile for a canefield locomotive. (Taken from [1] with permission of the Australian Lead Development Association.)

for canefield operations [1]. The test schedule is a scaled-down version of the actual power requirements of the locomotive as, in the interests of both economy and ease of operation, testing is conducted on 24-V battery packs rather than on the full-scale, 450-V, systems used in the locomotives themselves. The power peaks in Fig. 4.1 correspond to acceleration of the locomotive, the plateaux to constant-velocity cruising, and the zero regions to rest periods.

4.3. Simulating the Service of Road Vehicles

4.3.1. Standard driving schedules

The duty cycles of road EVs have a certain commonality in that each involves a substantial amount of driving under similar conditions as dictated by the general traffic situation. Standard driving schedules have been developed to describe vehicle operation in urban and suburban traffic flows, and then to allow performance comparisons between EVs and components (especially batteries) on the same basis.

After choosing an appropriate standard driving schedule, the next step towards testing EV batteries is to obtain the corresponding power profile. This profile can be determined by running the vehicle either on a dynamometer or on the road. The former is mostly preferred as the vagaries of weather and terrain are eliminated. Alternatively, the power demand can be calculated from equations which represent the forces that must be overcome in order to propel the vehicle over each phase of the driving schedule (see Section 3.1, Chapter 3).

No single standard schedule has been adopted universally for the representation of EV usage patterns. The early work was carried out using the J227a series recommended by the Society of Automotive Engineers (SAE) in the USA [2] and this became widely applied in the evaluation of EVs and batteries at that time. Figure 4.2 presents the SAE J227a/B, /C, and /D schedules; these were intended to simulate fixed-route urban, variable-route urban, and variable-route suburban travel, respectively. Each of these three schedules consisted of the following five successive periods: (i) acceleration to a velocity v within a time t_a; (ii) cruising at v for a time t_{cr}; (iii) coasting for a time t_{co}; (iv) braking to bring the vehicle to rest in a time t_b; (v) remaining idle for a time t_i. The schedules specified different values for v; the various time intervals are listed in Table 4.1. In battery testing programmes, the selected driving schedule was repeated continuously until the battery could no longer deliver the required power.

The SAE J227a procedures specified only the cruise velocity and the time of transmission from one mode to the next. The complete velocity profile was not fixed and hence the distance travelled during the acceleration, coasting and braking periods was variable; it was dependent upon the design characteristics of the vehicle under consideration.

The efficacy of the SAE J227a schedules in addressing EV service was questioned by US testing authorities in the light of data collected from EV field trials [3]. In particular, it was discovered that, unlike the conditions set down in the SAE J227a procedure (Table 4.1), the times spent in acceleration and deceleration were significantly longer than those in cruise

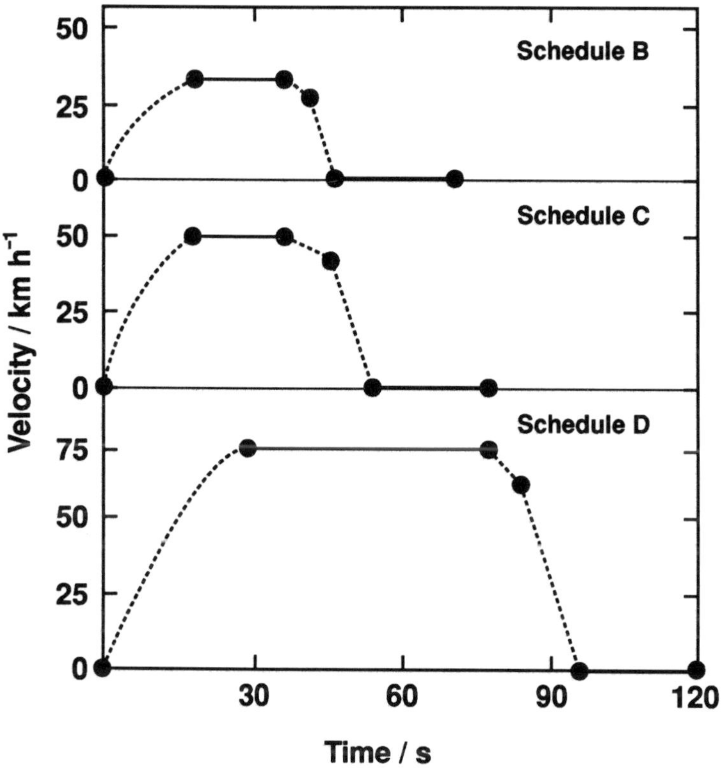

Fig. 4.2. SAE J227a standard driving schedules [2].

Table 4.1. Specifications of SAE J227a standard driving schedules.

		Schedule J227a		
Operation		**/B**	**/C**	**/D**
v	cruise velocity (km h^{-1})	32	48	72
t	total time (s)	72	80	122
t_a	accelerate (s)	19	18	28
t_{cr}	cruise (s)	19	20	50
t_{co}	coast (s)	4	8	10
t_b	brake (s)	5	9	9
t_i	idle (s)	25	25	25

at all velocity levels. A new procedure was therefore formulated [4] and was based on the Federal Urban Driving Schedule (FUDS) which was originally developed for evaluating both the exhaust emissions and the fuel economies of ICEVs [5].

The FUDS [5] is illustrated in Fig. 4.3. In terms of both time and distance, this schedule is considerably longer than the former SAE J227a procedure. Furthermore, the FUDS is much more random in character. It has average and maximum speeds of 31.5 and 89 km h^{-1}, respectively; an average acceleration of 0.79 m s^{-2}; a distance of 11.9 km; and idle periods which amount to 18% of the total time.

In assessing battery performance, the end of discharge is usually taken as the point at which the battery can no longer deliver the power necessary to follow the load profile of the vehicle. The single high-velocity peak in the early stages of the FUDS, which is preceded by a sharp acceleration, constitutes by far the greatest power demand (see Fig. 4.4, below). If this feat is achieved, the battery will most probably fulfil the requirements of the schedule until the high-velocity peak is again encountered. Thus, the measured range over the FUDS is likely to be multiples of ~12 km and could vary by this amount from one test to another.

A simplified version of the FUDS was developed [4]: (i) to reduce the cost of test hardware; (ii) to extract information which was difficult, or impossible, to obtain because of the continuously transient nature of the FUDS; (iii) to shorten the test cycle and thus reduce the uncertainty in range prediction. The simplified schedule was derived for a 'standard' vehicle — the 'Improved' Dual-Shaft Electric Propulsion (IDSEP) van. This was a hypothetical version of the Eaton DSEP van (see Section 1.11, Chapter 1) that assumed certain improvements in road load losses and drive-train efficiencies compared with the DSEP vehicle. The peak specific power limit was defined as the minimum power required to accelerate the IDSEP van from zero to 80 km h^{-1} in 20 s. This requirement was found to correspond to a battery specific power of 79 W kg^{-1}. Hence, the FUDS was modified in order to limit the peak power to that value; the new reference

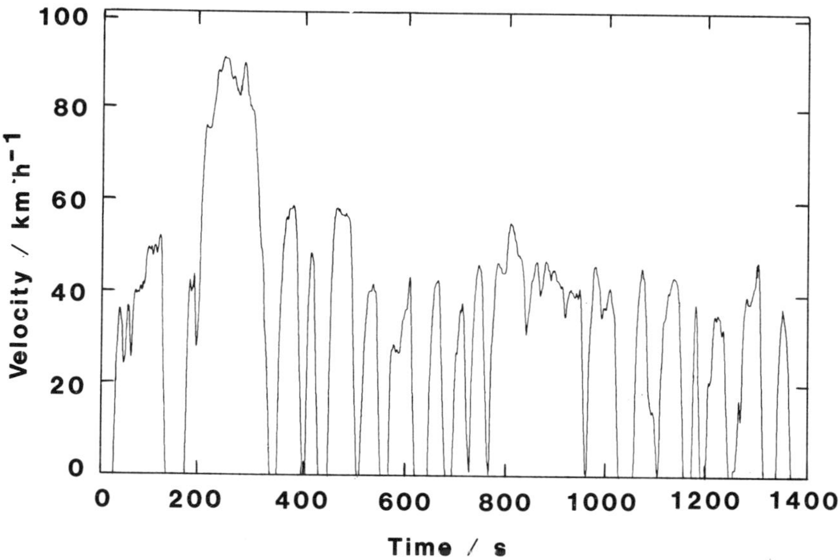

Fig. 4.3. Federal Urban Driving Schedule (FUDS) [5].

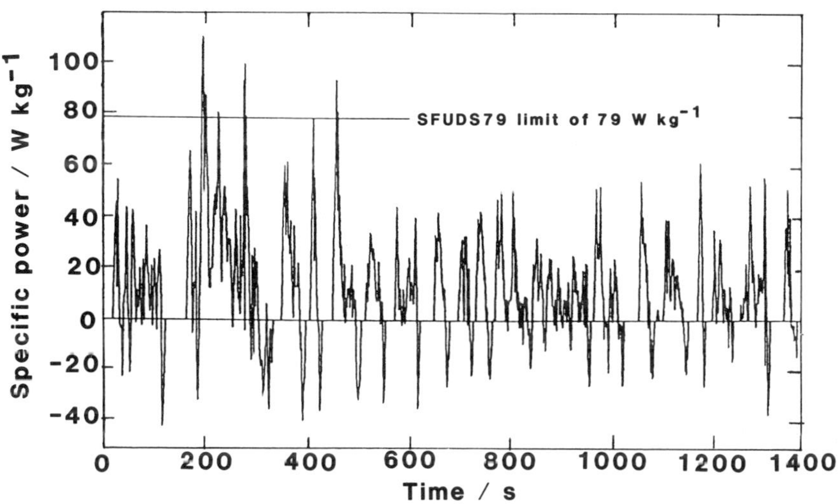

Fig. 4.4. Derived specific-power profile for the IDSEP vehicle operating on the FUDS [4].

driving schedule was referred to as the FUDS79. The impact of this change was minimal due to the few occurrences of higher power levels in the schedule (Fig. 4.4). Formulation of the simplified schedule involved an iterative process to sort the battery power at each second on the FUDS79 into a limited number of power steps that approximated the FUDS79 power spectrum. The resulting specific-power profile is shown in Fig. 4.5 and was first known as the SFUDS79 (i.e., 'Simplified' FUDS limited to 79 W kg^{-1} battery power), but is now generally termed the 'SFUDS'. The corresponding velocity profile for the IDSEP van is presented in Fig. 4.6. During the 8-s period over which the battery must deliver a peak power of 79 W kg^{-1}, the discharge will be at a relatively high current if the battery has a large weight and a small cell voltage. For example, a heavy 2-V lead/acid unit would be discharged at a higher current than a light 12-V counterpart. The profile includes a series of charges which simulates a contribution from regenerative braking, and the average discharge current during the profile is approximately equal to the $C_3/3$ rate.

From simulation tests, a comparison of the SFUDS and FUDS79 parameters shows that (Table 4.2) for the former schedule: (i) the profile time is reduced from ~23 to 6 min; (ii) the average velocity is 2% lower and the maximum varies by only 3.6 km h^{-1}; (iii) the length of the profile is reduced from 11.9 to 3.1 km; (iv) the average power differs by only 2%, the peak power agrees by definition; (v) the energy is within 1 Wh km^{-1} with regeneration, or 3% less without regeneration. A practical demonstration of the equivalence of the two schedules was carried out by the Argonne National Laboratory (ANL) using a nickel/iron module (Table 4.2).

The FUDS79 stipulates a 10-min, open-circuit, rest period for the battery after each application of the 1372-s profile. It is appropriate, therefore, to incorporate approximately the same percentage of rest time in SFUDS. From tests on both nickel/iron and lead/acid modules, ANL showed that there was little effect of open-circuit periods on FUDS79 and SFUDS discharges (in terms of both vehicle range and battery temperature), provided a rest period of 10 min was included after every four applications of the SFUDS.

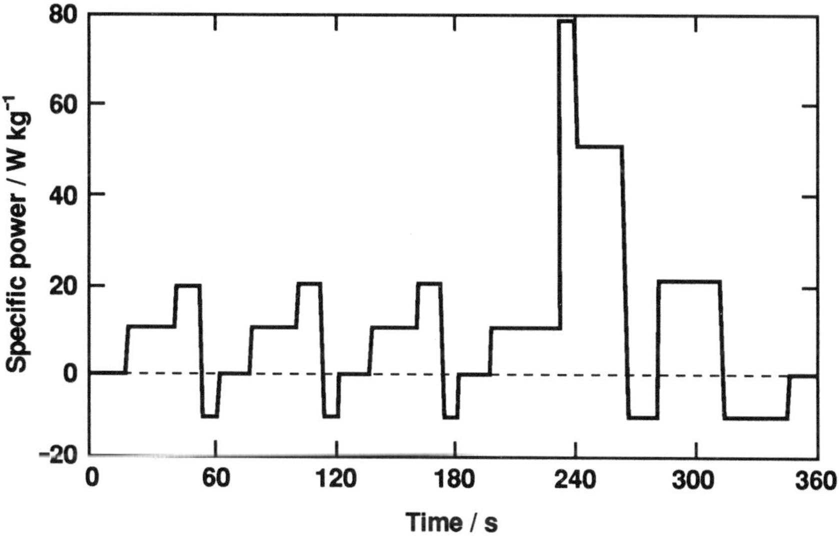

Fig. 4.5. SFUDS power profile [4].

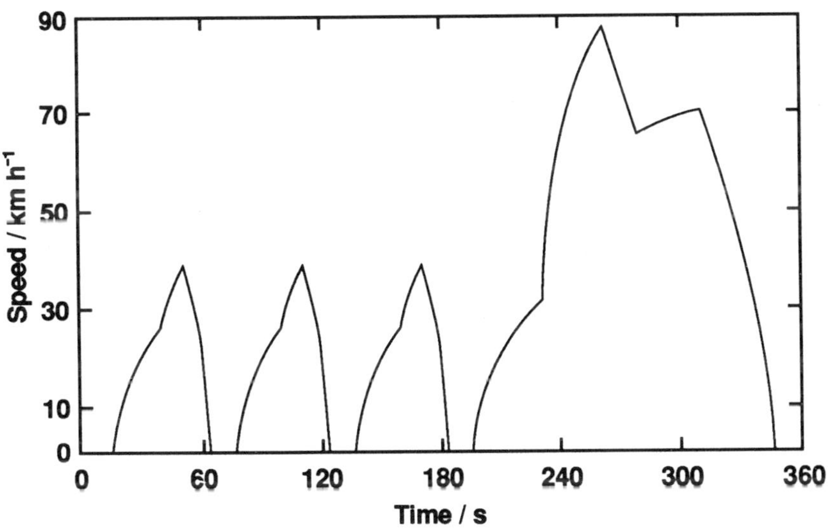

Fig. 4.6. Derived velocity profile for IDSEP vehicle using the SFUDS power profile [4].

Table 4.2. Comparison of FUDS79 and SFUDS [4].

Parameter	Simulation		Using Ni/Fe module	
	FUDS79	SFUDS	FUDS79	SFUDS
Time (s)	1372	360	1372	360
Velocity (km h^{-1})				
– average	31.2	30.6	31.2	30.6
– maximum	91.1	87.5	91.1	87.5
Distance (km)	11.9	3.1	11.9	3.1
Power (W kg^{-1})				
– average	10.1	9.9	10.0	9.9
– peak	79.0	79.0	79.0	79.0
Energy (Wh km^{-1})				
– w regeneration	225	224	221	221
– w/o regeneration	277	270	273	264

The minimum battery power required for the IDSEP vehicle to complete the SFUDS was found to be 50 W kg^{-1}. The portion of the FUDS at which, below this power limit, the vehicle could not deliver this minimum level of performance occurred during the acceleration immediately after the fourth stop (i.e., between the 447-s and 555-s marks, Fig. 4.3). The vehicle test was terminated at the point where the battery could not deliver 50 W kg^{-1} when demanded by the SFUDS power profile (Fig. 4.5). The estimated distance travelled by the vehicle was obtained by multiplying the discharge time (excluding rest periods) by the average speed for the schedule (30.6 km h^{-1}, Table 4.2).

In order to make SFUDS applicable to all classes of road EV, a 'generic' SFUDS — called the GSFUDS — has been proposed [6], Fig. 4.7. In the GSFUDS version, the ordinate scale in Fig. 4.5 is converted to the ratio of the power to the average power for a given vehicle. Adjustments have been made to the schedule so that the various power levels are integer multiples of the average power. These modifications do not significantly change the power profile of the SFUDS, yet they produce a more generic test schedule which is characterized by a single parameter, namely, the

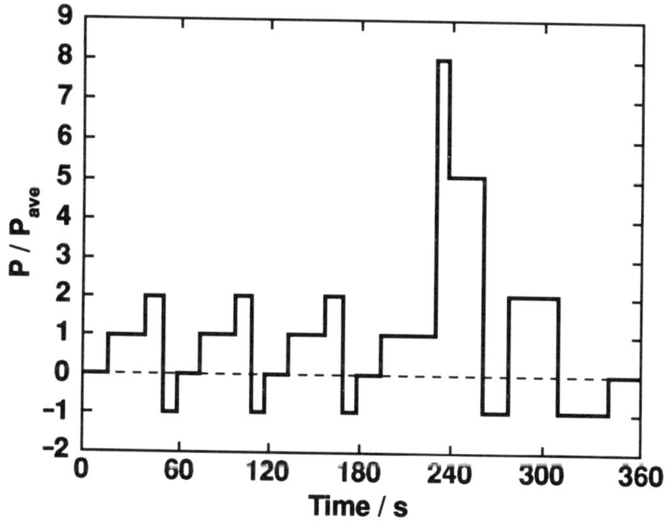

Fig. 4.7. The GSFUDS power profile [6].

average power. The minimum performance criteria for an EV battery when operating on the GSFUDS is a capability to supply five times the average power for the schedule. The GSFUDS has been employed [7] to generate data from which the performance of various types of batteries can be determined in EVs which have greatly different design characteristics, e.g., the IDSEP van, and the GM Impact (see Section 1.10, Chapter 1). The resulting battery performance is presented in terms of constant power, as well as Ragone plots (see Section 4.3.2) which display the specific energy (Wh kg^{-1}) as a function of the average specific power (W kg^{-1}) for the test schedule.

In a later development, the US Advanced Battery Consortium (USABC), see Section 3.4, Chapter 3, modified [8] the SFUDS into the Dynamic Stress Test (DST). The DST is scaled to a percentage of either the maximum rated power or the USABC peak-power goal, and also requires higher regeneration levels than the SFUDS. A graphical representation of a generic DST schedule is presented in Fig. 4.8. In general, a peak discharge power of 100% on this graph is intended to be 80% of the USABC peak-power

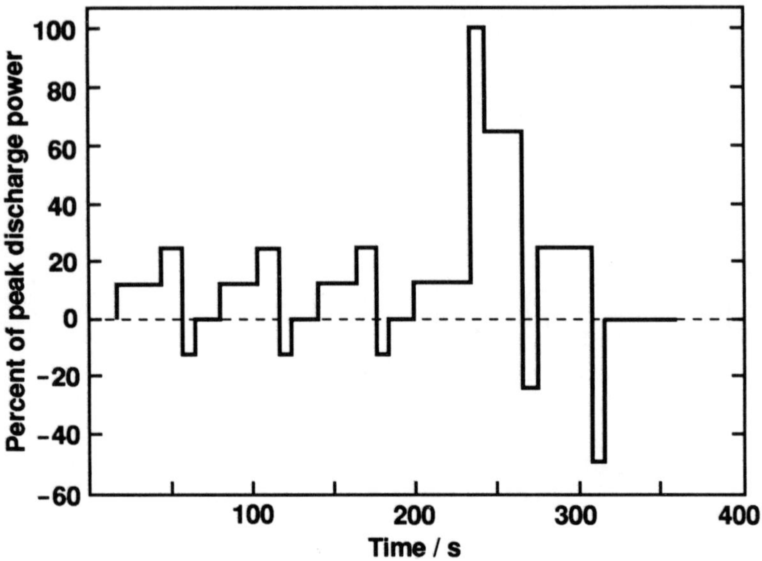

Fig. 4.8. DST power profile [8].

goal (at 80% DoD) for a given battery technology. In some situations, however, the test may be performed strictly in accordance with the rating specified by the manufacturer. As an example, if the DST profile is scaled to 80% of the USABC medium goal of 150 W kg^{-1} (see Table 3.4, Chapter 3), it would have a peak power of 120 W kg^{-1} and an average power of 15 W kg^{-1}.

The development of variable power discharge profiles has also been pursued [9] in Europe by the Joint Research Committee (JRC) – Traction Battery Working Group with the help of a number of local test institutes. The JRC consists of a consortium of eight major European automotive companies, namely, BMW, Daimler-Benz, Fiat, Peugeot, Renault, Rover, Volkswagen, and Volvo. A standard European test schedule (speed vs. time) — the ECE-15 — has been adopted and the corresponding battery power profile — the ECE15-L — has been calculated using a JRC standard vehicle. The ECE-15 is comprised of two parts, one urban part which is repeated four times, without rest periods, followed by one suburban part (Fig. 4.9). The total profile takes 1180 s (urban part: 4 × 195 s; suburban

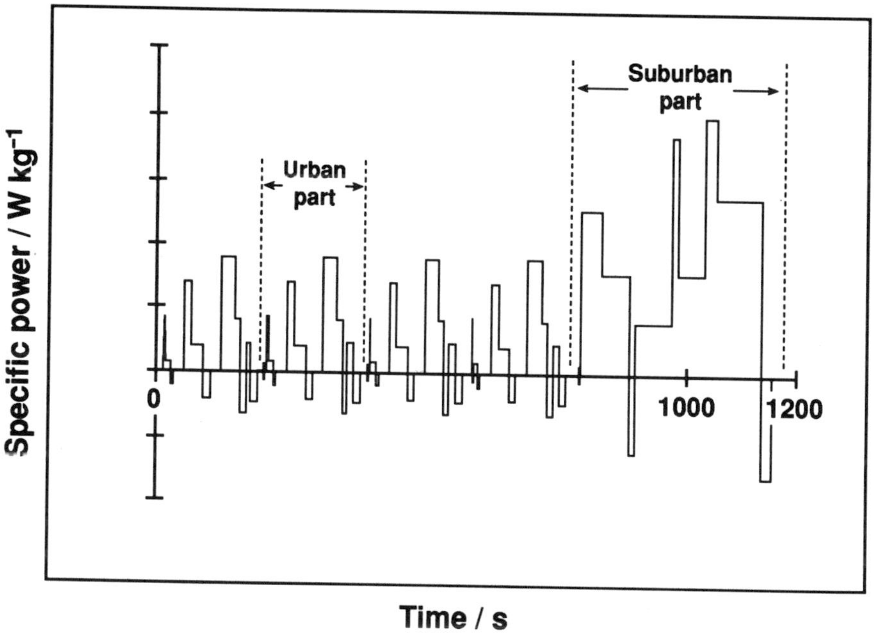

Fig. 4.9. ECE-15 power profile for the JRC reference vehicle [9].

part: 400 s) and is repeated without rest periods until the end of battery discharge is reached. The schedule of the ECE15-L is the same as that of the ECE-15, but the power levels of the individual stages differ according to the battery type. Separate schedules have been formulated for lead/acid, alkaline batteries, and high temperature batteries.

4.3.2. Determination of battery performance

Figure 4.10 shows the power profile of the 'Improved' ETV-1 vehicle (IETV-1) driven over the SAE J227a/D procedure [10]; this was a hypothetical vehicle that assumed certain improvements in performance compared with the ETV-1 (see Section 1.10, Chapter 1). It is clear that the segment in the profile corresponding to acceleration requires considerably more power than the segment for cruising. Power is returned to the battery during both the coasting and the braking modes through regenerative charging.

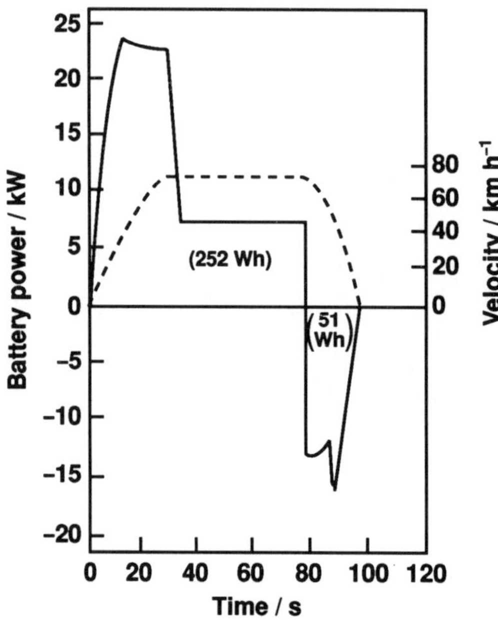

Fig. 4.10. Power (solid curve) required by the improved ETV-1 to negotiate the SAE J227a/D driving schedule (dashed curve). (Taken from [10] with permission of Elsevier Science Limited, UK.)

The discharge profile in Fig. 4.10 was applied by the US National Battery Test Laboratory (NBTL) to determine the ranges of EVs under simulated urban driving conditions when driven by different battery systems [11]. Figure 4.11 presents the ranges obtained for the IETV-1, together with the impact of regenerative braking. The charge returned to the battery during braking amounted to ~20% of that extracted for acceleration and cruising of the vehicle. For some types of battery, the corresponding increase in range was noticeably above 20%; in those cases, therefore, the regeneration procedure enhanced the utilization of plate materials.

Hornstra [10] developed an alternative method for the estimation of vehicle range. This involved the determination of specific-energy/specific-power relationships (i.e., Ragone plots) for batteries, together with the

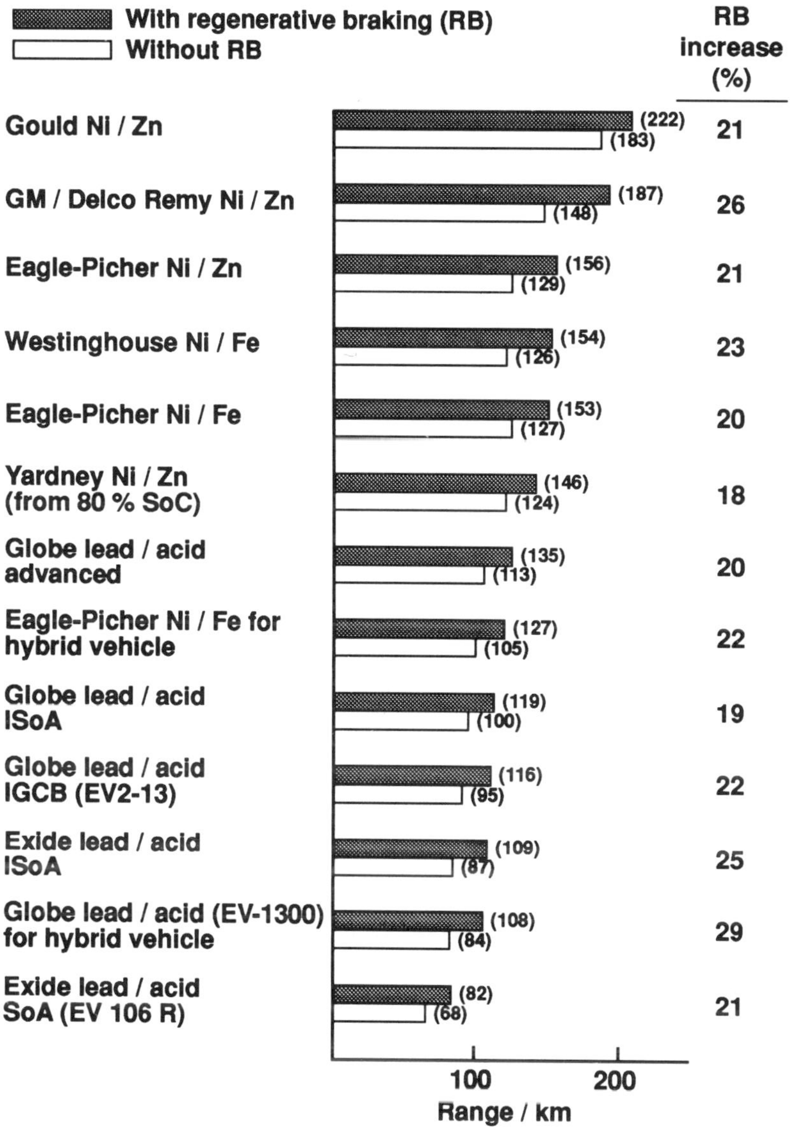

Fig. 4.11. Projected number of km travelled in one complete discharge of a 448-kg battery used to power the IETV-1 over the SAE J227a/D driving schedule. (Taken from [11] with permission of Elsevier Science Limited, UK.)

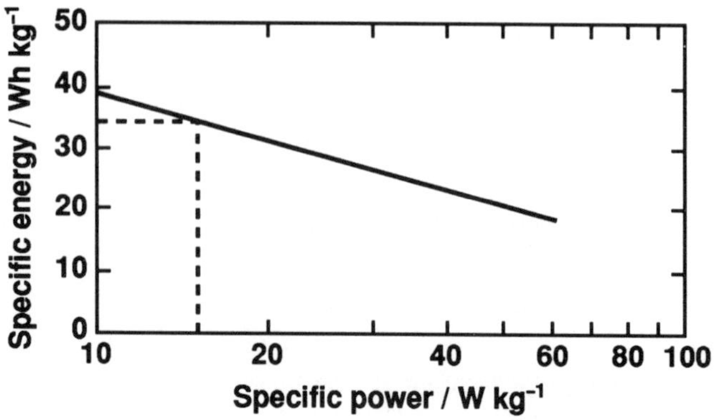

Fig. 4.12. Ragone plot for a EV-2300 lead/acid battery. The vertical dashed line corresponds to the average power requirement of the IETV-1 over the SAE J227a/D driving schedule. (Taken from [10] with permission of Elsevier Science Limited, UK.)

dependence of peak power on battery state-of-charge. By way of example, the Ragone plot for the EV-2300 lead/acid battery is shown in Fig. 4.12; this battery was developed by Johnson Controls, Inc. The average power required from the battery for driving the IETV-1 over the SAE J227a/D schedule is 15 W kg^{-1}. Inspection of Fig. 4.12 shows that the specific energy corresponding to this power rating is 34 Wh kg^{-1}. Thus, the battery discharge will last for 34/15 = 2.27 h, unless the battery becomes peak-power-limited before the discharge is completed. The latter condition can be determined by considering the measured variation in peak power as a function of the depth-of-discharge (DoD), Fig. 4.13. The peak specific power over the SAE J227a/D schedule is 47 W kg^{-1} and it can be seen that this power can be obtained for up to 93% DoD. Consequently, the battery will provide adequate power for $2.27 \times 0.93 = 2.11$ h. Since the average speed over each schedule is 47 km h^{-1}, the vehicle range will be 99 km. This derived range is close to the value of 91 km that was obtained from the same batteries when tested under simulated driving. Good agreement between derived and measured ranges was also obtained for nickel/iron and nickel/zinc batteries.

According to the literature which is available on the implementation of the SFUDS [4], cells/batteries for testing are subjected to consecutive

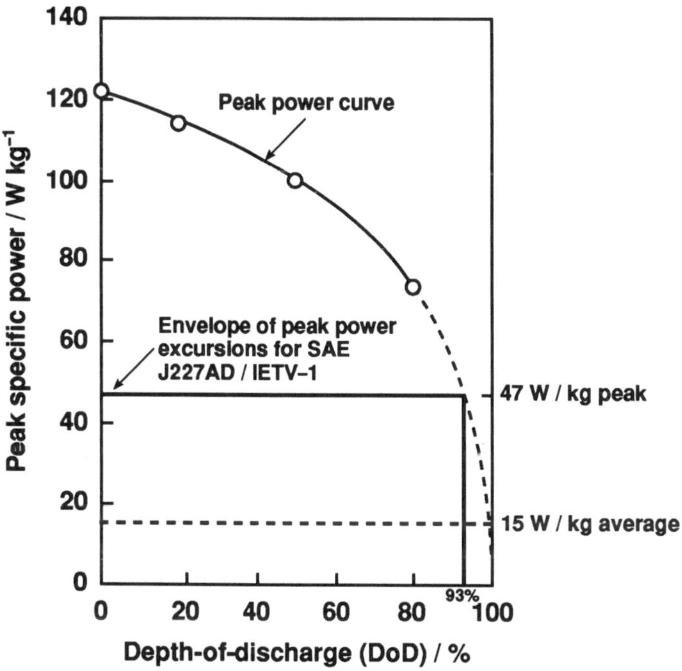

Fig. 4.13. Peak power versus depth-of-discharge (DoD) for an EV-2300 lead/acid battery, together with peak and average requirements of the IETV-1 over the SAE J227a/D driving schedule. (Taken from [10] with permission of Elsevier Science Limited, UK.)

iterations of the power profile (Fig. 4.5) until one of three criteria is met: (i) the unit cannot provide sufficient power for the 79 W kg^{-1} stage of the schedule; (ii) the unit cannot deliver the power required at any other stage of the schedule; (iii) the unit cannot operate at, or above, certain imposed conditions, such as the minimum (cut-off) voltage experienced under load. When the end-of-cycle limit is reached, the cell/battery is returned to 100% state-of-charge, according to the manufacturer's instructions, and is said to have completed one SFUDS cycle.

In completing one SFUDS cycle, a cell/battery will make a number of passes, N, through the SFUDS. Each of these passes through the 6-min profile is defined as a 'sub-cycle'. With each sub-cycle, there is a net discharge of 0.987 Wh kg^{-1}, i.e., total discharge = 1.187 Wh kg^{-1}, total charge due to simulated regenerative braking = 0.2 Wh kg^{-1}. Hence, the terminal voltage falls steadily until it reaches the lowest allowable value (as set by the manufacturer). In order to illustrate how the performance of a cell/battery changes during SFUDS duty, it is helpful to plot the lowest voltage registered during each sub-cycle versus the number of sub-cycles. Figure 4.14 provides an idealized version of such a plot for three successive SFUDS cycles; actual data for a lead/acid battery are given in Fig. 4.15.

In seeking to quantify the performance of a cell/battery under the SFUDS, the total (cumulative) discharge during one cycle (N sub-cycles) has been defined as the 'SFUDS capacity' [12]. The average value of this quantity over the first three cycles is the 'useful capacity', in line with the practice adopted for other test schedules [9]. This is the capacity which is available between charging periods and, therefore, provides a direct indication of the driving range of the vehicle. Consequently, a true evaluation of the performance of a particular cell/battery must consider the useful capacity, as well as the number of SFUDS cycles to the end of life. In general, the SFUDS is continued until the delivered capacity is 80% of the useful capacity. Note, the same methodology is adopted for the DST.

Controlled-power testing regimes like the SFUDS are more difficult to implement than conventional constant-current regimes because power cannot be measured directly. Rather, to determine the power load on a cell/battery, both the voltage across the load and the current through the load must be measured. Power is the product of these two parameters and is regulated by varying the voltage and/or the current. As voltage control of batteries is neither practical nor applicable for EV applications, the required power is set by adjusting the current. An alternative approach would be to assume an average battery voltage and then meet each of the constant-power steps by setting the current. Such a strategy, however, constitutes a poor approximation to true constant-power duty. Most importantly, the actual load only matches the required load when the terminal voltage equals the assumed average value. Tests on lead/acid traction batteries have shown [9], however, that values of the terminal voltage recorded during SFUDS duty span a considerable range.

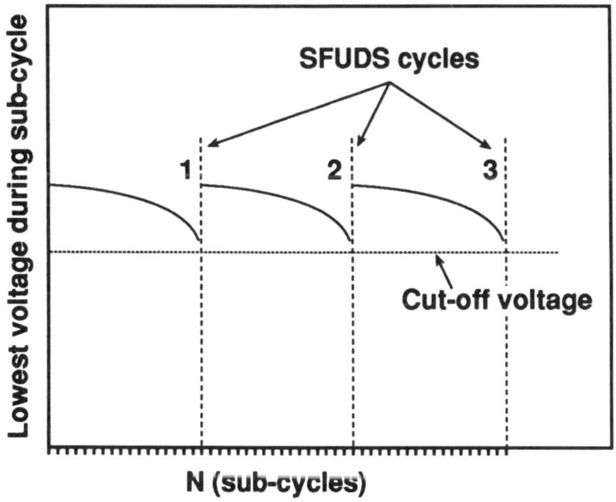

Fig. 4.14. Idealized representation of relationship between SFUDS cycles and sub-cycles. (Taken from [12] with permission of Elsevier Science Limited, UK.)

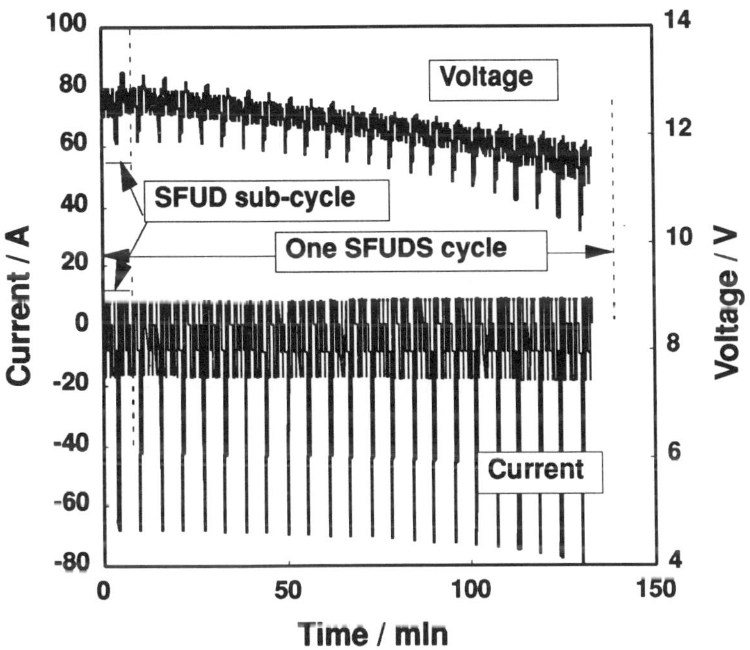

Fig. 4.15. Change in both current and voltage of a lead/acid battery during one SFUDS cycle (23 sub-cycles). (By courtesy of CSIRO, Australia.)

Table 4.3. Projected vehicle range (km) for simulated driving profiles [13].

Battery technology	Developer	Driving range (schedule : vehicle)		
		SFUDS : IDSEP	J227a/D : ETV-1	J227a/C : G-Van
Na/S (10-Ah PB cell)	CSPL	159	193	—
Na/S (60-Ah module)	CSPL	168	202	196
Lead/acid (flow-through)	JCI	83	111	—
Lead/acid (3ET 205, 6-V tubular)	Chloride	47	54	65
Ni/Fe (NIF-220, 220-Ah module)	EPI	75	98	93
Fe/air (80-Ah cell)	Westinghouse	66	113	—

CSPL = Chloride Silent Power Limited EPI = Eagle-Picher Industries, Inc.
JCI = Johnson Controls, Inc.

Therefore, for most of the test schedule, the battery will be subjected to a load which differs appreciably from that required. As a consequence, the important correspondence between duty under the laboratory test schedule and actual duty in the vehicle is lost.

Tests on traction batteries under the SFUDS (for the IDSEP) were reported in 1989 by the Analysis & Diagnostic Laboratory at ANL [13]. Table 4.3 summarizes the performances of four different technologies, viz., sodium/sulfur, lead/acid, nickel/iron and iron/air, and compares these with those achieved under both the J227a/D (for the IETV-1) and the J227a/C (for the G-Van) driving schedules. At that time, the maturity of the technologies ranged from research cells (flow-through lead/acid and electrolyte-flow iron/air) to production-type modules (advanced nickel/iron and tubular lead/acid). The Ragone plots presented in the ANL report [13] are given in Fig. 4.16.

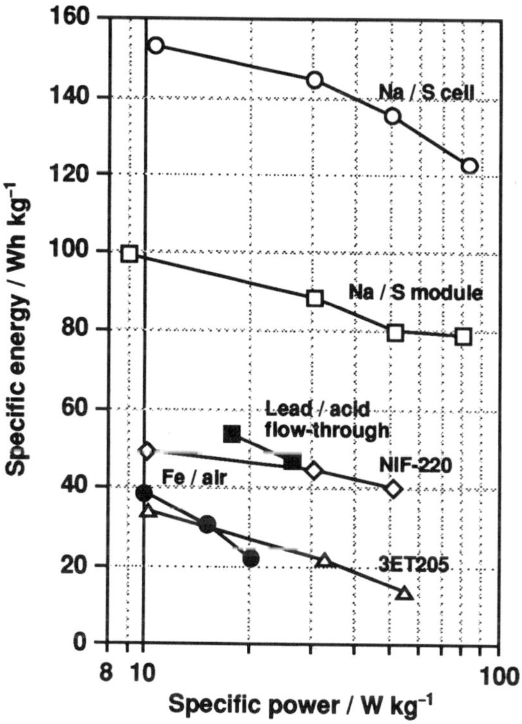

Fig. 4.16. Ragone plot for four EV battery technologies evaluated by ANL. (Reprinted with permission from SAE Paper No. 890820 ©1989, Society of Automotive Engineers, Inc. [13].)

4.3.3. Determination of battery life

In addition to range forecasts, laboratory testing also provides valuable information on the service lives of EV batteries. Life-cycle testing is typically conducted with 80 or 100% capacity discharges at a constant current (e.g., $C_3/3$ rate for lead/acid batteries), or with a simulated driving power profile (e.g., SFUDS). Since batteries do not usually fail catastrophically in EV operation, but lose capacity gradually, an arbitrary decision has to be made on what constitutes 'end of life'. For constant-current discharge tests, this state is usually defined in terms of the capacity level beyond which EV service is unacceptable. The DOE has taken 75% [14] or 80% [15] of the rated capacity as the end-of-life limit for lead/acid batteries. By contrast, Japanese practice adopted 60–70% of this quantity, as determined by the application [16]. Service lives of advanced batteries are often quoted to 50% of rated capacity (see, for example, [17]). Thus, different definitions of service life have to be taken into account when comparing the battery-life data reported by different workers. Unfortunately, these definitions are not always given.

An additional problem in comparing the service lives of batteries is that the discharge conditions applied by research groups in testing programmes are not always the same. For example, the life of a lead/acid battery decreases logarithmically with increase in the DoD [18], and is shorter at higher than at lower rates of discharge. Thus, it is possible to achieve artificially long service-lives by taking a low current and switching to charge at a shallow DoD, but the derived information will have little relevance to actual EV operation. As pointed out earlier, battery testing must be conducted at discharge rates which simulate vehicle driving conditions and must be taken to deep DoDs. Here again, there is variance between the testing philosophies of different research groups.

The service life of a battery is generally stated as the number of charge–discharge cycles performed. An alternative approach is to express life in terms of the total equivalent distance which the EV would have travelled. Such a measure is essentially one of cumulative capacity and is more meaningful in that it takes into account not only the number of cycles achieved, but also the work performed within each of these cycles. The latter can vary appreciably with different batteries.

Good agreement is not always obtained between the results of laboratory and field testing of batteries. Laboratory studies are usually conducted on single cells or battery modules, whereas field testing naturally involves full battery packs. The differences in measured performance largely arise from variability in the manufacture of individual cells. With battery packs, the discharge is terminated when the total voltage is the product of the cell voltage at the selected end-of-discharge and the number of cells present. If the cells are not identical, some will be discharged more deeply than others. Since service life decreases with DoD, the effect on individual cells will be cumulative as cycling proceeds and will cause the premature demise of the weakest. Thus, data obtained from tests on battery packs reflect the degree of quality control exercised in cell manufacture. The results from testing single cells and battery modules represent the performance which can be achieved if these units are properly matched in battery packs.

References

1. McWHINNEY, W. *A Review of Electric Traction Applications for Australian Cane Railway Operations*, Australian Battery Technologists Group, Paper No. 113, December 1983 (Australian Lead Development Association, Melbourne, Australia).

2. *Electric Vehicle Test Procedures, SAE J227a* (Society of Automotive Engineers, Inc., Warrendale, Pennsylvania, USA, 1976).

3. DOWGIALLO, E.J., and KEVALA, R.J. *Proc. 8th International Electric Vehicle Symposium (EVS-8)*, 20–22 October 1986, Washington, DC, USA, pp. 166–171.

4. *A Simplified Version of the Federal Urban Driving Schedule for Electric Vehicle Battery Testing*, Electric & Hybrid Propulsion Battery Test Working Task Force, DOE Report Number DOE/ID-10146, August 1988.

5. US Environmental Protection Agency, *Code of Federal Regulations, 40 CFR*, paragraph 86.115-78, 1 July 1985, p. 83.

6. COLE, G.H. *Recent Advances in Electric Vehicle Technology* (Society of Automotive Engineers, Inc., Warrendale, Pennsylvania, USA, 1989), Paper No. 891664, pp. 59–63.

7. BURKE, A.F., and COLE, G.H. *Proc. 10th International Electric Vehicle Symposium (EVS-10)*, 3–5 December 1990, Hong Kong (International Research Centre for Electric Vehicles, Department of Electrical and Electronic Engineering, University of Hong Kong, 1990), pp. 411–420.

8. *Electric Vehicle Battery Test Procedures Manual, Revision 1* (United States Advanced Battery Consortium, July 1994).

9. *Specification of Test Procedures for Traction Batteries for Electric Vehicles* (Joint Research Committee (JRC) – Traction Battery Working Group, October 1993).

10. HORNSTRA, F. *J. Power Sources*, 1986, **17**, pp. 284–294.

11. HORNSTRA, F. *J. Power Sources*, 1986, **17**, pp. 81–87.

12. HOLLENKAMP, A.F., LAM., L.T., PHYLAND, C.G., and WILSON, N.C. *J. Power Sources,* 1996, **59,** pp. 177–183.

13. DeLUCA, W.H., KULAGA, J.E., HOGREFE, R.L., TUMMILLO, A.F., and WEBSTER, C.E. *Proc. SAE International Congress and Exposition*, 27 February – 3 March 1989, Detroit, Michigan, USA (Society of Automotive Engineers, Inc., Warrendale, Pennsylvania USA, 1989), Paper No. 890820.

14. *Electric and Hybrid Vehicle Program, 9th Annual Report to Congress for Fiscal Year 1985* (U.S. Department of Energy, Washington, DC, USA, March 1986).

15. MILLER, J.F., CHRISTIANSON, C.C., HORNSTRA, F., GAY, E.C., ARNTZEN, J., and YAO, N.P. *Extended Abstracts, Proc. Vol. 85-2, Fall Meeting of the Electrochemical Society,* 13–18 October 1985, Las Vegas, Nevada, USA (The Electrochemical Society, Pennington, New Jersey, USA, 1985), pp. 53–54.

16. TAKAHASHI, S., and HIRAMATSU, T. *J. Power Sources*, 1986, **17**, pp. 55–63.

17. BROADHEAD, J., TRUMBORE, F.A., and BASU, S. *In* RAND, D.A.J., POWER, G.P., and RITCHIE, I.M. (Eds.) *Progress in Electrochemistry* (Elsevier, Amsterdam, The Netherlands, 1981), pp. 241–250.

18. KORDESCH, K.V., and TOMANTSCHGER, K. *In* BULLOCK, K.R., and PAVLOV, D. (Eds.) *Advances in Lead–Acid Batteries* (The Electrochemical Society, Pennington, New Jersey,. USA, 1984), *Proc.* Vol. 84–14, pp. 323–325.

CHAPTER 5

Lead / Acid Batteries

5.1. General Characteristics

Since the 1920s, the popular choice of power source for EVs has been the lead/acid battery. This system employs lead dioxide (PbO_2) as the active material of the positive electrode and metallic lead, in a sponge form, as the active material of the negative. The PbO_2 is present in two crystalline forms, namely, α-PbO_2 (orthorhombic) and β-PbO_2 (tetragonal), the latter is the predominant phase and there is evidence to suggest that the relative abundances of the two polymorphs influence battery performance. A grid of lead, or lead alloy, serves as both the current-collector and the support for the active material. In the traditional 'flooded' design, the electrodes (or plates) are immersed in an electrolyte which consists of a concentrated aqueous solution of H_2SO_4 (~35 wt.%). Thin sheets of porous insulators ('separators') are used to isolate the positive and negative plates from each other. The active material of both electrode polarities is converted to lead sulfate during discharge of the battery. Thus, the sulfuric acid electrolyte takes part in the cell reaction and forms the third active material.

Although the basic electrochemistry of the system remains the same, lead/acid batteries can differ widely in design, choice of component materials, and details of manufacture. These differences allow the batteries to meet the specific service demands of many applications other than EVs, e.g., ICEV starting, emergency power supplies, and electricity load management. Indeed, by virtue of its great versatility, the lead/acid system has captured the lion's share of the world's secondary battery market. In practical terms, the lead/acid battery has the following major advantages: (i) the highest cell voltage of all battery systems using aqueous electrolyte; (ii) the ability to supply both high and low currents over a comparatively wide range of temperature; (iii) a high degree of reversibility with a satisfactory energy efficiency (>80%); (iv) an acceptable charge retention during storage or inoperative periods in a charged condition; (v) a low cost compared with other secondary batteries; (vi) established manufacturing and recycling facilities.

For EV applications, the great disadvantages of the lead/acid system lie in the heavy weight of the basic material of construction (i.e., lead), a relatively low cycle life, and a long recharging time. (Note, the hitherto need for regular water addition has been overcome through the introduction of 'valve-regulated' technology, see Section 5.2.7.) Furthermore, the theoretical specific energy of the lead/acid system is the second lowest of all the candidate batteries for EVs listed in Table 3.3, Chapter 3. At the rates of discharge experienced in EVs, both the weight of the construction materials and the low utilization of the reactants themselves result in practical values of the specific energy being only a minor percentage of the theoretical value (171 Wh kg^{-1}). Traditional lead/acid EV batteries have had specific energies in the range of 20 to 30 Wh kg^{-1}, and this has translated into practical driving ranges of only 80 to 100 km at best. It is not surprising, therefore, that research [1,2] into the development of lead/acid batteries is directed principally towards achieving better electrical output and longer service life, together with reduced weight and faster recharging. In fact, the great scope for improvement offered by the lead/acid battery argues strongly that this technology will continue to be used to power EVs for some time to come.

5.2. Principles of Operation

5.2.1. Electrochemical reactions

The processes taking place during the discharging of a lead/acid cell are shown in schematic/equation form in Fig. 5.1(a). It can be seen that the HSO_4^- ions migrate to the negative electrode and react with the lead to produce $PbSO_4$ and H^+ ions. This reaction releases two electrons and, thereby, gives rise to an excess of negative charge on the electrode that is relieved by a flow of electrons through the external circuit to the positive

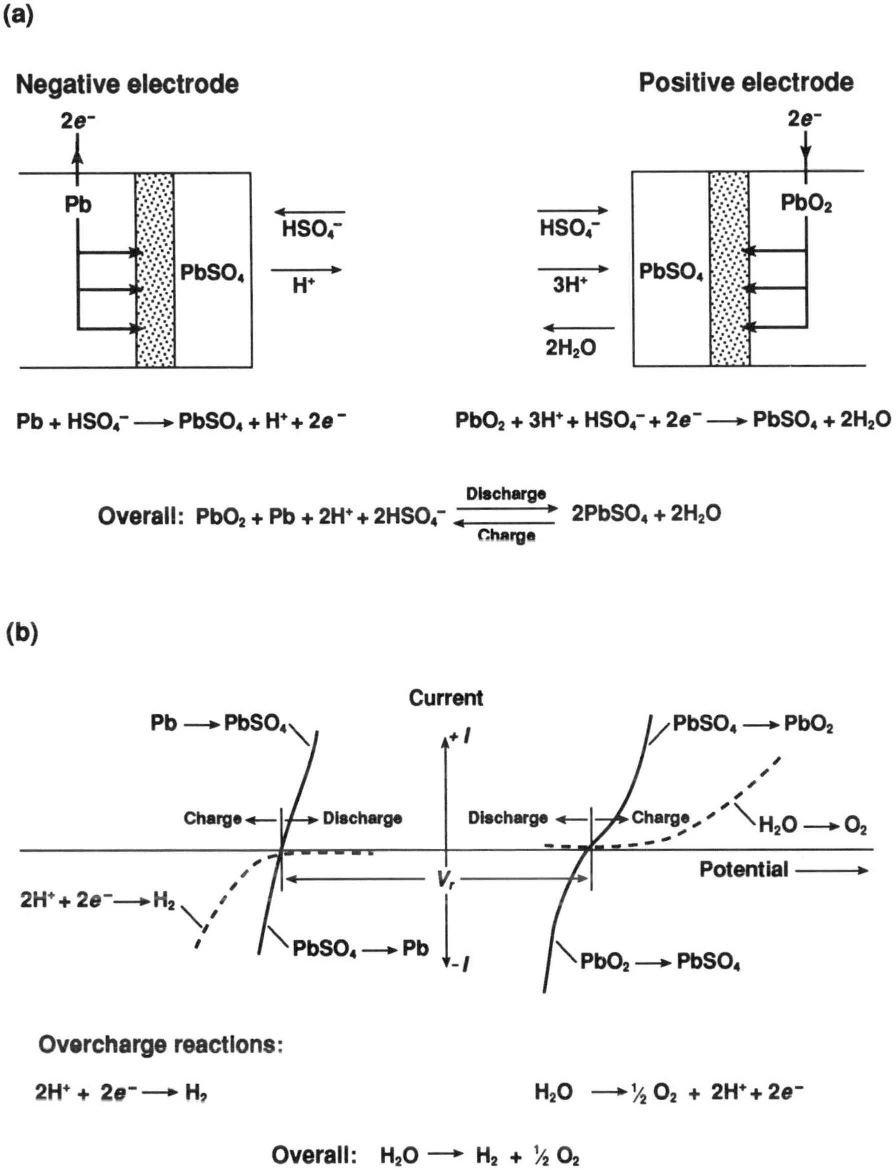

Fig. 5.1. (a) Electrochemical processes during discharge of the lead/acid cell. (b) Schematic representation of current–potential characteristic of each electrode.

electrode. At the positive electrode, the lead of the PbO_2 is also converted to $PbSO_4$ and, at the same time, water is formed. Thus, $PbSO_4$ progressively develops in equal quantities at both electrode polarities, with a concomitant weakening of the electrolyte. The decrease in the electrolyte concentration (or 'relative density' — traditionally termed: 'specific gravity', sp.gr.) is a convenient means for determining the degree of discharge which has taken place or, conversely, for checking the state-of-charge of the cell.

The above two electrochemical processes are reversed during the charging process. As the cell approaches full charge, the majority of the $PbSO_4$ will have been converted back to lead or PbO_2 and further passage of current will give rise to the evolution of hydrogen at the negative electrode and oxygen at the positive (Fig. 5.1(b)). The gases are released in stoichiometric proportions and, with traditional cell designs, this results in a loss of water from the cell electrolyte. In contrast to the discharge process, the specific gravity cannot be correlated with the state-of-charge of the cell until near the end of the charge when gassing provides good mixing of the electrolyte.

The fundamental discharge–charge reactions of the lead/acid cell involve dissolution–precipitation mechanisms which, collectively, are known as the 'double-sulfate theory', as first suggested by Gladstone and Tribe in 1882 [3]. In other words, discharge and charge of the electrodes may be considered as dissolution into, and electroplating out of, dilute lead-ion solutions. It should be noted that the lead/acid cell is able to operate effectively as an energy-storage device by virtue of the following three critical factors. First, contrary to thermodynamic expectations, the liberation of hydrogen from acids by lead takes place at only a negligible rate, i.e., there is a high hydrogen overpotential (Fig. 5.1(b)). Second, the high oxygen overpotential at the positive electrode allows $PbSO_4$ to be converted to PbO_2 before appreciable evolution of gas commences (Fig. 5.1(b)). Third, although $PbSO_4$ has a solubility in the electrolyte that is sufficient to promote the electrode dissolution–precipitation reactions, the value is so low that there is little migration of the material during charge–discharge cycling and, hence, a high degree of reversibility is maintained and the porous structures of the plates are largely retained.

5.2.2. Open-circuit voltage

The open-circuit potentials of the positive and the negative electrodes (and, therefore, the open-circuit voltage of the cell) depend on both the electrolyte concentration and the temperature. For a 1,1 dissociation of H_2SO_4, the potential-determining processes which occur on the electrodes may be written as follows.

Positive electrode:

$$PbO_2 + 3H^+ + HSO_4^- + 2e^- \rightleftarrows PbSO_4 + 2H_2O \qquad (5.1)$$

$$E_r = E^o + (3RT/2F) \ln a_{H^+} + (RT/2F) \ln a_{HSO_4^-} - (RT/F) \ln a_{H_2O} \qquad (5.2)$$

Using published electrochemical [4–6] and thermodynamic [7] data, it can be calculated that the E^o for α-PbO_2 and β-PbO_2 in eqn. (5.2) is 1.639 and 1.631 V, respectively.

Negative electrode:

$$Pb + HSO_4^- \rightleftarrows PbSO_4 + H^+ + 2e^- \qquad (5.3)$$

$$E_r = -0.302 + (RT/2F) \ln a_{H^+} - (RT/2F) \ln a_{HSO_4^-} \qquad (5.4)$$

Combining eqn. (5.2) and eqn. (5.4) yields the following expression for the open-circuit voltage of the lead/acid cell.

$$V_r = V^o + (RT/F) \ln a_{H^+} + (RT/F) \ln a_{HSO_4^-} - (RT/F) \ln a_{H_2O} \qquad (5.5)$$

or:

$$V_r = V^o + (RT/F) \ln (a_{H_2SO_4} / a_{H_2O}) \qquad (5.6)$$

Thus, V^o has a value of 1.941 and 1.933 V for α-PbO_2 and β-PbO_2, respectively (but, as pointed out above, positive active-material is composed mainly of β-PbO_2).

Note, the open-circuit voltage is often quoted for a 2,1 dissociation of the electrolyte, even though this dissociation does not take place over the range of H_2SO_4 concentrations employed in lead/acid batteries. Thus [8]:

$$H_2SO_4 \rightleftarrows 2H^+ + SO_4^{2-} \tag{5.7}$$

and:

$$V_r = 2.047 + (RT/F)\ln(a_{H_2SO_4} / a_{H_2O}) \tag{5.8}$$

The new value for V^o, viz., 2.047 V, is due to a change in the definition of the activity of H_2SO_4. For example, consider a strong electrolyte $C^{z+}_{\nu+} A^{z-}_{\nu\_}$ dissociated in solution according to:

$$C^{z+}_{\nu+} A^{z-}_{\nu\_} \rightleftarrows \nu_+ C^{z+} + \nu\_ A^{z-} \tag{5.9}$$

in which the condition for electrical neutrality is:

$$\nu_+ z+ = -\nu\_ z- \tag{5.10}$$

The activity, a_{CA}, is given by:

$$a_{CA} = a_+^{\nu_+} a\_^{\nu_-} = a_\pm^{\nu} = (\gamma_\pm m_\pm)^{\nu} \tag{5.11}$$

where: a_+, $a\_$ are the individual ion activities; $a_\pm$ is the mean activity of the ions; $\nu = (\nu_+ + \nu\_)$; $\gamma_\pm$ is the mean activity coefficient and is given by $\gamma_\pm^{\nu} = \gamma_+^{\nu_+} \gamma\_^{\nu_-}$; m is the molality (mole kg^{-1}); $m_\pm$ is the mean ionic molality and is given by $m_\pm = m(\nu_+^{\nu_+} \nu\_^{\nu_-})$.

Thus, for 1,1 dissociation of H_2SO_4:

$$a_{H_2SO_4} = (\gamma_{H^+} \gamma_{HSO_4^-})\, m_\pm^2 = \gamma_\pm^2 m^2 \tag{5.12}$$

and for a 2,1 dissociation of H_2SO_4:

$$a_{H_2SO_4} = (\gamma_{H^+}^2 \gamma_{SO_4^{2-}})\, m_\pm^3 = 4\gamma_\pm^3 m^3 \tag{5.13}$$

It should be understood that the $\gamma_\pm$ term takes on a different significance in eqns. (5.12) and (5.13). Values of $\gamma_\pm$ for H_2SO_4 are scattered throughout the literature; there appears to be no unifying text. Bode [9], for example, reports a_{H_2O} and $\gamma_\pm$ for 1,1 and 2,1 dissociated H_2SO_4 at various molalities, but the only common molality is m = 5.55 (i.e., sp. gr. ~1.440). At this molality, a_{H_2O} = 0.649, $\gamma_\pm$ (1,1 H_2SO_4) = 4.67, and $\gamma_\pm$ (2,1 H_2SO_4) = 0.212. By calculation from eqns. (5.6) and (5.8), the corresponding values of V_r are 2.111 and 2.106 V, respectively. Of course, the same value should have been obtained, irrespective of the method of derivation. The slight difference in V_r values observed here is, in fact, a consequence of the limitations of the Debye-Hückel theory for the dissociation of strong electrolytes.

The variation of the open-circuit voltage with electrolyte concentration is presented in Fig. 5.2. This plot has been constructed from various reported data [8,10–12]. The voltage increases uniformly above an acid sp. gr. of 1.05 and the following rule-of-thumb can be assumed:

$$V_r = \text{sp. gr.} + 0.84 \tag{5.14}$$

At electrolyte concentrations that are typically used in traction batteries (i.e., sp. gr. = 1.25–1.30), the open-circuit voltage is around 2.1 V at room temperature. For simplicity, the nominal or 'rated' value is usually taken to be 2.0 V. It should be remembered, however, that the open-circuit voltage is determined solely by the concentrations of acid at the plates. Thus, the voltage will not agree with sp.gr. readings unless the acid is uniform everywhere in the cell. The time required to reach such a state can be hours. The unstabilized voltage will always be higher than the true value immediately after charging of the cell, and lower after discharging. As mentioned above, there is a lag in sp. gr. during most of the charging period due to incomplete mixing of the electrolyte.

The open-circuit voltage is also influenced by temperature; this is due to minor changes in the interfaces between the active materials and the electrolyte, as well as solution activity effects at different temperatures. Experimental data for the temperature coefficient of the open-circuit voltage

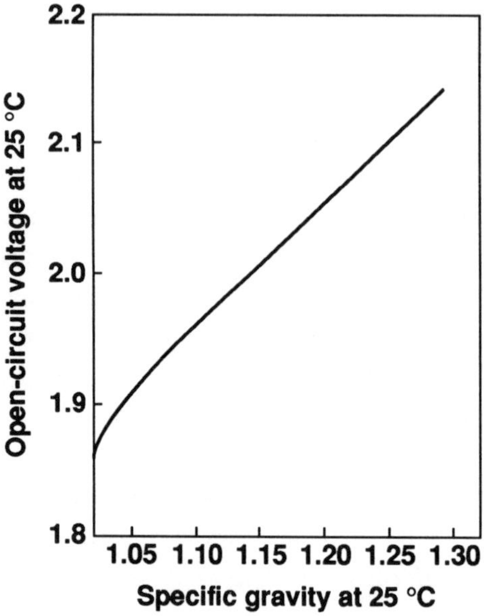

Fig. 5.2. Dependence of open-circuit voltage of the lead/acid cell on electrolyte sp. gr.

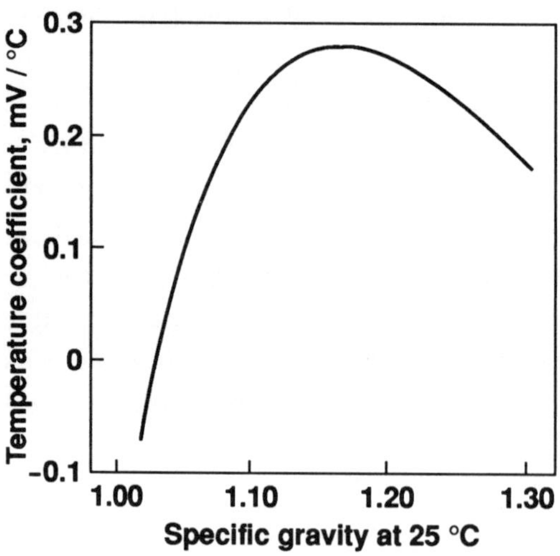

Fig. 5.3. Dependence of temperature coefficient of open-circuit voltage of the lead/acid cell on electrolyte sp. gr. (Taken from [12] with permission of McGraw-Hill, Inc.)

as a function of electrolyte sp. gr. are given in Fig. 5.3 [12]. The coefficient is positive for sp. gr. >1.04, i.e., the open-circuit voltage increases with increasing temperature. A thermal coefficient of about +0.2 mV / °C is generally assumed for EV batteries.

Finally, it should be understood that the dependence of the open-circuit voltage on pressure can be neglected in practical battery applications [13]. For example, at 25 °C and an acid concentration of 3.74 M, the pressure coefficient is only -3.32×10^{-5} μV / Pa [1].

5.2.3. Voltage during discharge and charge

The discharge curves for positive and negative electrodes in a lead/acid cell are illustrated schematically in Fig. 5.4. Immediately on applying a load, there is an instantaneous drop in cell voltage (region A). This is due to resistive and polarization losses in the cell. The sloping portion of the curve (region B) arises from discharge of the double-layer capacity. The internal resistance of a lead/acid cell in the fully-charged condition is of the order of milliohms; the value depends on the design and size of the cell, the methods used for manufacturing the plates, and the temperature. Cell resistance increases during discharge due to the plate materials being converted to lead sulfate and the electrolyte solution becoming depleted in acid. The former is deposited in the pores of the active material and thereby inhibits the access of acid. The pore-blocking effect of lead sulfate is a particular problem at high rates of discharge, e.g., at high power demands from EV battery packs during acceleration and/or hill-climbing operations. Decrease in temperature also hinders the diffusion of acid and, in addition, slows down the convective flow by increasing the viscosity of the electrolyte.

The potential minimum in the discharge curve (region C, Fig. 5.4) is known as the 'coup de fouet' or the 'whip crack'. Studies have shown [13–15] that the phenomenon is confined to the behaviour of the positive plate and, more specifically, to the β-PbO_2 polymorph. Moreover, the resulting dip in the cell voltage is particularly pronounced at high discharge rates, i.e., the period taken to reach the minimum becomes shorter, and the minimum

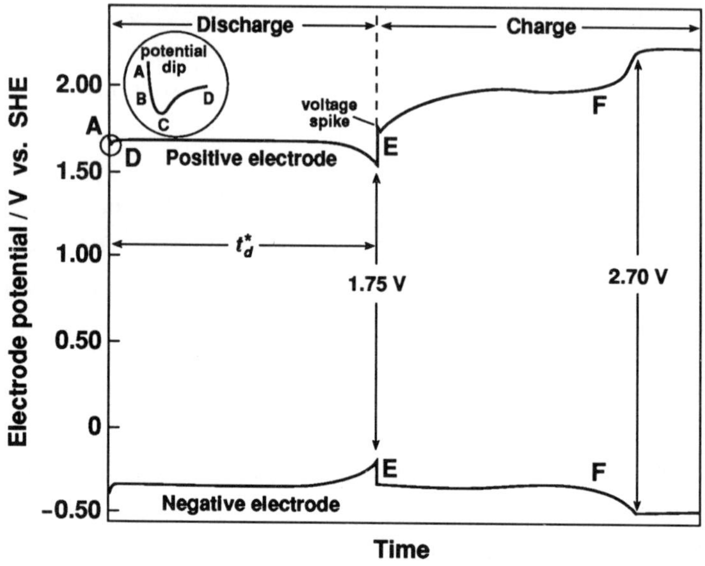

Fig. 5.4. Schematic of discharge–charge characteristics of the lead/acid cell.

itself is deeper. Berndt and Voss [15] claim that the voltage dip is due to impeded nucleation of $PbSO_4$. As a result, the solution in the pores of the active material becomes supersaturated with Pb^{2+} ions, produced by the electrochemical reaction, and the potential of the positive electrode is decreased. The Pb^{2+} ions are ultimately removed at the minimum value of the electrode potential and the $PbSO_4$ nuclei which are formed start to grow. The growth rate of the crystals will now be faster than that of the formation of ions so that the electrode potential (and hence the cell voltage) increases. A state is eventually reached where the electrochemical reaction and the nucleation–growth of $PbSO_4$ crystals proceed at equal rates and a potential plateau is observed (region D, Fig. 5.4).

A maximum (or spike) in potential is similarly observed on the positive electrode at the beginning of the recharge. This is considered to be due to the electrical resistance of an insulating $PbSO_4$ layer which covers the internal surface of the discharged active material [15]. Both effects — the potential minimum during discharge and the maximum during recharge —

are small compared with the terminal voltage of the cell, and are generally not regarded to be important in EV operations.

Further discharge of the cell results in a progressive decrease in the acid concentration and a concomitant fall in the voltage (Fig. 5.4). The voltage decreases slowly at first, but then drops abruptly near the end of discharge; this signifies that exhaustion of the cell is approaching. Beyond this point, it is impractical (and harmful) to extract further energy from the cell. It is standard practice to assume that the cell has reached 100% depth-of-discharge (DoD) at the 'knee' of the voltage curve (see Section 2.3.4, Chapter 2). The value of the cut-off voltage for the cell is dependent upon both the discharge rate and the temperature; for EV operations, it is usually taken to be 1.75 V (i.e., for batteries: 1.75 V / cell, on average). The electrical output of a cell is generally limited by one or other of the electrodes; or, in some cases, by the electrolyte. A 'negative-limited' design is generally recommended for EV applications [16] as it serves to protect the positive which is more susceptible to degradation (see Section 5.4.1).

The changes in potential that take place during the recharging of the positive and negative electrodes are also shown in Fig. 5.4. It can be seen that immediately after the discharge current is switched off, the potential of the positive electrode rises while that of the negative falls (region E) until the open-circuit voltage of the cell reaches ~2.0 V. A steady increase in the cell voltage is observed on charging with a constant current. At ~2.4 V (region F), practically all of the $PbSO_4$ is converted back to PbO_2 and Pb at the positive and negative electrodes, respectively. The potentials of the two electrodes now increase rapidly to values where the electrolytic decomposition of water becomes the predominant process (see Fig. 5.1), and the cell begins to 'gas'. As charging nears completion, the cell voltage stabilizes at ~2.70 V. The value of the final voltage depends on the design of the cell, the composition of the grid alloy, the type of negative-plate additives, the rate of charging, and the temperature. Note, in order to minimize both gassing and positive-grid corrosion, the cell voltage is usually regulated to lower values (typically, 2.35–2.45 V) during the final stages of charging.

5.2.4. Cell capacity

The 'rate of discharge' of a lead/acid cell, or indeed any other type of cell, is usually expressed in terms of the current flow (amperes) or the time (hours) that elapses before the cut-off voltage is reached (e.g., t_d^* in Fig. 5.4). As stated in Section 2.3.4, Chapter 2, the 'capacity' (C) of a cell refers to the quantity of electricity which can be delivered during discharge at constant current (I_d) to the cut-off voltage, i.e., $C = I_d t_d^*$. The capacity realized is determined by several factors which include the method of production, construction, maintenance and age of the cell, as well as the mode/rate of discharge, and the temperature. There can also be an influence from both the previous charge–discharge history (i.e., a 'memory effect') and the stand time. In most cases, the initial capacity is lower than the rated value. This is usually because the final step in the plate-manufacturing process, called 'formation' (see Section 5.3.7), does not normally proceed to completion. The conversion is completed, however, during the early use of the cell, and the capacity increases until it reaches a stabilized value. Under normal operating conditions, the capacity should remain at or near the stabilized value for most of the cell's useful life. Cells will then begin to suffer some capacity degradation due to ageing effects and the duty to which they have been subjected. This permanent loss usually increases slowly with age until the capacity falls below 80% of the rated value, which is the common definition of 'end-of-life'. Note that, the initial, stabilized capacity can be above the rated value because battery manufacturers often 'under-rate' their products. This design strategy is an added protection to ensure that each battery meets the promised life specification.

The capacity obtained from a lead/acid cell is strongly dependent upon the rate of discharge. Increase in this rate results in voltage depression, shortening of the discharge time, and a discharge characteristic with a slope as opposed to a plateau (Fig. 5.5). During rapid discharge, the electrochemical reactions take place mainly on the surface of the plates because of the limited time available for adequate diffusion of acid into the pores of the active material. Moreover, the reaction product ($PbSO_4$) tends to close off the pores, a process which further restricts the ingress of acid. Thus,

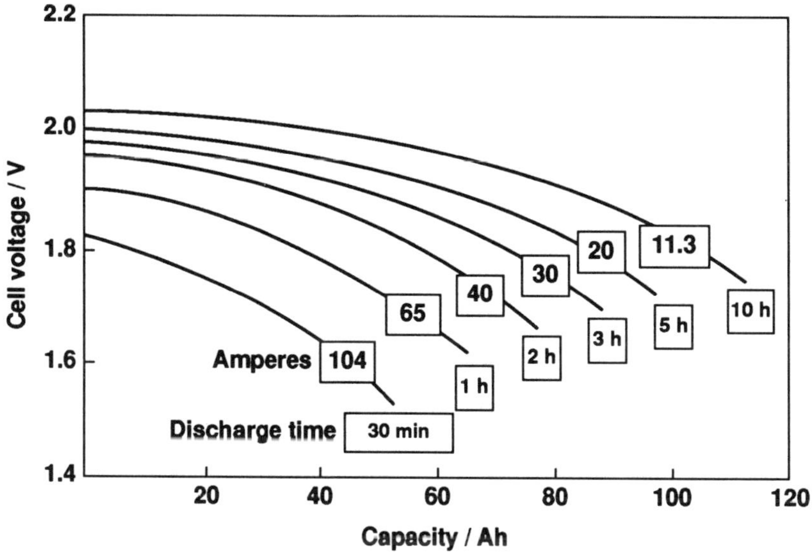

Fig. 5.5. Typical discharge characteristics of lead/acid EV batteries, base: 100 Ah at $C_5/5$ rate. (Taken from [17] with permission of the Australian Lead Development Association.)

capacity is severely reduced at heavy rates of discharge. For this reason, international and domestic standards have been introduced in order to compare the performance of different batteries. The capacity ratings of EV batteries are stated by manufacturers on the standard basis of a 3-h or 5-h constant-current discharge under specified conditions of cut-off voltage (usually, 1.75 V / cell) and temperature (usually, 25 °C). This means that a battery specified to have a 100-Ah capacity at the 5-h rate will deliver a constant current of 20 A for 5 h at the given temperature without the battery voltage falling below the cut-off value. For convenience, the term C/t (where t = discharge time in hours) is often used to define the rate at which the capacity is calculated. More specifically, (see Section 2.3.4, Chapter 2), the discharge rate should be normalized to the rated capacity and expressed as C_X/t, where X is the hour rate for the rated capacity and, as before, t is the discharge time in hours. Thus, the above battery has a 'nominal' capacity of 100 Ah at the $C_5/5$ rate. By way of example, taking a current of 50 A from this battery can be expressed as discharge at the $C_5/2$

rate. The latter is simply an expression of the current used and does not signify that the battery can be discharged at this same current for 2 h. As shown by the data given in Fig. 5.5, if the battery is to operate effectively for 2 h, then it must be discharged at a current of no more than 40 A, i.e., the $C_2/2$ rate. The manufacturer's rating cannot provide for the widely varying conditions of discharge that are encountered in typical EV operations. The nominal, or rated, capacity is intended only as a convenient standard by which the capacities of various batteries may be compared. Further, to obtain maximum EV utilization and to protect battery life, experience has shown that the battery should not be discharged beyond 80% of the rated capacity, regardless of the value of the discharge current.

As long ago as 1897, Peukert [18] determined empirically the following relationship for the dependence of capacity on discharge current:

$$C = K / I_d^{\,(p-1)} \tag{5.15}$$

where K and p are constants that are determined from discharge data. If the nominal capacity is known, then only the value of p is required in order to determine the capacity at other rates of discharge. This value generally lies in the range $p = 1.3$–1.4 for discharge rates experienced during typical EV duties. For example, the data given in Fig. 5.5 yield $p = 1.36$. From studies of the separate discharge curves of positive and negative plates over a wide range of discharge currents, Gillibrand and Lomax [19] have found that both polarities of plate have the same p value, but different K values.

Empirical observations have shown [20] that the Peukert equation is also valid for batteries subjected to current profiles that simulate the operations of road EVs (e.g., the SFUDS power profile, see Fig. 4.5, Chapter 4). Under such schedules, the battery undergoes essentially 'continuous' discharge to the nominated cut-off voltage, and the capacity can be calculated from the corresponding average current and nominal capacity. In practice, however, both road and plant EVs will not be in continuous operation but will be parked for long periods while their drivers pursue other activities. Accordingly, the battery discharge will be more intermittent and, typically,

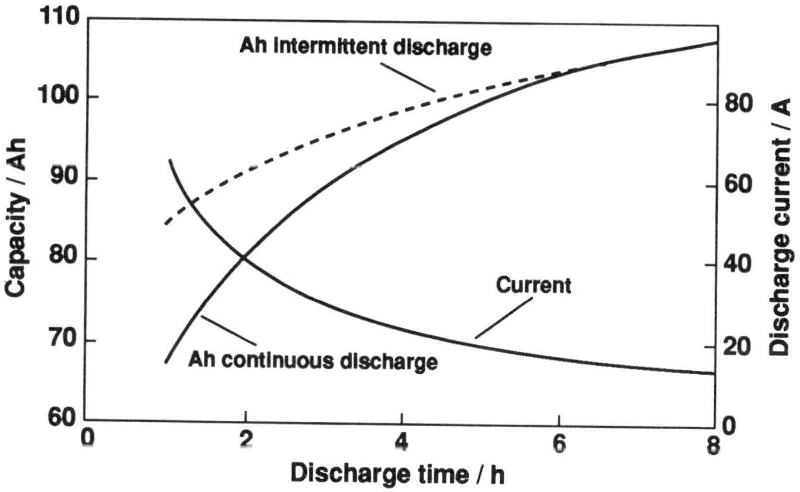

Fig. 5.6. Available capacity of an EV battery on continuous and intermittent discharge, base: 100 Ah at $C_5/5$. (Taken from [21] with permission of Pitman Publishing.)

will be spread over a working day, i.e., 6–8 h. The data presented in Fig. 5.6 demonstrate that there is a significant gain in useful capacity, and hence EV driving range, under intermittent operations. It is interesting to note that the reverse holds for ICEVs — these vehicles perform more efficiently under continuous driving conditions. With EVs, the improved capacity under intermittent discharge is referred to as the battery 'recuperation' or 'recovery' phenomenon, and is due to the diffusion of sulfuric acid into the active materials of the plates during the rest periods.

The major limitations of the Peukert equation are: (i) temperature effects are not considered; (ii) reliable predictions of capacity are obtained only for discharge currents which have values between the two rates used to determine the value of p; (iii) the accuracy decreases markedly at both very low and very high discharge rates. Kaushik and Mawston [22] have developed approaches towards overcoming these fundamental problems in order to give a greater degree of accuracy in forecasting the cell capacity for a particular discharge rate at a given temperature.

Modern EV design favours the use of solid-state switches for stepless speed control (also known as 'chopper control'). These electronic devices have the effect of discharging the battery under pulsed-current conditions in which the peak discharge current can be many times greater than the average discharge current. The effect of pulsating currents on lead/acid battery capacity has been studied in detail [23–26]. If changes in battery temperature are accounted for (note, battery heating, $I^2\mathcal{R}'$, is considerably higher for pulsed discharges), the mean battery voltage is taken as the test cut-off criterion, and the results are compared with a constant-current discharge test which has the same time-average discharge current, then the effect of pulsating currents on battery capacity is small (<4% increase). This finding has been confirmed by tests conducted both in the laboratory [27] and in EVs themselves [28].

The discharge capacity of the lead/acid cell increases with increasing temperature. This temperature effect is mainly the result of a reduction in the viscosity of the electrolyte and a concomitant increase in the rate of diffusion of sulfate ions to the active sites within the bulk of the electrodes. In addition to this decrease in concentration polarization, there is also a decrease in each of the other two components of electrode polarization (viz., activation and resistance effects, see Section 2.3.2, Chapter 2) at higher operating temperatures so that the battery cut-off voltage is reached later and, hence, more charge is delivered at a given current. Conversely, the capacity starts to decrease rapidly when the operating temperature is lowered, especially below 0 °C. Thus, unless a thermal management system is provided, EV range will be severely restricted in cold climates — the General Motors EV-1, for example, could achieve only 20 km on a full charge in the harsh Michigan winter.

The development of a quantitative relationship between capacity and temperature is difficult because, as stated above, the capacity also depends upon the design, method of manufacture, length of storage, maintenance and service history of the cell, as well as upon the discharge current-density. In particular, it has been found [29] that variations in battery production, although generally acceptable in commercial practice, can have an appreciable influence on the early capacity performance of batteries at elevated temperatures. Furthermore, studies at later stages of service life are not meaningful since cell ageing processes can exert significant adverse effects on capacity performance. Finally, there is also the difficulty in determining the true discharge temperature; the latter is defined as that measured at the reaction sites, i.e., the battery plates. Studies have shown that the external surface which best represents the temperature inside a cell is the surface of the container wall, parallel to the plates [30].

Despite the above qualifications, attempts have been made to predict the influence of temperature on the discharge capacity of lead/acid cells [31]. For automotive batteries operated at low discharge rates, Bode [9] has given as a rule-of-thumb a 1% increase in capacity (Ah) for each 1 °C rise in temperature. A number of empirical equations have been devised for the dependence of capacity on temperature, and according to the International Electrochemical Commission [32], the capacity C (Ah) – temperature T (°C) dependence within a restricted temperature range close to 30 °C can be expressed as:

$$C_T = C_{30}\,(1 + 0.008\,(T - 30)) \qquad (5.16)$$

where C_{30} is the capacity at 30 °C. This relationship is in good agreement with a theoretical treatment presented by Micka and Koudelka [33] in which the temperature coefficient of the capacity of the positive plate is expressed as a function of the temperature coefficient of the conductivity of 35% H_2SO_4 solution. A study has also been made [34] of the relationship between temperature and the capacity of positive and negative plates at different, constant, current densities. For both types of plates over the temperature range −18 to +16 °C, the capacity–temperature dependence is found to be linear; the temperature coefficient of capacity is greater at the lower discharge rates, and is also greater for the negatives than for the positives. The effect exerted by the concentration of sulfuric acid on the capacity–temperature relationships of both types of plates appears to be significant only at temperatures below 20 °C and at specific gravities below 1.24 [8].

There have been comparatively few detailed investigations of the effect of temperature on the capacity of lead/acid traction cells and batteries when discharged at EV rates. Under constant-current conditions, Chapman [35] found that the capacity increased linearly with temperature at ~0.9% / °C between 27 and 55 °C. On the other hand, Nowak [36] observed a capacity maximum at around 40 °C. Under variable-current discharge of the type normally encountered in EV operations, it was shown [31] that batteries exhibited a linear trend of increasing capacity with increasing temperature up to 50 °C with a temperature coefficient of ~2% / °C. Above 50 °C, the batteries experienced a decrease in initial capacity with further increase in temperature, but the values obtained at 60 and 70 °C were still greater than those achieved below 50 °C. In other words, a capacity maximum was observed and was similar to that reported by Nowak [36] for constant-current discharging, albeit at a higher temperature (50 vs. 40 °C).

The sensitivity of the capacity of lead/acid cells towards production and temperature variations emphasizes the well-recognized need for good battery matching, whereby the voltage under load at the end of discharge is the same for all cells. If the temperature is not controlled, battery matching is rendered difficult. Without adequate safeguards, operation at an uncontrolled temperature for an extended number of charge–discharge cycles could promote the overdischarge of individual cells and, hence, could result in battery damage.

In summary, the effect of temperature on initial capacity performance is determined by cell design, quality control in manufacture, and mode of discharge. Given that a wide range of battery production technologies and applications exist, it is clear that, at best, only generalized expressions can be formulated to describe the capacity–temperature dependence of lead/acid batteries. For traction batteries with appreciable positive temperature coefficients, it is clear that running at elevated temperature would increase the vehicle range for each discharge cycle. This would offer a decided advantage provided such high-temperature operation is not detrimental to battery service-life through, for example, an accelerated rate of positive-grid corrosion, an increased rate of self-discharge, and/or degradation of the negative-plate expander. Such problems are discussed in Section 5.2.5 and Section 5.4.1.

5.2.5. Self-discharge

Any charged battery tends to lose capacity over time; the capacity loss is referred to as 'self-discharge'. This process takes place both during the storage of freshly manufactured batteries, and during subsequent service life when the batteries are left under open-circuit conditions. The lost capacity is normally recoverable on recharge, but this may not be possible if the battery has remained at a low state-of-charge for prolonged periods (the discharge product, lead sulfate, tends to grow into large crystals which, on charge, cannot be readily converted back to the active materials, i.e., so-called 'refractory' or 'hard' sulfate is formed). The self-discharge in lead/acid batteries occurs as a consequence of: (i) the inherent thermodynamic instability of lead dioxide (positive plate) and lead (negative plate) in the presence of battery-strength sulfuric acid; (ii) extraneous reactions which occur within the cell. On open-circuit, the following reactions can take place.

Self-discharge at the positive electrode

(i) Oxygen evolution:

$$PbO_2 + H_2SO_4 \rightarrow PbSO_4 + H_2O + \tfrac{1}{2}O_2 \quad (5.17)$$

(ii) Oxidation of organics, X^R (these substances usually originate from the separators, or are decomposition products of the negative-plate expander, see Section 5.4.1):

$$2PbO_2 + X^R + 2H_2SO_4 \rightarrow 2PbSO_4 + 2H_2O + CO_2 + X^O \quad (5.18)$$

(iii) Corrosion of lead in the current-collector:

$$Pb + PbO_2 + 2H_2SO_4 \rightarrow 2PbSO_4 + 2H_2O \quad (5.19)$$

(iv) Corrosion of alloying constituents in the current-collector (grid), e.g., antimony [37]:

$$5PbO_2 + 2Sb + 6H_2SO_4 \rightarrow (SbO_2)_2SO_4 + 5PbSO_4 + 6H_2O \quad (5.20)$$

The rates of the above self-discharge reactions depend, variously, on several factors, e.g., oxygen overpotential, acid concentration, temperature, and plate design. Nevertheless, the overall rate of self-discharge of positive plates is slow; typically, it is less than 0.2% per day at 25 °C with new, fully-charged cells, and it decreases as the reactions proceed. The lead sulfate films do afford a certain degree of passivation, but impose the penalty of increased internal resistance. Furthermore, the selective permeability of the $PbSO_4$ corrosion layer to solution species (it is non-permeable for HSO_4^- ions, but permeable for H_2, H^+ and OH^- ions) can result in the formation of a highly resistive 'barrier' layer of α-PbO at the grid/corrosion-layer interface [38–40], particularly with non-antimonial grids (see Section 5.5.3).

Note, the following reaction can also take place at the positive electrode in newly formed cells, i.e., in those cells in which the lead monoxide (PbO) used to prepare the active material has not been converted completely to lead dioxide (see Section 5.3.7):

$$PbO + H_2SO_4 \rightarrow PbSO_4 + H_2O \quad (5.21)$$

Self-discharge at the negative electrode

(i) hydrogen evolution:

$$Pb + H_2SO_4 \rightarrow PbSO_4 + H_2 \quad (5.22)$$

(ii) oxygen recombination:

$$Pb + H_2SO_4 + \tfrac{1}{2}O_2 \rightarrow PbSO_4 + H_2O \quad (5.23)$$

Overall, the self-discharge of negative plates is generally more rapid than that of positive plates. The rate of reaction (ii) is limited by the diffusion rate of oxygen to the negative plate. The latter is deliberately enhanced in valve-regulated designs of lead/acid battery ('sealed' units) in order to allow the internal removal of oxygen produced during charging and to prevent the loss of water (see Section 5.2.7). The evolution of hydrogen (eqn. (5.22)) increases with increase in either the concentration of sulfuric acid or the temperature. It is also influenced by the presence of alloying additives in the grid metal and by organic components in the expander, as well as by any accidental contamination of the cell components with certain metallic ions (e.g., chromium, iron, manganese, nickel). A particular problem arises from the use of positive plates with antimonial grids. During battery charge, antimony dissolves from the positive grid, diffuses through the electrolyte, and deposits on the negative plate and, thereby, enhances hydrogen evolution (eqn. (5.22)). This results in a reduction of the hydrogen overpotential at the negative plate. The subsequent effect on self-discharge increases as the battery ages. The detrimental gassing effect of antimony can be ameliorated by using grids of lower antimony content or can be avoided by employing pure-lead or lead–calcium substitutes. Pure-lead grids have inferior strength, however, and batteries using these grids would be vulnerable to the vibration and mechanical shocks experienced during EV service. Batteries with lead–calcium grids are more robust, but they encounter cycle-life problems when operated under deep-discharge EV duties (see Section 5.5.3). Surfactants, such as α- and β-naphthol, have been introduced into negative plates in attempts to eliminate the problems caused by alloying additives (especially antimony) on the self-discharge processes. Unfortunately, such 'inhibitors' tend to lower the discharge capacity of negative plates.

5.2.6. Charging

As with all rechargeable batteries, the charging process is the reversal of the discharge. The type and degree of charging is usually critical to both discharge performance and life. Insufficient input will result in reduced discharge capability (which often becomes irreversible), whereas charging without adequate controls or for prolonged periods may reduce battery life by any one of several processes. With lead/acid batteries, the process is further complicated by the parasitic decomposition of water to produce oxygen at the positive plate and hydrogen at the negative plate. This reduces the charging efficiency, that is, the battery is not fully charged when the amount of charge returned is equal to that previously removed — a certain amount of 'overcharge' is required. The rates of gas evolution are relatively low at low states-of-charge, but increase markedly as full charge is approached. At this latter stage, essentially all of the charge current is being applied to the generation of gas, because most of the usable active material on the plates has been converted to the charged state, i.e., to lead dioxide in the positives and lead in the negatives.

The 'charge-acceptance' (C_A) represents the part of the current flowing through the cell that is utilized for the charging processes, namely, the reverse of eqn. (5.1) at the positive plate and the reverse of eqn. (5.3) at the negative plate. The charge-acceptance is defined as the ratio, expressed as a percentage, of the number of ampere hours usefully accepted during a small increment of time to the total number of ampere hours supplied during that time. Thus:

$$C_A = 100\,((I - I_g)\,/\,I) \qquad (5.24)$$

where: I is the total current flowing through the cell; I_g is the current wasted in the decomposition of water. The value of C_A for both plate polarities decreases as the charging temperature is decreased, the current density (i.e., charge rate) is increased, or the state-of-charge is increased. As with self-discharge (see Section 5.2.5), the charge-acceptance can be influenced significantly by the presence of additives and/or impurities, particularly those which induce a low hydrogen overpotential. Obviously, the charge-acceptance is also dependent upon the age of the battery.

The work of Peters et al. [41] demonstrated that positive and negative plates of conventional (flooded-electrolyte) batteries have different charge-acceptances, Fig. 5.7. During increase in the cell voltage to 2.4 V, the

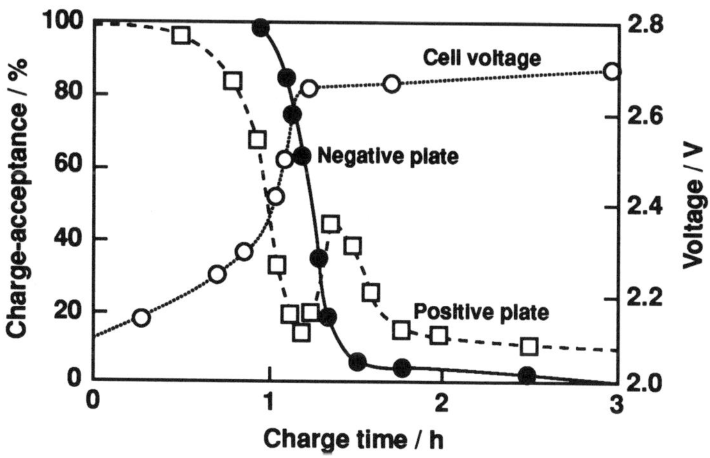

Fig. 5.7. Charge-acceptance of positive and negative plates at 40 °C. (Taken from [41] with permission of Elsevier Science Limited, UK.)

charge-acceptance of the negative plate is almost 100%. Then, it decreases abruptly and approaches zero when the plate is fully charged. By contrast, the charge-acceptance of the positive plate is more difficult and complex. After an initial slow decrease, it falls rapidly and then passes through a minimum followed by a maximum (60–80% state-of-charge region), prior to tapering off to a value above zero at the end of charge. It was concluded that this charge behaviour is associated with the formation and subsequent decomposition of persulfuric acid. Given the observed difference in the charge-acceptance characteristics, oxygen is evolved in increasing amounts as the charging of a lead/acid cell proceeds, but very little hydrogen is produced until the cell is nearly charged. This feature is used to advantage in valve-regulated designs (see Section 5.2.7).

It is difficult to determine precisely when a battery is fully charged. Thus, a certain degree of overcharge is applied to ensure that the battery does in fact reach full charge. On the other hand, uncontrolled charging is undesirable since it leads to copious gassing as well as excessively high temperatures. Vigorous gas evolution can weaken the mechanical integrity of the positive plate and remove small particles of the active material. In turn, these particles can be transported to the negative plates and reduced to lead. The resulting deposit grows as a 'moss' (or 'sponge') on the upper edges and sides of the plates and may give rise to short-circuiting (see Fig. 5.24). During heavy gassing, the separators can suffer attack from oxygen, while the negative-plate expanders can be chemically destroyed or mechanically removed by hydrogen. High temperatures will also cause degradation of these two battery components in addition to accelerating the corrosion of the positive grid. When exposed to sulfuric acid and elevated temperature during overcharge, the grid metal oxidizes to form, principally, lead dioxide. This reaction takes place quite slowly because the grid is surrounded by active material which tends to isolate or shield the grid from the electrolyte. As the grid is gradually oxidized, the cross-section is diminished and the current-carrying capability is impaired because lead dioxide is not such a good conductor. Eventually, the oxidation may become so severe that the grid ceases to provide electrical continuity between the positive active-material and the cell terminal. Corrosion of the grid can also create other problems. Lead dioxide occupies a greater volume than the lead it replaces. Thus, as the positive grid oxidizes, the positive plate expands. The resulting plate 'growth' can give rise to electrical shorts, or to physical damage (such as buckling) of the positive plate. In summary, therefore, it is essential to select charging strategies that minimize gas evolution and avoid enhancing the corrosive attack of the positive grid.

To ensure the right balance of charge and discharge, correct charging should involve control of both the current and the time of charging, together with the battery voltage, to suit both the size and the design of battery, as well as the operating conditions, e.g., state-of-charge, temperature, EV duty cycle. (Note, in valve-regulated designs, the pressure within the cell is also important during charging, see Section 5.2.7.) Although it is well recognized that a correct charging procedure is most important for satisfactory battery life and electrical performance, there is no clear consensus of opinion on the 'best' method of charging. Consequently, using regimes of current and/or voltage control, numerous charging strategies have been developed for different types of lead/acid technology according to the requirements of the corresponding applications. Common procedures for flooded-electrolyte lead/acid batteries include the following (Fig. 5.8): (i) constant-current

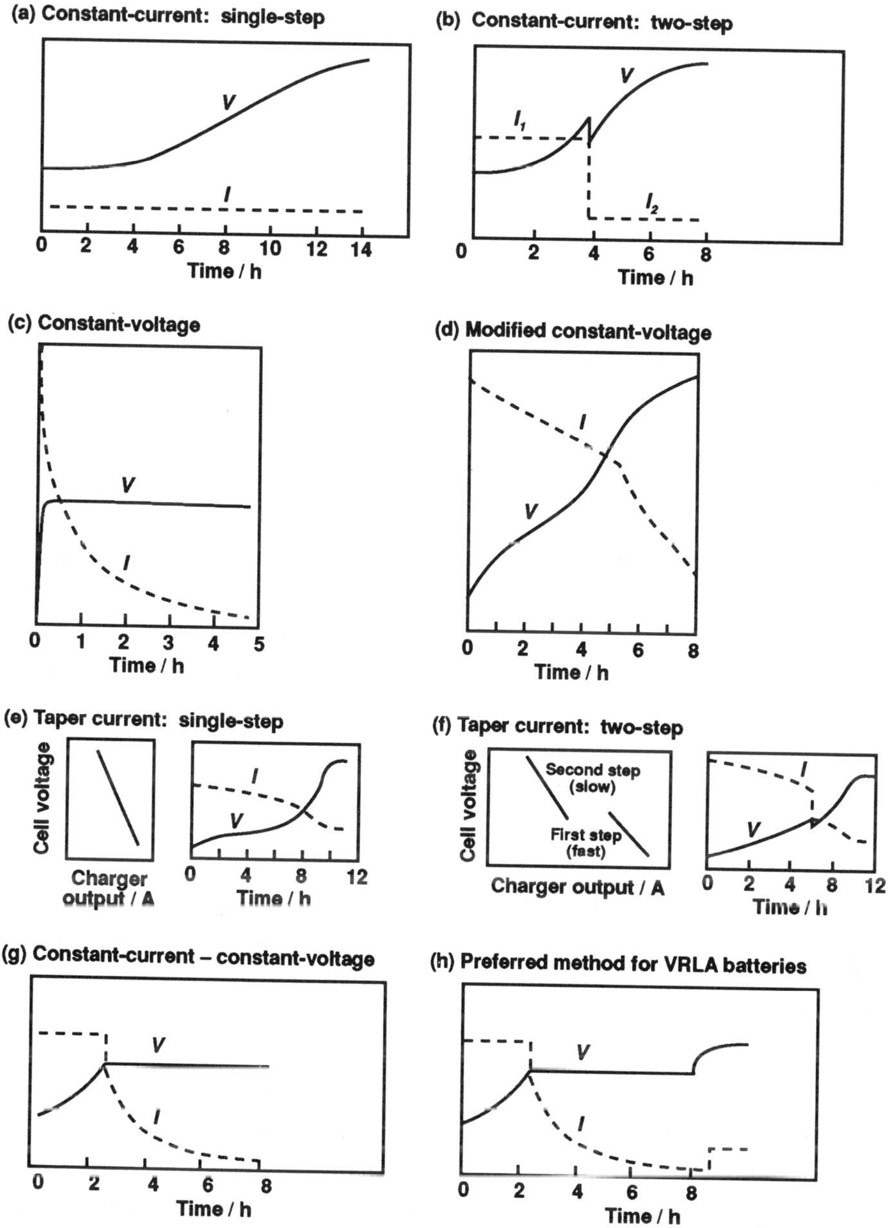

Fig. 5.8. Charging procedures and cell characteristics for lead/acid batteries. I = cell current, V = cell voltage.

(single- or two-step); (ii) constant-voltage; (iii) taper current (single-or two-step); (iv) constant-current–constant-voltage. Valve-regulated designs of batteries may use slight variants of these procedures (see Section 5.2.7).

Constant-current charging

In this form of charging, the charger furnishes a relatively uniform current, regardless of the battery state-of-charge or temperature. This type of charger is both inexpensive and very reliable. On the other hand, single-step, constant-current charging (Fig. 5.8(a)) is slow and is therefore appropriate for cyclic operations where a battery is only required to attain a full charge overnight, e.g., in fork-lift truck, golf-cart and milk float operations. The concept of two-step, constant-current charging (Fig. 5.8(b)) is very simple. For the first period of charging, the current is set at a medium-to-high rate, and is then switched to a very low 'trickle' rate as the cells approach full charge. The charger may then be left connected to the battery indefinitely. A two-step charger entails extra cost and complexity, but these factors are outweighed by the accompanying benefits of improved battery performance and life. In detail, the following advantages are offered by a two-step charger: (i) the battery can be charged more rapidly; (ii) cells receive only a limited amount of overcharge as the trickle charge rate is very low; (iii) cells in a long series string can be kept reasonably balanced throughout the life of the battery. Nevertheless, true constant-current recharging, at one or more current rates, is generally considered to be too slow a procedure for road EVs except, perhaps, for the initial charging ('conditioning') of the packs.

Constant-voltage charging

Constant-voltage chargers are most often used in two very different modes: as a fast charger to restore a high percentage of charge in a short time, or as a float charger to minimize the effects of overcharge on batteries which have infrequent discharges. The current starts out extremely high (Fig. 5.8(c)), but then decays rapidly as the voltage difference between the charger and the battery decreases. As a consequence, even though the battery reaches partial charge quickly, obtaining a full charge requires prolonged charging. The initial current is determined by the previous discharge. If the latter has been deep, then the current may attain very high values. Thus, true constant-voltage charging requires a charger with a very high output that is used only for short periods. This is an expensive proposition and, therefore, constant-voltage charging is generally restricted to batteries which experience shallow discharges (i.e., <25% DoD), such as those discharges commonly encountered in stationary applications. Moreover, the charging characteristics are not well suited to the cycling duties of EVs (i.e., repeated discharges with intervening times which are usually insufficient to reach asymptotically 100% capacity), and will cause 'cycling down' of the battery. Consequently, modified constant-voltage charging is preferred for road EVs (Fig. 5.8(d)). In this method, a higher voltage is applied and the current output is limited by including a series resistor in the charger circuit.

Taper-current charging

The taper charger is an unregulated constant-voltage charger, i.e., it is a variation of the modified constant-voltage method. The charger circuitry is inexpensive and is designed so that the battery, at low states-of-charge, receives charge at a high rate; as the battery approaches full charge, the charge rate decreases. The typical characteristic of a single-step taper is shown in Fig. 5.8(e). The steepness of the taper in the charging current determines the cost of the charger, its sensitivity to fluctuations in the mains voltage, and the time required to complete the charging. If shorter recharge times are required, then the two-step taper charger is employed (Fig. 5.8(f)). In this system, a higher charging current is used for the first part of the recharge cycle. When the voltage reaches ~2.4 V/cell (i.e., the start of gassing), the charging current is reduced. The taper in the second step is similar to that of a single-step charger. During the gassing phase, the current from either type of charger must be limited, whatever the taper, to prevent both excessive loss of electrolyte and overheating. For flooded-electrolyte traction batteries, a typical current acceptable to battery manufacturers is one-twelfth of the nominal C_5 capacity in amperes at 2.5 V/cell. Nevertheless, the charge rate is still quite high and may degrade battery life.

Constant-current–constant-voltage

This is the charging method which is best suited to road EVs, see Fig. 5.8(g). Initially, the current is set to about 15–20% of the battery capacity ($C_5/5$ rate), and is held constant until the average cell voltage in the battery rises to the gassing value of about 2.35 V/cell. The voltage is then maintained at this level and the current decays rapidly as with true constant-voltage charging. The total charge time is controlled by a timer and is selected to ensure the return of a predetermined percentage of the ampere-hour output of the previous discharge, e.g., 105–110%, or 5–10% overcharge. This method also permits the charging of EV battery packs that are made up of cells connected in parallel, irrespective of their capacity and/or state-of-charge (note, constant-current charging may result in an uneven distribution of the charging current among parallel-connected batteries).

It is important to note that because of inherent differences in cell characteristics, such as self-discharge rates, material composition, etc., a method for charge equalization is required to prevent the systematic undercharging of individual cells during the operational life of the battery. For example, if a low-capacity cell is present, it will be more heavily drained than the remainder of the battery pack. Consequently, even though the other cells are charged to capacity and the charge current has diminished significantly, the low-capacity cell still requires more charging. If the charge time is inadequate and the battery is discharged again, then the low-capacity cell will be even more heavily discharged. Thus, the problem is compounded and the battery can quickly develop one or more 'dead' cells; in extreme situations, some weak cells can be driven into reversal during discharge. This cycling-down behaviour can be exacerbated by: (i) large differences in the initial state-of-charge of the cells; (ii) variable product quality, e.g., uneven pasting of plates; (iii) the use of very long strings; (iv) deep and continuous cycling. The problem can be managed by subjecting the battery periodically to an 'equalizing' or 'levelling' charge through either adjusting charge voltages or setting charge times to ensure adequate charging for all cells. In the constant-current–constant-voltage charging procedure, for example, the charging time after attainment of the set voltage limit is usually doubled every few cycles in order to equalize variations in the charge conditions of the individual cells within the battery [16].

In summary, proper selection of the charging procedure and the charge parameters is essential for achieving maximum battery performance and life. Excessive overcharging gives rise to increased gassing, enhanced component corrosion and elevated temperatures, all of which adversely affect maintenance, performance and cycle life. Consistent undercharging produces a gradual decrease in battery capacity, which often becomes irreversible. Among the various charging techniques available for lead/acid batteries: (i) two-step, constant-current chargers are appropriate for most applications; they use a high-rate current to restore charge quickly and then switch to a lower-rate current in overcharge; (ii) constant-voltage chargers restore a large fraction of the discharged capacity quickly, but then require quite long charging times to bring the battery to full charge; (iii) taper chargers are inexpensive but, because of their high overcharge currents, are not conducive to long battery life; (iv) the constant-current–constant-voltage technique offers the most practical method of charge control for EVs through which battery efficiency and discharge performance may be optimized by proper selection of the current and voltage levels. Recent developments in the area of fast-charging of lead/acid traction batteries are discussed in Section 5.5.4.

5.2.7. Valve-regulated batteries

Two fundamental drawbacks of flooded (i.e., conventional) lead/acid batteries are the heavy gassing levels on overcharge and the tendency to leak electrolyte. The former shortcoming can be ameliorated by the use of lead–calcium and low–antimony grids (see Section 5.3.2), but gassing is still substantial and the presence of free electrolyte remains a safety hazard in the event of accidents. Clearly, these restrictions limit the practicality of flooded lead/acid batteries as EV power sources; for example, gassing causes water loss (and hence introduces the need for regular battery maintenance), produces explosive hydrogen–oxygen mixtures, and carries acid (which is corrosive) outside the battery.

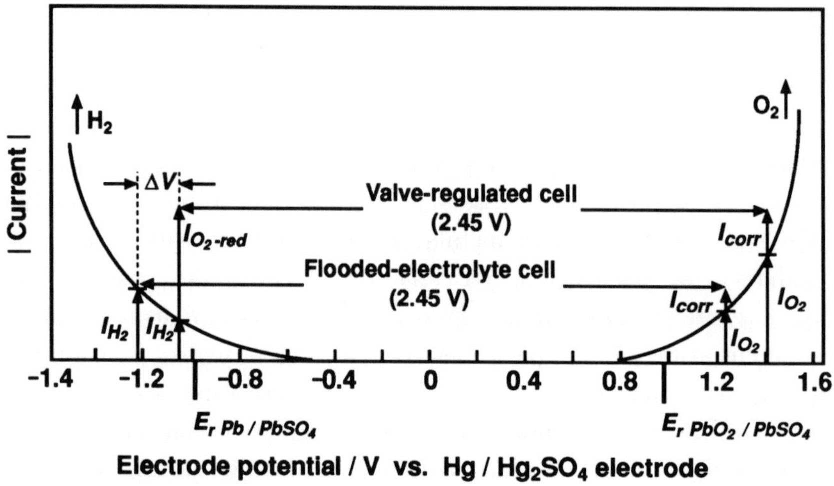

Fig. 5.9. Current distribution in flooded-electrolyte and valve-regulated batteries during constant-voltage charging at 2.45 V/cell. The current for oxidation of hydrogen at the positive electrode is negligible and is therefore not shown. (Note, anodic and cathodic currents are opposite in sign but, for convenience, are both presented on the same side of the potential axis.)

The generation of hydrogen and oxygen in lead/acid batteries can never be completely suppressed. This is because hydrogen evolution starts above the equilibrium potential of the negative ($Pb/PbSO_4$) electrode, and oxygen evolution below the equilibrium potential of the positive ($PbSO_4/PbO_2$) electrode, see Fig. 5.9 (further discussion of the current distribution shown in Fig. 5.9 is given later in this Section). Starting in the 1950s, however, many attempts were made to 'recombine' (i.e., to convert back to water) the hydrogen and oxygen gases given off during the overcharging of flooded batteries, and thus to produce 'sealed' or 'maintenance-free' units, i.e., units which require no topping up with water. The methods have included the use of catalyst plugs [42–46] and of auxiliary electrodes [47–52]. The catalysts are typically palladium, or another platinum-group metal, dispersed in an alumina or activated carbon host that is pelletized to give a material with a high surface area. The pellets are treated with either silicone or fluorocarbon resin to impart hydrophobic properties to the catalyst, and are housed in a device which is usually inserted in place of the vent plug in the lid of the battery container. The reaction of hydrogen and oxygen with the plug catalyst is purely a chemical one and, therefore, the gases will combine in stoichiometric amounts. During battery operation, however, hydrogen and oxygen can be evolved in a non-stoichiometric ratio (see, for example, Fig. 5.7). Gas recombination under such conditions can be achieved through the use of sets of auxiliary electrodes of suitable materials, e.g., platinum-group metals [49,50], tungsten carbide [51,52]. The auxiliary electrodes are electrically connected to the positive and negative plates in such a manner (e.g., via solid-state diodes) as to ensure a suitable potential region for the respective reactions of hydrogen oxidation and oxygen reduction to proceed at the required rates [49]. Thus, combination of the over-stoichiometric gases can take place on the auxiliary electrodes via electrochemical reactions which proceed in parallel with the catalytic chemical recombination between the two gases.

Despite the above studies, neither catalyst plugs nor auxiliary electrodes have proved to be a practical means for gas recombination. Shortcomings with catalyst plugs include: (i) added cost; (ii) complexity of cell manufacture; (iii) formation of electrolyte concentration gradients due to the return of condensed water and, consequently, the possibility of excessive corrosion of the top-lead and/or upper parts of the grids; (iv) early failure through inactivation of the catalyst by the flooding of its pores with product water. Corresponding disadvantages with auxiliary electrodes are: (i) added cost; (ii) complexity of cell manufacture; (iii) partial dissolution which, especially with noble metals, increases both self-discharge and sulfation of the plates; (iv) loss of electrochemical activity on prolonged usage; (v) decrease in cell specific energy through added weight; (vi) increase in cell volume; (vii) limited effectiveness at high overcharge.

In the late 1960s and early 1970s, research conducted in both Europe and the USA resulted in the development of commercially-feasible, sealed, lead/acid technology. In such batteries, the cell is designed to promote the chemical recombination of oxygen at the negative electrode via the so-called 'internal oxygen cycle' or 'oxygen-recombination cycle', i.e.,

At the positive electrode:

$$H_2O \rightarrow 2H^+ + \tfrac{1}{2}O_2 + 2e^- \qquad (5.25)$$

At the negative electrode:

$$Pb + \tfrac{1}{2}O_2 + H_2SO_4 \rightarrow PbSO_4 + H_2O \qquad (5.26)$$

$$PbSO_4 + 2H^+ + 2e^- \rightarrow Pb + H_2SO_4 \qquad (5.27)$$

In this cycle, the oxygen discharges chemically the negative plate and, thus, 'depolarizes' the potential to a value at which the rate of hydrogen evolution is very low, i.e., the rate is much less than that in a flooded battery. Since the plate is simultaneously on charge, however, the discharge product is immediately reduced electrochemically to lead and the chemical balance of the cell is restored. As long as the overcharge current remains moderate, the charge and recombination reactions can remain in equilibrium and little net gas is generated. For example, consider a flooded-electrolyte cell held at a constant voltage, e.g., 2.45 V as in Fig. 5.9. The current at the positive electrode is consumed principally by oxygen evolution (I_{O_2}) and by grid corrosion (I_{corr}), and is balanced by that consumed by hydrogen evolution at the negative electrode (I_{H_2}). With a valve-regulated cell, however, oxygen reduction ($I_{O_2\text{-}red}$) shifts the potential of the negative electrode (ΔV) and the rate of hydrogen evolution (I_{H_2}) becomes much reduced, but is not eliminated completely, compared with that experienced in the flooded system. With a high recombination efficiency and with no loss of oxygen from the cell, the rate of hydrogen evolution is balanced by the rate of positive-grid corrosion. It should be recognized, however, that for a cell to be maintained in a fully-charged state, the hydrogen-evolution current at the negative plate must be sufficiently high for the electrode potential to be more negative than the associated reversible value of the Pb/$PbSO_4$ couple, otherwise self-discharge of the plate will occur. In general, the rate at which oxygen recombines at the negative plate is a complex function of the cell design, the operating conditions, and the overcharge regime. For a given type of cell or battery, the efficiency of oxygen recombination begins to fall at some level of charge current and it continues to decline as the rate of charge increases. Finally, it should be noted that certain impurities can give rise to appreciable emissions of hydrogen and, consequently, the lead materials used to manufacture the plates must be of high purity. The specifications for these materials are still the subject of considerable debate within the industry [53].

Given that no significant overall chemical change is occurring, the electrical energy input into the cell is converted to heat which must be dissipated to safeguard against accelerated cell degradation and to avoid the possibility of thermal runaway (see Section 5.4.2). A corresponding recombination cycle for hydrogen does not exist because the oxidation of the gas at the positive electrode is too slow. This, together with the fact that the oxygen-recombination efficiency is <100% in most cases (typically, 95–99%), requires each cell to be fitted with a one-way valve (located in the container lid) as a safeguard against excessive pressure build-up and subsequent distortion of the container. (Note, the valve must not allow any ingress of air into the cell, otherwise the negatives would suffer serious self-discharge.) By virtue of this design feature, the product has become known as the 'valve-regulated' lead/acid battery, or VRLA battery (other, less appropriate, terms include: 'immobilized-electrolyte', 'maintenance-free', 'recombinant-electrolyte', and 'starved-electrolyte').

For the VRLA battery to function, the transport of oxygen from the positive to the negative plates must be facilitated. This is achieved by 'immobilizing' the electrolyte between the plates. Two techniques are employed, namely: (i) gelatinization with silicon dioxide — the 'gel' type; (ii) containment within absorptive glass-microfibre separators — the 'AGM' type. (Note, the term RBSM, i.e., recombinant-battery separator mat, has also been used [54] to denote the AGM design; there is still debate over which is the better terminology.) Gas transport can be further enhanced by operating the batteries under a slight excess of internal pressure. The recombination reaction is promoted by arranging for the positive-to-negative stoichiometric ratio of the active material to be close to unity, so that gassing at the positive plate occurs before that at the negative plate [55].

To reduce gassing levels, the grids are usually made from lead–calcium rather than lead–antimony alloys (see Section 5.3.2); the former alloys have a higher hydrogen overpotential. Gel batteries are produced with either flat or tubular positive plates. By contrast, AGM units are only available with flat positives, although efforts are underway to develop batteries with tubular plates [56]. Flat negative plates are used in both designs.

The silicon dioxide, or 'fumed silica', used in the construction of the gel batteries has a very small average particle size (approximately 10 nm) and, consequently, has a high surface area (up to 400 $m^2 g^{-1}$). On mixing with sulfuric acid, the fumed silica (typically, 3–5 wt.%) forms a viscous solution which develops into a gel on standing. The thickening action can be reversed by subjecting the material to a significant impact, although the gel will reform if the solution is not disturbed. It is thought that weak hydrogen-bonding is responsible for this thixotropic behaviour. Oxygen transfer occurs via fissures in the substrate that are formed during the early stages of battery life through partial drying out (water loss) and concomitant shrinkage of the gel (it is also understood that some transport of oxygen occurs via bonding and movement along the silica chains of the gel [57]). As a result of this behaviour, cells connected in series can exhibit at the beginning of service: (i) different recombination rates; (ii) variable top-of-charge voltages (i.e., by up to 250 mV [58]); (iii) significant water consumption. After the initial 'running-in' period, these features usually decrease to much lower levels, although the variation in the top-of-charge voltage is always considerably higher than that of flooded batteries. Despite the fact that the process of making a gelled electrolyte is an old concept [8], it was not until the mid-1960s that Sonnenschein GmbH in Germany became the first to translate the technology into commercial practice [59–61]. Since then, these batteries have been marketed under the 'Dryfit®' brandname. Later work by both Sonnenschein and Johnson Controls, Inc. (in the USA) extended the development of gel batteries to solar [62], large stationary [63], and EV [64,65] applications.

The glass-microfibre separator material employed in AGM batteries consists of a blend of glass fibres of varying lengths and diameters. Good

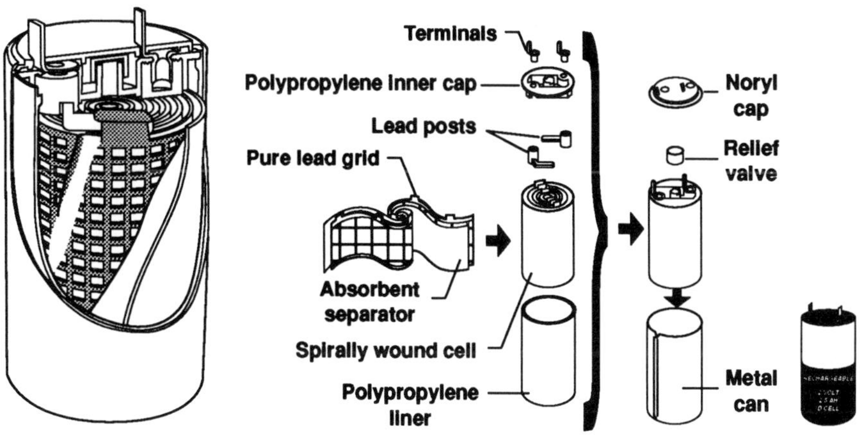

Fig. 5.10. The Gates' D-cell and its method of construction. (Taken from [67] with permission of Hawker Energy Products, Inc.)

movement of the acid (known as 'wicking') along these glass strands is essential for satisfactory operation of the batteries. Initially it was thought that thin fibres with a high surface area (1.6–2.0 $m^2 g^{-1}$) were required to achieve this action, but it was later established that strands with low surface area (1.1–1.3 $m^2 g^{-1}$) can provide acceptable performance at a much reduced cost. The AGM is usually saturated with acid to between 90 and 95% of its maximum, while the remaining volume allows the passage of gas so that recombination can occur. The AGM separators must be in intimate contact with the plates in order to allow fast transfer of ions during both charge and discharge. To achieve the 'stack pressure' required to give this contact, a tight fit is required between the cell group and the battery container. Drying out of the electrolyte, with subsequent stabilization of top-of-charge voltages, also occurs in these batteries, but to a lesser extent than that experienced in gel units. Nevertheless, after the 'conditioning' water loss, the deviation in the individual voltages in strings of both AGM and gel cells is similar. The first, commercially successful, AGM cells were made in 1971 by Gates Energy Products, Inc. in the USA [66]. The cells were later marketed under the 'Cyclon®' brandname. In the Gates approach [67], a single pair of positive and negative plates (using soft, pure-lead grids) is interleaved with AGM separators, and the assembly is then spirally wound

Table 5.1. Advantages/disadvantages of valve-regulated lead/acid batteries with respect to flooded-electrolyte lead/acid batteries [69].

Advantages	Disadvantages
• No water addition	• Careful charging required
• No acid spillage	• Thermal management is more critical, especially for AGM batteries
• Negligible acid fumes	• Significant variation in top-of-charge voltages
• Easy transportation	• Increase in overcharge required at elevated temperatures
• No special ventilation requirements	• Deep-cycle life often inferior under optimum operating conditions
• Can be operated horizontally	• Cannot measure specific gravity
• Smaller footprint (batteries can be multi-stacked)	• Not available in dry-charged state
• Negligible acid stratification (gel type only)	• Shelf-life of two years maximum
• Less overcharge required at room temperature	
• Good high-rate, discharge capacity (AGM type only)	

to form a D-cell element (Fig. 5.10). The principal patent [68] on Gates' technology has been licensed world-wide, and forms the basis for the AGM technology which is now also produced in many sizes of rectangular cells. The Gates battery business was acquired by Hawker Batteries in December 1993 and was renamed Hawker Energy Products, Inc.

The advantages and disadvantages of VRLA batteries are presented in Table 5.1 [69]. Although most of these are common to both types of cell design, the gel cells experience little stratification of the electrolyte compared with AGM counterparts. (Note, electrolyte stratification is particularly prevalent in flooded cells and refers to the development of a vertical concentration gradient of sulfuric acid within the cell container that is caused by the continual flux of acid during charge–discharge operations; stratification can reduce appreciably the performance of the cell, see Section 5.4.1, and therefore should be minimized or avoided.) The mild level of stratification experienced by gel technology means that, at normal temperatures, less overcharge is required to recharge and equalize the cells [63]. On the other hand, AGM batteries have a very low internal resistance and, therefore, display excellent high-rate performance. The higher internal resistance of gel batteries is mainly the result of having to use a conventional microporous separator in order to suppress short-circuiting that could otherwise arise through lead dendrites bridging the space between the electrodes.

As stated above, VRLA batteries are more susceptible to cell-to-cell imbalances than their flooded counterparts. This is because, apart from the inherent differences in cell characteristics, there can also be differences in the oxygen-recombination efficiencies. The latter give rise to variability in the polarization behaviour of the negative electrode and, hence, to variability in the top-of-charge voltages of the individual cells, especially in new batteries. For example, under either constant-current or constant-voltage charging, the negative plates of each cell can have a potential which can lie between values set by the highest and the lowest oxygen-recombination efficiency and, typically, the cell-to-cell voltage can vary by up to 100 mV (ΔV in Fig. 5.11). As a consequence, during charging, some cells will not become fully charged and on the next discharge will be discharged to a greater depth. This will result in a greater degree of undercharging on the next recharge. The process continues with each cycle until, eventually, such cells in the battery become grossly overdischarged. Clearly, such cell-imbalance problems are exacerbated when VRLA batteries are operated in strings. The problem can be minimized by the judicious application of appropriate charge-equalization procedures.

Typical pressure, temperature and voltage characteristics for a VRLA cell during constant-current charging are shown in Fig. 5.12(a). The cell pressure remains low until the cell approaches 80% state-of-charge and the positive plates begin to generate oxygen. At this point, or shortly thereafter, the negative plates start to evolve hydrogen. Under conditions where the majority of the charge current goes into gas generation, venting may occur as dictated by the overcharge rate. Eventually, the gassing rate

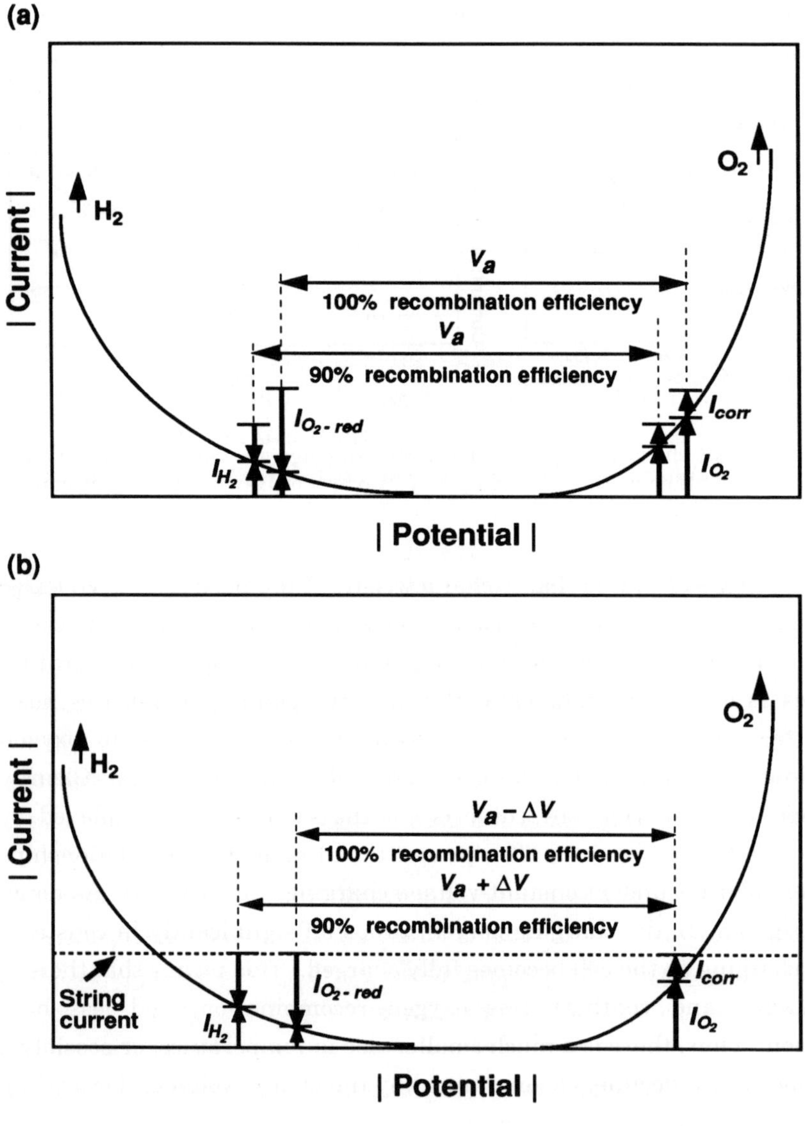

Fig. 5.11. Schematic representation of how VRLA cells/batteries with different oxygen-recombination efficiencies exhibit variable top-of-charge voltages during constant-voltage charging: (a) two cells each charged at V_a, current through cell with 100% recombination efficiency is greater than that through cell with 90% recombination efficiency; (b) cells from (a) charged in series at $2V_a$, the individual voltages adjust by $\pm \Delta V$, as shown, to equalize the current through each cell.

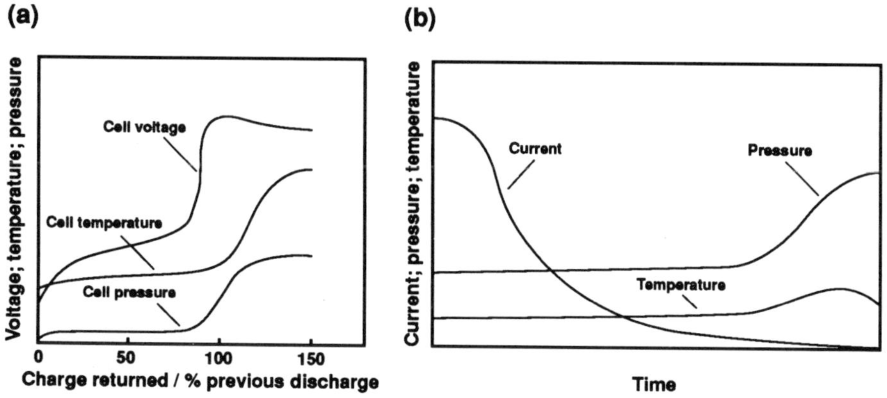

Fig. 5.12. Typical characteristics of VRLA cells during: (a) constant-current and (b) constant-voltage charging. (By courtesy of Hawker Energy Products, Inc., USA.)

approaches a steady value that is characteristic of the particular overcharge conditions which are being employed. The temperature profile of the cell is similar to the pressure profile, but lags it in time. The cell temperature increases as oxygen recombines at the negative plates and releases heat. The voltage reaches a peak level and then drops off due to oxygen recombination 'pulling down' the potential of the negative plates. Clearly, the shape of this part of the curve is not the same as that displayed by flooded cells (refer Fig. 5.4). The pressure and temperature relationships are quite different under constant-voltage charging (Fig. 5.12(b)). As noted above (Fig. 5.8(c)), the charging current drops off significantly in constant-voltage charging as the cell becomes fully charged. This means that there is less gassing (and venting), less oxygen recombination, and less heat generation. Thus, there is a much smaller rise in temperature, or possibly a rise followed by a decline, as determined by the charge voltage. For cycling applications, such as those experienced in EV service, a constant-current–constant-voltage charging method is preferred and is followed by an equalizing charge at an increased cell voltage for a set time (Fig. 5.8(h)). To avoid undesirable overdischarge, the charging current during this last stage is limited to a fairly low level.

5.3. Principles of Manufacture

5.3.1. Plate designs

Three varieties of positive plate are used in the manufacture of lead/acid batteries, namely, the Planté, the flat pasted (or Faure), and the tubular configurations. By contrast, all negative plates are constructed as pasted types.

The Planté version of a positive plate is made from a casting of pure-lead that has numerous thin vertical grooves, strengthened by a series of horizontal cross-ribs, in order to increase the surface area. The active material (lead dioxide: PbO_2) is formed electrochemically from the lead itself by carrying out a series of charge–discharge cycles in a mixture of dilute sulfuric and perchloric acids, and is held within the grooves cast in the plate. Planté batteries are flooded types and are used solely in standby-power service.

The pasted type of positive plate is produced by machine application of a stiff paste of the active material on to a rectangular, lattice-type grid which is fabricated from a suitably selected lead alloy (Fig. 5.13(a)). The grid has a lug (or tab) for connection to the strap to which all plates of the same polarity (in this case, positives) are joined. Thus, in addition to providing a support for the active material, the grid conducts the electrical current to and from the plate. The paste consists of a mixture of 'leady oxide' (see Section 5.3.3), water and sulfuric acid blended together in proportions that are specific to each battery manufacturer. Minor percentages of red lead (Pb_3O_4) and certain proprietary additives are often included as these are considered to benefit the process and/or are claimed to improve battery performance. After pasting, the plates are passed rapidly through a high-temperature tunnel oven (so called 'flash drying') in order to reduce the surface moisture so that the plates will not stick together during subsequent handling. In the next process stage, the plates are 'cured' for a given period, preferably in an oven which is operated under controlled conditions of humidity and temperature. These two parameters are

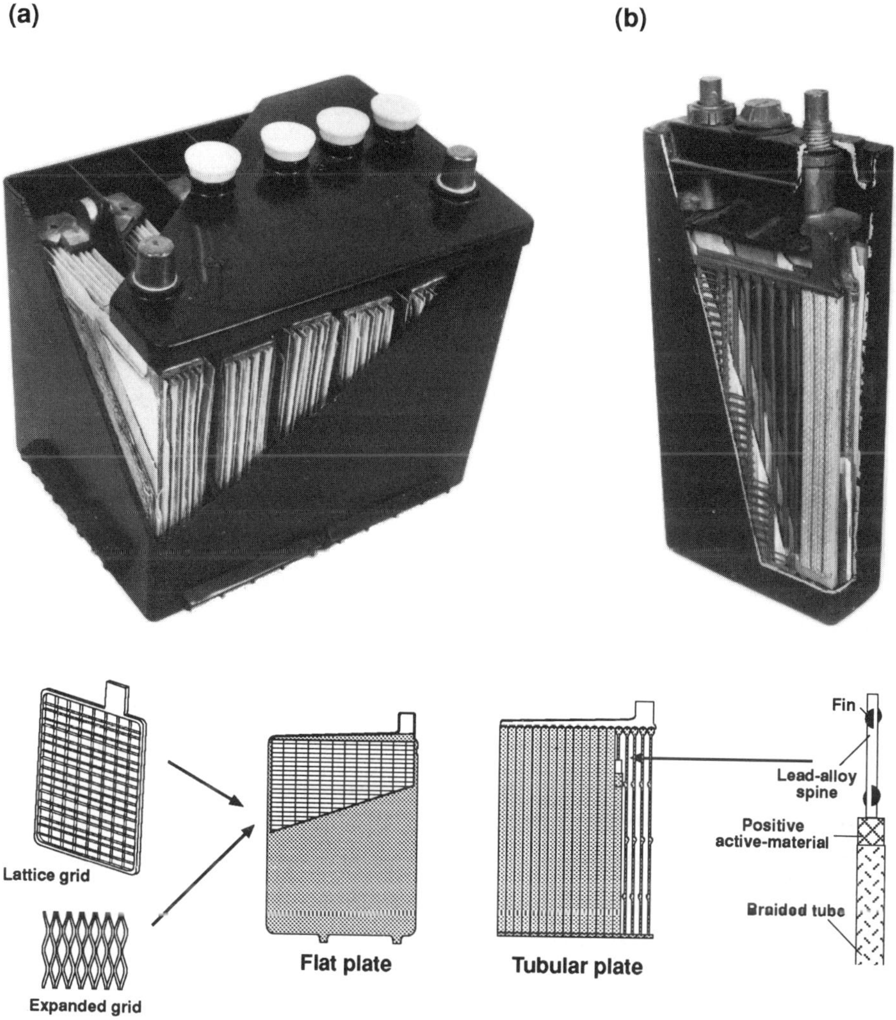

Fig. 5.13. Lead/acid batteries with: (a) flat plates; (b) tubular plates.

regulated [70–72] to provide cured plates rich in either tribasic lead sulfate ($3PbO{\cdot}PbSO_4{\cdot}H_2O$ = 3BS) or tetrabasic lead sulfate ($4PbO{\cdot}PbSO_4$ = 4BS). The consolidation of the plate material is assisted by the production of these basic lead sulfates. Finally, during the 'formation' process, the cured material is converted to PbO_2 by electrochemical oxidation in dilute sulfuric acid for a fixed period under controlled conditions of charging rate and temperature. Pasted-plate designs are employed over the complete range of flooded and VRLA battery applications.

The tubular form of positive plate consists of a row of tubes which contain coaxial lead-alloy rods (also known as 'spines') surrounded by active material (Fig. 5.13(b)). Two types of tubes are commonly used: single braided glass-fibre or woven polyester-fibre, multi-tube 'gauntlets'. After filling with oxide (usually a mixture of leady oxide and red lead), the tubes are soaked in sulfuric acid (the so-called 'pickling' process) in order to convert the majority of the lead oxide into lead sulfate. This extensive sulfation assists the subsequent formation of the tubular plates. Flooded batteries with tubular plates are used predominantly in standby-power and motive-power applications.

5.3.2. Grid alloys

Conventional grid alloys for flooded traction batteries are based on the lead–antimony alloy system. Antimony strengthens the lead through the formation of an eutectic phase at the grain boundaries that consists of a fine structure of lead and antimony platelets. These platelets are not only considerably stronger than the lead-rich phase contained between them, but also do not deform as the antimony content is raised. In fact, by virtue of the rigidity of the platelets, the tensile strength, yield strength and creep resistance of the alloy system all increase with antimony content. This strengthening of the grid alloy by the eutectic phase is augmented by precipitation of fine particles of antimony (as well as other minor alloying elements, particularly arsenic). The precipitates inhibit the movement of imperfections or 'dislocations' (which weaken the alloy) both in the crystals and more especially in the grain boundaries. These beneficial mechanical properties enable lead–antimony grids to resist major deformation under the action of the stresses which are imposed by corrosive attack and by volume changes in the active-material during the charge–discharge cycling of battery plates.

Due to differences in the ease of fabrication, the spines of tubular plates have traditionally required an alloy with a higher antimony content (up to ~10 wt.%) than that appropriate for flat plates (up to ~6 wt.%). Similar alloys are employed in the manufacture of standby-power batteries, but the use of pure-lead in both Planté and pasted positive plates is also quite common. Traditionally, flat grids are produced in pairs (to aid subsequent handling) by the gravity casting, or 'book-mould', method. This involves feeding a metered amount of molten alloy to the top or 'gate' of a pair of vertically supported half-moulds. When the alloy has solidified, the half-moulds open and the double grid is ejected. (Note, pressure casting is employed in the manufacture of spines for tubular plates.) The production process for flat grids is non-continuous, labour intensive, and gives low productivity. Accordingly, advanced techniques have been developed for the continuous casting of low-antimony, flat grids [73–76]. The improved technology has largely overcome the problems of excessive corrosion and grid growth that had hitherto hindered attempts to produce commercially viable grids via a continuous casting process.

Antimony serves to improve the castability, hardness, strength, creep resistance and corrosion resistance of the grid. It also exerts a subtle, beneficial effect on both the structure and the life of the positive active-material, particularly under severe conditions of charge–discharge cycling. As mentioned above in Section 5.2.5, however, there is a major disadvantage with using antimony. During battery charge, antimony dissolves (corrodes) from the positive grid, diffuses through the electrolyte, and deposits on the negative plate. Antimony lowers the overpotential for hydrogen evolution and thus causes greater rates of gassing, water loss (and, consequently, the need for more frequent maintenance), and self-discharge. The severity of this problem is determined by the amount of antimony deposited on the negative plates and, hence, the rate of water consumption increases with battery age. Such features are in opposition to the growing demand for maintenance-free operation in all types of battery applications. This, together with obvious cost savings, has resulted in the development and widespread use of 'low-antimony alloys' in flooded batteries. At first, casting problems (e.g., cracking) were experienced with low-antimony alloys. These have been largely overcome, however, by the addition of grain-refining agents such as arsenic, copper, selenium, sulfur and tellurium [77] which enhance both the stiffness and the tensile strength of the grid alloys. Castability can also be improved by the presence of a small amount of tin (~0.1 wt.%) which increases the fluidity of lead alloys. Moreover, these additional elements can enhance both the mechanical strength and the corrosion resistance of the alloys.

Low-antimony alloys used to fabricate the positive grids of present-day automotive batteries generally contain ~2 wt.% Sb. In the USA — the world's largest market for automotive batteries — most units are built with either lead–calcium or lead–calcium–tin alloys for the negative grids (so-called 'hybrid' units). There is also a growing proportion of batteries in service with positive and negative grids that are both made from lead–calcium alloys. These alloys exhibit a higher hydrogen overpotential than antimonial types (and, thus, suffer a lower water loss during charging), have increased electrical conductivity, and are sufficiently malleable to allow grid fabrication by rapid and continuous techniques. On the other hand, lead–calcium alloys are significantly weaker (by about two orders of magnitude) than lead–antimony counterparts, especially in terms of creep resistance [78]. Grids using calcium-containing alloys are produced by one of three processes: (i) direct gravity casting (book-moulding or continuous strip); (ii) slitting and expansion, into a diamond-shaped mesh, of thin strip cast on a rotating, water-cooled drum which is immersed in a bath of the molten alloy; (iii) slitting and expansion of rolled (wrought) cast strip. The last two techniques produce a double-width grid form with a solid central section. Metal is punched from the latter to leave opposing lugs that connect the two halves of the strip. The strip is pasted, clad on both sides with paper, cut into plates, and then transferred to the flash-drying and curing ovens [79]. The different methods for the fabrication of lead–calcium grids can give rise to variations in the metallurgy of the alloy that can be expected to influence performance in the battery. For example, wrought manufacturing techniques produce alloys with mechanical properties which are comparable with those of high-antimony alloys, together with finer, defect-free grain structures which offer a marked improvement in creep resistance and less susceptibility to intergranular corrosion.

The calcium content is important in determining the mechanical properties, grain structure, rate of strengthening ('age-hardening') and corrosion resistance of the alloy. In particular, the tensile strength peaks at about 0.07 wt.% Ca and then exhibits a marked decline at higher calcium contents. The rate of corrosion of lead–calcium alloys is also accelerated above 0.07 wt.% Ca. Preserving the correct calcium content is a particular problem during the casting of lead–calcium grids or strip — calcium has a propensity to oxidize and form a dross. Under such circumstances, the cast alloy has a large grain size and the resulting grids are relatively weak. Moreover, the enlarged grain size and lowered calcium content brings battery grids into a composition range where cellular precipitation of calcium, as the inter-metallic compound Pb_3Ca, occurs at the grain boundaries. When this occurs, severe corrosion can take place at the grain boundaries during battery service. In addition, the oxides generated during calcium oxidation from the molten alloy can give rise to pockets of trapped dross in the grids and these, in turn, can cause very rapid corrosion of cast lead–calcium grids. Drossing can be minimized either by maintaining an inert gas atmosphere above the molten metal or by adding aluminium to the alloy. In the latter case, the aluminium forms a protective skin of oxide over the surface of the melt pot, and if there is sufficient aluminium (viz., ~0.03 wt.%), any ruptures in this skin are quickly repaired.

Lead–calcium–tin alloys are also preferred by most manufacturers of VRLA batteries (both AGM and gel types) and great emphasis is placed upon alloy purity to minimize water loss [80]. The addition of tin has been shown: (i) to enhance the mechanical properties of the alloys [81], by changing the mode of precipitation from Pb_3Ca to the more stable $Pb_3(Ca,Sn)$ compound; (ii) to increase the electrical conductivity of the grid/active-material interface and thus reduce the tendency towards passivation and improve recharge performance [82]; (iii) to increase the corrosion resistance [81,82]. Silver also increases both the creep and the corrosion resistance of lead–calcium–tin alloys [83]. Research studies have suggested that the optimum composition for positive-grid alloys contains 0.05 to 0.07 wt.%Ca–1 to 2 wt.%Sn–0.05 wt.%Ag. The main exception to the use of lead–calcium-based alloys is the Gates' AGM design (Fig. 5.10) which employs grids punched from pure-lead. Such grids require careful processing because of the low strength of the metal. Glass-microfibre VRLA batteries have also appeared on the market with low-antimony positive grids [84]. These are based on a lead–antimony–cadmium system which contains approximately equal amounts (viz., 1.5 wt.%) of the two alloying additions. It would appear that the degree of antimony poisoning of the negative plate is minimal and that the alloys are suitable for deep-cycling service. There are, however, toxicity problems associated with the use of cadmium.

Clearly, the use of the different alloy compositions described above will affect grid corrosion rates and mechanisms, grid growth, and the leaching of impurities into solution. In turn, these features will influence the life and failure modes of batteries — dependent, of course, upon the cell design. Of particular importance was the early recognition that the deep-discharge cycle life of batteries made with lead–calcium positive grids is inferior to that of batteries using lead–antimony alloys. Such behaviour (initially referred to as the 'antimony-free effect', but later as 'premature capacity loss') is discussed in detail in Section 5.5.3.

5.3.3. Leady oxide production

Leady oxide (also called 'grey' oxide) is made by either the Barton-pot or the ball-mill process [85–89]. With either method, the oxide usually contains about one-part unreacted, fine lead particles (so-called 'free-lead') and three-parts lead monoxide (predominantly α-PbO). It is also possible that minor amounts of Pb_3O_4 (red lead, or minium) are produced. The oxidation is performed at a much higher temperature in a Barton pot and as a consequence, the conversion of lead to oxide is correspondingly higher, i.e., Barton-pot oxide usually has a lower free-lead content (18–25 wt.%) than its ball-mill counterpart (25–30 wt.%). There are established standards of lead purity for making leady oxide for use in flooded-electrolyte batteries. Not surprisingly, however, there are differences of opinion between suppliers of lead derived from ores which are mined in different parts of the world. Also, there is continuing debate on the relative merits of using primary and/or secondary lead for leady-oxide production. A particular difficulty relates to the purity of lead required to make leady oxide for use in VRLA batteries, in which it is vital to restrict hydrogen generation to a low level. Experience shows that certain trace elements, present either alone or in combination with other trace elements, can seriously increase the level of gas generation [53]. This has given rise to a marked shift by many battery companies towards tighter purity controls for the lead that is used in oxide manufacture.

In the Barton-pot process for manufacturing leady oxide (first patented by G.B. Barton in 1898), molten lead is partially oxidized by humidified air at a specific temperature (typically, ~400 °C) in a crucible or 'reaction pot' (Fig. 5.14(a)). Usually, a spray of molten lead droplets is fed into the reaction pot and any accumulated bulk molten metal is broken up again into droplets by a revolving paddle which forces the lead against a fixed baffle arrangement inside the pot. Sometimes, the lead is melted directly in the reaction pot itself. In both procedures, a stream of air, blown over the molten droplets, oxidizes the lead and, once this reaction is in progress, the temperature is maintained by the oxidation itself. The resulting oxide powder is carried by the air stream into a size-classification apparatus ('cyclone classifiers') where the required particle-size range is separated out and the coarser material is returned to the pot. Battery oxide of the desired chemical composition can be obtained by careful control of: (i) the size of the operating load (determined by the rate of solid or molten lead feed); (ii) the pot temperature; (iii) the speed and humidity of the air flow; (iv) the design and rotation speed of the paddle. Generally, the particle-size distribution of the resulting oxide is satisfactory for good reaction with sulfuric acid in the paste-mixing process (see Fig. 5.16, later). In order to avoid pasting and plate-making difficulties, however, some battery manufacturers pass the Barton pot oxide through a hammer mill to produce particles with greater reactivity. The fracturing action of the hammer mill both flattens the spherically shaped oxide particles and increases the amount of β-PbO. Typically, this treatment increases the reactivity by

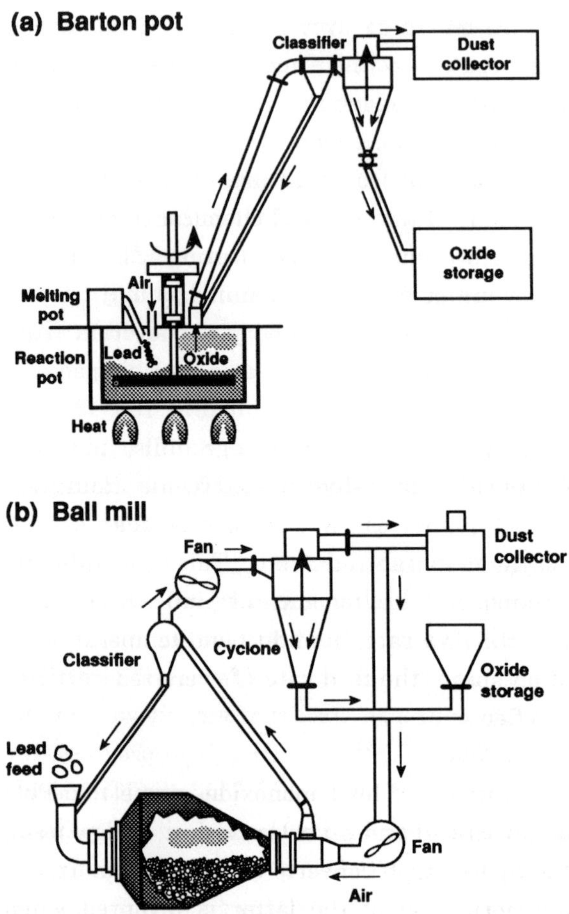

Fig. 5.14. Schematic representation of the production of leady oxide: (a) Barton-pot process; (b) Hardinge ball-mill process.

about 12% over that of standard Barton-pot oxide, but improvements up to 25% are achievable. Over the years, the basic Barton-pot process has been refined by Balox, Chloride, Hammond, Heubach, Linklater, Minimet, Pennaroya, and others.

The original ball-mill process was developed by Shimadzu in Japan in 1926, and by Tudor in Europe. In this method of leady oxide production, lead balls, cylinders, or part/whole lead ingots are tumbled in a steel drum which is rotated (typically, at 25–30 rpm) with its central axis in the horizontal plane. The heat generated by the friction between the lead pieces is sufficient to start oxide formation in the presence of a stream of humidified air which is passed through the drum. This reaction, in turn, generates more heat and the particles of lead that are rubbed off by the abrasion are converted to leady oxide of the required chemical composition. Cooling the drum with air or water prevents excessive heating. The oxide is removed by entrainment in the air stream (e.g., Shimadzu mill), or by discharge through small holes located in the periphery of the drum (e.g., Tudor mills, operated at 70–90 °C), or by overflowing at a cone-shaped exit which is fitted with screens (e.g., Hardinge mill, operated at 160–190 °C). The air-sweep principle is now applied also to Hardinge mills, and both these and Shimadzu mills are usually fitted with cyclone classifiers (Fig. 5.14(b)). Oxide of consistent quality can be obtained through careful control of the residence time of the material in the mill. This is achieved by regulating the following variables: the amount of mill charge; the temperature of the mill charge; the flow rate, humidity and temperature of the air stream; the mill rotation speed; the feed rate of oversized particles returned from the cyclone classifier.

There are two crystal modifications of lead monoxide, namely, α-PbO (litharge, red form, tetragonal structure) and β-PbO (massicot, yellow form, orthorhombic structure). The α-polymorph converts to the β-polymorph at around 480 °C and, therefore, formation of the latter is favoured when using the higher-temperature, Barton-pot process (Fig. 5.15). Control of the relative proportions of α-PbO and β-PbO is essential. This is because there is convincing evidence [71] that the polymorphs promote the formation of different basic lead sulfate compounds during the plate-curing process, and that these sulfate phases, in turn, can strongly influence the performance characteristics of battery plates (see Section 5.3.6).

In addition to differences in phase chemistry, Barton-pot and ball-mill oxides also exhibit dissimilarities in both morphology and particle size. The free-lead particles are roughly spherical in Barton-pot material, but elongated (or 'flaky') in ball-mill oxide [86]. The smaller surface area of the

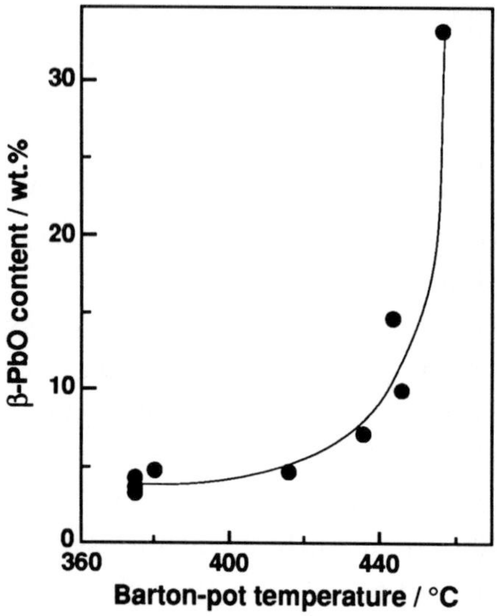

Fig. 5.15. Influence of process temperature on β-PbO content of Barton-pot leady oxide. Note, the absolute value of the temperature depends on the location of the temperature sensor. (Taken from [88] with permission of Elsevier Science Limited, UK.)

free-lead particles in Barton-pot material decreases the rate of secondary oxidation during storage, that is, Barton-pot oxide has a longer storage life than its ball-mill counterpart. On the other hand, the higher surface area of the free-lead particles in ball-mall oxide enables paste made from this material to be cured faster (note, oxidation of free lead is the driving force in curing, see Section 5.3.6). It is also claimed that ball-mill oxide provides plates with greater mechanical strength and higher initial capacity.

The Barton-pot process produces the larger oxide particles. This feature renders the oxide less prone to aggregation and, therefore, it is easier to handle. The larger particle size also invests the oxide with good flow characteristics and, as a consequence, favours its use in tubular-plate production. By contrast, plates filled with ball-mill oxide tend to exhibit poor uniformity in weight and a lower packing density. Leady oxide has often been blended with red lead (Pb_3O_4) due to the latter's ability to assist in plate formation (about one-third of the oxide is converted directly into

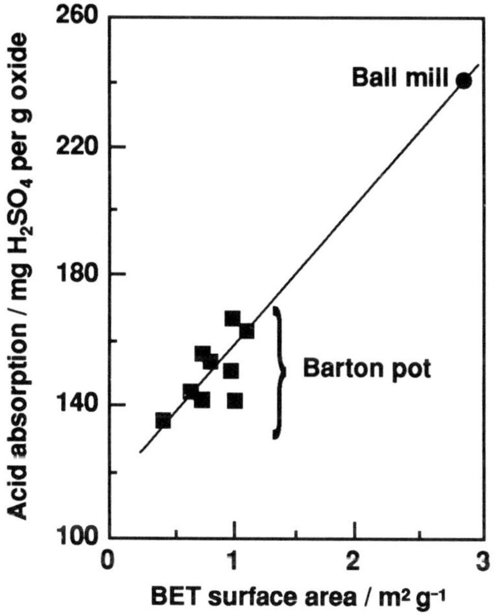

Fig. 5.16. Reactivity of Barton-pot and ball-mill leady oxides. (Taken from [88] with permission of Elsevier Science Limited, UK.)

PbO_2 before the formation charge is started). In recent times, however, the application of red lead has slowly declined because of its extra cost and handling (i.e., occupational health and safety) problems.

The larger size of the Barton-pot oxide particles also results in the material having a lower reactivity with sulfuric acid (i.e., a lower 'acid-absorption value', defined as the weight, in milligrams, of sulfuric acid absorbed per gram of oxide) than ball-mill oxide, see Fig. 5.16. As outlined above, however, Barton-pot oxide can be made more reactive by mechanical treatment in a hammer mill. Oxide reactivity is especially important in the manufacture of flat plates as it provides a useful indicator of the paste-mixing attributes of the oxide. Differences in reactivity will give rise to variations in paste properties and, thus, will necessitate modifications to paste formulations. Naturally, such problems will be transmitted through to the subsequent, curing stage. It is claimed that the mixing and application of battery paste is easier when using Barton-pot oxide. Nevertheless, slight

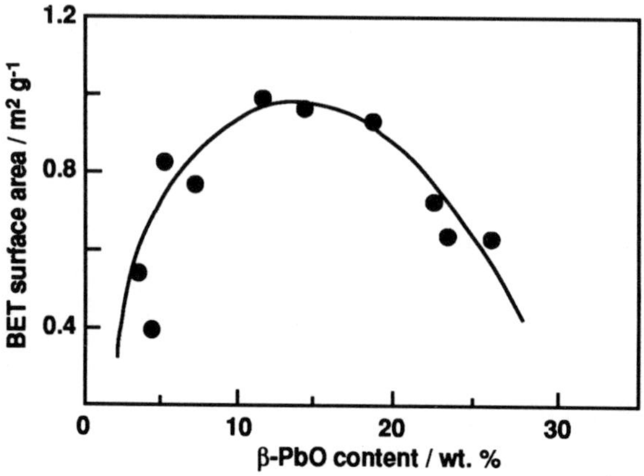

Fig. 5.17. Influence of β-PbO content on surface area of leady oxide. (Taken from [88] with permission of Elsevier Science Limited, UK.)

adjustments in either, or both, of the quantities of water and acid will yield perfectly satisfactory pastes from either type of leady oxide.

The surface area (and, hence, the reactivity) of Barton-pot oxide is dependent upon the β-PbO content; maximum reactivity is displayed at about 15 wt.% β-PbO (Fig. 5.17). Therefore, in order to obtain optimum results, it is imperative to control the level of β-PbO in the starting battery oxide through regulation of the oxide-making process itself (principally, the temperature of operation), through milling, or through blending with β-PbO that has been synthetically prepared.

The purity of both the leady oxides is almost exclusively dependent on that of the lead feedstock. Tolerance limits for impurities in lead have been set, but are presently under serious review [53]. Contamination with the constituents of steel is a major concern during the oxide manufacturing process itself, particularly in the case of the Barton-pot method where excessive temperatures can cause the lead to attack the reaction vessel.

Although the Barton-pot process is regarded to be the more difficult to control, the equipment has faster start-up and response to change, is less capital intensive, produces more oxide per unit space (i.e., it has a smaller

Table 5.2. Comparison of Barton-pot and ball-mill oxides.

Parameter	Barton-pot	Ball-mill
Particle size	Large particles (4–8 μm diameter)	Small particles (2–3 μm diameter)
Reactivity in air	Generally more stable i.e., longer storage life	Generally high; can cause storage and long-distance transport problems
Polymorphism	60 wt.% α-PbO 15 wt.% β-PbO	60–70 wt.% α-PbO 1–4 wt.% β-PbO
Acid absorption (mg H_2SO_4/g oxide)	140–160 (200 with milling)	220–240
Surface area ($m^2\ g^{-1}$)	0.4–1.8 (influenced by β-PbO)	2.4–2.8
Free-lead content (wt.%)	18–25 (roughly spherical particles)	25–35 (elongated particles)
Ease of making battery paste	Makes a soft paste which can result in easier pasting	Makes a stiff paste which can require careful control
Paste curing	Average curing rate	Capable of fast curing
Battery performance	Batteries have lower initial capacity, but possibly longer life	Batteries have higher initial capacity, but possibly shorter life
Deep-cycleability	Sometimes good	Usually good
Process control	Can be more difficult, but recent computer controls are helping	Easier; more consistent oxide; slower start-up
Typical production rate (kg h^{-1})	300–800	Up to 1000
Investment considerations	Lower initial cost; compact in size; relatively quiet; costs less to maintain; uses less energy to run	Higher initial costs; requires more space; costs more to maintain
Energy use (kWh t^{-1})	Up to 100	100–300+
Environmental aspects	With well-engineered environmental systems (including baghouse and storage), existing emission standards can normally be met; typically, in the range of less than 0.1 mg oxide per N m^3. The ball mill has a higher noise level.	

footprint), and consumes less energy per tonne of leady oxide produced than a ball mill. In many cases, economic considerations are often the main determinant in choosing between the two methods of manufacturing leady oxide. Despite the advantages and disadvantages of the two oxides and their respective production technology (summarized in Table 5.2), the key requirement of the battery manufacturer is to have an oxide system that will deliver a consistent and reproducible product with good handling characteristics. In terms of the effect(s) on the performance of the resulting battery product, the selection of either the Barton-pot or the ball-mill route to leady oxide preparation will depend upon the pasting, curing and formation strategies that are in practice at the plant.

5.3.4. Paste production and application

The basic ingredients of the paste used to prepare flat plates are leady oxide, sulfuric acid and water. Often, chopped plastic fibres are added to bond the material together and thus enhance its mechanical strength, while special additives (so-called 'expanders') are incorporated in the paste used to make negatives in order to maintain plate porosity during charge–discharge service (see Section 5.4.1). The main objectives in paste mixing are to produce an active material which will: (i) exhibit good plastic-flow properties and readily fill the lattice structure of the grid; (ii) retain sufficient moisture for effective curing; (iii) restrict shrinkage during drying and, thereby, avoid local sinkage and excessive cracking; (iv) produce consistently, after plate formation, a uniform gravimetric density which meets specification. Various mixing actions are employed, e.g., dough-type paddles, (e.g., Ross mixer), Muller wheel (e.g., Simpson mixer), and shear blades (e.g., Beardsley-Piper mixer).

During pasting, a significant proportion of the total PbO converts to basic lead sulfates of which there are three possible varieties, namely, the monobasic ($PbO{\cdot}PbSO_4$ = 1BS) and the aforementioned tribasic (3BS) and tetrabasic varieties (4BS). These compounds serve to consolidate and strengthen the paste through the development of interlocking, crystalline networks. The addition of water and the degree of sulfation must be controlled precisely in order to obtain a paste with the desired density and texture. By virtue of its higher surface area and its greater degree of mechanical strain, ball-mill oxide generates more heat of wetting than its Barton-pot counterpart, but the effect is small for both varieties of oxide. If the same initial amount of water is added, the heat of acid dilution will be the same for both oxides. On the other hand, the heat of reaction of ball-mill product with sulfuric acid is higher because of the greater surface area of this oxide compared with that of the Barton-pot material (Table 5.2). The heat added to the system by the mechanical work expended during mixing is greater for ball-mill paste because the greater reactivity of this oxide drives off water and gives rise to a stiffer paste. In addition, the finer particles of the ball-mill oxide pack more densely than those of Barton-pot oxide. Therefore, to achieve the same paste density with each variety of oxide, further water has to be added to the ball-mill paste towards the end of the mixing. This acts as a coolant so that, overall, there is very little difference in the final temperature of the two types of paste.

The addition of acid is regulated to provide the paste with the required thixotropic properties. With too much acid, the paste mixture becomes 'sticky'; with too little acid, it becomes 'soft'. In either case, grid pasting is made difficult, the cured material is fragile, and the finished plates are susceptible to high levels of active-material shedding during battery service. When the appropriate quantity of acid is used, not all of the PbO is converted to basic lead sulfate on mixing (Fig. 5.18). Furthermore, although β-PbO is thermodynamically less stable than α-PbO below 480 °C, there is evidence that the transformation of some α-PbO to β-PbO occurs during paste mixing. It has been suggested that mixing encourages this phase change via a strain-induced mechanism, although others have argued that β-PbO may result from localized and 'hot' oxidation of the free lead. By the end of the paste-making process, the free-lead content is reduced to ~10 wt.%.

Under well-controlled mixing conditions, the level of 1BS is determined by the acid-to-oxide ratio specified in the paste formulation; the content increases when large amounts of acid are used (Fig 5.18). With lower (i.e., normal) ratios, the acid must be added slowly and with good dispersion,

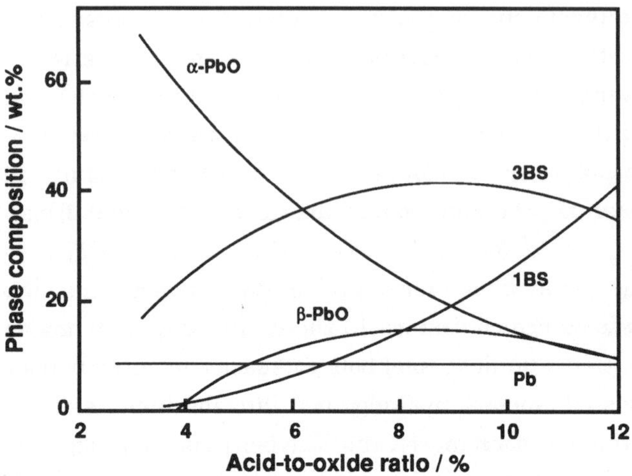

Fig. 5.18. Changes in paste composition with acid-to-oxide ratio for the same starting leady oxide. (Taken from [71] with permission of *Batteries International.*)

otherwise localized decreases in the pH of the paste will promote an abnormally high yield of 1BS. This degrades the consistency of the paste and thus introduces problems during grid pasting. It is also known that 1BS is harmful to plate stability. The material consists of fine, acicular crystals which have a poor cohesive strength. Accordingly, the presence of appreciable amounts of 1BS will promote shedding and/or spalling of active material during both plate formation and subsequent battery service. For this reason, it is advisable to restrict the level of 1BS to <10 wt.%.

The relative quantities of 3BS and 4BS produced in the paste are influenced by the properties of the starting oxide (i.e., phase composition, particle size, morphology), the volumes of both water and sulfuric acid added, the maximum mixing temperature, the mechanics of mixing (i.e., method of acid addition, mixer action), and the duration of the process. For 4BS production, the temperature of the process is the most important determinant: the rate of formation of 4BS becomes appreciable only above ~70 °C, provided that the material is maintained in a moist state. The appearance of 4BS at such high temperatures (and of 3BS at lower temperatures) is also favoured by an acid-to-oxide ratio of ~6%. It is usual, however, to keep the temperature of the paste mixing at a value below ~50 °C. This is because 4BS-rich material has a gritty/sandy texture which renders it difficult to work and prevents the establishment of a firm bond with the grid. Temperature control is best achieved by surrounding the mixer with a water-cooled jacket. In summary, at the end of processing, the paste should consist of a mixture of unreacted starting materials and 3BS.

Two types of machine are available for applying paste to grids, namely, the belt paster and the fixed-orifice paster. In the former, the grids travel on a porous cloth or a slitted, stainless-steel belt and pass beneath a hopper from which paste is pressed, spread and smoothed into the upper sides of the grids. Water is forced out of the paste, into the belt, and collected in a sump. By contrast, the orifice paster pushes a relatively soft paste into both sides of the grid simultaneously and, thereby, gives added protection against grid corrosion and provides considerable improvement in the control of both paste weight and plate thickness [75]. It is necessary, however, to have a very uniform grid thickness and flatness, otherwise, costly jams and production stops are likely to occur..

5.3.5. Flash drying

After pasting, it is common practice to pass the plates through a high-temperature, 'flash drying', tunnel oven. The aim of this heat treatment is to remove the surface moisture (but to preserve the internal absorbed moisture) so that plates can be stacked in contact with one another without risk of sticking together, ready for the next processing stage. Traditionally, mains gas has been used as the heating source; the burners are baffled from direct contact with the plates. Nowadays, both electric radiant and infra-red heaters are gaining in popularity as they allow greater control of the drying process and minimize the possible formation of unwanted basic lead carbonates (see Section 5.3.6).

The rate of plate travel and the degree of heating must be balanced to provide a uniform and precise removal of moisture. Generally, flash drying should be just sufficient to evaporate all the surface moisture, and no more.

Over-drying will not only reduce the bulk water content to a level below that required for efficient curing (viz., ~10 wt.%), but will also promote cracks in the material that will become accentuated during the curing process. In addition to the obvious effects of lowering the mechanical integrity of the plate and increasing the electrical resistance (hence, hindering plate formation), cracking may also result in the preferential discharge of the grid itself and the concomitant development of pernicious α-PbO layers (see Section 5.5.3). The latter problem is especially prevalent with grids made from lead–calcium alloys. Apart from incorrect drying, features such as the water-absorption capability of the oxide, the degree of sulfation, and differential heating between grid and plate, also increase the probability of crack formation. The incidence of cracking in the material can be reduced markedly by subjecting the plates, on exit from the pasting machine, to a fine spray of dilute sulfuric acid. A thin crust of lead sulfate (and, possibly, basic lead carbonates) is produced on the plate surface. This coating ameliorates the drying rate and thus allows the optimum level of internal moisture to be retained during passage through the tunnel oven.

5.3.6. Curing

In essence, curing is the process by which satisfactory levels of cohesion and adhesion are imparted to the active material prior to plate formation, and are maintained thereafter. In addition, curing must ensure that a highly-developed surface area is maintained.

Cohesion is determined by the nature and degree of interlocking between the crystals (or particles) of the paste. Adhesion is dependent upon the strength of the bond which is formed between the paste and the grid. These two essential properties are best accomplished by conducting the curing procedure in the following sequence: (i) a steaming period (so-called 'hydroset'), to equalize the moisture content between all the plates and to encourage the development of basic lead sulfates; (ii) a lower-humidity treatment, to effect the oxidation of free-lead; (iii) a drying process, to reduce the water content to the very low level that is essential for the active-material to have good stability.

Efficient and consistent plate curing is usually the most challenging technical goal that manufacturers face in the fabrication of lead/acid batteries. An effective procedure must allow close control of the chemical, physical and morphological changes that take place within the active material. These changes are both subtle and intricate, and are brought about by at least four reactions, namely:

- progressive growth of basic lead sulfate crystals
- development of basic lead carbonates (principally, hydrocerussite (HC): $2PbCO_3 \cdot Pb(OH)_2$), especially in the surface zones of the plate material
- conversion of free-lead particles to α-PbO with the liberation of heat
- corrosive attack of the metal/alloy grid.

Overall, these reactions cause the active material to harden and strengthen, and bond well to the grid.

The development and growth of the basic lead sulfates provides the essential particle-to-particle cohesion that enables the active material to withstand the regular expansion–contraction forces which are encountered during the course of battery discharge–charge service. Rather than being deleterious, the unavoidable (but mild) corrosion of the grid is beneficial in that it helps to establish a firm contact between the grid surface atoms and the lead oxides or basic lead sulfates in the paste. In simple terms, the effectiveness of the curing process can be evaluated both by the degree of free-lead oxidation and by the amount of water remaining in the material after the process has complied with the specified conditions.

The further generation and growth of basic lead sulfates are both largely dependent on the temperature and humidity conditions that are applied during the initial steaming period. Conversion of 3BS to 4BS will take place if the curing temperature is raised above 70 °C, if the material is maintained in a sufficiently moist condition, and if adequate time is allowed for the reaction to develop. The change in phase chemistry from 3BS to 4BS is accompanied by a corresponding change in the morphological character of the active material: the finely-divided crystals of 3BS are replaced by larger crystals of 4BS (Fig. 5.19). Both types of crystal have an elongated,

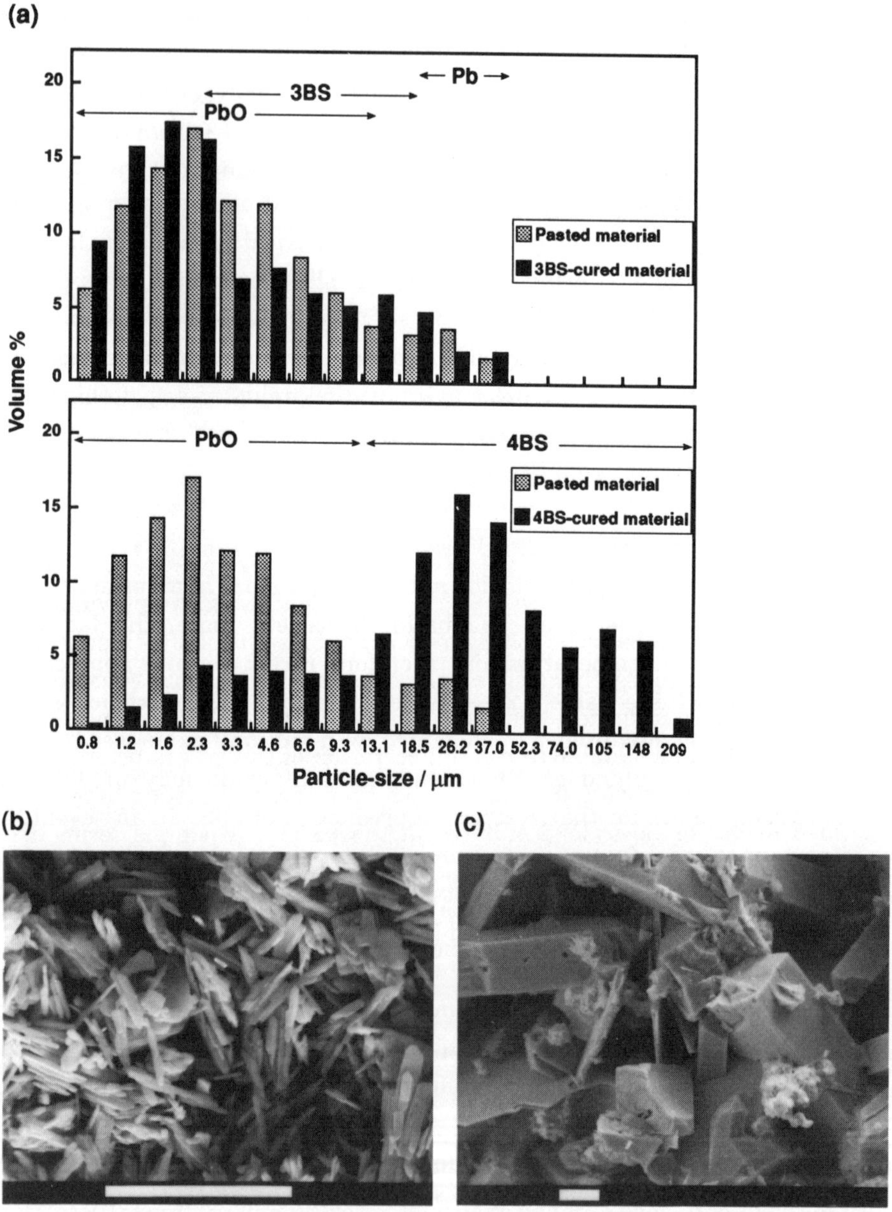

Fig. 5.19. (a) Comparison of particle-size distribution in paste and 3BS- and 4BS-cured materials. (b) Electron micrograph of 3BS crystals. (c) Electron micrograph of 4BS crystals. Magnification bar in (b),(c) = 100 μm. (By courtesy of CSIRO, Australia.)

prismatic form. With the development of 4BS, there is, as expected, a decrease in specific surface area and, subsequently, in discharge capacity (see Section 5.3.7). This disadvantage is offset, however, by the fact that the 4BS crystals interconnect strongly to provide a rigid network (i.e., they give the active material a backbone). The porous matrix is, therefore, more robust than the 3BS-counterpart which is composed of loose aggregates of smaller crystals. Furthermore, the broad geometrical features of the 4BS structure are retained during the following plate-formation process (see Section 5.3.7). Both these attributes serve to enhance the mechanical strength and, hence, the cycle life of the plates. It has been suggested that the favourable effect on cycle life may lie in the ability of 4BS to bind the active material more securely to the grid, and/or to reduce the incidence and severity of crack formation in the corrosion layer [71,72].

Studies have shown that β-PbO promotes the formation of 4BS [71,72]. Thus, if this basic lead sulfate is desired, then it is more appropriate to use Barton-pot oxide rather than the ball-mill variety (note, the latter is produced at a lower temperature and, therefore, rarely contains appreciable amounts of β-PbO, see Table 5.2). On the other hand, such a choice would remove the curing advantages to be gained with ball-mill oxide (see below). Clearly, a suitable blend of β-PbO and ball-mill oxide may provide an effective compromise for the preferential production of 4BS. In any event, it is important to manipulate, control and monitor the levels of 3BS and 4BS because, as highlighted above, these two compounds can exert a substantial influence on battery performance and life.

The oxidation of free-lead is the major process that takes place during the second stage of curing. The oxidation reaction actually proceeds via an electrochemical mechanism and, therefore, requires the presence of moisture. Once enclosed in a sheath of water, each lead particle can develop anodic and cathodic sites (Fig. 5.20). The lead surface is corroded at the former site, while oxygen is reduced at the latter. Overall, the lead is converted to lead hydroxide which, in turn, probably loses some water and becomes partially-hydrated, lead monoxide. The oxidation reaction proceeds at its maximum rate over a relatively narrow range of moisture levels in the

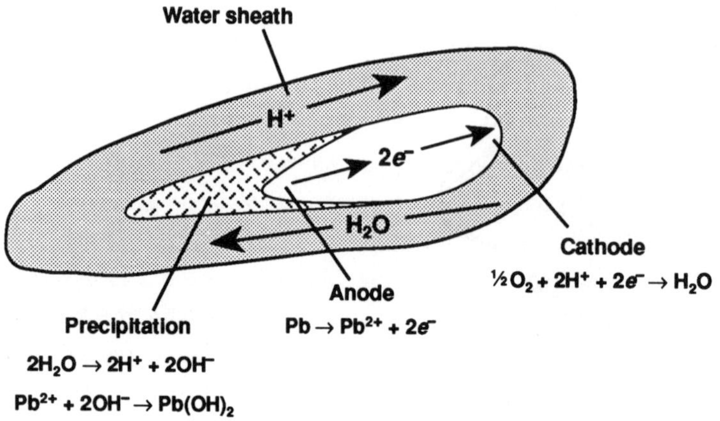

Fig. 5.20. Schematic representation of the oxidation of free-lead particles during the second stage of plate curing.

material, i.e., 7 to 8.5 wt.%. Too little moisture would be insufficient to sustain water films around all the lead particles, while too much moisture would diminish the rate of oxygen diffusion to the cathodic sites. The moisture is supplied initially from within the plate material itself. This emphasizes the importance of delivering plates from the flash-drying oven with a well-controlled and precise amount of internal moisture. On the other hand, lead oxidation generates heat and this will tend to dry out the plate. It is therefore desirable to set the flash-drying conditions so that the plates are dispensed from the oven with a moisture content that errs on the high side of the optimum range, i.e., ~10 wt.%. Spray treatment assists in meeting this requirement and ensures that maximum advantage is taken of the full band-width of moisture change. As the oxidation reaction proceeds, the water content of the plate should be maintained at the optimum level by careful regulation of the ambient humidity conditions.

It should also be recognized that the oxidation reaction is assisted by the lead particles having an elongated, as opposed to an equidimensional, shape — the extended dimensions encourage a better-defined separation of the anodic and cathodic sites. This explains why pastes prepared from ball-mill oxide (flaky lead particles) cure faster than those made from Barton-

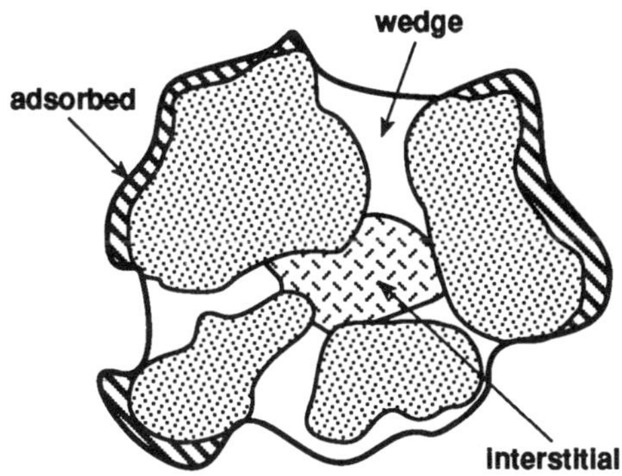

Fig. 5.21. Schematic representation of the distribution of water in plate material after the second stage of curing.

pot oxide (spherical lead particles). In practice, fragments of lead will remain at the termination of curing; these are the cathodic ends of the particles. Nevertheless, the overall residual free-lead content should be kept below 5 wt.%. Under-curing generally results in the swelling/warping of plates, the shedding/spalling of active material and the development of short-circuits — both during the ensuing formation stage and during battery operation. Such debilitating effects are due to the preferential evolution of oxygen on buried lead remnants and subsequent build-up of gas pressure to a level which exceeds the cohesive strength of the active-material pellets.

Immediately prior to the third and final stage of the curing process — the drying period — the active material is essentially still a paste. It is considered that the entrapped water is distributed in three distinct forms (Fig. 5.21): (i) interstitial water, which is drawn by capillary forces into the pore system; (ii) wedge water, which accumulates at the contact points of the particles; (iii) adsorbed water, which covers the exposed surfaces of the paste components.

The importance of thorough drying cannot be over-emphasized — it enhances the mechanical strength of the active material and allows plates to be stored without risk of degradation (see below). The water is removed in the order given above, i.e., interstitial → wedge → adsorbed. During evaporation of the wedge water, dissolved lead species precipitate at the contact points of the particles and act as a cement which consolidates the matrix into a sturdy and continuous structure.

Inevitably, drying will be accompanied by some shrinkage of the material. This must be kept to a minimum in order to retain sufficient porosity without the introduction of large voids and/or cracks. The latter would reduce the mechanical strength of the plate and impose resistance to the passage of electrical current during plate formation and cycling service. Thus, for successful drying, there must be a balance between the rate of surface evaporation and the rate of capillary movement of water from the interior of the active material. If the former rate is the faster, then shrinkage will be greater at the surface of the plate than in the bulk, and differential stresses will promote the formation of cracks. In other words, rapid capillary flow is essential for rapid drying. For this reason, cured plates should be subjected to 'humidity drying'. With this approach, the required capillary flow is achieved by raising the temperature of the material (and, thereby, lowering the viscosity of water), while a controlled evaporation rate is assured by maintaining an environment with a high relative humidity. As drying and shrinking proceed, the humidity is reduced gradually at a constant high temperature until the plates are free of moisture and devoid of cracks. Such thorough drying is mandatory. High residual moisture complicates the plate-formation procedure (see Section 5.3.7) and, for plates cured via the 3BS route, this can result in the degradation of the basic lead sulfate during battery storage. For example, under certain ambient conditions, 3BS decomposes to 1BS and HC. Since plates with large quantities of these compounds have a tendency to shed excessively (see Section 5.3.4), battery manufacturers must be alert to the potential instability of 3BS when adopting a low-temperature procedure for plate curing (note, 4BS does not exhibit the same behaviour as 3BS).

The factors which can determine the curing process of battery plates are summarized in Fig. 5.22. In essence, the key to successful curing is the

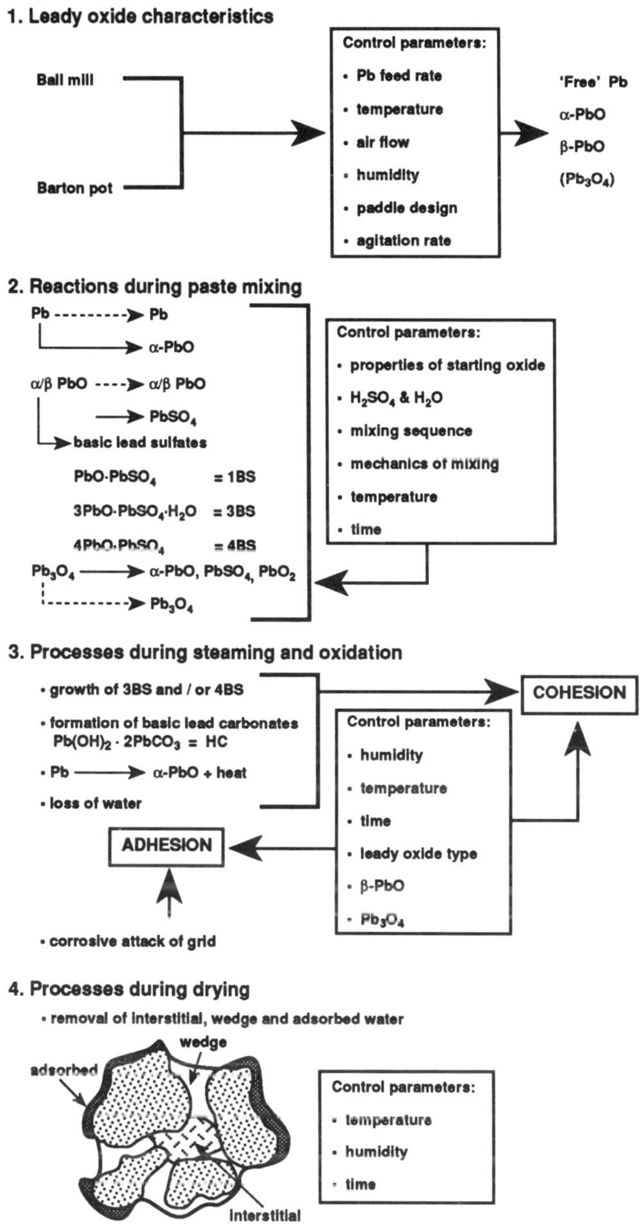

Fig. 5.22. Determinants of the curing of lead/acid battery plates.

maintenance of an exact balance between the moisture content and the temperature of the plate material. It is essential, therefore, to conduct the process within an enclosed chamber which allows precise regulation of the ambient conditions. In order to achieve good control, high energy efficiency, minimum handling of materials and optimum use of floor space, it is necessary to perform all three stages (i.e., hydroset, lead oxidation, drying) within the same chamber. This requires the chamber to be fitted with heaters, humidification controls, and an air supply–recirculation system. The design, location and operation of these facilities is critical if all plates are to be processed in the same manner [70]. Such considerations are extremely important when plates containing high contents of 4BS are being prepared. Quality control is monitored by regular measurements of the water, free-lead and basic lead sulfate contents, as well as by periodic evaluations of plate strength.

5.3.7. Formation

During formation, the lead monoxide and basic lead sulfates in the cured paste are converted electrochemically to lead dioxide in the positive plate and lead sponge in the negative plate. This is achieved either by 'tank' formation or by 'container' formation. The former, older process is more applicable to the formation of plates for automotive batteries and involves assembling positive and negative plates alternately in large tanks; plates of the same polarity are connected in parallel. The tanks are filled with a liberal excess of dilute sulfuric acid (typically, 1.05–1.10 sp. gr.) and charging is conducted to the gassing state. The charging rate can be either a fixed value or a sequence of several values for different periods of time. After the process is completed, the plates are removed from the tank, washed to remove the sulfuric acid solution, and then dried under controlled conditions (note, negatives will spontaneously oxidize in air). The plates are now ready for assembly into batteries.

Flooded-electrolyte lead/acid batteries designed for EV applications are usually subjected to container formation. (Note, container formation is not generally recommended for valve-regulated designs since it is difficult to obtain good uniformity of acid distribution in freshly filled cells [90].) In this procedure, the cured plates are assembled in their final container, acid is added (~1.24 sp. gr.) and, after a short period to allow acid diffusion into the plate materials (the 'soaking' time), the battery is put on charge with others in series. When the formation process is complete, the final electrolyte concentration reaches that of the fully-charged battery. Compared with tank formation, container formation offers the major advantages of lower process costs (i.e., less maintenance, less direct labour, and less scrap), improved environmental conditions (i.e., reduced problems with lead-in-air, gassing, acid spray and waste effluent), and improved product quality.

Some manufacturers employ a 'two-shot' process for container formation, more particularly for automotive batteries. In this procedure, the cells are first charged in dilute acid (1.05–1.10 sp. gr.), the acid is then dumped and replaced by a fresh quantity of higher strength (1.30–1.32 sp. gr.) and, finally, a short mixing charge is applied so that the electrolyte can attain the working concentration of the fully-charged battery. Preference for the two-shot method is based on the greater efficiency that arises from the use of weak electrolyte. By contrast, the one-shot method tends to be less efficient and requires a tighter control of the processing of the pastes, to ensure a minimum variation in the sulfate contents of the plates and a balanced amount of sulfate between positive and negative cured materials. Nevertheless, two-shot formation does involve the messy discarding of the initial electrolyte and generally takes longer to complete than the alternative one-shot procedure.

Temperature control during container formation poses something of an operational problem. To avoid damage to the plates (e.g., active-material shedding), temperatures should not exceed 50 °C. Therefore, to prevent overheating, the batteries are held in shallow tanks of water, or sometimes sprayed with chilled water or subjected to forced, chilled-air cooling. In addition to temperature, other factors which influence the efficiency of plate formation include: (i) the concentration of the sulfuric acid electrolyte; (ii) the duration of the soaking period to which plates are subjected prior to formation; (iii) the phase composition of the cured paste; (iv) the density of

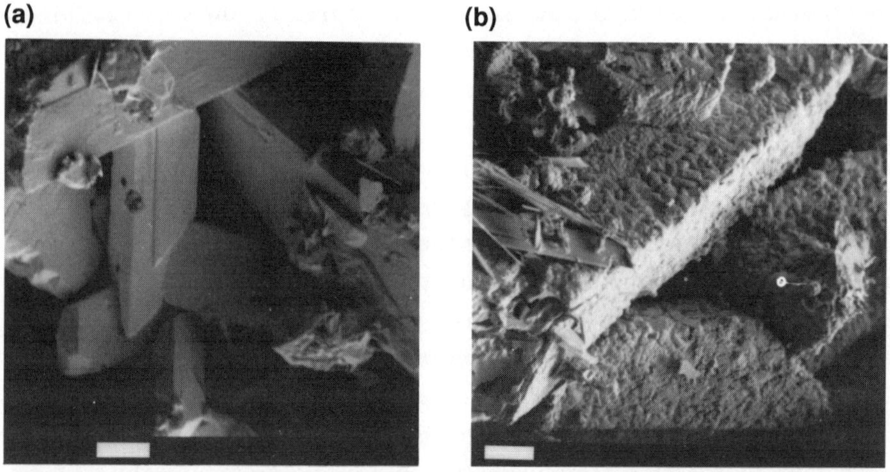

Fig. 5.23. Electron micrographs of 4BS crystals (a) before and (b) after formation. Magnification bar in (a), (b) = 100 μm. (By courtesy of CSIRO, Australia.)

the dry, cured paste; (v) the formation current density. Even when these conditions are optimum, and for practical formation times, positive plates generally contain little more than 85 wt.% PbO_2, while negative plates are transformed to 95 wt.% Pb; in both types of plate, the remaining material is mainly lead sulfate.

As mentioned above, 4BS crystals are much coarser than their 3BS counterparts and, therefore, are more difficult to convert to PbO_2 during the formation process; the oxidation occurs predominantly on the surfaces of the 4BS crystals (Fig. 5.23). As a consequence, batteries prepared from 4BS-rich plates exhibit a lower initial capacity and an inferior charge-acceptance compared with those which employ plates containing 3BS. Several charge–discharge cycles are required to bring the batteries to full capacity, i.e., to convert the inner parts of the 4BS crystals to PbO_2. On the credit side, however, the reticular structure of the 4BS crystals is mostly retained by the PbO_2 in the formed plates, that is, the PbO_2 is gathered into aggregates that follow the overall geometry of the precursor 4BS network (Fig. 5.23). This 'metasomatic' process serves to increase the mechanical integrity and, consequently, the cycle life of the plate. It should be noted, however, that formation at high potentials, under vigorous gassing conditions, causes severe attack of the 4BS reticular network [91] and, consequently, may remove the advantage of greater resilience to deep-discharge cycling that is generally achieved with plates that are cured under high-temperature (4BS) conditions.

Clearly, the initial capacity and charge-acceptance problems of 4BS-rich plates would be lessened if curing techniques could be found to reduce the size of the 4BS crystals without weakening their interlocking structure. In one study [92], the addition of red lead to the paste mix has been used to modify 4BS crystal size without impairing plate strength, while another investigation [72] has shown that manipulation of the phase composition of the leady oxide itself (principally through regulation of the β-PbO content) can accelerate the 4BS growth-rate and provide a matrix of smaller, finer crystals with good mechanical integrity. With decrease in the crystal size, a greater proportion of 4BS will be required to provide an adequate network throughout the active mass. Thus, for each plate design, it is necessary to determine precisely the 4BS morphology–content relationship that will yield a maximum-benefit cured material. Further studies have shown [93] that pulsed-current charging can improve the efficiency of 4BS oxidation to PbO_2. Indeed, several battery manufacturers have implemented pulsed-current technology in their formation circuits, despite the extra cost of the more complicated electrical equipment that is required. In its simplest form, such a formation schedule involves the application of a square-pulsed current of constant amplitude.

Two crystal modifications of lead dioxide are found in the positive-plate active material of lead/acid batteries, namely, α-PbO_2 (orthorhombic) and β-PbO_2 (tetragonal). Although these two polymorphs differ in structure only by the spatial arrangement of the oxygen octahedra, there are marked differences in the respective physical properties (Table 5.3). Compared with β-PbO_2, α-PbO_2 has a more compact morphology, greater mechanical strength, larger pore/crystallite size and higher electrical conductivity, but lower surface area and less discharge capacity. The α form of PbO_2 is deposited electrochemically in a neutral or alkaline environment, the β form in an acidic environment. Studies of the distribution of the two polymorphs

Table 5.3. Physical and electrical characteristics of α-PbO_2 and β-PbO_2 [91].

Parameter	α-PbO_2	β-PbO_2	Reference
Density (g cm^{-3})	9.1–9.4	9.1–9.4	[94]
Pore size (nm)	<60	<10	[95]
Crystallite size (nm)	55	20	[95]
BET surface area (m^2 g^{-1})	0.48	9.53	[4]
Discharge capacity (Ah g^{-1})	0.041	0.113	[96]
	0.110	0.172	[97]
	0.135	0.164	[98]

in positive battery plates have revealed [99,100] that whereas β-PbO_2 predominates in outer surface layers of the plates, α-PbO_2 resides in the inner zones. This finding suggests that a pH gradient develops within the active material during plate formation.

Attempts have also been made to detect relationships between the phase compositions of plates in the cured and formed states. The findings have given rise to the 'tetrabasic lead sulfate $\rightarrow$ α-PbO_2' debate [101]. The origin of this debate has been the long-standing suggestion [102,103] that the reticular structure in plates formed from 4BS-rich cured material is composed of α-PbO_2. This, together with evidence that α-PbO_2 improves plate stability and has lower intrinsic capacity than β-PbO_2 (Table 5.3), has given rise to the common assumption that 4BS promotes the formation of the α polymorph and, by analogy, that 3BS converts preferentially to β-PbO_2. More recent work has shown [93], however, that the phase chemistry of the cured material exerts little influence on the relative proportions of α-PbO_2 and β-PbO_2 that are produced during plate formation. Thus, since the development of α-PbO_2 is favoured in weak-acid–alkaline media, physical factors such as increasing paste density and increasing plate thickness that progressively hinder the penetration of sulfuric acid into the plate are more important determinants of the α-PbO_2 content.

5.3.8. Separators

In lead/acid batteries of conventional design, the positive and negative plates are welded to the straps in parallel groups, generally with one extra negative plate. The plate groups are interleaved with sheets of insulating material known as 'separators'. The separator fulfills three basic functions: (i) to provide electrical insulation between each adjoining positive and negative plate and, hence, prevent short-circuits; (ii) to act as a mechanical spacer which holds the plates in the prescribed position; (iii) to help retain the active materials in firm contact with the grid. The main technical requirements for separators used in flooded designs of battery are as follows:

- a non-conductive, microporous structure (~60% porosity) with good wettability, together with a small, uniform pore-size (<1 μm) which will permit both the free diffusion of electrolyte and the migration of ions, but will restrict the penetration of particulate matter that could lead to short-circuits
- adequate strength and flexibility to maintain quality during storage, and to withstand handling during cell assembly as well as mechanical and/or thermal stresses that are likely to be encountered in service
- good chemical stability in the presence of sulfuric acid
- high resistance to oxidation over a wide temperature range
- low electrical resistance (~5% of total internal resistance)
- minimum displacement of acid
- high purity.

In recent years, there has been an increasing trend away from traditional 'leaf' separators towards the use of 'envelope' separators which enclose either the positive or the negative plates. These serve to prevent dendritic growth of lead around the plate edges (and thus shorting, see Fig. 5.24), and to provide improved retention of the plate active-material, especially during the latter stages of battery life when the active material tends to 'soften' and shed from the grid (see Section 5.4.1). Envelope separators are particularly important for batteries which employ positive plates with lead calcium grids because of the propensity for such plates to exhibit high rates of dendritic growth, appreciable plate expansion, and poor adhesion and cohesion of active material.

The materials used for separators can be usefully classified in terms of those made from naturally occurring materials, e.g., wood veneers, cellulose (paper), and those made from synthetic materials, e.g., rubber, polyvinyl chloride (PVC), phenolic resins, polyethylene, and glass-microfibre. Fibrous products are made via the customary paper-making process and are usually stiffened with a phenol–formaldehyde resin binder. Random alignment of the fibres produces a porous structure and thus there is no need for special pore-forming agents. The PVC separator is manufactured by the light fusion of spherical particles of resin. The void space between the particles provides the porosity. Often, a surfactant is added to improve wettability and to lower the electrical resistance. Most types of resinous or polymeric separator utilize a purified, precipitated silica as the carrier for a liquid pore-forming agent, such as starch or oil. The precipitated silica, which remains in the separator, has a skeletal structure and is highly absorptive and naturally acid-resistant. It is also hydrophilic and, therefore, is easily wetted and thereby offers a low electrical resistance.

The use of glass-microfibre ('glass-mat') separators became important with the introduction of VRLA technology. Indeed, this type of separator is the eponym for one of the two forms of VRLA battery — the AGM design (see Section 5.2.7). The microfibres are made from borosilicate glass of very high purity. These are processed into sheet form by a wet-laying process on a conventional paper-making machine. The resulting product has a three-dimensional anisotropy by which most of the fibres lie in the plane of the sheet. The anisotropy promotes differences in both the tensile properties and the pore structure in the three directions. The influence of pore-size anisotropy is considered [104] to be important in the mechanism of gas transport through the AGM separator at the high degrees of acid saturation which are required for VRLA operation. The glass-microfibres have diameters typically in the range 0.25 to 4 μm, and a length of ~1 mm. These are often blended with finer fibres (diameter: ~0.8 μm; length: ~0.3 mm) as the latter, though more costly, can cause improvements in battery performance, e.g., reduced acid stratification and increased cycle life [105]. Nevertheless, the optimum ratio of coarse-to-fine fibre has still to be determined. The addition of polymeric fibres enables the separator mat to become sealable, but such modification lowers the porosity and, hence, reduces the volume of acid that can be retained [106]. Glass microfibre separators have several attributes, namely: (i) good wettability; (ii) small pore size and high porosity (90–95%); (iii) high compressibility, so that they can conform to irregularities in the plate surface with subsequent good acid transfer and low internal resistance (but even higher compressibility is required, see Section 5.5.3); (iv) very low solubility in acid; (v) high oxidation and thermal stability.

Given the growing demand for VRLA technology, considerable attention is continuing to be focused on the development of improved separators for both gel [107,108] and AGM [106,109] designs of batteries. Many of these efforts are directed towards obtaining new materials which can sustain accurate, uniform and high levels of plate-group compression (sometimes called 'stack pressure') throughout battery lifetime. This is because control of the expansion of positive active-material has been identified [110] as the key to combating the problem of premature capacity loss (see Section 5.5.3).

Separators used in the construction of conventional gel batteries are of a 'ribbed' design. This consists of a backing sheet (termed, 'the backweb') to which is moulded a series of ribs. A sheet of glass mat is also placed between the positive plate and the separator. In addition to adding strength to the separator, the ribs determine and maintain the distance between the positive and negative plates and, hence, the supply of acid. In contrast to AGM designs, flat-plate, gel batteries have traditionally been assembled with relatively low levels of plate-group compression. Moreover, the existing separator technology is inadequate to withstand the much higher levels of compression that are required to produce significant gains in battery life. First, the ribbing does not provide even support across the entire positive plate — areas of the positive plate not in direct contact with the ribs rely only on the sheet of glass mat for compression. Second, under high compressive loads, the ribs can deform and fold to bring about collapse of the plate group. Such problems can be ameliorated by increasing the number of ribs and/or by strengthening the backweb. Any increase in backweb thickness, however, raises the electrical resistance of the separator and, in turn, decreases the high-rate performance of the battery (a feature not conducive to meeting EV power demands). These limitations have led to the development of a novel separator system [111,112] that can apply and sustain high levels of compression, as well as preserve or improve the high-rate discharge performance of the battery.

The main body of the new separator system comprises two sheets of acid-resistant material that are joined by equally spaced, vertical ribs. These ribs form a series of compartments and provide the main body of the separator with most of its mechanical strength. The thickness of the sheets/ribs and the spacing of the ribs are determined by the strength of the material used. Holes are drilled through the face of the main body, including the areas backed by the ribs, to permit the movement of electrolyte both through the separator and from one compartment to another. The location and size of the holes depend upon the intended application of the battery and the required separator strength. To provide support to regions of the active material that are not constrained directly by the separator (i.e., where holes occur in the main body), a thin sheet of microporous separator, glass mat, polyester cloth, or some combination of these, is placed on either side of the main body. Overall, the separator system does not significantly affect acid diffusion, occupies minimal volume, and guarantees even compression across the entire face of the positive plate. Remarkable improvements in cycle life under simulated EV duty have been displayed by prototype gel cells fitted with this separator system.

Similar attempts are being made to increase the strength of AGM separators, as well as to improve their ability to maintain high levels of compression. The latter property relies on the AGM separator behaving as a spring, i.e., the separator material must display good resilience such that the necessary rebound force is exerted uniformly on the positive mass [113]. Unfortunately, separator resilience can be diminished by: (i) the absorption of acid (i.e., wetting), which can cause relaxation/contraction of the AGM; (ii) inordinate compression, which can induce permanent 'crushing' of the separator material, either immediately or with time. Hence, given that very minor variations in plate-group thickness can result in significant changes in stack pressure, the compressive force within AGM batteries after acid filling and/or extended operation can be considerably less than that at the time of assembly. The Advanced Lead-Acid Battery Consortium (ALABC) has instigated a research programme to develop methods and equipment for characterizing the resilience and fatigue of AGM separator materials [113]. The aim is to gain a complete understanding of those separator properties that are influential in enabling constant pressure at the required level to be imposed on active material throughout its working life.

5.4. Failure Modes

Before discussing recent and ongoing efforts to enhance the performance of lead/acid batteries for EV applications, it is instructive to consider the common types of failure that limit the performance of present designs. In order to do this, it is necessary to distinguish between 'catastrophic' battery failure, as characterized by a sudden inability of the battery to function, and 'progressive' failure, as demonstrated by a gradual decline in capacity to an unacceptable level during charge–discharge cycling.

In general, catastrophic failure is easily diagnosed and characterized. The following problems are commonly encountered:

- incorrect cell design and component selection, e.g., short-circuits due to plate and separator faults
- poor quality control during battery manufacture, e.g., open-circuits (arising from loose plates/terminals, and/or from the fracture of weak plate-to-strap, strap-to-terminal or cell-to-cell connections), attack from contaminants
- incorrect treatment/operation of batteries ('abuse'), e.g., undercharging, overcharging, low electrolyte level (flooded systems), vibration, high ambient temperature, entry of harmful foreign species, prolonged storage with insufficient recharging

- external and/or internal damage, e.g., broken containers and covers, damaged terminals, electrolyte leakage.

Clearly, there is a certain degree of overlap between some of these categories of failure, and the demise of a battery may have well resulted from some combination of effects.

'Progressive' failure is generally more difficult to predict and explain. Often, it represents a subtle deviation from optimum performance that originates from microscale changes in plate characteristics brought about by the influence of manufacturing variables and service conditions. Parameters which depend upon the materials, processing and design of the battery are known as 'inner parameters', while those determined by the conditions of battery use are referred to as 'outer parameters'. The classification is as follows.

Inner parameters

- the chemical composition and physical properties of leady oxide
- the composition (including additives), density, method of application, loading and pellet size of the paste (for flat plates)
- chemical composition and structure of cured plates (flat type) or pickled plates (tubular type)
- chemical composition and structure of formed plates
- plate thickness
- composition and processing conditions of grids
- composition (including additives) of electrolyte
- choice of separator
- cell design.

Outer parameters

- storage time before use
- frequency and current density of discharge
- depth-of-discharge
- stand time in a state of partial or complete discharge
- current–voltage schedule of recharge

- rate of recharge
- degree of overcharge
- temperature
- uniformity of concentration and maintenance of electrolyte
- battery configuration (length of series string, series/parallel)

Any deleterious effect of the inner parameters on cycle life can be minimized, or perhaps avoided, through control and optimization of the materials' processing stages, whilst any problems emanating from the outer parameters can be reduced by selecting the battery design and system features to match the operating requirements of the intended application. Nevertheless, both sets of parameters will act (to varying degrees) to cause the eventual failure of the battery, principally via progressive degradation of the positive plate, the negative plate and the separator, or by some combination of all three.

5.4.1. Flooded batteries

Degradation of positive plates

The positive plate is the major life-limiting factor. From the outset, it should be remembered that the positive plate has to be sufficiently robust to withstand the mechanical stress which is caused by the discharge product ($PbSO_4$) having a molar volume that is ~70% greater than that of the active material (PbO_2). Under such forces, the structure of the active material tends to change with repeated charge–discharge cycling; the agglomerate structure is gradually and irreversibly broken down into individual crystallites. This process causes softening and shedding of the active material and, hence, degrades the mechanical integrity of the plate. Shedding, in turn, promotes the secondary problems of 'mossing' and 'leading through'. These problems refer to the transport of particles of positive active-material to the negative plate and their subsequent reduction to lead during charging to build up voluminous deposits ('mossing', see Fig. 5.24) and/or dendrites ('leading through') that can, respectively, bridge plates and puncture separators to produce short-circuits. The loosening of the inter-particle contact, together with the fact that not all of the $PbSO_4$ is necessarily converted back to PbO_2 during charging, also causes a reduction in electrical performance.

Fig. 5.24. Examples of 'moss' formation in lead/acid cells. (By courtesy of CSIRO, Australia.)

The corrosion and growth of positive grids are also life-determining degradation phenomena. These processes have been the subject of a vast number of investigations. In summary, the results show that the rate of grid corrosion is influenced by: grid composition and microstructure, plate potential, electrolyte composition, and temperature. These parameters, together with the geometry of the grid and the creep properties of the alloy, also determine the rate/extent of the growth (or elongation) of grid members during battery service.

A particular problem arises with grids made from lead–calcium alloys; namely, batteries can exhibit a rapid walkdown in deep-discharge capacity with cycling. This degradation in performance is more severe than that observed for batteries with lead–antimony alloys containing >1 wt.% Sb and, accordingly, it came to be known as the 'antimony-free effect'. It is now recognized, however, that this is but one manifestation of the much broader problem of 'premature capacity loss' (see Section 5.5.3).

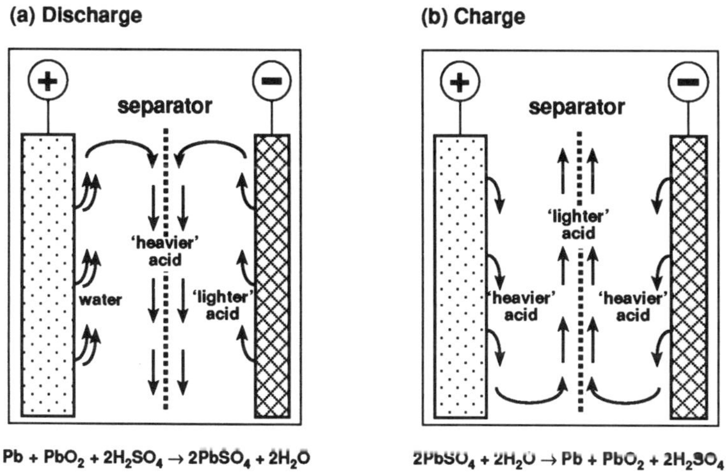

Fig. 5.25. Schematic representation of the processes and electrolyte flow that contribute to acid stratification during: (a) discharge and (b) charge. (Taken from [114] with permission of Elsevier Science Limited, UK.)

During the discharge and recharge of lead/acid batteries, sulfuric acid is, respectively, consumed and produced. These processes can result in the development of a vertical concentration gradient of acid (Fig. 5.25). This, in turn, gives rise to non-uniformity in both current distribution and active-material utilization — factors which shorten service life through the irreversible formation of $PbSO_4$ [115] — as well as to irregular positive grid corrosion and growth. Acid stratification is a particular problem in tall cells and can be avoided in flooded systems by agitating the electrolyte, e.g., by 'gassing' during extended overcharge, air bubblers, or circulation with pumps.

Degradation of negative plates

Negative plates require the presence of additives, known as 'expanders', to prevent the individual crystals of lead from growing and combining (or packing) into a dense structure with low surface area and, therefore, low electrical capacity. Expanders usually comprise a mixture of barium sulfate, carbon black, and a sulfonate derivative of lignin. The lignin has surfactant properties and is adsorbed on the spongy lead, as well as on the surface of lead sulfate crystals which are formed during discharge. This has the effect of inhibiting crystal growth during the charge and discharge processes. The barium sulfate particles act as nuclei for the formation of lead sulfate crystals during discharge, and the end result is a uniform distribution of lead sulfate throughout the spongy lead active mass. The actions of lignin and barium sulfate are considered to be synergistic [116]. Carbon black has a lower hydrogen overpotential than spongy lead and thus, during charge, the hydrogen gas accumulates around the carbon particles to cause an advantageous, mild expansion of the plate material. In summary, the expander counteracts the tendency for the spongy lead to coalesce, contract and densify. Through such action, expanders impart desirable low-temperature and high-rate capabilities to the negative plate, in addition to inhibiting capacity-loss processes.

Degradation of the lignin fraction of the expander, and hence failure of the negative plate, may occur via one or more of the following three processes: (i) dissolution in the electrolyte (and, consequently, transport to the positive plate and destructive oxidation thereupon); (ii) thermal decomposition during battery manufacture and/or service; (iii) chemical reaction with sulfuric acid. Of these, thermal stability has emerged as the most important criterion for expanders as certain applications (e.g., automotive and remote-area power- supply duties) are requiring batteries to operate under increasingly aggressive conditions of elevated temperature. The same problem would be encountered by lead/acid-powered EVs when operating in hot climates, particularly when subjected to rapid recharging (see Section 5.5.4). An example of negative-plate failure caused by thermal decomposition of the expander is shown in Fig. 5.26. In this case, large crystals of lead have developed (Fig. 5.26(a)) and have consolidated to such an extent that the density of the active material resembles that of the grid itself (Fig. 5.26(b)). As a result, the active material undergoes severe shrinkage and becomes detached from the grid members (Fig. 5.26(c)). Naturally, the resulting cracks will introduce electrical resistance losses and, thereby, will severely lower the high-rate performance of the battery.

Fig. 5.26. Densification of active material in negative plates caused by expander degradation: (a) electron micrographs of lead crystals; (b) polished cross-section of plate; (c) shrinkage of active material. (By courtesy of CSIRO, Australia.)

Degradation of separators

Apart from obvious faults associated with mechanical defects, construction errors and damage during battery manufacture, common problems arising with separators include: degradation under conditions of persistently high temperature; attack from contaminants (introduced, for example, via top-up water); failure to resist penetration by lead dendrites which may be growing from the negative plate, i.e., 'leading-through'. In recent years, however, advances in polymer science have made significant improvements in separators so that thin, low-cost materials with small pore-size and high porosity can now be used with a high degree of reliability.

5.4.2. Valve-regulated batteries

Grid corrosion

Grid corrosion is an important life-limiting process when VRLA batteries are used in cycling duties. This is not unexpected as the polarization characteristics of electrodes in VRLA batteries are different to those of electrodes in flooded-electrolyte designs. In particular, the float current is higher because of the depolarizing effect of oxygen recombination on the negative electrode (see Fig. 5.9). Furthermore, for the same applied voltage, the positive electrode in a VRLA battery will experience a higher potential. Obviously, these characteristics will influence the rate of grid corrosion.

During overcharge, water is lost at the positive. This gives rise to an increase in acidity ($2H_2O \rightarrow O_2 + 4H^+ + 4e^-$) which, in turn, will affect the corrosion rate. The problem is compounded by the gradual loss of water during service life as the recombination efficiency is <100%. Finally, the partial saturation of the system and the oxygen environment around the positive also serve to enhance grid corrosion, particularly later on in life when the saturation falls below the initial level.

The above corrosion effects are mitigated to a certain degree by the cell design. In this, the plates are held under compression so that mechanical support of the grid is less of a problem than in flooded batteries.

Grid growth

Positive-grid growth is closely associated with corrosion. The incidence and severity of the problem is determined by the creep resistance and tensile strength of the alloy used, the extent of the corrosion, and the morphology of the corrosion product. VRLA batteries can be designed to accommodate some grid growth. Excessive corrosion will result in unacceptable growth which will promote loss of contact between the grid and the active material and, in severe cases, will give rise to the development of a short-circuit between the expanded positive plate and the negative-plate strap.

Water loss

Water is lost from VRLA batteries via three main mechanisms: (i) the result of oxygen-recombination inefficiencies (particularly under cycling duty); (ii) permeation through the container material (the driving force is the difference in water vapour pressure between the inside and outside); (iii) grid corrosion ($Pb + 2H_2O \rightarrow PbO_2 + 4H^+ + 4e^-$). Water loss will reduce the degree of saturation of the system and, consequently, will increase the resistance of the separator. This is particularly important in situations where the batteries are required to deliver high currents, as in EV applications. For low-rate applications, a much greater loss of water can be tolerated before the deterioration in battery performance becomes significant. At this stage, the problem will be compounded by an increase in the rate of grid corrosion that is brought about by the development of a higher acid concentration. In this situation, failure of the battery will result from a combination of both effects.

Acid stratification

With VRLA batteries, it is obviously not possible to equalize the acid concentration through mixing the electrolyte. As with other lead/acid technologies, the severity of acid stratification in VRLA batteries depends on both the cycling regime and the cell design. The problem is more common with AGM cells than with gel cells [117]. In tall AGM designs, however, the problem can be alleviated by mounting and operating the cells in a horizontal position.

Cell imbalance

Oxygen recombination can result in an irregular distribution of voltage between cells, especially in new batteries. This is because there are two possible reactions at the negative plate (hydrogen evolution and oxygen reduction) and, therefore, the potential of the plate can lie between the two values (i.e., ΔV in Fig. 5.9) and will depend on the level of oxygen recombination (Fig. 5.11). This gives rise to cell-to-cell imbalances which, in series strings, can cause premature system failure due to some cells being persistently undercharged and then deep discharged so that, as time goes on, the weak cells become progressively weaker. The possibility of cell failure by this mechanism is increased if the quality of the product is variable, the string is very large, and/or the cycling is deep and continuous. On the other hand, if the cells experience long recharges, the problem is alleviated by allowing the weak cells to reach top-of-charge before the next discharge. Such a strategy is not practical for EV operations when duty requirements often demand a rapid recharge (see Section 5.5.4). Therefore, traction packs of VRLA batteries require complex management systems.

Leading-through of separator

In VRLA batteries, the electrolyte is completely immobilized and the circulation of particulate matter in the electrolyte cannot occur. Thus, the development of leading-through via such a mechanism is not possible. Despite this and given the acid-starved conditions, a different type of leading-through can occur if the acid concentration reaches very low levels, e.g., as a result of overdischarge or prolonged stand on open-circuit in the discharged condition. In such situations, the pH can increase and give rise to soluble lead species which promote the growth of dendrites from the plates. As this involves a solution mechanism, it is generally unaffected by separator characteristics and can be largely avoided by careful cell design.

Thermal runaway

Thermal runaway is a possible problem with many battery systems and is caused by the heat, produced during charge and overcharge, depressing the cell voltage and producing an increase in current. In VRLA batteries, the recombination reaction results in an increase in heat generation and this enhances the possibility of thermal runaway. In addition, VRLA batteries tend to be smaller and lighter than flooded types and have reduced gassing levels — features which encourage heat build-up and also hinder heat dissipation. Nevertheless, as long as the battery layout is arranged correctly, failure via thermal runaway should not be a common occurrence.

5.5. Advanced Designs for EVs — Technical Challenges

Although the lead/acid battery is a mature product and the basic electrochemistry remains the same, significant advances have been made in recent times. These advances have included: improved lead alloys and grid-fabrication techniques; more reactive oxides for pastes; more effective plate-curing procedures; improved microporous separator systems; valve-regulated technology; a better understanding of the properties of the positive and negative electrode systems. Nevertheless, the duty requirements of modern EVs are placing even greater demands on battery performance. The most pressing need is to develop more energetic, more powerful, long-lived, deep-discharge, inexpensive VRLA batteries which can both co-operate and communicate with the electrical circuits that they are powering, and which can be recharged rapidly. The technical challenges are discussed in the following Sections.

5.5.1. Greater specific energy

Compared with its competitors, lead/acid has a low specific energy. The reason is two-fold: (i) the obvious weight penalty which is imposed by the high density of lead and its compounds; (ii) an exceptionally low utilization of active material in the positive plate.

Minimizing the inert components

About 30% of the total battery weight consists of grids and top-lead (i.e., straps, cell connectors, posts, terminals) that do not directly provide electrical output (see, for example, Fig. 3.2, Chapter 3). Clearly, it is tempting to minimize such 'dead weight', but this must be done without diminishing energy/power levels, without shortening cycle life through complete corrosion of less substantial positive grids and top-lead, or without accelerating failure through the effects of vibration. Weight reduction has been explored via the use of lighter grid materials, thinner grids, and/or novel cell configurations.

Numerous attempts have been made to produce inexpensive, lightweight grids of lead combined with materials of lower density, and without losses in corrosion resistance, mechanical properties or conductivity. Composites have included lead-coated aluminium, copper, steel, nickel or titanium, as well as lead-coated graphite, fibreglass, plastics or ceramics. With the possible exception of co-extruded metal–fibreglass grids (see below), none of these alternatives has proved satisfactory.

A complementary approach is to reduce the thickness of traditional designs of grid. This necessitates the development of stronger and more corrosion-resistant lead alloys, especially calcium-based systems for VRLA batteries. For the latter alloys, tin additions both increase the mechanical properties (yield strength, tensile strength, creep resistance, etc.) and improve the electrochemical characteristics by: (i) preventing passivation; (ii) permitting recharge of batteries from deeply-discharged states; (iii) increasing corrosion resistance. Mechanical properties and corrosion resistance can be enhanced further by incorporating minor amounts of silver. Thin grids are best made from wrought strips formed by rolling cast ingots of the alloys; reduction ratios of up to 10 : 1 are acceptable [118]. The rolling process endows alloys with superior mechanical properties (the improvement is dependent on the amount of reduction imposed by rolling), and with uniform grain structures which retard both corrosion and grid growth (Fig. 5.27) and, thereby, provide better adhesion between the grid and the active material. The major problem is the conversion of the rolled material into battery grids. Expanded mesh is employed in some battery designs, but this can introduce weakening of the metal at the nodes. It is more appropriate, therefore, to use either a punch or a rotary cutter for grid fabrication.

A more adventurous approach to weight reduction is to minimize the current pathways between cells through reconfiguration of the vertical, interleaved, plate assembly used in conventional units. Bipolar and quasi-

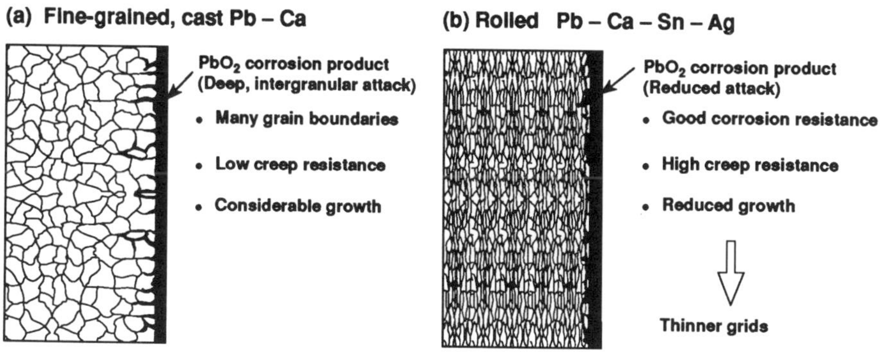

Fig. 5.27. Schematic representation of the effect of rolling on grain refinement and orientation. Rolling reduces corrosive attack and also improves resistance to growth. These attributes allow the manufacture of thinner grids.

bipolar designs are promising examples of such imaginative engineering. As discussed in Section 2.5, Chapter 2, in a bipolar configuration, each interior plate serves as the positive of a cell on one side, and as the negative of the adjacent cell on the other, and the intervening spaces are filled with the electrolyte and separators (Fig. 5.28). The term 'quasi-bipolar' describes intercell connections that are not true bipolar connections. In general, the design involves a non-conducting partition (e.g., plastic) between cells, together with various types of connections similar to those employed in conventional (monopolar) batteries to join cells in series. The Electrosource Horizon® battery has a quasi-bipolar configuration and is being developed for EV propulsion [119]. This valve-regulated, AGM battery comprises several layers of woven, co-extruded, lead–tin/fibreglass wires built up in side-by-side, double-polarity plates ('biplates') that are arranged horizontally and compressed in a cage (Fig. 5.29). The wires have superior corrosion resistance, and the mesh into which they are woven permits a more frequent distribution of wires in the active materials than can be practically achieved with cast or expanded grid technologies. The overall effect is a lower resistance and more uniform utilization of active material. The horizontal orientation provides a convenient low profile, and is said to eliminate active-material shedding and intra-plate electrolyte concentration gradients (i.e., acid stratification) — effects that contribute to loss of cycle

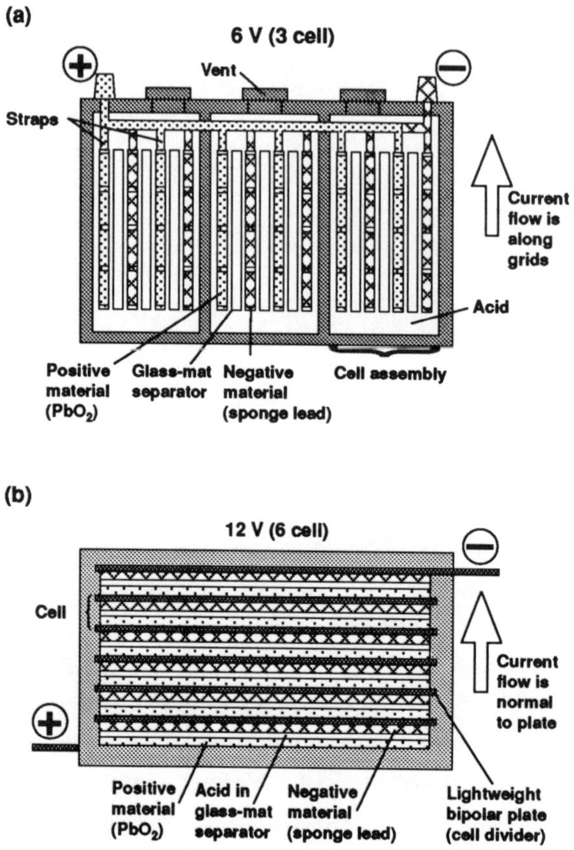

Fig. 5.28. Schematic representation of (a) monopolar (conventional) and (b) bipolar battery designs.

life in most lead/acid batteries. The cell configuration also improves heat distribution and oxygen recombination. By virtue of these design features, the Horizon® technology promises greater specific energy (>50 Wh kg^{-1}, $C_3/3$ rate), higher specific power (>300 W kg^{-1}, 80% DoD) and longer cycle life (900 cycles, $C_2/2$ rate) than conventional systems. At present, batteries are powering both electric and hybrid-electric vehicles in many locations in the USA, as well as in other countries. In two notable projects, prototype battery modules are being produced for hybrid-electric versions of the Orion transit bus and the 'smart Swatchmobil'. The latter is an innovative, two-seater car which is being developed by SMH Automobile in Switzerland.

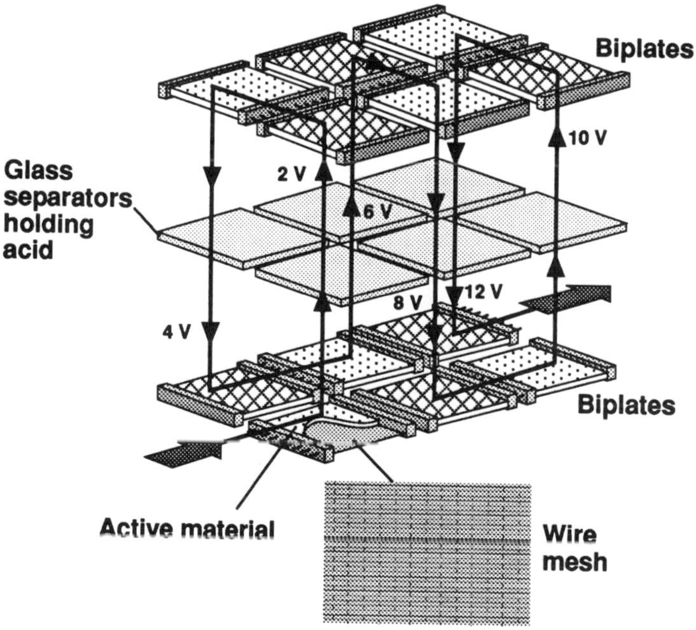

Fig. 5.29. Schematic representation of quasi-bipolar design of Horizon® lead/acid battery. (By courtesy of Electrosource, Inc., USA.)

Improving the active-material utilization

Incomplete utilization of positive active-material is caused by the slow diffusion of acid into the plate interior, and by the continuous decrease in the conductivity of both the positive and negative plates during discharge. These discharge problems become more acute as the discharge rate is increased. In EV (high-rate) operations, for example, only about 20 to 30% of the positive active-material in flooded systems can be accessed to provide electrical energy. Obviously, this limitation restricts vehicle range. VRLA technology has raised the utilization level to ~40%, but further significant improvements must be made to achieve, for example, the target specific energy (50 Wh kg^{-1}, $C_3/3$ rate) set for EVs by the ALABC [120].

Acid supply can be enhanced by using thinner or more porous electrodes, or by the use of external pumps. To date, both these methods have been

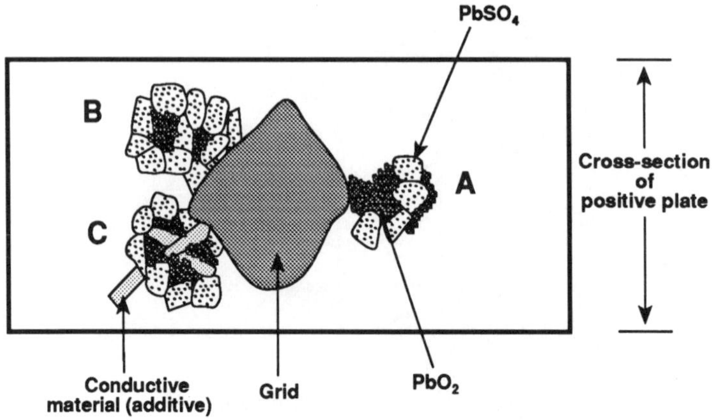

Fig. 5.30. Schematic illustration of the cross-section of a lead/acid positive plate in the presence/absence of a conductive additive.

constrained by adverse effects on cycle life. The future development of thin, wrought grids is expected to ameliorate this problem. Pumping does not appear to be a viable option, however, as it increases system complexity and constitutes a parasitic load.

The second impediment to the utilization of positive active-material is the progressive decrease in plate conductivity during discharge. The PbO_2 particles of the active material serve two functions: (i) to participate in the discharge reaction; (ii) to act as electron pathways from each reaction-site in the active mass to the grid. The non-reacting material provides structure and conductivity to the plate, but does not contribute to the reaction and, clearly, should be minimized. As discharge proceeds, the conductivity decreases with the formation of increasing amounts of $PbSO_4$ (region A, Fig. 5.30). In some locations, areas of PbO_2 can become encapsulated by the non-conductive $PbSO_4$ and, thereby, are rendered inaccessible to further reaction (region B, Fig. 5.30). This is because the electron pathways, as well as the flow of solution reactants through the pores of the active mass, are both hindered by the $PbSO_4$.

Increasing the porosity of the plate is an obvious approach to reducing the problem of $PbSO_4$ encapsulation (and, incidentally, to improving acid supply). This has encouraged numerous studies of the influence of the plate-processing stages on active-material morphology, and how the growth and the distribution of the crystals during these stages decrease material availability. The use of low paste-densities has been explored, but this tends to reduce cycle life through lowering of both the mechanical strength and the corrosion resistance of the plates. The incorporation of additives to promote porosity (e.g., carboxymethyl cellulose, carbon black, graphite, glass microspheres) has also been investigated in detail. Except for the glass microspheres, all the materials tested to date undergo oxidative attack and, consequently, any gains in plate performance are not lasting.

The use of an electrically-conducting additive is the alternative means to overcoming the insulating effects of $PbSO_4$. If a conductive material can be deployed to connect regions of isolated PbO_2 with their reactive neighbours, and thus with the grid (region C, Fig. 5.30), the discharge reaction will proceed at the otherwise inactive locations and the degree of active-material utilization will be increased. Doubtless, a wide range of prospective conductive additives has been examined in-house by battery manufacturers. Published reports, however, are surprisingly few. To date, none of the additives tested, namely, carbon fibres, barium metaplumbate, titanium suboxides (Magnéli phases: Ti_xO_{2x-1}) and tin-dioxide-coated glass-flakes, has produced sufficient and enduring improvements to become commercially practical [121]. Nevertheless, the search still continues for appropriate materials with the following attributes: high conductivity; good chemical/ electrochemical stability; an ability to form a strong network with the PbO_2; negligible enhancement of gassing rates during battery charging. Materials which consist of particles with a high aspect ratio (i.e., fibrous particles) are likely to be more effective in building up a conductive network than spherical alternatives. In addition, the dosage should be kept to a minimum in order to prevent any favourable conductivity effects from being suppressed by a concomitant decrease in the amount and/or pore volume of (potentially) available active material.

5.5.2. Higher specific power

Compared with other systems, the lead/acid battery has good specific power ratings, despite its weight disadvantage. Moreover, VRLA technology (especially, AGM designs) has advanced further the high-rate capability. Nevertheless, new applications are placing ever-increasing power demands on batteries. This is particularly true for EVs, where it is not only the intrinsic power capabilities of the battery system that are important, but also the performance window in which the battery is operated.

Increase in the power demand of an EV drive-train, and hence in the current drawn from the battery, will result in a decrease in voltage due to the $I\mathcal{R}$ drop caused by the internal resistance. Obviously, this results in energy losses which will increase with the DoD of the battery. Thus, high power demands necessitate batteries with minimal internal resistance down to low DoDs (at least 80%). The internal resistance can be reduced by: (i) developing thinner plates with high surface areas; (ii) improving current-collection with new grid designs; (iii) increasing grid/top-lead conductivity through substitution with lead composites; (iv) developing low-resistance separator materials; (v) ensuring good inter-particle contact within the plate materials through optimizing both phase chemistry and architecture, adding binders and dopants, etc.; (vi) minimizing current pathways through innovative cell/battery configurations. Clearly, such advances will also assist attempts to increase specific energy and to enhance recharge capabilities (see Section 5.5.4), but they are generally incompatible with efforts to extend cycle life.

The difficulty of simultaneously maximizing a battery's energy/power and cycle-life capabilities has given rise to the dual-battery concept for EVs in which one unit would be optimized for range (specific energy), and the other for acceleration and hill-climbing (specific power). This is the battery–battery hybrid EV (see Section 10.4.1, Chapter 10). During less-demanding driving, the vehicle's energy management system would charge the power battery from the range unit. This has rekindled interest in bipolar and quasi-bipolar designs for high-power battery modules.

In a bipolar configuration, the cell-to-cell current flow is perpendicular through the whole area of each plate, and is transmitted only through its thickness (Fig. 5.28(b)). Thus, resistive losses are much reduced and the distribution of both current and potential across the plate is uniform. Moreover, the posts and straps that carry the current in conventional (monopolar) designs are eliminated and this further reduces resistance, as well as weight. Consequently, the bipolar design can have a significantly higher peak power compared with a traditional monopolar design. By way of example, the state-of-the-art Arias bipolar battery is a 90-V, 24-Ah unit that delivers 300 W kg^{-1} and 50 Wh kg^{-1} for 3000 cycles at 50% DoD [122]. The practical realization of the bipolar lead/acid battery has been delayed by difficulties in maintaining the integrity of both the electrodes themselves and the seals which they must make with the container. Complete sealing is essential in order to prevent the possible formation of electrolytic short-circuits when gases carrying acid spray are vented. Other problems include finding a stable conductive substrate, removing heat from the battery during high-rate charge–discharge, and establishing low-cost, high-volume methods of manufacture.

Early bipolar designs employed lead sheet as the substrate material. To minimize weight, a thin sheet is desirable but corrosion can render this porous and, thereby, create short-circuits between cells. Furthermore, self-discharge rates increase markedly when only thin layers of lead dioxide are applied to a lead substrate. Various alternative substrate materials have been examined, namely: carbon, conductive ceramics, transition metal oxides, barium metaplumbate, and glass-fibres coated with tin dioxide. None of these has proved entirely satisfactory and the search continues for a material which is stable both electrochemically and chemically, has high oxygen and hydrogen overpotentials, has good conductivity, bonds well with the active materials, withstands mechanical stresses imposed by changes in the volume of the active materials during cycling, and is inexpensive.

The Bolder TMF® cell (Fig. 5.31) presents a different approach to the bipolar concept for achieving high power. The cell is a VRLA design made by the spiral-wrapping of ultra-thin (0.13 mm), lead-alloy, foil plates with

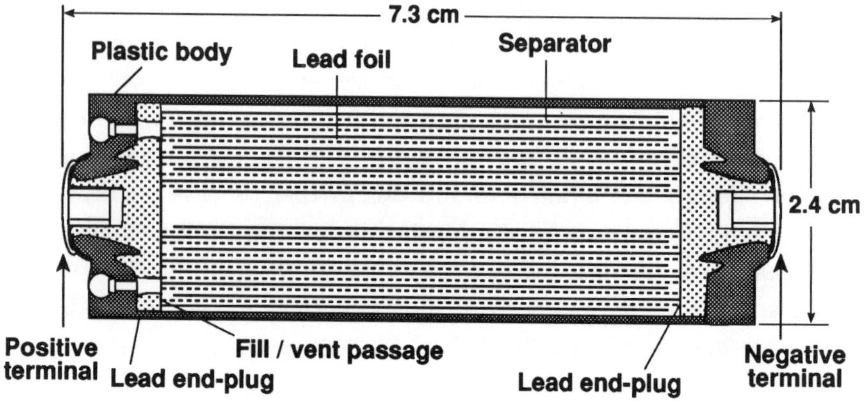

Fig. 5.31. Cross-sectional view of Bolder TMF® cell. (Taken from [123] with permission of Elsevier Science Limited, UK.)

glass-microfibre separators [123]. A key feature is the continuous bonding of the positive plate to a lead plug at one end of the cell, and similar bonding of the negative plate at the other end. The use of opposite-end current collection with continuous contact between the plugs and the electrode foils results in uniform current distribution over the electrode surfaces. Combined with the high plate surface area, this provides a low-resistance path for current flow across the full electrode surfaces. The extremely low internal resistance and good form factor for heat dissipation enables the Bolder technology to accept–deliver current at very high rates. Power levels of 800 and 400 W kg^{-1} can be delivered over 70 and 90% of the discharge lifetime, respectively. Thus, while the specific energy is no better than for contemporary lead/acid designs (typically, 30–40 Wh kg^{-1}), the specific power is outstanding. Indeed, the Bolder cell promises to provide a better power-to-energy ratio at a lower cost and with greater safety than alternatives such as flywheels, ultracapacitors, and other types of battery system (Fig. 5.32). By virtue of these characteristics, it appears to be a promising high-power-density device for both acceleration and uptake of regenerative braking in hybrid electric vehicles.

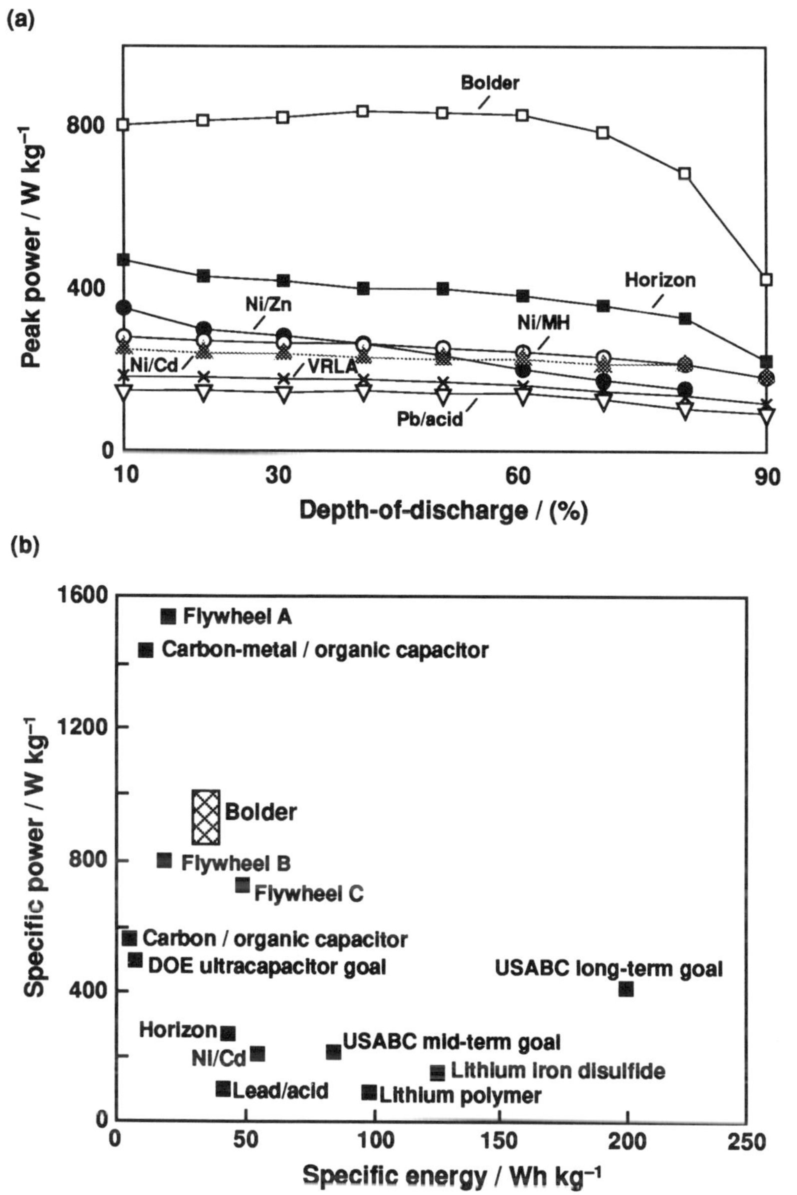

Fig. 5.32. Comparison of the performance of the Bolder TMF® cell with some other technologies: (a) peak power as a function of depth-of-discharge; (b) power–energy relationship. (Taken from [123] with permission of Elsevier Science Limited, UK.)

5.5.3. Longer deep-discharge service life

In most cases, the cycle life of lead/acid batteries is mainly limited by the performance of the positive plate, the capacity of which declines on cycling, especially under deep discharge. The above discussion of failure modes of positive plates (Section 5.4.1) noted that common degradation mechanisms include: (i) loss of inter-particle contact due to morphological changes; (ii) shedding of material due to morphological changes and grid corrosion; (iii) grid deterioration and growth; (iv) irreversible plate sulfation due to acid-stratification effects. It was also noted that batteries with lead–calcium or low-antimony (<1 wt.% Sb) positive grids suffer very poor cycle life under deep-discharge conditions. Early research resulted in a general consensus of opinion that the inferior capacity performance is due to the absence of antimony, rather than the presence of calcium, hence, the term 'antimony-free effect' came into vogue. Eventually, there emerged three main explanations of this phenomenon [124].

First, early studies [125–127] attributed the antimony-free effect to the irreversible formation of an insulating $PbSO_4$ corrosion layer between the grid and the active material. In later work [82,128–130], performance losses were associated with the development of a highly resistive layer of α-PbO on the grid. To explain the presence of this compound, which is stable only in an alkaline environment, it was suggested [38–40] that the $PbSO_4$ discharge product is impermeable to sulfate species. During battery charging, therefore, deposition of SO_4^{2-} ions in the interior layer as lead sulfate or mixed basic sulfates, together with the net flow of H^+ ions to the negative electrode, will result in a rise in the local pH beneath the $PbSO_4$ layer and, hence, will permit the formation of α-PbO. The mechanism is shown schematically in Fig. 5.33 along with the analogous process of α-PbO formation during battery discharging. At high positive-plate potentials, these conditions will also promote the formation of α-PbO_2 [131]. Thus, it is concluded that the passivation, or 'barrier', layer has a mixed $PbSO_4/PbO_x$ structure ($1 \le x < 2$) where PbO_x is probably α-PbO and/or α-PbO_2 [132]. The presence of α-PbO has been detected by different workers using a wide

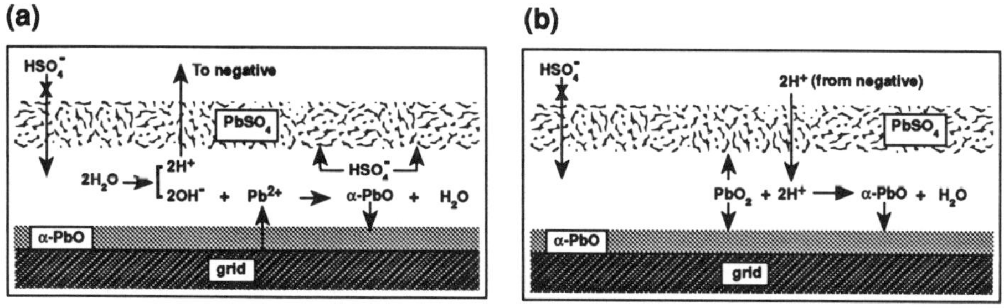

Fig. 5.33. Mechanism whereby α-PbO 'barrier' layers can form on positive grids during (a) battery charging and (b) battery discharging.

range of techniques, e.g., photoacoustic spectroscopy [133], photocurrent spectroscopy [134–136], and scanning laser imaging microscopy [137].

The characteristics and influence of the barrier layer are thought to be moderated by the incorporation of tin — either alloyed into the positive grid, or as a laminated [138] or electrodeposited [139] layer. When present in the alloy, tin has been shown to be effective in providing recovery from deep discharge [82,140,141], particularly during the early stages of battery life. Thus, the development of a passivating layer on non-antimonial grids has also been attributed to a 'tin-free effect' [82]. There still remains considerable debate over the mechanism of the proposed tin effect [142]; suggestions range from removal of the α-PbO layer (via a simple redox process, i.e., $SnO + PbO \rightleftarrows Pb + SnO_2$ [143]) to improved conductivity via complex semi-conductor processes [144,145]. By contrast, recent detailed studies have shown that, under certain operational conditions, tin exerts either only a transitory effect [132] or no clear benefit at all [146].

A second explanation for the antimony-free effect has been based on perceived differences in the nature and degree of cracking in the corrosion layers formed on antimonial and calcium alloys. The cracks develop under the action of mechanical stresses which are introduced because the oxide of the grid corrosion product has a greater volume than the lead from which it has been formed. The presence of cracks, together with poor adherence of the corrosion layer to the grid, will introduce electrical resistance losses. It appears that there is a reduced tendency for stress-cracking in the presence of antimony [147] and this has been related both to the additional space resulting from the preferential leaching of antimony-rich regions from the grid [148], and to a decreased hardness of the corrosion layer [149]. It has also been suggested [150] that the compound $PbSb_2O_6$ is formed as a grid corrosion product and, by virtue of its layered-type structure, helps to relieve the mechanical stresses which accompany transformations of the corrosion product during battery charge and discharge. Cracking of the corrosion layer occurs with prolonged cycling of batteries, irrespective of the choice of grid alloy. Lead–antimony alloys tend to exhibit radial fracture, whereas lead–calcium alloys tend to suffer concentric fracture [151]. Thus, under such conditions, there will be a better conduction path for electrons in the antimonial corrosion layer and, hence, less electrical resistance at the grid/active-material interface.

As opposed to $PbSO_4/\alpha$-PbO resistive-layer formation or corrosion-layer cracking, a third school of thought advocated that the antimony-free effect is due to structural changes which take place in the porous active-material during battery charge–discharge cycling. This behaviour degrades the contact between the individual PbO_2 particles. In an extreme situation, the particles become separated from each other and from the grid (so-called 'material softening') and, consequently, are easily shed from the plate. Prior to this catastrophic event, it is argued that more subtle changes take place within the PbO_2 matrix. It is assumed [152] that the positive active-material consists of a three-dimensional network of spherical PbO_2 particles (the 'Kugelhaufen' electrode) and that the properties of the narrow contact zones (the 'necks') determine the electronic conductivity of the total mass. In particular, it is claimed that, during cycling, the formation of smaller neck zones causes an irregular increase in the electrical resistance of the active material. The corresponding changes in the apparent conductivity lead to an increasingly inhomogeneous distribution of the current density within the plate. As a result of this modification, large zones of the material become excluded from the electron-conducting network, i.e., there is an 'insufficient mass utilization'. Such behaviour occurs under certain unfavourable battery-cycling conditions and is aggravated by the absence of

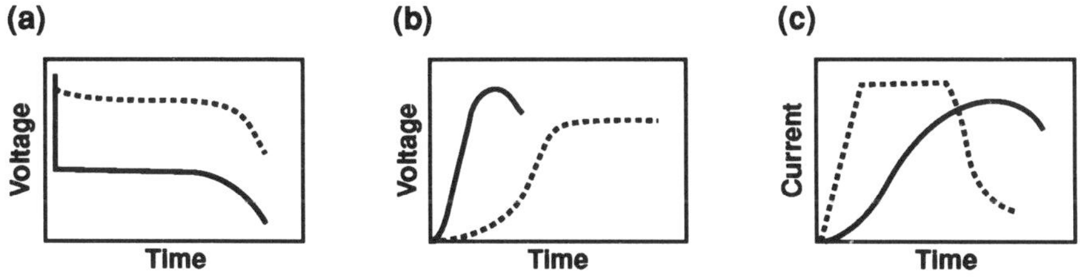

Fig. 5.34. Manifestation of positive-plate passivation during: (a) discharge; (b) constant-current charging; (c) constant-voltage charging. Broken lines represent ideal behaviour.

antimony in the grid alloy. On the other hand, various types of treatment (e.g., washing, drying, open-circuit stand) have been shown [153,154] to broaden the neck zones and, hence, to 'relax' the debilitating effect. Thus, according to the Kugelhaufen Model, the antimony-free effect is described as 'relaxable insufficient mass utilization', or RIMU.

In summary, the above three phenomenological explanations of the antimony-free effect, as well as the more obvious positive-plate degradation processes of shedding, irreversible sulfation and grid corrosion/growth, all serve to decrease the discharge capability by reducing active-material availability and by introducing large resistance losses. When the collective effects cause a battery to fall significantly short of its design life, the behaviour has now come to be described as 'premature capacity loss', PCL. This is a more appropriate term as batteries with high antimony grids (~6 wt.%) also suffer PCL at high levels of active-material utilization. Moreover, the occurrence of PCL appears to be independent of plate design — pasted, tubular and Planté plates can all succumb to the effect.

Unravelling the mysteries of PCL has been hampered by the fact that this phenomenon has rarely been recognized as being quite distinct from that of positive-plate passivation. The latter refers to situations where there is some demonstration of an additional energetic barrier at the grid/active-material interface. This barrier creates an initial sharp fall in voltage during battery discharge, a rapid rise in voltage during constant-current charging, or a low current during constant-voltage charging (Fig. 5.34).

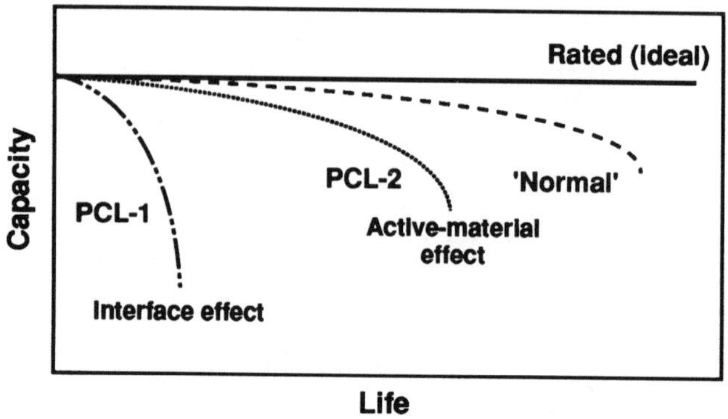

Fig. 5.35. Typical capacity–cycle plots for PCL-1, PCL-2, and ideal cycling service.

Positive-plate passivation can occur when:

- formed plates are dried for extended periods at high temperatures ('thermopassivation')
- batteries are stored for extended periods ('storage passivation')
- non-antimonial batteries are floated without any discharge ('floating passivation'), or are deeply discharged with a very small current ('squeeze passivation').

In summary, positive-plate passivation is related more to difficulties which are experienced during the initial charging or discharging of batteries, while PCL is associated with repetitive deep discharge at appreciable rates.

Two distinct forms of premature capacity loss are now recognized, namely, PCL-1 and PCL-2 (Fig. 5.35). Simply stated, the more severe form, PCL-1, is thought to arise from the development of 'barrier layers' close to the grid (e.g., $PbSO_4$ and/or α-PbO, as discussed earlier in this Section), particularly if the positive-plate potential is driven too low by deep discharge. Batteries that are constructed with lead–calcium grids (i.e., VRLA batteries) are particularly susceptible to this problem. As mentioned above, passivation can also cause sudden loss of capacity, but can be distinguished from PCL-1 through the application of simple electrical tests. Following research and technological advances, the problem of very early positive-plate failure, caused either by passivation or by PCL-1, has been virtually

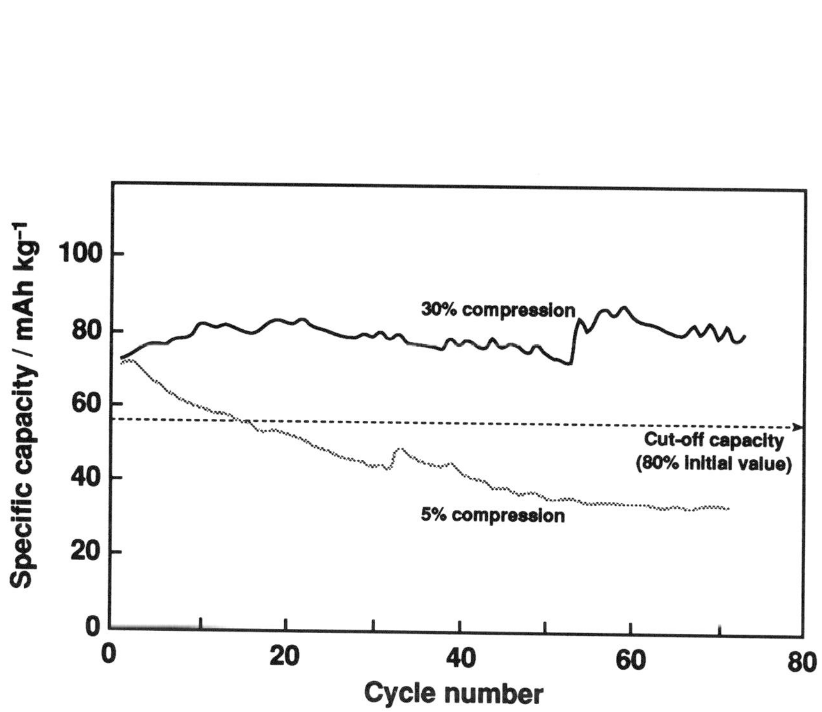

Fig. 5.36. Influence of compression on the performance of positive plates. (By courtesy of CSIRO, Australia.)

solved. Remedial strategies include: addition of tin to grid alloys; improved manufacturing techniques; use of phosphoric acid (with gel batteries).

The longer-term manifestation of premature capacity loss, PCL-2, applies when positive plates lose capacity at lower rates, but the common and readily identifiable causes of plate failure (e.g., active-material shedding, grid corrosion) are largely absent. PCL-2 is far more subtle in nature and, therefore, considerably more difficult to rectify. It is thought to be active-material related, and studies have shown that the life of a VRLA cell can be enhanced greatly by maintaining a certain degree of compression on the plate/separator assembly [113]. The data shown in Fig. 5.36, for example, provide clear evidence of the benefit to be gained. This important benefit originates from the effect of the compressive load on the 'apparent density' of the porous positive material. Apparent density is defined as the mass of PbO_2 divided by the total volume occupied by the porous structure (i.e., pore volume is included); it is the complement of porosity. Hollenkamp [110] has proposed that a decrease in apparent density is, in fact, the principal

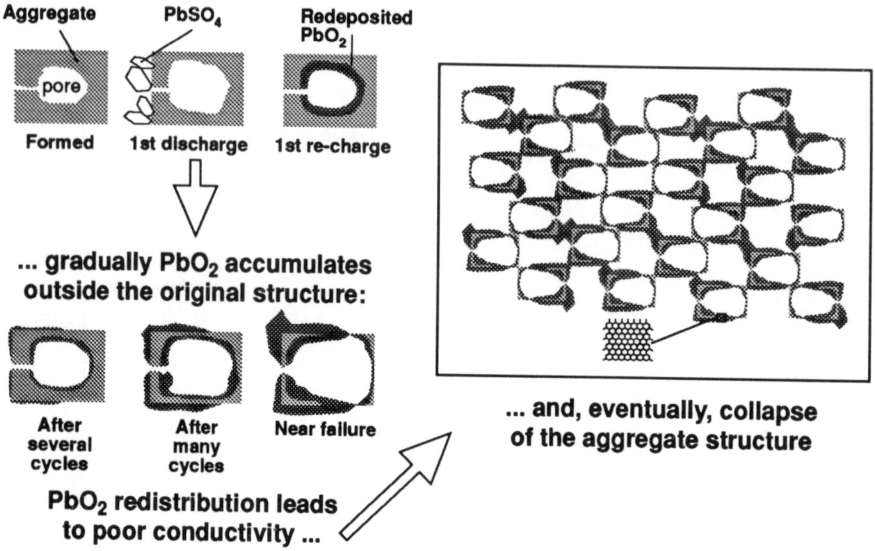

Fig. 5.37. Two-dimensional representation of changes in positive electrode structure during charge–discharge cycling. Redistribution of lead dioxide during cycling leads ultimately to collapse of the aggregate structure. (By courtesy of ALABC and CSIRO Australia.)

determinant of active-material degradation and, hence, capacity loss during battery service.

The pivotal role of apparent density arises from its direct relationship to active-material conductivity. Two conductivity-related factors conspire to limit the utilization of positive-plate material. First, the transformation of PbO_2 into insulating $PbSO_4$ reduces electrode conductivity during each discharge. Second, positive active-material expands with cycling service; this emanates from the volume changes which accompany PbO_2–$PbSO_4$ inter-conversion. Accordingly, the apparent density decreases and the concomitant increasing disruption to the PbO_2 network (Fig. 5.37) continuously lowers the amount of active material that can be utilized before plate conductivity limits the discharge. The region close to the current-collector is particularly important. Here, an increase in resistance can effectively prevent discharge of the bulk material. Typically, such a situation is found when severe capacity loss develops within a few cycles (i.e., PCL-1). Virtually all of the $PbSO_4$ is produced near to the corrosion layer, while the bulk positive material exhibits no signs of discharge.

Rapid loss of capacity is promoted by low-density pastes, the employment of relatively weak grids, and an ineffective level of compression. All of these features contribute to the expansion of the porous material, but if they are attenuated to maintain the apparent density above the minimum acceptable level, the capacity will be preserved for longer periods of service. A progressive decrease in density will still occur, however, but at a lower rate. The eventual failure is described as PCL-2.

From these observations, Hollenkamp [110] has developed a 'unified mechanism' of capacity loss by which a continuous scale of positive-plate degradation is established, rather than two distinct groups of failure modes (i.e., PCL-1 and PCL-2). Previous interpretations of PCL-2 are in line with the unified mechanism because they focus on changes in the structure and/or function of the positive material. By comparison, explanations advanced for rapid capacity loss (PCL-1) have created confusion since they invoke the concept of some form of permanent barrier which electrically isolates the current-collector from the porous material. As intimated above, however, the development of a low apparent density can localize discharge and, hence, can cause the formation of high concentrations of $PbSO_4$ close to the grid. The result of this preferential discharge is indistinguishable from the resistive effects of a 'barrier layer' and, importantly, no separate explanation is required for cases of severe capacity loss.

Basic design strategies for maintaining the apparent density of the positive active-material above the critical value are widely known, but only at a rudimentary level. New methods and materials are required to guarantee the three-dimensional stability of positive plates throughout service. For VRLA batteries with pasted plates, this demands the development of: (i) new alloys and designs to increase the toughness of grid frames, i.e., to restrict expansion in the plate plane; (ii) improved separators with greater strength and elasticity, i.e., to provide sufficient compression at right angles to the plate.

5.5.4. Faster recharge time

Methods for very fast charging of lead/acid batteries have long been sought. With the growing demand for EVs, this need has intensified. Clearly, fast charging offers a promising approach towards overcoming the somewhat limited driving range of vehicles powered by lead/acid batteries. For practical purposes, this involves shortening charging times from the hitherto usual period of hours to a few minutes; the ALABC, for example, has set a target of 80% capacity return within 15 min (see Section 3.4, Chapter 3). Accordingly, considerable research is being directed towards the development of procedures and equipment that will achieve fast-charging goals without degrading battery performance. An example of a modern fast-charging procedure (or 'algorithm') is given in Fig. 5.38. With this particular procedure, the battery is charged via a series of current pulses, and the average current for consecutive sets of pulses is reduced by decreasing simultaneously both the current amplitude and the duty cycle. The change in these two parameters is triggered either by the average, resistance-free voltage or by the amount of charge return. The $I\mathcal{R}$ component is subtracted from the pulsed voltage during the on-time to give the resistance-free, pulsed voltage. The average value of this voltage for each pulse is then calculated and compared with the recommended voltage limit. The final, constant-voltage charging stage is initiated when the average pulsed current reaches a low value, typically ~0.5 A. Such a charging strategy keeps overcharge (and, hence, water loss) to a minimum. Concurrent with algorithm development, cell designs (and, most probably, the cell materials themselves) will have to be modified for minimal resistive/chemical heat generation, as well as effective heat dissipation.

There have been surprisingly few investigations of the electrochemical, chemical and physical effects that fast-charging might exert on battery plates. This fact, together with the recent demonstrations that very fast and efficient recharging of lead/acid batteries is possible [155], and that

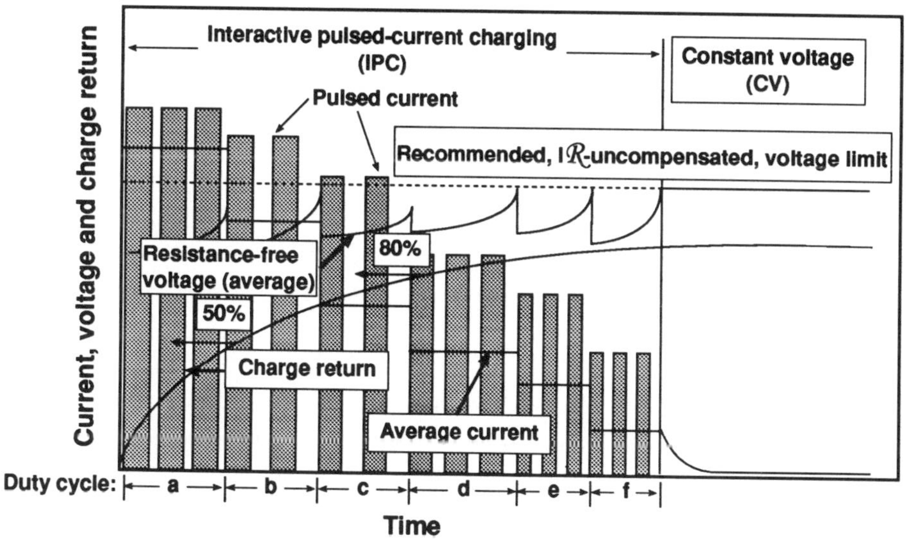

Fig. 5.38. Algorithm for fast-charging VRLA batteries in EVs. (By courtesy of ALABC and CSIRO, Australia.)

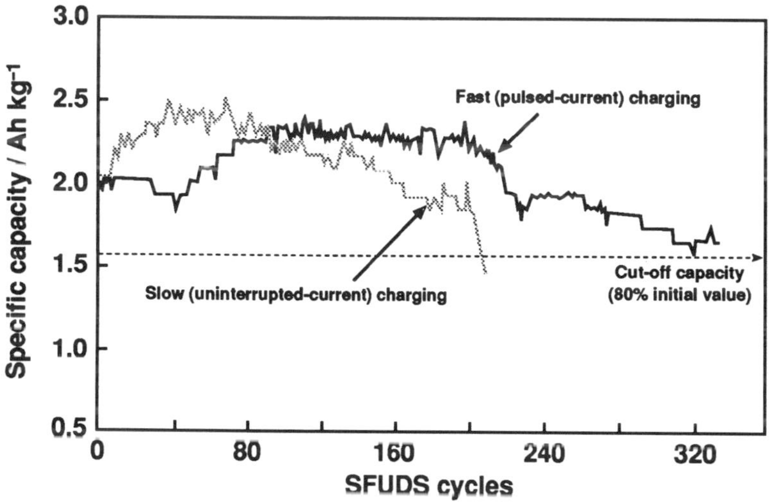

Fig. 5.39. Demonstration of enhanced cycle life obtained under fast-charging conditions. (By courtesy of ALABC and CSIRO, Australia.)

pulsed-current techniques can also extend battery life (Fig. 5.39) — possibly through depressing the crystal growth of the particles of the positive active-material and, hence, restricting plate expansion [156] — opens up rapid charging as an exciting new area of research. Apart from optimization of charging algorithms and battery design, there are challenges to find new materials for plates and grids that have good strength, good heat-transfer properties, and manageable gassing characteristics.

5.6. Summing Up

Although the lead/acid battery is a mature product, with a long and well established record of commercial performance, the search continues for the most appropriate design, the best raw materials of construction, and the optimum charge–discharge management system for EV applications. The incentive is that present designs offer considerable scope for improvement in terms of greater specific energy and faster recharging, as well as the obvious operational advantages of the neoteric valve-regulated technology.

Ongoing materials research is critical to the advancement of any battery system. For lead/acid, with weight a central issue, the development of lighter, better-conducting designs/materials for grids, straps and terminals is a priority. Weight reductions, together with active-material utilization enhancement, translate immediately into improvements in specific energy. Conducting plastics, metal composites and novel processes for creating thinner grids with greater strength and lower corrosion rates will also assist efforts to maximize specific power and recharging capabilities. Finally, the incorporation of radically new materials and/or the adoption of new cell-construction techniques and geometries will enhance the structure and containment of the positive active-material and, thereby, will improve cycle life under the most demanding of EV duties.

References

1. BULLOCK, K.R. *J. Power Sources,* 1994, **51,** pp. 1–18.

2. RAND, D.A.J. *Phil. Trans. R. Soc. Lond. A*, 1996, **354,** pp. 1529–1544.

3. GLADSTONE, J.H., and TRIBE, A. *Nature*, 1882, **25**, p. 221 and p. 461.

4. RÜETSCHI, P., SKLARCHUK, J., and ANGSTADT, R.T. *Electrochim. Acta*, 1963, **8**, pp. 333–342.

5. BONE, S.J., SINGH, K.P., and WYNNE-JONES, W.F.K. *Electrochim. Acta*, 1961, **4**, pp. 288–293.

6. COVINGTON, A.K., DOBSON, J.V., and WYNNE-JONES, W.F.K. *Trans. Faraday Soc.*, 1965, **61**, pp. 2050–2056; 2057–2062.

7. WAGMAN, D.D., EVANS, W.H., PARKER, V.B., SCHUMM, R.H., HALOW, I., BAILEY, S.M., CHURNEY, K.L., and NUTTALL, R.L. *The NBS Tables of Chemical Thermodynamic Properties, J. Physical and Chemical Reference Data*, 1982, **11**, Supplement 2.

8. VINAL, G.W. *Storage Batteries* (John Wiley & Sons, Inc., New York, USA, 1940).

9. BODE H. *Lead–Acid Batteries,* translated by BRODD, R.J. and KORDESCH, K. (John Wiley & Sons, Inc., New York, USA, 1977).

10. VINAL, G.W., and CRAIG, D.N. *J. Research*, 1940, **24**, p. 475.

11. HARNED, H.S., and HAMER, W.J. *J. Am. Chem. Soc.*, 1935, **57**, p. 33.

12. SALKIND, A.J., KELLY, J.J., and CANNONE, A.G. *In* LINDEN, D. (Ed.) *Handbook of Batteries, Second Edition* (McGraw–Hill, New York, USA, 1995), pp. 24.1–24.89.

13. WILLIHNGANZ, E. *J. Electrochem. Soc.*, 1955, **102**, p. 99.

14. MARK, H.B. *J. Electrochem. Soc.*, 1962, **109**, pp. 634–638.

15. BERNDT, D., and VOSS, E. *In* COLLINS, D.H. (Ed.) *Batteries 2, Research and Development in Non-Mechanical Electrical Power Sources* (Pergamon Press, Oxford, UK, 1965), pp. 17–26.

16. HILL, R.J., RAND, D.A.J., and WOODS, R. *In* PEARCE, L.J. (Ed.) *Power Sources 11, Research and Development in Non-Mechanical Electrical Power Sources* (International Power Sources Symposium Committee, Leatherhead, UK, 1987), pp. 103–124.

17. BAIN, C.J. *The Electric Industrial Truck* (Australian Lead Development Association, Melbourne, Australia, 1972), p. 39.

18. PEUKERT, W. *Electrotech. Z.*, 1897, **18**, p. 287.

19. GILLIBRAND, M.I., and LOMAX, G.R. *Electrochim. Acta*, 1963, **8**, pp. 693–702.

20. BALDSING, W.G.A., GARDNER, J.R., HILL, R.J., RAND, D.A.J., and WOODS, R. *Optimization of an Electric-Vehicle Lead Battery System (NERDDC Project No. 79/9317. Final Report)*, CSIRO Div. Min. Chemistry, Melbourne, Commun. MCC 631, April 1985, 62 pp.

21. SMITH, G. *Storage Batteries* (Pitman Publishing Ltd., London, UK, 1980), p. 77.

22. KAUSHIK, R., and MAWSTON, I.G. *J. Power Sources*, 1989, **28**, pp. 161–169.

23. JAYNE, M.G. *In* COLLINS, D.H. (Ed.) *Power Sources 6, Research and Development in Non-Mechanical Electrical Power Sources* (Academic Press, London, UK, 1977), pp. 35–44.

24. CATALDO, R.L. *Proc. 13 Intersoc. Energy Conv. Eng. Conf.*, 20–25 August 1978, San Diego, USA (Society of Automotive Engineers, Inc., Warrendale, PA, USA, 1978), pp. 764–768.

25. AHAMED, T.U., and SHEPHERD, W. *IEE Proceedings*, 1980, **127 B**, pp. 120–123.

26. HILL, R.J., RAND, D.A.J., and WOODS, R. *In* PEARCE, L.J. (Ed.) *Power Sources 10, Research and Development in Non-Mechanical Electrical Power Sources* (The Paul Press Ltd., London, UK, 1985), pp. 459–480.

27. DeLUCA, W.H. *SAE Paper 820179*, Detroit, USA, 1982.

28. BURKE, A.F. *In* McNICOL, B.D., and RAND, D.A.J. (Eds.) *Power Sources for Electric Vehicles* (Elsevier, Amsterdam, The Netherlands, 1984), pp. 931–974.

29. KURTZ, D.W., PRICE, T.W., and BRYANT, J.A. *SAE Paper 810418*, Detroit, USA, 1981.

30. VACCARO, F.J., RHOADES, J., and LE, B. *Eighteenth International Telecommunications Energy Conference, INTELEC '96*, 6–10 October 1996, Boston, Massachusetts, USA (The Institute of Electrical and Electronics Engineers, Piscataway, New Jersey, USA), pp. 39–44.

31. CONSTABLE, D.C., GARDNER, J.R., HARRIS, K., HILL, R.J., RAND, D.A.J., and ZALCMAN, L.B. *J. Electroanal. Chem.*, 1984, **168**, pp. 395–414.

32. International Electrochemical Commission, *Publication 254*, 1967.

33. MICKA, K., and KOUDELKA, V. *J. Power Sources*, 1981, **6**, pp. 319–325.

34. BAIKIE, P.E., GILLIBRAND, M.I., and PETERS, K. *Electrochim. Acta*, 1972, **17**, pp. 839–844.

35. CHAPMAN, P. *Performance Characteristics of an Electric Vehicle Lead–Acid Battery Pack at Elevated Temperatures* (Jet Propulsion Laboratory, Pasadena, California, USA, April 1982), JPL Publication 82-85, DOE Contract CS 54209-8.

36. NOWAK, D. *Extended Abstracts, Fall Meeting of the Electrochemical Society,* 9 October 1983, Washington, DC, USA (The Electrochemical Society, Pennington, USA, 1983), Proc. Vol. 83-2, pp. 594–595.

37. RÜETSCHI, P., and ANGSTADT, R.T. *J. Electrochem. Soc.,* 1958, **105**, pp. 555–563.

38. PAVLOV, D., and POPOVA, R. *Electrochim. Acta,* 1970, **15,** pp. 1483–1491.

39. PAVLOV, D., and IORDANOV, N. *J. Electrochem. Soc.,* 1970, **117**, pp. 1103–1109.

40. RÜETSCHI, P. *J. Electrochem. Soc.,* 1973, **120,** pp. 331–336.

41. PETERS, K., HARRISON, A.I., and DURANT, W.H. *In* COLLINS, D.H. (Ed.) *Power Sources 2, Research and Development in Non-Mechanical Electrical Power Sources* (Pergamon Press, Oxford, UK, 1968), pp. 1–14.

42. DYSON, J.I., and SUNDBERG, E. *In* COLLINS, D.H. (Ed.) *Power Sources 4, Research and Development in Non-Mechanical Electrical Power Sources* (Oriel Press, Newcastle upon Tyne, UK, 1973), pp. 505–523.

43. Accumulatorenwerke Hoppecke Carl Zoellner & Sohn, *German Patent 20 08 218* (issued 30 September 1976).

44. SHIMIZU, K., and TSUBUKI, A. *In* MIYAKE, Y., and KOZAWA, A. (Eds.) *Rechargeable Batteries in Japan* (JEC Press, Cleveland, Ohio, USA, 1977), Ch. 7.

45. WATANABE, M., and YONEZU, K. *Prog. Batteries Sol. Cells,* 1979, **2**, pp. 153–160.

46. LEDJEFF, K., and WINSEL, A. *J. Power Sources,* 1982, **8**, pp. 211–227.

47. RÜETSCHI, P. *U.S. Patent 2,951,106* (issued 1960); *U.S. Patent 3,080,440* (issued 1963).

48. VOSS, E., and DEHMELT, K. *U.S. Patent 3,170,816* (issued 1965).

49. RÜETSCHI, P., and OCKERMAN, J.B. *Electrochem. Technol.,* 1966, **4**, pp. 383–393.

50. SHIMIZU, K., and OHKAWAY, Y. *In* MIYAKE, Y., and KOZAWA, A. *Rechargeable Batteries in Japan* (JEC Press, Cleveland, Ohio, USA, 1977), p. 257.

51. NIKOLOV, I., PAPAZOV, G., PAVLOV, D., VITANOV, T., and NAIDENOV, V. *J. Power Sources,* 1990, **31**, pp. 69–77.

52. PAPAZOV, G., NIKOLOV, I., PAVLOV, D., VITANOV, T., ANDREEV, P., and DOJINOV, M. *J. Power Sources,* 1990, **31**, pp. 79–88.

53. RICE, D.M., and MANDERS, J.E. *J. Power Sources,* 1997, **67**, pp. 251–255.

54. ZGURIS, G.C., KLAUBER, D.W., and LIFSHUTZ, N.L. *In* KEILY, T., and BAXTER, B.W. *Power Sources 13, Research and Development in Non-Mechanical Electrical Power Sources* (International Power Sources Symposium Committee, Leatherhead, England, UK, 1991), pp. 45–57.

55. NELSON, R.F. *Batteries International,* 1993, **17**, pp. 64–65.

56. MERZ, K.-D. *J. Power Sources,* 1997, **67**, pp. 121–123.

57. BRECHT, W.B., and O'LEARY, N.F. *Tenth International Telecommunications Energy Conference, INTELEC '88,* 30 October – 2 November 1988, San Diego, California, USA (The Institute of Electrical and Electronics Engineers, Piscataway, New Jersey, USA, 1988), pp. 35–42.

58. TUPHORN, H. *Rev. Gen. Electr.,* 1990, **3**, pp. 45–49.

59. JACHE, O. *U.S. Patent 3,257,237* (issued 1966).

60. EBERTS, K., and JACHE, O. *Lead 65, Proc. 2nd International Conference on Lead* (Pergamon Press, Oxford, UK, 1967), pp. 199–210.

61. EBERTS, K. *In* COLLINS, D.H. (Ed.) *Power Sources 2, Research and Development in Non-Mechanical Electrical Power Sources* (Pergamon Press, Oxford, UK, 1970), pp. 69–92.

62. TUPHORN, H. *Prog. Batteries Sol. Cells,* 1987, **6,** pp. 145–149.

63. MAHATO, B.K., and BULLOCK, K.R. *Prog. Batteries Sol. Cells,* 1987, **6,** pp. 136–140.

64. McKINNEY, B.L., MAHATO, B.K., and BULLOCK, K.R. *In* BULLOCK, K.R., and PAVLOV, D. (Eds.) *Proc. Symposium on Advances in Lead–Acid Batteries* (The Electrochemical Society, Pennington, USA, 1984), Proc. Vol. 84–14, pp. 426–440.

65. BULLOCK, K.R., MAHATO, B.K., BRILMYER, G.H., and WIERSCHEM, G.L. *In* BULLOCK, K.R., and PAVLOV, D. (Eds.) *Proc. Symposium on Advances in Lead–Acid Batteries* (The Electrochemical Society, Pennington, USA, 1984), Proc. Vol. 84–14, pp. 451–465.

66. DEVITT, J. *J. Power Sources,* 1997, **64**, pp. 153–156.

67. *CYCLON: Battery Application Manual* (Gates Energy Products, Inc., Denver, Colorado, USA, 1982), p. 7.

68. McCLELLAND, D.H. and DEVITT, J.L. *U.S. Patent 3,862,861* (issued 1975).

69. NEWNHAM, R.H. *J. Power Sources,* 1994, **52,** pp. 149–153.

70. RAND, D.A.J., HILL, R.J., and McDONAGH, M. *J. Power Sources,* 1990, **31,** pp. 203–215.

71. LAM, L.T., and RAND, D.A.J. *Batteries International,* 1992, **13,** pp. 52, 53, 56, 57.

72. RAND, D.A.J., and LAM, L.T. *The Battery Man,* 1992, **34,** pp. 18–22, 24, 27, 28.

73. LAMBERT, D.W.H. *Batteries International*, 1992, **13,** pp. 32, 33, 36, 37.

74. SKLARCHUK, J., DEWAR, M.J., VALERIOTE, E.M., and VINCZE, A.M. *J. Power Sources,* 1993, **42,** pp. 47–53.

75. WIRTZ, J.O. *Batteries International,* 1996, **26,** pp. 56–57.

76. CATTANEO, E., STUMPF, H., TILLMAN, H., and SASSMANNSHAUSEN, G. *J. Power Sources,* 1997, **67**, pp. 283–290.

77. BAGSHAW, N. *J. Power Sources,* 1995, **53**, pp. 25–30.

78. PRENGAMAN, R.D. *J. Power Sources,* 1991, **33,** pp. 13–20.

79. MELNIK, D.C. *The Battery Man,* 1980, **22,** pp. 2, 4–6.

80. CULPIN, B., PILLING, M.W., and FLEMING, F.A. *J. Power Sources,* 1988, **24,** pp. 127–136.

81. PRENGAMAN, R.D. *J. Power Sources,* 1997, **67**, pp. 267–277.

82. GEISS, H.K. *In* BULLOCK, K.R., and PAVLOV, D. (Eds.) *Proc. Symposium on Advances in Lead–Acid Batteries* (The Electrochemical Society, Pennington, USA, 1984), Proc. Vol. 84–14, pp. 241–251.

83. BAGSHAW, N.E. *In* KEILY, T., and BAXTER, B.W. (Eds.) *Power Sources 12, Research and Development in Non-Mechanical Electrical Power Sources* (International Power Sources Committee, Leatherhead, UK, 1989), pp. 113–129.

84. COOK, G.M., and SPINDLER, W.C. *J. Power Sources,* 1991, **33,** pp. 145–161.

85. CORINO, G.L., HILL, R.J., JESSEL, A.M., RAND, D.A.J., and WUNDERLICH, J.A. *J. Power Sources,* 1985, **16**, pp. 141–168.

86. RAND, D.A.J. *J. Power Sources,* 1989, **28**, pp. 107–111.

87. BUSDIEKER, E.E., and SCHECK, M.C. *Batteries International,* 1993, **16**, pp. 56–57.

88. MAYER, M.G., and RAND, D.A.J. *J. Power Sources,* 1996, **59**, pp. 17–24.

89. BLAIR, T. *Batteries International,* 1997, **30**, pp. 37, 38, 40, 42.

90. NELSON, R.F. *Proc. 7th Asian Battery Conference (7ABC),* 3–5 September 1997, Penang, Malaysia (Pasminco Metals, Melbourne, Australia, 1997); to be published in *J. Power Sources,* 1988.

91. LAM, L.T., VECCHIO-SADUS, A.M., OZGUN, H., and RAND, D.A.J. *J. Power Sources,* 1992, **38**, pp. 87–102.

92. PAVLOV, D., and KAPKOV, N. *J. Power Sources,* 1990, **31**, pp. 189–201.

93. LAM, L.T., OZGUN, H., CRANSWICK, L.M.D., and RAND, D.A.J. *J. Power Sources,* 1993, **42**, pp. 55–70.

94. *Lead Chemicals*, (International Lead Zinc Research Organization, Inc., New York, 1975), p. 64.

95. SEXTON, B.A., COTTERILL, G.F., FLETCHER S., and HORNE, M.D., *J. Vac. Sci. Technol. A,* 1990 **8,** pp. 544–548.

96. RAND, D.A.J., HARMER, P.B., HILL, R.J., and WUNDERLICH, J.A. *In* TRAN, T., and SKYLLAS-KAZACOS, M. (Eds.) *Electrochemistry: Current and Potential Applications, Proc. Seventh Australian Electrochemistry Conference (7AEC),* 15–19 February 1988, Sydney, Australia (Royal Australian Chemical Institute, Melbourne, Australia, 1988), pp. 20–23.

97. DODSON, V.H. *J. Electrochem. Soc.,* 1961, **108**, pp. 406–412.

98. VOSS, E., and FREUNDLICH, J. *In* COLLINS, D.H. (Ed.) *Batteries, Research and Development in Non-Mechanical Electrical Power Sources* (Pergamon, Oxford, UK, 1963), pp. 73–87.

99. BAGSHAW, N.E., and WILSON, K.P. *Electrochim. Acta*, 1965, **10**, pp. 867–873.

100. BUTLER, W.O., VENUTO, C.J., and WISLER, D.V. *J. Electrochem. Soc.,* 1970, **117**, pp. 1339–1342.

101. CULPIN, B. *J. Power Sources,* 1989, **25**, pp. 305–311.

102. SIMON, A.C., and JONES, E.J. *J. Electrochem. Soc.*, 1962, **109**, pp. 760–770.

103. PIERSON, J.R. *Electrochem. Technol.*, 1967, **5**, pp. 323–327.

104. CULPIN, B., and HAYMAN, J.A. *In* PEARCE, L.J. (Ed.) *Power Sources 11, Research and Development in Non-Mechanical Electrical Power Sources* (International Power Sources Symposium Committee, Leatherhead, UK, 1987), pp. 45–66.

105. COOPER, A.J. *J. Power Sources,* 1996, **59**, pp. 161–170.

106. ZGURIS, G. *The Battery Man,* 1997, **39**, pp. 14, 16, 18, 20, 21, 24.

107. HOLLENKAMP, A.F., and NEWNHAM, R.H. *J. Power Sources,* 1997, **67**, pp. 27–32.

108. FERREIRA, A.L., and LINGSCHEIDT, H.A. *J. Power Sources,* 1997, **67**, pp. 291–297.

109. ZGURIS, G. *Batteries International,* 1997, **31**, pp. 87–89.

110. HOLLENKAMP, A.F. *J. Power Sources,* 1996, **59**, pp. 87–98.

111. HOLLENKAMP, A.F., and NEWNHAM, R.H. *International Patent Application, PCT/AU97/00080.* (14 February 1997).

112. WOOLF, G. *Batteries International,* 1997, **30**, pp. 23, 24, 26–28.

113. McGREGOR, K., HOLLENKAMP, A.F., BARBER, M., HUYNH, T., OZGUN, H., PHYLAND, C.G., URBAN, A.J., VELLA, D.G., and VU, L.H. *Proc. 7th Asian Battery Conference (7ABC),* 3–5 September 1997, Penang, Malaysia (Pasminco Metals, Australia, 1997); to be published in *J. Power Sources,* 1988.

114. APÂTEANU, L., HOLLENKAMP, A.F., and KOOP, M.J. *J. Power Sources,* 1993, **46**, pp. 239–250.

115. SHINPO, M., NAKASHIMA, H., SASABE, S., and KASAI, Y. *Yuasa Jiho,* 1990, **68**, pp. 26–34.

116. SIMON, A.C., CAULDER, S.M., GURLUSKY, P.J., and PIERSON, J.R. *Electrochim. Acta,* 1974, **19**, pp. 739–743.

117. TUPHORN, J. *J. Power Sources,* 1990, **31**, pp. 57–67.

118. PRENGAMAN, R.D. *The Battery Man,* 1997, **39**, pp. 23,24, 26–31.

119. MOORE, T. *EPRI Journal,* 1994, April–May Volume, pp. 7–13.

120. COLE, J.F. *J. Power Sources,* 1992, **40**, pp. 1–15.

121. McGREGOR, K. *J. Power Sources,* 1996, **59**, pp. 31–43.

122. ARIAS, J.L. *Proc. 9th Annual Battery Conference on Applications and Advances,* 1994, Long Beach, Florida, USA, pp. 121–123.

123. JUERGENS, T., and NELSON, R.F. *J. Power Sources,* 1995, **53**, pp. 201–205.

124. RAND, D.A.J., HOLLENKAMP, A.F., and NELSON, R.F. *The Battery Man,* 1993, **9**, pp. 16–18, 20–23.

125. TUDOR, S., WEISSTUCH, A., and DAVANG, S.H. *Electrochem. Technol.,* 1965, **3**, pp. 90–94; 1966, **4**, pp. 406–411; 1967, **5**, pp. 21–25.

126. HATTORI, S., YAMAURA, M., KONO, M., YAMANE, M., NAKASHIMA, H., and YAMASHITA, J. *ILZRO Project LE-253, Report for Period: March–November, 1977* (International Lead Zinc Research Organization, Inc., Research Triangle Park, North Carolina, USA, 15 December 1977).

127. KOMAKI, A., KAWAMURA, G., and MOCHIZUKI, S. *Prog. Batteries Sol. Cells,* 1982, **4**, pp. 167–171.

128. TAMURA, H., YONEYAMA, H., IWAKURA, C., and IKEDA, O. *ILZRO Project LE-254, Final Report* (International Lead Zinc Research Organization, Inc., Research Triangle Park, North Carolina, USA, 31 December, 1977).

129. FUCHIDA, K., OKADA, K., HATTORI, S., KONO, M., YAMANE, M., TAKAYAMA, T., YAMASHITA, J., and NAKAYAMA, Y. *ILZRO Project LE-276, Final Report (Progress Report 8)* (International Lead Zinc Research Organization, Inc., Research Triangle Park, North Carolina, USA, 31 March 1982).

130. KITA, A., MATSUMARU, Y., SHINPO, M., and NAKASHIMA, H. *In* PEARCE, L.J. (Ed.) *Power Sources 11, Research and Development in Non-Mechanical Electrical Power Sources* (International Power Sources Symposium Committee, Leatherhead, UK, 1987), pp. 31–44.

131. BULLOCK, K.R., and BUTLER, M.A. *J. Electrochem. Soc.,* 1986, **133**, pp. 1085–90.

132. NELSON, R.F., and WISDOM, D.M. *J. Power Sources,* 1991, **33**, pp. 165–185.

133. BRILMYER, G.H. *In* BULLOCK, K.R., and PAVLOV, D. (Eds.) *Proc. Symposium on Advances in Lead–Acid Batteries* (The Electrochemical Society, Pennington, USA, 1984), Proc. Vol. 84–14, pp. 142–153.

134. PAVLOV, D., ZANOVA, S., and PAPAZOV, G. *J. Electrochem. Soc.,* 1977, **124**, pp. 1522–1528.

135. FLETCHER, S., and MATTHEWS, D.B. *J. Electroanal. Chem.,* 1981, **126**, pp. 131–144.

136. BUCHANAN, J.S., and PETER, L.M. *Electrochim. Acta,* 1988, **33**, pp. 127–136.

137. PEAT, R., MOSELEY, P.T., HOLLENKAMP, A.F. and RAND, D.A.J. *J. Power Sources,* 1993, **42**, pp. 119–135.

138. TAKAHASHI, K., HOSHIHARA, N., YASUDA, H., ISHII, T., and JIMBO, H. *J. Power Sources,* 1990, **30**, pp. 23–31.

139. DÖRING, H., GARCHE, J., FISCHER, W., and WIESENER, K. *J. Power Sources,* 1989, **28**, pp. 367–380.

140. WEISENER, K., GARCHE, J., and ANASTASIJEVIC, N. *In* THOMPSON, J. (Ed.) *Power Sources 9, Research and Development in Non-Mechanical Electrical Power Sources.* (Academic Press, London, UK, 1983), pp. 17–37.

141. BARTON, R.T., MITCHELL, P.J., and FLEMING, F.A. *In* KEILY, T., and BAXTER, B.W. (Eds.) *Power Sources 13, Research and Development in Non-Mechanical Electrical Power Sources* (International Power Sources Symposium Committee, Leatherhead, UK, 1991), pp. 25–33.

142. CULPIN, B., HOLLENKAMP, A.F., and RAND, D.A.J. *J. Power Sources,* 1992, **38**, pp. 63–74.

143. TERADA, M., SAITO, S., HAYAKAWA, T., and KOMAKI, A. *Prog. Batteries Sol. Cells,* 1989, **8**, pp. 214–216.

144. PAVLOV, D., MONAKHOV, B., MAJA, M., and PENAZZI, N. *J. Electrochem. Soc.,* 1989, **136**, pp. 27–33.

145. DÖRING, H., GARCHE, J., DIETZ, H., and WIESENER, K. *J. Power Sources,* 1990, **30**, pp. 41–45.

146. VALERIOTE, E.M.L., HEIM, A., and HO, M.S. *J. Power Sources,* 1991, **33**, pp. 187–212.

147. MAHATO, B.K. *J. Electrochem. Soc.,* 1979, **126**, pp. 365–374.

148. SIMON, A.C., *J. Electrochem. Soc.,* 1967, **114**, pp. 1–8.

149. HATTORI, S., YAMAURA, M., KONO, M., YAMANE, M., NAKASHIMA, N., YAMASHITA, J., and NAKAYAMA, Y. *ILZRO Project No. LE-276, Progress Report 5, 1 December – 30 May 1980* (International Lead Zinc Research Organization, Inc., Research Triangle Park, North Carolina, USA, 1980).

150. HILL, R.J. *J. Solid State Chem.,* 1987, **71**, pp. 12–18.

151. HARDMAN, A.M., HOPWOOD, R.T., MANDERS, J.E., RAND, D.A.J., REITZ, J.W., and TUPHORN, H. *J. Power Sources,* 1988, **23**, pp. 257–277.

152. HULLMEINE, U., WINSEL, A., and VOSS, E. *J. Power Sources,* 1989, **25**, pp. 27–47.

153. WINSEL, A., VOSS, E., and HULLMEINE, U. *J. Power Sources,* 1990, **30**, pp. 209–226.

154. MEISSNER, E., and VOSS, E. *J. Power Sources,* 1991, **33**, pp. 231–244.

155. GRIBBLE, D. *Batteries International,* 1994, **19**, pp. 60–61.

156. LAM, L.T., OZGUN, H., LIM, O.V., HAMILTON, J.A., VU, L.H., VELLA, D.G., and RAND, D.A.J. *J. Power Sources,* 1995, **53**, pp. 215–228.

CHAPTER 6

Alkaline Batteries

6.1. Electrodes Operating in Alkaline Media

The development of storage batteries based on alkaline electrolytes was initiated in the late 1890s. Since then, these systems have generated significant markets in a variety of applications. Electrodes which are used in alkaline media are listed in Table 6.1, together with their respective standard potentials. Several combinations of these positives and negatives have been applied in practical cells (see, Table 3.3, Chapter 3); the resultant cell voltage is the difference between those of the two component electrodes (see Section 2.3.1, Chapter 2).

In alkaline batteries, the electrical energy is essentially derived from the reaction of a metal with oxygen. The latter reactant can either be used on its own, as in metal/air systems, or be held in the form of a metal oxide. The efficacy of supplying oxygen from an oxide is clearly demonstrated by the fact that the potentials of the silver oxide and nickel oxide positives are both greater than that of the oxygen electrode itself (Table 6.1). It must be remembered, however, that since a carrier metal is involved, weight is added to the battery and hence the specific energy is lowered.

The usual convention when designating a battery is to cite the negative electrode first (e.g., zinc/air, sodium/sulfur, lithium/manganese dioxide), but with alkaline-electrolyte batteries that employ metal oxide positives, it has been traditional to reverse the order and omit the world 'oxide'. Thus, for example, the cadmium/nickel oxide battery is usually referred to as 'nickel/cadmium'. This practice is so widely used that it is adopted in the subsequent discussion of such batteries.

Table 6.1. Electrodes in alkaline batteries.

Electrode	Half-cell discharge reaction			E^o (V)*
Positive				
Ag	$2AgO + H_2O + 2e^-$	$\rightarrow$	$Ag_2O + 2OH^-$	0.57
	$Ag_2O + H_2O + 2e^-$	$\rightarrow$	$2Ag + 2OH^-$	0.34
Ni	$NiOOH + H_2O + e^-$	$\rightarrow$	$Ni(OH)_2 + OH^-$	0.49
O_2	$O_2 + 2H_2O + 4e^-$	$\rightarrow$	$4(OH)^-$	0.40
MnO_2	$MnO_2 + H_2O + e^-$	$\rightarrow$	$MnOOH + OH^-$	0.27
Negative				
Al	$Al + 4OH^-$	$\rightarrow$	$Al(OH)_4{}^- + 3e^-$	–2.33
Zn	$Zn + 2OH^-$	$\rightarrow$	$ZnO + H_2O + 2e^-$	–1.25
Fe	$Fe + 2OH^-$	$\rightarrow$	$Fe(OH)_2 + 2e^-$	–0.88
H_2	$H_2 + 2OH^-$	$\rightarrow$	$2H_2O + 2e^-$	–0.83
Cd	$Cd + 2OH^-$	$\rightarrow$	$Cd(OH)_2 + 2e^-$	–0.81

* Referred to the hydrogen electrode at pH = 14

On an energy-to-weight basis, the best negative active material is hydrogen. Both silver/ and nickel/hydrogen batteries (with hydrogen stored in gaseous form) have been shown to perform excellently in space applications where unit cost is not a major factor [1]. Designs for terrestrial energy-storage markets, including EVs, have been made by Johnson Controls, Inc. [2] and Eagle-Picher Industries [3]. In order to minimize production costs, both these designs employ a common-pressure vessel for the complete battery stack. The proven cycle life of the nickel/gaseous-hydrogen system confers on such batteries the ability to outlive the EV itself. Nevertheless, there remain the cost and safety problems associated with storing hydrogen gas. The need for a pressure vessel can be eliminated by using certain alloys to store the hydrogen as a hydride rather than as a gas. Indeed, following recent improvements in the alloy technology, the nickel/metal hydride system has become a promising candidate for EVs and other energy-storage applications. Note, the hydrogen electrode can be used in conjunction with the oxygen electrode; such an arrangement is a 'fuel cell', and not a secondary battery. Fuel cells are also being considered [4] as power sources for EVs and are discussed briefly in Section 10.3, Chapter 10.

The silver/zinc battery has a high practical specific energy (namely, ~150 Wh kg^{-1} [1]) and, consequently, is employed in satellite technology and rocketry, as well as in supplying propulsion power to torpedoes and submersibles [5]. There have also been instances of the use of this battery for driving road EVs, particularly in range and speed competitions. Nevertheless, batteries using silver electrodes cannot be considered as serious contenders for EV traction because of the high cost and limited availability of the metal, as well as the short cycle lives.

The other practical combinations of electrodes, namely, nickel/iron, nickel/cadmium, nickel/zinc, manganese/zinc, iron/air, aluminium/air and zinc/air have all received considerable attention as candidate power sources for EVs.

6.2. Nickel / Iron Batteries

The nickel/iron system was commercialized during the early years of the twentieth century following research and development work by Jungner in Sweden [6] and Edison in the USA [7] and, for a time, became the prominent secondary battery in industrial use. Applications included: fork-lift trucks, mine locomotives and shuttle cars, railway locomotives, and motorized hand-trucks.

The system comprises a nickel(III) hydroxy-oxide (NiOOH) positive electrode and a metallic iron negative. The electrolyte is a concentrated solution of potassium hydroxide (typically, 240 g l^{-1}) containing lithium hydroxide (50 g l^{-1}). The inclusion of lithium in the electrolyte exerts a stabilizing effect on the capacity of the nickel electrode during charge–discharge cycling. The cell reaction is given in Table 3.3, Chapter 3 and the electrode potentials in Table 6.1. The voltage exhibited by a practical cell in the fully-charged state is 1.4 V, while the average value during discharge is 1.2 V.

The complete discharge of an iron electrode proceeds in two steps; the successive products are $Fe(OH)_2$ and $Fe(OH)_3$. Nickel/iron cells are always designed, however, to be 'positive limited'; the excess negative material is sufficient to restrict the discharge of iron to the formation of $Fe(OH)_2$ (Table 6.1). The positive material can be present in different crystalline forms in both the charged and discharged states; cobalt oxide is usually added (in a Co-to-Ni ratio of 1 : 20) to assist formation of the preferred crystalline structures.

In traditional designs of nickel/iron cells [8] (and nickel/cadmium cells, see Section 6.3), the positive material is located in ribbons of perforated nickel-plated steel (Fig. 6.1(a)) that are wound into tubes, while the negative material is contained in flat rectangular pockets made also from perforated nickel-plated steel (Fig. 6.1(b)). In other cell designs, both the positive and negative materials are contained in flat pockets. The positives are filled with alternate layers of nickel(II) hydroxide and nickel flake. The latter component is added to provide a conducting matrix since both the charged and discharged nickel oxides have low conductivity. Powdered and flaked graphite can be used in place of nickel flake in order to obtain good conductivity. The nickel hydroxide is prepared by spraying nickel sulfate solution into hot sodium hydroxide; the resulting precipitate is filtered, dried, washed, and crushed into particles with sizes which range between 0.07 and 0.5 mm. The negative material is prepared from pure iron by a process in which iron slugs are dissolved in 20–25% sulfuric acid. The ferrous sulfate product is re-crystallized, roasted, reduced, and then partially re-oxidized. These steps further purify the material and produce a powder of iron and Fe_3O_4 that has both good conductivity and high surface area.

In recent years, new types of nickel oxide positive electrode have been introduced, particular for use in premium, high-rate batteries. One type is based upon sintered, porous plaques of nickel powder (~80% porosity) that

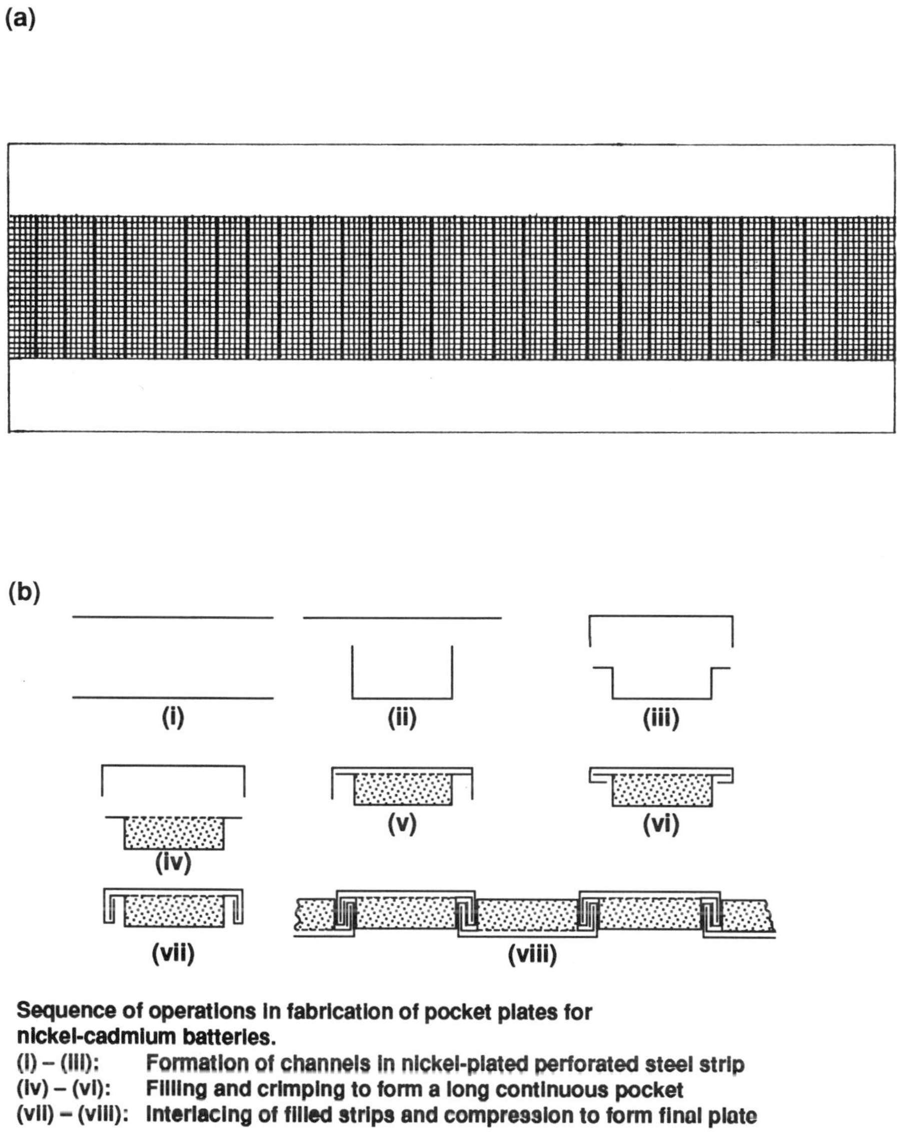

Fig. 6.1. (a) Perforated steel strip with large total hole area used for nickel electrodes. (Re-drawn from illustration of Sonesson, Sweden.) (b) Sequence of operations in fabrication of iron pocket plates: (i)–(iii) formation of channels in nickel-plated perforated steel strip; (iv)–(vi) filling and crimping to form a long continuous pocket; (vii), (viii) interfacing of filled strips and compression to form final plate.

are chemically or electrochemically impregnated with active material. Others utilize nickel felts or nickel foams as the substrate for the active material. Such batteries are generally more costly than those with pasted or pocket plates and are reserved for premium applications. For example, nickel/cadmium and nickel/hydrogen cells in satellite batteries often use sintered nickel positives.

Negatives have been developed that contain copper as a conducting matrix. To prepare these plates, copper and iron powders in acicular or arborescent crystalline forms are mixed and then coated as a uniform layer over a fine wire mesh. Under very high levels of compression at ambient temperature, the metal particles become inter-nested to yield a material with a high porosity.

In general, nickel/iron cells are electrochemically formed and then conditioned by applying charge–discharge cycles which bring the capacity to the rated value. During service, battery charging is usually carried out under constant-current control; a recommended procedure is to charge for 7 h at $C_5/5$.

Traditional designs of nickel/iron cells have specific energies which are 1.5–2.0 times those of conventional lead/acid cells, but have considerably higher materials and manufacturing costs. Further, the alkaline system has a higher self-discharge rate (due to the reaction between iron and water to release hydrogen) and this results in cell heating and, by necessity, in more complex thermal management. Nickel/iron also has the disadvantages of inferior low-temperature performance and lower energy efficiency on account of hydrogen evolution on charging. This leads to an increased need for battery maintenance to replenish lost water, or to an automatic watering system. As a consequence of such deficiencies, nickel/iron has been displaced by its lead/acid counterpart in most traction battery applications since the 1920s. Only in the former Soviet Union does the alkaline system still command a sizeable fraction of the traction market.

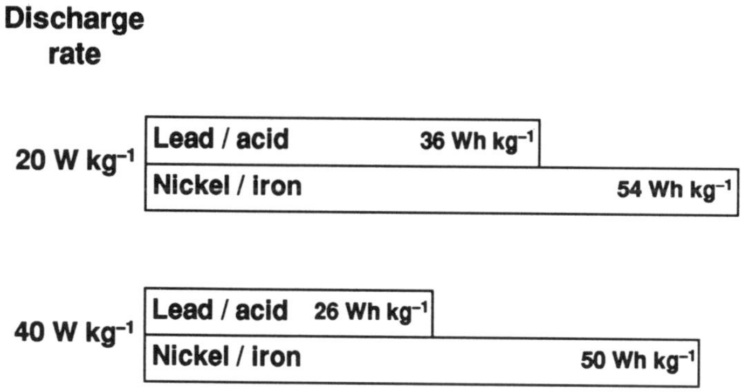

Fig. 6.2. Comparative specific energies of nickel/iron and lead/acid batteries at two discharge rates.

The nickel/iron battery has certain advantages. In particular, the system is noted for its ruggedness and long life (2000 cycles at 80% DoD). These features counteract to some extent the price penalty and allow the nickel/iron system to have a market edge over lead/acid in a number of specialist applications.

The theoretical specific energy of nickel/iron is 50% greater than that of lead/acid (Table 3.3, Chapter 3) and thus considerable scope should exist for enhancing the practical value. Furthermore, the decrease in specific energy with increase in discharge current (i.e., the Ragone relationship, see Section 4.3.2, Chapter 4) is much less pronounced with nickel/iron than with lead/acid (Fig. 6.2). The prospect of good specific energy, together with the fact that the basic technology was already in place, focused attention in the late 1970s and early 1980s on nickel/iron as an EV battery technology which both would give a greater vehicle range than that offered by lead/acid and could be brought into production quickly.

In the USA, nickel/iron was one of the systems identified by the US Department of Energy (DOE) as having clear potential for meeting near-term EV requirements. As part of the DOE programme, Eagle-Picher Industries developed a battery in which the nickel electrode was based on sintered plaque. The technology had been established for nickel/cadmium systems and involved the sintering of pure nickel powder on to a centrally located current-collector in the form of a metallic wire screen or a perforated metal sheet. Such plaques have high porosity, large surface area, good conductivity, and favourable mechanical properties. Loading of nickel hydroxide is carried out by impregnating the plaques with nickel nitrate solution under vacuum, followed by either electrochemical or thermal treatment. The Eagle-Picher battery achieved a specific energy of 57 Wh kg^{-1} ($C_3/3$ rate) and accomplished 1500 cycles; the specific power was >100 W kg^{-1} [9]. An improvement in the purity of the iron component led to a decrease in the self-discharge rate and a corresponding increase in the energy efficiency from 50 to 67% [10]. Nevertheless, plaque technology is expensive; the estimated cost in 1986 was 2.5 times that of advanced lead/acid [11]. A pack of Eagle-Picher batteries (28 $\times$ 6 V modules) served as the power source for an electric lightweight van which was developed under the Dual-Shaft Electric Propulsion (DSEP) Program of the DOE (see Section 1.11, Chapter 1). The DSEP design concentrated on lowering the internal resistance of the battery in order to make transfer of energy to and from the pack more efficient. It was claimed that the improved efficiency not only acted to improve range, but also solved the problems of high water consumption and overheating that are often associated with the nickel/iron system. Plans were made in the late 1980s to commence the commercial production of batteries to be installed in the TEVan from Chrysler (see Section 1.8, Chapter 1). In 1992, it was reported [3] that Eagle-Picher was producing nickel/iron batteries in several sizes. The most popular was the NIF-200-5 module which weighed 30 kg and delivered 200 Ah ($C_1/1$ rate) at a nominal 6 V.

Parallel DOE-sponsored research into nickel/iron batteries was carried out by Westinghouse with the objective of developing low-cost electrodes. Cells using plate materials contained in steel wool delivered 53 Wh kg^{-1} ($C_3/3$ rate) and 1000 cycles [9]. The projected cost of the system did not, however, meet the DOE programme goal and remained well above that of lead/acid [9,10].

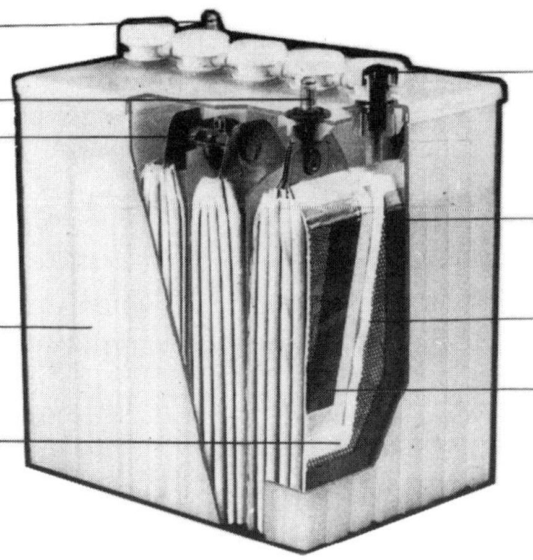

Fig. 6.3. SAFT nickel/iron battery. Voltage: 6 V; specific energy: 60 Wh kg^{-1}; energy density: 120 Wh l^{-1}; specific power: 170 W kg^{-1}. (Taken from [15] with permission of SAFT, France.)

Improvements in nickel/iron battery technology have also been made by: Heyrovsky Institute in the Czech Republic; VARTA, Daimler-Benz and Deutsche Automobilgesellschaft GmbH (DAUG) in Germany; SAFT in France; Furukawa and Matsushita in Japan; Swedish National Development Company and SAB NIFE in Sweden; Tsing Hua University of Taiwan; INCO Europe in the UK; and Leningrad Industrial Amalgamation Istochnik in Russia. Specific energies of 60 Wh kg^{-1} ($C_5/5$ rate) were reported for both the SAFT [12] and the Matsushita [13] batteries; these values are at least 50% greater than the baseline performance given by conventional lead/acid batteries.

In most cases, the performance of the advanced nickel/iron batteries developed by the above companies was assessed in actual EV operation. Life in vehicle tests matched that demonstrated in the laboratory [14]. As expected, increases in vehicle range reflected improvements in specific energy, and the range of a VW Rabbit powered by Eagle-Picher batteries was 2–3 times that obtained with a conventional lead/acid pack [9]. Peugeot fitted SAFT nickel/iron batteries (Fig. 6.3) to an experimental EV, the 205 Electrique (Fig. 6.4). The specific energy of the battery as a

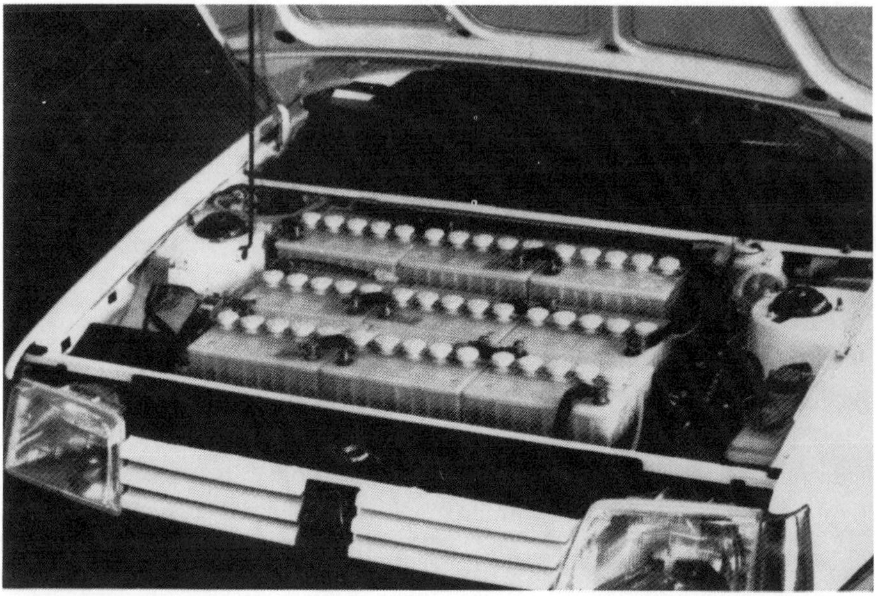

Fig. 6.4. SAFT nickel/iron battery pack (12 cells of type shown in Fig. 6.3) fitted to the Peugeot 205 Electrique. (Taken from [15] with permission of SAFT, France.)

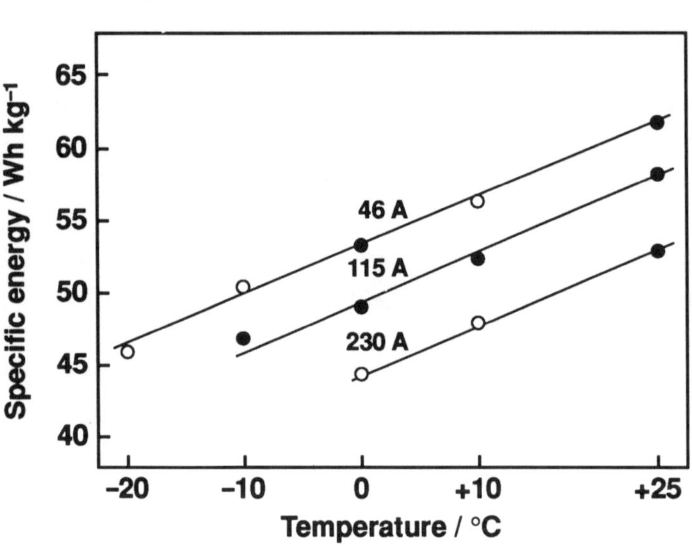

Fig. 6.5. Influence of temperature on the specific energy of SAFT nickel/iron battery for various constant-current charge rates. (Taken from [15] with permission of SAFT, France.)

function of discharge rate and temperature is shown in Fig. 6.5. SAFT reported a driving range of 140 km under stop–go conditions, a maximum range of 200 km, a top speed of 100 km h^{-1}, an estimated battery life of 1500 deep-discharge cycles (~200 000 km), and adequate performance down to –20 °C. SAFT has now terminated its work on nickel/iron batteries for EV applications and is focusing on nickel/metal hydride batteries (see Section 6.6) and lithium-ion batteries (see Section 9.4.4, Chapter 9).

6.3. Nickel / Cadmium Batteries

The nickel/cadmium system employs the same positive electrode as the nickel/iron system, in combination with a metallic cadmium negative. The electrode and cell reactions are presented in Table 6.1 and in Table 3.3, Chapter 3, respectively. The voltage in the fully-charged condition is 1.3 V; the average value on discharge is 1.2 V. Historically, the development of the battery has coincided with that of nickel/iron. Conventional versions of nickel/cadmium batteries are based on either the pocket or the sintered electrode technologies that have been discussed in Section 6.2.

In many respects, the nickel/cadmium battery is comparable with its nickel/iron counterpart. For example, similar technology is employed in its manufacture. The practical specific energy, traditionally ~40 Wh kg^{-1}, has been raised to 60 Wh kg^{-1} in recent designs. Like nickel/iron, the battery has long cycle life and displays good resistance to both mechanical and electrical abuse. On the credit side, nickel/cadmium operates better than nickel/iron at low temperatures, has a much slower rate of self-discharge, a higher electrical efficiency, and requires much less maintenance. On the debit side, the charge efficiency decreases rapidly above 35 °C [16]. Also, cadmium is more expensive than iron, is toxic, and has limited availability — problems which may be overcome in part by efficient battery recycling.

The comparatively high cost of traditional nickel/cadmium batteries (~US$ 1000 per kWh) has caused many EV manufacturers, particularly in Japan and the USA, to reject this battery out of hand. It has also stimulated a number of research and development programmes that are directed towards producing less-expensive electrode materials. For example, in the DOE programme, the Energy Research Corporation (ERC) aimed to perfect a roll-bonded fabrication process for electrodes which utilize a lower content of inert metallic nickel than sintered-plaque types. It was claimed that such an approach would produce an economic, long-lived battery with a specific energy significantly greater than that of conventional designs. In Germany, Hoppecke [17] developed an innovative fibre-structured electrode which the Company contended would make nickel/cadmium competitive in EV operations. This technology is being exploited by a new joint company that has been formed with DAUG (DAUG-Hoppecke) [18]. Similarly, Sorapec in France has manufactured a nickel-felt electrode, electrochemically impregnated, which shows good charge retention and charge-acceptance at high temperatures (60 °C) where traditional nickel/cadmium does not perform well [19].

Improved designs of nickel/cadmium battery for EV traction have been developed by SAFT [16] and these have been selected as the power source for three French passenger cars, all of which are being put into production, initially, at the 5000–10 000 per year level. The cars are the Peugeot 106, the Citroën AX and the Renault Clio (see Fig. 1.19, Chapter 1). The justification for this choice is seen in Table 6.2 which compares the performance of the latest SAFT nickel/cadmium batteries with the USABC mid-term goals (see Table 3.4, Chapter 3).

It is apparent that the SAFT battery meets, or exceeds, most of the USABC mid-term goals; the principal difficulties are stored energy and capital cost. The battery displays exceptional cycle life and power, good recharge characteristics, and a wide range of operating temperatures. In addition, the power/capacity degradation, self-discharge and continuous discharge characteristics are all acceptable. The failure to meet the stored-energy goals may not be too inhibiting for urban vehicles where elimination of pollution is more important than vehicle range.

Table 6.2. Comparison of SAFT nickel/cadmium battery performance with USABC mid-term goals [20].

Characteristic	SAFT	USABC mid-term goal	Pass/fail
Primary objectives:			
• Specific energy (Wh kg^{-1})	60	80	Fail
• Energy density (Wh l^{-1})	115	135	Fail
• Specific power (W kg^{-1})	225	150	Pass
• Power density (W l^{-1})	400	250	Pass
• Recharge time (h)	1.0	<6	Pass
• Fast recharge time (h)	0.3	0.25	Pass (?)
• Life, years	8.5	5	Pass
• Life, cycles @ 80% DoD	2500	600	Pass
• Ultimate cost (US$ / kWh)	350	<150	Fail
• Operating environment (°C)	–40 to +85	–30 to +65	Pass
• Power and capacity degradation (% rated specification)	13	20	Pass
Secondary objectives:			
• Efficiency (%)	74	75	Pass (?)
• Self-discharge (% in 48 h)	~1	<15	Pass
• Maintenance	Low	zero	Fail
Derived goal:			
• Ultimate cost (US$ / kWh / cycle)	0.14	0.25	Pass

This leaves capital cost as the major problem outstanding. Production costs for prototype batteries are claimed to be US$ 900–1000 per kWh [20]. These are reduced to US$ 500 per kWh for 2000–3000 EV traction batteries per year, and are expected to fall to US$ 350 per kWh for 10 000 units per year. This is still much greater than the target production cost of US$ 150 per kWh for lead/acid that has been set by the Advanced Lead-Acid Battery Consortium. It should be noted, however, that because of the exceptional cycle life of the nickel/cadmium battery, its cost per cycle compares favourably with the USABC mid-term goal (Table 6.2). There is even a prospect of meeting the long-term goal of US$ 0.10 per kWh per cycle, provided the manufacturing costs can be reduced to US$ 250 per kWh for mass production. It is expected that the problem of capital cost will be addressed through battery leasing. Overall, it is estimated [21] that the total life-cycle costs will be at least as good as those for EVs with lead/acid batteries, and considerably better than those for ICEVs.

The confidence of the French industry in the future of urban EVs powered by nickel/cadmium batteries is such that SAFT has built a new plant in Bordeaux to manufacture these batteries. This facility is capable of producing 15 000 traction batteries per year, and there are plans to expand to 50 000 batteries per year. A recycling plant for spent batteries is also under construction.

6.4. Nickel / Zinc Batteries

The nickel/zinc cell combines the same positive electrode which is used in the nickel/iron and nickel/cadmium systems with a metallic zinc negative. The electrode and cell reactions are given in Table 6.1 and in Table 3.3, Chapter 3, respectively. The voltage in the fully-charged state is 1.7 V; the average value on discharge is 1.6 V. In Ireland during the late 1920s and early 1930s, Drumm used nickel/zinc batteries to supply power for the propulsion and lighting of passenger trains [8]. Short cycle life, however, discouraged further practical application.

The theoretical specific energy of the nickel/zinc system is double that of lead/acid and 30% greater than that of nickel/iron. A practical nickel/zinc cell can achieve 80–100 Wh kg^{-1} and a battery pack 70 Wh kg^{-1}. These are attractive specific energies for EV traction and would double the range of present EVs fitted with lead/acid batteries. The zinc system has other attractive properties: low-temperature performance is admirable and self-discharge is slow. In both these latter features, the battery approaches the cadmium system and is superior to the iron counterpart. Nickel/zinc shares with other alkaline nickel batteries the characteristics of a good specific power as well as a specific energy that does not decrease appreciably with the discharge rate. Because of these favourable properties, nickel/zinc was pinpointed in the 1970s as the most likely challenger to lead/acid for the perceived near-term EV market. But the question of battery cost had to be addressed and the problem of short service-life had to be overcome.

The nickel electrode is the major determinant of cost despite the fact that, because of the higher cell voltage, the energy stored per unit weight of nickel is 25% greater than in nickel/cadmium cells. Some success in decreasing the cost of nickel electrodes was achieved in the DOE programme by containing the nickel hydroxy-oxide in either nickel-plated steel wool or in plastic-bonded structures [22]. Nevertheless, the cost of batteries with these electrodes remained above the DOE project goal [23].

Much research effort has been devoted to extending the service life of nickel/zinc batteries. The life-limiting problem is associated with the zinc electrode and is a result of the partial solubility of the discharge product, ZnO, in the alkaline electrolyte to form zincate, i.e.,

$$Zn + 4OH^- \underset{\text{charge}}{\overset{\text{discharge}}{\rightleftarrows}} Zn(OH)_4{}^{2-} + 2e^- \tag{6.1}$$

The zinc electrode is usually prepared by pasting zinc oxide on to a metal current-collector and then electrolytically reducing the oxide to the metal. Heavy metals (bismuth, cadmium, indium, lead, thallium, tin) are incorporated in the zinc electrode in order to increase the hydrogen overpotential [24]. During subsequent charging, zinc metal is deposited from the zincate solution in the form of dendrites. Penetration of the separators by these zinc dendrites can lead to short-circuiting and, consequently, to catastrophic failure. Research studies have shown that internal shorting of cells can be checked by the use of separators which reject zincate ions, and by the addition of reagents to the electrolyte that inhibit dendritic growth. Sometimes, multiple separators are employed. In these designs, a zincate-rejecting separator is sandwiched between two reservoir separators which hold the electrolyte.

In view of such developments, there was considerable optimism in the late 1970s with respect to the early commercialization of nickel/zinc batteries for EVs. Zincate formation, however, gives rise to another problem, and one that has proved to be particularly intractable. Movement of zincate within the cell can lead to redistribution of active material on the negative plate, and even to complete removal from some parts of the grid, as shown in Fig. 6.6 [25]. This phenomenon is commonly known as 'shape change' and results in an appreciable loss of capacity. The decline in performance is, however, not due to the decrease in the projected area of the material *per se*, but rather to the accompanying densification of the remaining zinc (Fig. 6.7).

The adverse effect of shape change can be alleviated to some extent, albeit at the expense of battery specific energy, by the incorporation of a large excess of zinc in the negative plate over that required to balance the capacity of the positive counterpart. Cycle life has also been improved by the addition of specific reagents to the zinc material [22]. In this respect, bismuth oxide has proved particularly effective and forms a conducting matrix of bismuth metal throughout the electrode structure [25]. Additives that lower zincate solubility, such as calcium hydroxide, inhibit shape change through retarding the dispersion of zincate within the cell [22]. A decrease in solubility enhances zinc passivation, but the deleterious effects of this process on capacity can be ameliorated by the addition of cellulose fibres to maintain a porous structure throughout the plate and to avoid drying out of the electrode by serving as a wicking agent. The zinc electrode developed by General Motors (GM) had an active material with a composition of 56.7 wt.% ZnO, 38.5 wt.% $Ca(OH)_2$, 3.4 wt.% Pb_3O_4, and 1.4 wt.% newsprint (cellulose fibres) [27], and was pasted on to a copper current-collector which was plated with a lead–tin coating. Electrodes of

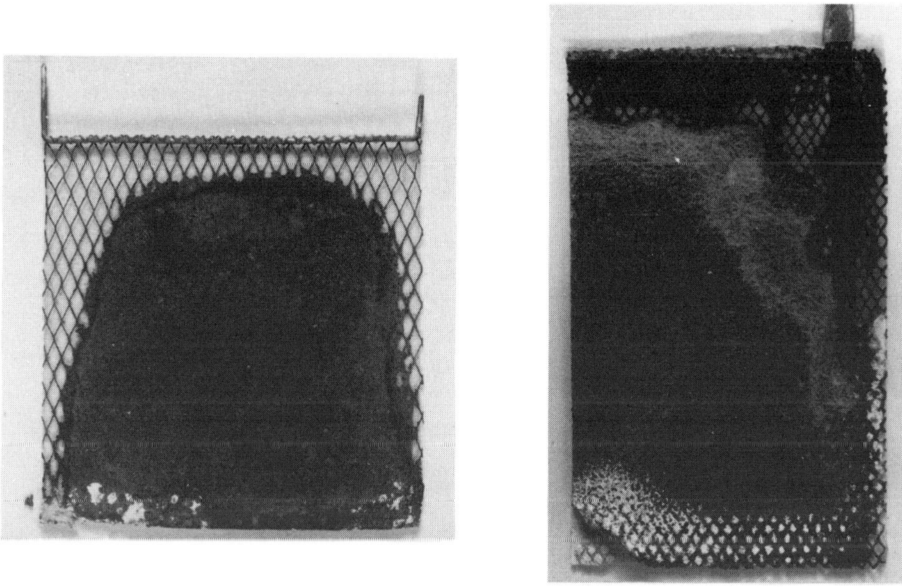

Fig. 6.6. Examples of shape-change in a zinc electrode after charge–discharge cycling. (Taken from [25] with permission of CSIRO, Australia).

(a) (b)

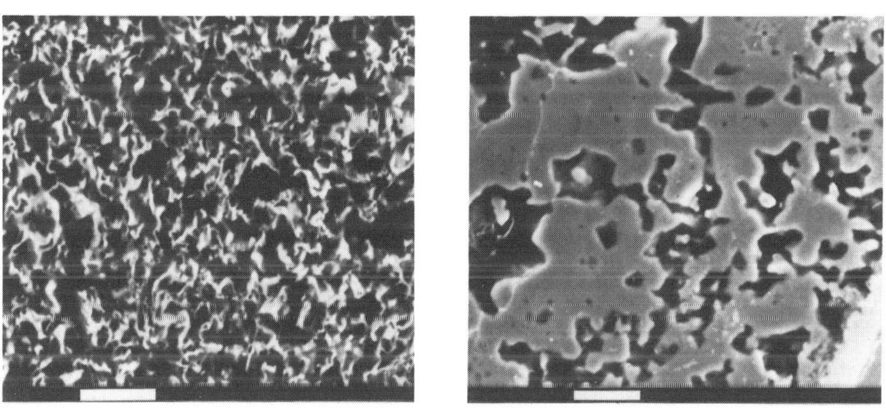

Fig. 6.7. Electron micrographs of cross-sections of zinc electrodes: (a) 1st charge; (b) 105th charge. Densification of zinc is observed with charge–discharge cycling. (Taken from [26] with permission of The Electrochemical Society, Inc.)

Table 6.3. Performance of Samsung nickel/zinc EV batteries [31].

Property	USABC mid-term goal	Samsung battery
• Specific energy (Wh kg^{-1})	80	70
• Energy density (Wh l^{-1})	135	150
• Specific power (W kg^{-1})	150	150
• Power density (W l^{-1})	250	330
• Recharge time (h)	<6	<8
• Cycle life (@ 80% DoD)	600	300
• Capacity degradation (% rated specification)	20	10
• Ultimate cost (US$ / kWh)	<150	150

this type were pressure sealed in talc–calcium acetate separators [28]. The GM battery gave the best life performance of all the nickel/zinc batteries examined under the SAE J227a/D driving schedule (see Section 4.3.1, Chapter 4) by the US National Testing Laboratory. Nevertheless, the recorded life of only 280 cycles was well below the goal of 800 cycles that had been set by DOE for near-term batteries. The GM system is illustrated in Fig. 6.8. General Motors scaled down nickel/zinc EV battery development in the late 1980s and, instead, focused attention on other battery options. In 1995, the Company considered nickel/metal hydride to be the most viable battery technology in the mid-term [30].

The Samsung Corporation in Korea continues to work on nickel/zinc batteries and has developed a range of sealed modules specifically for EV traction applications (180 Ah, 1.6 V; 50 Ah, 8 V; 80 Ah, 6.4 V) [31]. The performance characteristics of these batteries come close to meeting the USABC mid-term goals, except in the case of cycle life (Table 6.3).

Attempts to improve the service life of nickel/zinc batteries through lowering zinc solubility have continued at the Lawrence Berkeley Laboratory in the USA [24]. The best results (namely, a service life of about 500 cycles) were achieved with an electrolyte that had a composition

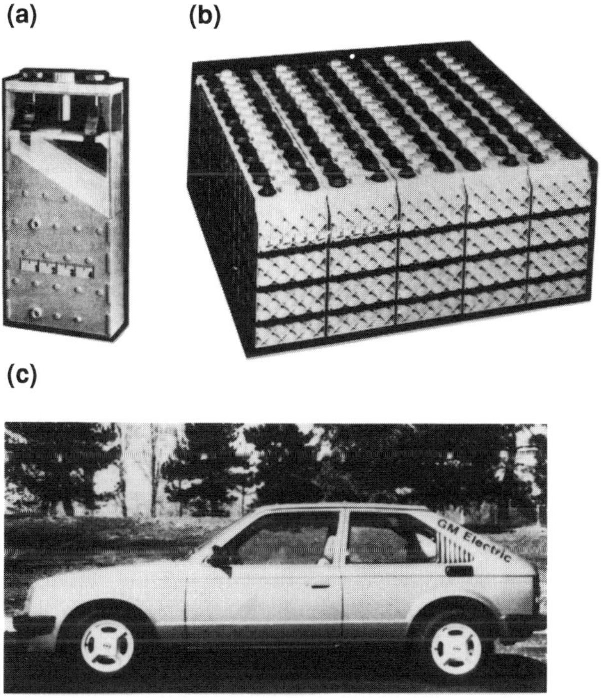

Fig. 6.8. Nickel/zinc battery system developed by General Motors: (a) 150-Ah cell; (b) EV battery pack; (c) test vehicle. (Taken from [29] with permission of General Motors, USA.)

of: 3.2 M KOH, 1.8 M KF, 1.8 M K_2CO_3. A series of similar studies has been conducted by Electrochimica, also in the USA [32]. With a typical electrolyte composition of 3 M KOH, 3 M K_3PO_4, over 600 cycles (80% DoD) were obtained for 225-Ah cells. By contrast, cells which employed a conventional KOH electrolyte failed after about 100 cycles.

It has been pointed out [25] that if the major mechanism of capacity loss is a decrease in plate porosity, then additives which diminish zincate solubility could be counter-productive. This is because a decrease in zinc solubility would promote passivation and hence decrease plate capacity. Consequently, it was suggested that an electrolyte which maximizes zinc solubility would be more useful. Such an approach has subsequently been adopted by the Sorapec Corporation in France [33]. The electrolyte in the

Sorapec system is in large excess and consists of 8 M KOH with 2 wt.% sodium silicate to stabilize supersaturation of zincate. A 60-Ah unit displayed a moderate specific energy of 57 Wh kg^{-1} and a service life in excess of 1000 cycles.

There appears to be scope for improving the service life of nickel/zinc batteries through the use of pulsed-charging regimes. Charging at low frequencies (<10 Hz) has been found [34] to give a superior cycle life to that obtained with either constant-current or constant-voltage charging. A significant increase in life was achieved [35] by charging at 0.5 Hz with a programmed pulse-height. The latter was initially set to a high value, to nucleate zinc throughout the electrode, and then was decreased in a controlled manner to avoid dendritic growth in the final stages of charging. When using this charging regime, the performance of 2-Ah Eagle-Picher cells was extended to 450 cycles from the 130 cycles obtained with constant-current charging. The optimum performance of Sorapec flooded cells was obtained with pulse charging [33]. The regime comprised pulses of 5 mA cm^{-2} for 2.75 s, each followed by a reverse pulse at the same current density for 0.15 s. This charging approach, together with the addition of lead ions to the electrolyte, maintained an acceptable morphology for the zinc electrode.

The 'Vibrocel' conceived by AB Tudor in Sweden, and taken up by Exide in the USA [36], adopted a radically different approach to combating the life problem of the nickel/zinc system. In this concept, the quantity of cell electrolyte was sufficient to dissolve all the oxidized zinc formed on discharge, and the negative electrodes were vibrated during charging to modify the hydrodynamic conditions so that the deposited zinc layer was smooth, adherent and uniform. In order to maintain good zinc distribution and morphology, it was necessary to discharge the cell fully at regular intervals. Long service lives (>1000 cycles) were claimed for the Vibrocel. On the other hand, the achieved specific energy of 47 Wh kg^{-1} [9] was well below that of conventional nickel/zinc designs, and was little better than that of advanced lead/acid designs. Furthermore, the energy density of 72 Wh l^{-1} did not even match the performance which was given by lead/acid. Such disadvantages, together with the higher materials cost and the greater complexity of the system, led to the abandonment of the Vibrocel concept [37].

There have been many other efforts to develop nickel/zinc batteries: in France by Compagnie Générale d'Electricité; in Germany by DAUG; in Italy by Fiat; in Japan by Furukawa, Japan Storage Battery Company, Sanyo, and Yuasa; in Sweden by AGA Innovation Centre; in the UK by Lucas; in the USA by Energy Research Corporation (ERC), Gould, NASA-Lewis, Rockwell, and Yardney Electric; in the former USSR by Volzhski. Specific energies of up to 75 Wh kg^{-1} (C_3/3 rate) with specific powers of up to 150 W kg^{-1} have been reported [38]. Most projects have been taken to the stage of constructing battery packs and then conducting tests in EVs. A sealed battery produced by ERC has achieved 11 000 cycles at 10% DoD and is considered to suit the discharge–charge patterns of HEVs. The status of nickel/zinc technology was reviewed by McBreen [39] in 1994, and by Jindra [40] in 1997. McBreen concluded that this system is appropriate for large EVs by virtue of its substantial economic advantage over the competing near-term alkaline battery technology of nickel/metal hydride. Using a figure-of-merit of Wh-cycles per kg per US$, it is claimed that nickel/zinc is now superior to all competing ambient-temperature traction batteries.

6.5. Manganese / Zinc Batteries

The manganese/zinc cell comprises a zinc negative electrode (as in nickel/zinc systems) and a manganese dioxide positive. The electrode and cell reactions are given in Table 6.1 and in Table 3.3, Chapter 3, respectively. It should be noted that the dioxide can undergo a two-electron discharge process to the manganese(II) state. The second electron-step causes major structural changes, however, that result in mechanical degradation of the positive material. Thus, the operation of secondary versions of the cells has been restricted to the one-electron discharge. The voltage of the cell in the fully-charged state is 1.5 V; the average value on discharge is 1.3 V.

Manganese dioxide is considerably less expensive than nickel, and the theoretical specific energy of the manganese-dioxide/zinc couple lies between that of nickel/iron and nickel/zinc (see Table 3.3, Chapter 3). The manganese/zinc system is the market leader amongst primary cells. Secondary types were introduced for small-scale applications in about 1959 but their consumer acceptance was inhibited initially by poor service life (<50 cycles) and a modest continuous-current rating (400 mA for AA-size cells) [8]. On the other hand, these drawbacks were offset by higher capacity, longer shelf-life and lower cost than nickel/cadmium. In the mid-1990s, there was a sharp rise in sales in the USA for moderate-drain consumer applications [41]. The endurance of rechargeable manganese/zinc cells can be limited by the characteristics of either electrode. In addition to shape-change and dendritic-growth problems at the zinc negative (see Section 6.4), there can be a marked decline in the capacity of the manganese dioxide positive on charge–discharge cycling. Early commercial cells used electrolytic manganese dioxide (EMD) and were rechargeable only if the DoD of the EMD was kept below 50%. To prevent the positive from discharging beyond this value, cells must be of the negative-limited design. Such an approach exacerbates the zinc electrode problems.

Co-deposition of nickel or titanium into EMD was shown [42] to improve the rechargeability of the positive electrode. Modification of the manganese dioxide structure by the incorporation of bismuth was found to be especially effective [43]; 200–300 deep-discharge cycles were obtained from laboratory cells. Through such improvements in performance, secondary manganese/zinc cells have begun to establish a market in small-scale operations [44]. Further advances are required, however, if this system is to become a serious candidate for EV applications.

6.6. Nickel / Metal Hydride Batteries

The nickel/metal hydride (Ni/MH) battery is a subsequent version of the nickel/hydrogen battery in which hydrogen is incorporated in an alloy as a metal hydride rather than stored in gaseous form. The positive electrode is the same as that used in both the nickel/iron and the nickel/cadmium batteries discussed in Sections 6.2 and 6.3, respectively.

It is most appropriate to employ the metal hydride directly as a reversible hydrogen electrode, as opposed to applying the hydride as an external storage medium. The reaction at the negative electrode is:

$$MH_x + OH^- \underset{\text{charge}}{\overset{\text{discharge}}{\rightleftarrows}} MH_{x-1} + H_2O + e^- \tag{6.2}$$

On discharge, OH^- ions are consumed at the negative hydride electrode and generated at the nickel oxide positive (Table 6.1). The converse is true for water molecules. Thus, there is a counterflow of OH^- ions and water molecules across the separator, which reverses direction on charge. The overall electrolyte concentration remains unchanged during both charge and discharge, although locally at the electrodes the concentration changes as water is generated or consumed. It is important to avoid dry-out of the positive and over-dilution of the negative on rapid discharge. Control is exercised through careful design of the pore size in the separator to effect the necessary wicking action.

The Ni/MH cell has an open-circuit voltage of ~1.4 V in the fully-charged state, as determined by the choice of alloy. This compares with a value of ~1.5 V for a typical nickel/hydrogen cell. Thus, there is only a small loss in voltage by storing the hydrogen as a hydride. The practical cell voltage (1.2–1.3 V) is almost the same as that of nickel/cadmium (this makes for ready interchangeability), while the specific energy is 1.5–2 times higher (70–80 Wh kg^{-1}).

The Ni/MH battery has a high discharge-rate capability and is resilient with regard to overcharge, overdischarge and cell reversal. During overcharge, oxygen is generated at the nickel positive and is reduced back to water at the alloy negative. During overdischarge, the potential of the nickel electrode drifts into the hydrogen evolution region; the hydrogen generated is re-absorbed at the alloy negative electrode and thus prevents

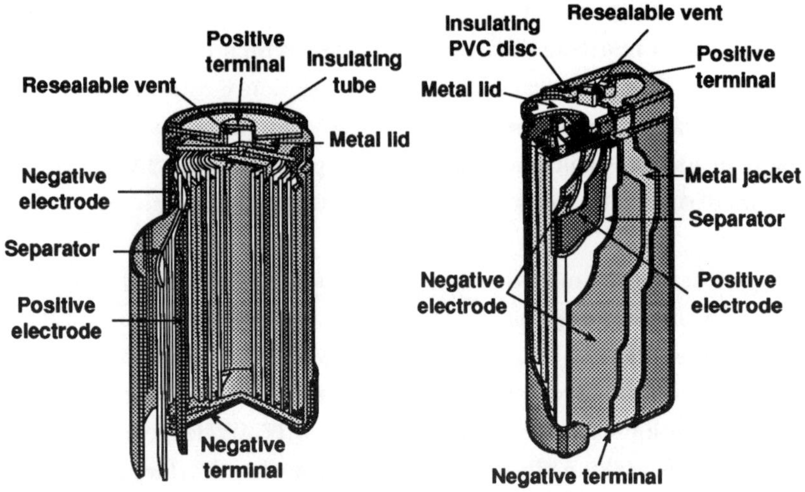

Fig. 6.9. Cross-sections of cylindrical and prismatic nickel/metal hydride cells. (Taken from [45] with permission of Research Studies Press Limited, UK.)

serious build-up of pressure. The requirements for a hydrogen-storage alloy for the Ni/MH battery are: a high hydrogen-absorption capacity; a suitable dissociation pressure for the hydride over the operating temperature range; ease of hydrogen absorption and desorption at ambient temperature; good stability in strongly alkaline aqueous solutions at the operating potential. In addition, the alloy constituents should be inexpensive, readily available, and environmentally benign.

The first Ni/MH technology to enter the market was based on single cells of both the cylindrical and the prismatic design (Fig. 6.9). The further development of Ni/MH batteries for a range of portable-power applications rapidly became a common goal of battery manufacturers throughout the world given the growing environmental concerns associated with nickel/cadmium batteries. By the end of 1995, the installed production capacity in Japan alone was 64 million cells per month [46]. At the same time, many companies started to tailor battery designs specifically for EV markets. For example, the Ovonic Battery Company [47,48] and Eagle-Picher Industries in the USA [3], ABB Hochenergiebatterie GmbH [49] in Germany, SAFT [50,51] in France, and Matsushita [52,53] in Japan reported on EV battery development at EVS-11 in Florence in 1992 and/or EVS-12 in Anaheim in 1994. The production line which SAFT has built in France for nickel/cadmium traction batteries (Section 6.3) is adaptable to metal hydride technology and, in mid-1997, the Company delivered its first Ni/MH batteries to Chrysler for testing in an Epic minivan. SAFT expects a vehicle range of 130 km and a battery life of 1000 cycles.

More recent developments have been reported by VARTA [54] who have Ni/MH batteries undergoing field trials in a number of electric and hybrid cars (Fiat, Daimler-Benz, Opel, Volvo, VW), as well as in hybrid trucks and buses. These traction batteries are based either on high-energy modules (75 Wh kg^{-1}, 160 W kg^{-1}) or on high-power modules (55 Wh kg^{-1}, 300 W kg^{-1}). The truck battery, at 75-kWh size, is claimed to be the largest Ni/MH battery yet. VARTA is also developing ultra high-power modules which are projected to have a specific power output in excess of 1000 W kg^{-1} with specific energy of up to 45 Wh kg^{-1}. In Japan, a Ni/MH traction battery is being marketed by Panasonic EV Energy, a joint venture company involving both Toyota and Matsushita, and is being fitted to the Toyota RAV4-EV which won outright the 1995 Scandinavian electric car rally. The vehicles are available in small numbers for purchase (Fig. 1.21, Chapter 1). Toyota has also announced the production of two further EVs powered by Ni/MH batteries: the 'E-com' is a two-seater, ultra-compact car designed for driving in busy areas; the 'Prius' is a five-seater hybrid jointly powered by a 1500-cc ICE. Compared with a similar Toyota ICEV, the Prius is expected to reduce emissions of carbon dioxide by half, and of hydrocarbons and nitrogen oxides to one-tenth of statutory requirements. In 1997, Honda announced that it will place 300 EVs, powered by Ni/MH batteries, in California, USA, as part of a three-year monitoring and testing programme. The Company is also preparing a further 20 vehicles for use in Japan, and is considering introducing several vehicles in Europe. General Motors has formed a joint venture with the Ovonic Battery Company (GM Ovonic) to develop and manufacture Ni/MH traction batteries, while Ford plans a Ni/MH version of its Ranger electric van. Meanwhile, Hyundai has reported that Ni/MH batteries will be offered in its production EVs.

Research on metal hydride materials focused initially on the family of intermetallic compounds that have the general formula AB_5, where A is either a rare earth element (such as lanthanum) or a mixture of rare earths, and B is partially substituted nickel. It was found that the storage capacity of $LaNi_5$ (as $LaNi_5H_{6.5}$) decreased drastically after repeated charge–discharge cycling. The decay in capacity is caused by degradation of the electrode due to the irreversible formation of lanthanum hydroxide. The problem can be partially prevented by encapsulating the alloy in a thin porous film of copper or nickel [55], or by coating the surface with platinum black [56]. This oxidation of the $LaNi_5$ electrode can also be minimized by adding appropriate alloying elements [57]. The partial substitution of the lanthanum by zirconium, and the nickel by aluminium, increased the cell life from 180 to over 800 cycles [58]. This improvement has been attributed to the formation of a surface oxide film which acts as a barrier to prevent hydride oxidation. The use of mischmetal (Mm), an unrefined mixture of rare earth elements, instead of lanthanum made it possible to produce a suitable alloy at a significantly lower cost. This promoted, in 1990, the commercial production of small sealed batteries. Most alloy compositions in use are proprietary; a typical composition is $MmNi_{3.55}Co_{0.75}Mn_{0.4}Al_{0.3}$ [59]. It has been reported that a majority of battery manufacturers are using AB_5-type alloys [60].

An alternative alloy system is based on the AB_2 stoichiometry, where component A is titanium and/or zirconium, and component B can be nickel, cobalt, vanadium, manganese, aluminium, chromium, iron, etc. The Ovonic Battery Company, one of the leading developers of Ni/MH batteries, favours the AB_2 system and contends [61] that this system has the advantages over AB_5-based materials of higher storage capacity, superior oxidation and corrosion resistance, less cost, and easier processing. The custom alloys have been designed with particular elements in order to emphasize one or more of the key performance criteria of hydrogen-storage capacity, thermodynamic behaviour, oxidation and corrosion resistance, kinetics of absorption and desorption, and primary metallurgical properties. One such alloy is $V_{15}Ti_{15}Zr_{20}Ni_{28}Co_5Fe_6Mn_6$ [62]. Ovonic is also proposing to develop magnesium alloys which offer the prospects of superior hydrogen-storage capacity, reduced weight, and lower cost. The alloys will be prepared by a rapid solidification process which yields special microstructures and tailored surface properties.

The Ovonic, multi-component, hydrogen-storage alloys are produced [48] by vacuum induction melting in an inert atmosphere and casting into moulds. The alloys are then reduced to a fine powder (~75 μm) by application of a single hydrogen absorption–desorption cycle, followed by mechanical grinding. Negative electrodes are manufactured from the powder in a continuous strip by compression rolling on to wire screen, expanded metal, or perforated foil.

Pasted nickel positive electrodes are used in the Ovonic battery with an electrolyte which comprises an aqueous solution of 30% KOH and a LiOH additive. The cells have a prismatic design and are sealed. Ovonic Ni/MH cells in commercial production have demonstrated a specific energy of more than 80 Wh kg^{-1}, while engineering prototype batteries have exceeded 95 Wh kg^{-1} [63]. Ovonic has also reported [48] that a 30-kWh pack, trialled in an EV, had a specific energy and energy density, both at the $C_3/3$ rate, of 71 Wh kg^{-1} and 172 Wh l^{-1}, respectively, together with a specific power and power density, both at 80% DoD, of 226 W kg^{-1} and 546 W l^{-1}, respectively. It was further disclosed that a module of 80-Ah cells had achieved over 850 cycles prior to failure; this would represent a service of over 200 000 km for a vehicle with a range of 320 km. Successful operation of EVs was demonstrated over ambient-temperature conditions that ranged from −29 to +40 °C. More recently, Ovonic has introduced [64] Ni/MH batteries for HEVs with a power capability in excess of 600 W kg^{-1} and 1700 W l^{-1}, while retaining a specific energy of more than 70 Wh kg^{-1} during a service life of over 10 000 cycles to 30% DoD. In a further development, 40-Ah and 50-Ah cells have been designed for electric scooter applications. When using such cells, commercial scooters converted to electric drive have attained ranges of 125 km on a single charge. This latter enterprise is targeted at the emerging market for electric scooters in Taiwan and elsewhere (see Section 1.7, Chapter 1).

Electro Energy, Inc. is developing a bipolar Ni/MH secondary battery for a wide spectrum of applications that include global EV markets [65]. It is considered that bipolar designs will provide enhanced specific energy and rate capabilities. Cost savings can also be gained, both through the simplicity of the wafer design as well as the synergistic use of low-cost, plastic-bonded electrode technology. Estimated performance parameters for present designs are 70 Wh kg^{-1} and 200 Wh l^{-1}; future improvements are expected to yield near-term and long-term values of 80 Wh kg^{-1}, 225 Wh l^{-1} and 100 Wh kg^{-1}, 300 Wh l^{-1}, respectively. In 1994, the cost was estimated at US$ 200 per kWh, but the Company is confident that it can meet the mid-term goal of US$ 150 per kWh that has been set by the USABC (see Table 3.4, Chapter 3). A bipolar Ni/MH battery is also being developed by DAUG in Germany [66]. From laboratory tests, this Company projects a more modest value for the specific energy (46 Wh kg^{-1}), but outstanding specific power (1400 W kg^{-1} at 80% DoD). These performance characteristics make the battery an ideal candidate for hybrid EVs.

In summary, the following advantages have been claimed for the Ni/MH battery: (i) comparatively high specific energy (70–95 Wh kg^{-1}); (ii) high discharge-rate capability (200–300 W kg^{-1}); (iii) good low-temperature performance (–20 °C); (iv) long cycle life; (iv) tolerance to overcharge and overdischarge; (v) sealed for maintenance-free operation; (vi) acceptable cost projections on a cycle-life basis (viii) non-toxic materials. Another attribute, of particular importance for small EVs where space is at a premium, is the high energy and power density levels for this battery; the levels greatly exceed those of lead/acid. Indeed, the power density quoted for the Ovonics EV battery (546 W l^{-1}) is well in excess of the USABC mid-term target of 250 W l^{-1} (Table 3.4, Chapter 3) and eclipses the power density of sodium batteries (Chapter 8). Negative features of the Ni/MH system are the poor charge retention on open-circuit stand and the decline in charge-acceptance at elevated temperatures. Amelioration of the latter problem, however, has been reported by Nagano et al. [67] who have demonstrated a significant improvement in the charge efficiency at high temperatures through the addition of a calcium compound to the nickel electrode.

6.7. Iron / Air Batteries

The iron/air battery is an attractive concept because its theoretical specific energy is nearly three times that of nickel/iron and more than four times that of lead/acid (see Table 3.3, Chapter 3). The system consists of an iron negative electrode of the type used in nickel/iron cells and a positive operating on oxygen present in air. The electrode and cell reactions of the iron/air system are presented in Table 6.1 and in Table 3.3, Chapter 3, respectively. The open-circuit voltage is 1.28 V, but the operating value is 0.7–0.8 V due to overpotential losses at the air electrode.

The air electrode in metal/air batteries is similar to that employed in alkaline fuel cells. (Indeed, metal/air systems are sometimes referred to as 'fuel cells' or 'fuel batteries'.) The electrode consists of an inert, porous, electronically-conducting matrix that contains a small quantity of catalyst to facilitate the discharge process of oxygen reduction. Air is fed into the matrix on the side opposite to that in contact with the electrolyte. Oxygen reduction takes place predominantly in the aqueous phase close to the meniscus which is formed at the gas/solution interface; the mass-transport rate of oxygen is greatest in this region. The boundary between gas and solution is maintained within the pores of the matrix by making the gas-side hydrophobic and the solution-side hydrophilic. The ideal site for the catalyst is the junction between the hydrophobic and hydrophilic layers.

The problems encountered with the air electrode are the same as those in alkaline fuel cells (viz., requirement for CO_2 removal, blanketing of the electroactive catalyst by nitrogen, overpotential losses, water make-up, compressor energy loss). The use of pure oxygen, rather than air, gives significantly higher current densities and avoids the problem of CO_2 poisoning of the electrolyte. Unfortunately, however, this defeats the prime advantage of the metal/air battery, namely, that air is freely available and does not have to be carried around.

The charging process of oxygen evolution can lead to degradation of the catalyst and corrosion of the substrate. To overcome this problem, separate positive electrodes can be used for charge and discharge. Such an arrangement is ruled out, however, by the increased volume and complexity, the additional circuitry, and the higher cost compared with bifunctional air electrodes. Since oxygen reduction can occur anywhere within the flooded portion of the electrode, it is possible to design electrodes in which the charge and discharge processes take place in different regions. This approach [68] is obviously more practical than one which employs an auxiliary electrode.

The substrate used for the air electrode is a porous matrix of nickel or carbon and the catalyst is usually silver or platinum. The cost of these catalysts is significant, e.g., silver constitutes one-third of the estimated production price of air electrodes which use this metal [69]. Therefore, the search for less expensive catalyst materials has attracted considerable attention, not only for the iron/air system, but for all metal/air batteries and fuel cells.

The iron/air battery suffers from those disadvantages associated with the iron electrode discussed in Section 6.2, namely, rapid self-discharge, poor energy efficiency, and inferior low-temperature performance. The energy efficiency is further diminished by overpotential losses at the air electrode during both charge and discharge operations, so that the overall value is less than 50%. On the other hand, the iron/air system has an attractive specific energy (even though this declines appreciably with increase in power demand), and the iron electrode has proven long-life properties. These advantages encouraged several companies to develop the battery to the demonstration stage, namely, Siemens in Germany [55], Matsushita in Japan [69], the Swedish National Development Company (SNDC) [70,71], and Westinghouse in the USA [72].

The SNDC battery was based on a monopolar design which comprised sintered iron electrodes and silver-catalyzed-nickel air electrodes; a 30-kWh module is shown in Fig. 6.10. The auxiliaries required for its operation included: a blower to force air through the positive electrodes; a scrubber to remove carbon dioxide from the air stream (to avoid carbonate

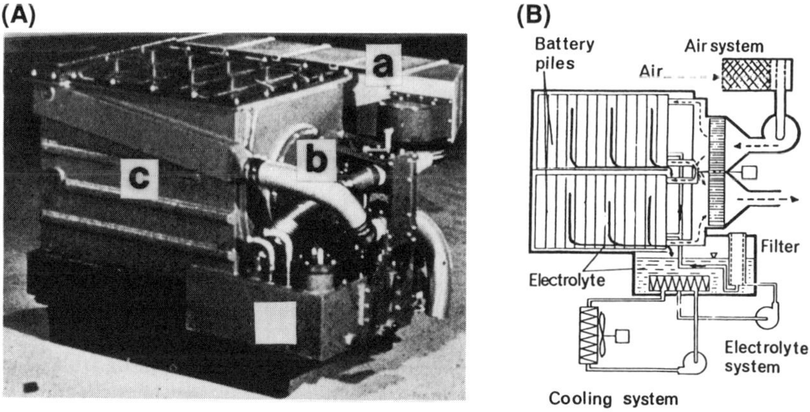

Fig. 6.10. SDNC iron/air battery. (A) 30-kWh system: (a) CO_2 scrubber; (b) moisture exchanger; (c) cell stack; (d) internal heat exchanger and electrolyte storage. (B) Schematic of battery and auxiliary system. (Taken from [71] with permission of Elsevier Science Limited, UK.)

formation by reaction with the electrolyte); a moisture exchanger to prevent excessive water loss from the electrolyte; an electrolyte cooling system to maintain the battery temperature at around 40 °C. The battery had a specific energy of 50 and 80 Wh kg^{-1} at the $C_2/2$ and $C_5/5$ rates, respectively. A service life in excess of 200 cycles was achieved [71] and the anticipated cost was US$ 160 per kWh [73]. The battery was tested both in a mine vehicle and in a delivery van. The SDNC programme was terminated in the mid-1970s because the market potential for the battery was uncertain. The project recommenced in 1984 with the aim of developing a cheaper system with an increased specific energy (target: 100 Wh kg^{-1}) [73]. The revised approach was to develop a bipolar system in which a conducting polymer substrate carried an iron electrode on one side and a bifunctional oxygen electrode on the other.

The Westinghouse iron/air design had each iron electrode coupled to two bifunctional air electrodes [9,72]. The negatives were prepared by pressing and sintering a mixture of sponge iron and urea between perforated nickel current collectors. The urea acted as a pore-former and gave an electrode with an apparent density which was about 22% that of solid iron. The air electrode consisted of sintered nickel-fibre grids which supported silver-catalyzed carbon. The latter was wet-proofed with fluorinated polymers. Inorganic additives (e.g., tungsten carbide with cobalt, iron tungstate, nickel sulfide) were included to reduce silver loss on charge–discharge cycling. The electrolyte was 25% KOH and was circulated during charge to facilitate thermal management of the battery. The performance of the air electrode was the life-limiting factor of the iron/air system; Westinghouse electrodes achieved 500 cycles when tested in isolation. Typical prototype systems delivered 100–120 Wh kg^{-1} at the $C_3/3$ rate [74], and 71 Wh kg^{-1} over the SFUDS driving schedule [75].

6.8. Aluminium / Air Batteries

The electrode and cell reactions of the aluminium/air system are presented in Table 6.1 and in Table 3.3, Chapter 3, respectively. The theoretical cell voltage is 2.7 V, but overpotential losses at both the aluminium and the air electrodes result in a lower operating value of 1.4–1.5 V.

Aluminium/air is the only battery system evaluated to date that offers the prospect of an EV with a range and a refuelling time that approach those of ICEVs. The range factor is derived from the high available specific energy — the theoretical value is more than sixteen times that of lead/acid and about three times that of zinc/air (see Table 3.3, Chapter 3). Rapid refuelling is achieved by recharging the battery mechanically. Indeed, aluminium cannot be electrodeposited from aqueous solution and hence it is not possible to recharge aluminium/air batteries in the same manner as that used for conventional secondary systems.

In the USA, DOE has supported the development of aluminium/air systems since the late 1970s. This work has been co-ordinated by the Lawrence Livermore National Laboratory (LLNL) and has involved contributions from Alcoa, Alcan, Case Western University, Diamond Shamrock, Eltech, Lockheed, Reynolds Aluminium, and the University of Akron (Ohio). Other research has been carried out by both Hoppecke and the University of Bonn in Germany, the Norwegian Defence Research Establishment, and the University of Belgrade in Serbia.

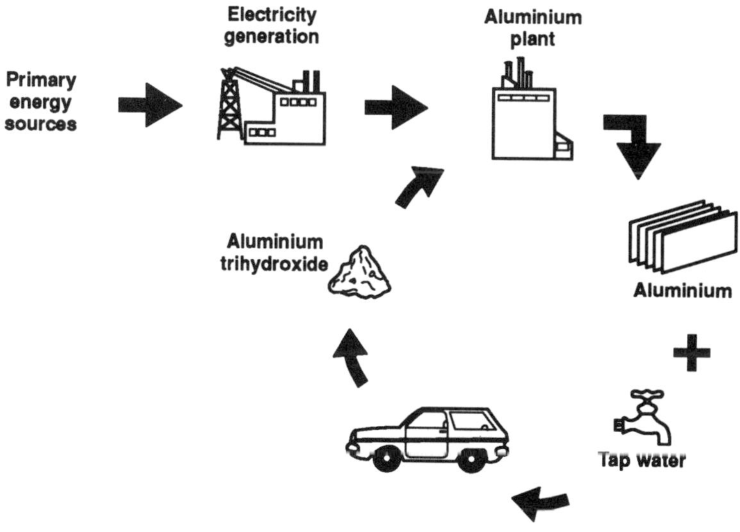

Fig. 6.11. Fuel cycle of aluminium/air vehicle as proposed by LLNL. (Taken from [76] with permission of the University of California, Lawrence Livermore National Laboratory, and the US Department of Energy.)

The fuel cycle proposed by LLNL for the aluminium/air system is illustrated in Fig. 6.11 [76]. As with all EV batteries, the primary energy source is first converted to electricity in a power station. In the aluminium/air concept, however, the electricity is not fed directly to the battery but to a smelter for the production of aluminium sheet. When placed in the battery, the aluminium reacts both with oxygen in the air and with water in the electrolyte to provide power for the EV. The product of the discharge reaction is aluminium hydroxide which is precipitated in the form of hydrargillite, $Al_2O_3{\cdot}3H_2O$. The latter material is extracted from the battery and returned to the smelter to regenerate aluminium metal. In effect, the 'charge cycle' of the aluminium/air battery is the production of metal by the established aluminium industry. Note, since the battery cannot be recharged electrically, it is not possible for the system, when operating EVs, to utilize directly the energy derived from regenerative braking. On the other hand, auxiliary secondary batteries are required for start-up, lighting, etc., and these could make use of regenerative charging.

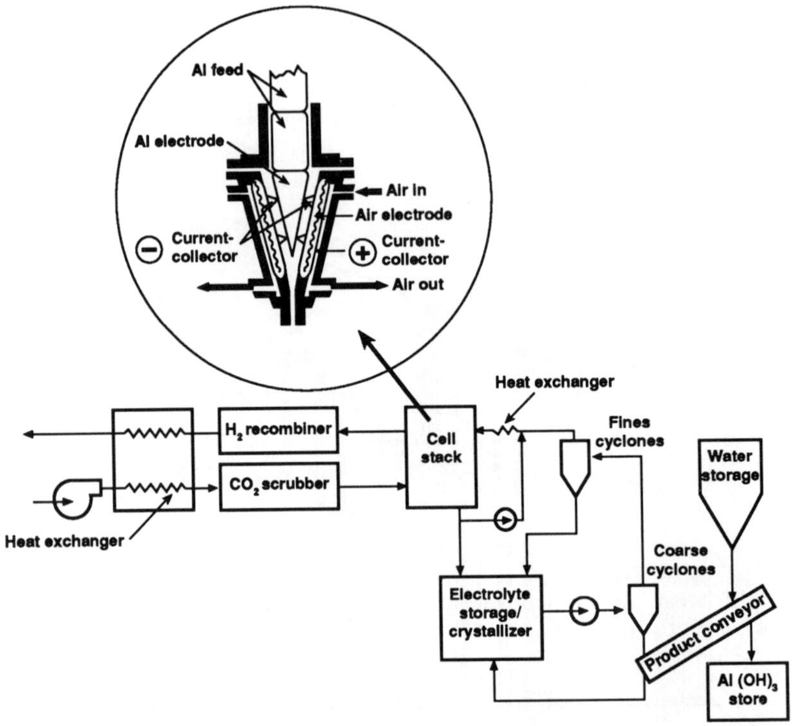

Fig. 6.12. Schematic flow diagram of aluminium/air cell. (Reprinted with permission from SAE Paper No. 859052 ©1985, Society of Automotive Engineers, Inc. [77].)

A plan of the various components of the aluminium/air system as envisioned by LLNL [77] is given in Fig. 6.12. The heart of the system is the cell stack. Each cell is triangular in cross-section so that aluminium plates fed into the battery adopt a self-perpetuating wedge-shape and are completely consumed. The cell is shown in more detail in Fig. 6.13. In order to allow the insertion of fresh aluminium plates without having to make complex electrical connections, the plates are put in contact with the 'wet-side' of the current-collectors. Air electrodes are sited on either side of each aluminium plate. The electrolyte is 4 M NaOH and is circulated through the battery. The operating temperature is maintained in the range 60 to 80 °C. Another developer of the system (Alupower, Canada) has employed 8 M KOH electrolyte [78].

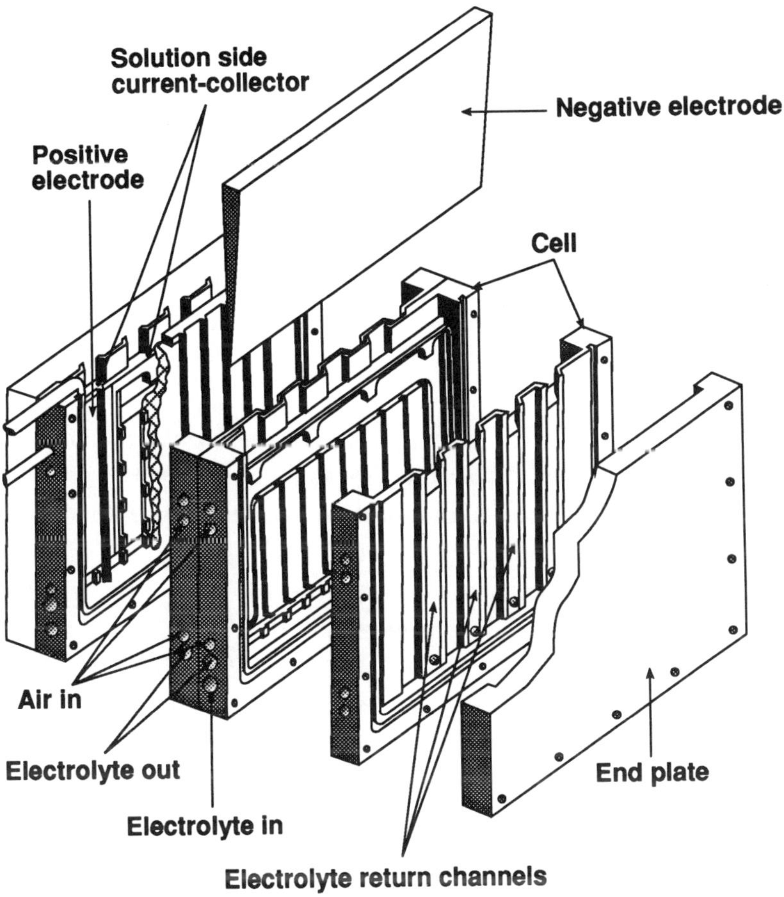

Fig. 6.13. Aluminium/air wedge cell. (Reprinted with permission from SAE Paper No. 859052 ©1985, Society of Automotive Engineers, Inc. [77].)

Aluminium reacts with strong alkaline solutions to generate hydrogen:

$$2Al + 6H_2O \rightarrow 2Al(OH)_3 + 3H_2 \quad (6.3)$$

This corrosion reaction results in self-discharge and, hence, a loss in stored energy. In the LLNL concept (Fig. 6.12), evolved hydrogen is reacted with oxygen in a 'hydrogen combiner' and the water product is returned to the cell. There was a major research effort into devising means for decreasing the rate of the parasitic corrosion reaction. Success was achieved by incorporating very small amounts of gallium, indium and/or thallium in the aluminium electrodes, and by including sodium stannate in the electrolyte. Eltech reported [79] a more than ten-fold decrease in corrosion current and a concomitant increase in open-circuit cell voltage of ~0.1 V with proprietary alloy systems. Parasitic corrosion, however, is significant even with the best systems so far identified and, to avoid excessive energy losses, the cell stack must be drained of electrolyte when the battery is standing on open circuit.

The air enters the cell stack after preheating and humidification in a heat exchanger and removal of carbon dioxide in a scrubber (Fig. 6.12). Since the air electrodes in the aluminium/air battery are unifunctional (recharge is conducted mechanically), they encounter less severe operating conditions than those experienced by bifunctional electrodes in other metal/air cells. This allows a wider choice of materials for the catalyst and, consequently, much research has been directed towards identifying inexpensive materials to replace noble metals. Heat-treated transition-metal macrocycles and related compounds have been found [77,80] to display high activity for oxygen reduction (comparable with that exhibited by platinum), as well as relatively long life under cycling conditions.

The electrolyte is circulated through the cells to achieve thermal management of the battery. Removal of heat is particularly important with the aluminium/air battery because about half the stored energy would be converted to heat during typical EV driving. Circulation of the electrolyte also extracts the reaction product as soluble sodium aluminate, $NaAl(OH)_4$. A fraction of the electrolyte is transferred to the crystallizer (Fig. 6.12) where, upon cooling, hydrargillite is precipitated. This process is facilitated by the addition of seed crystals. Large crystals of the precipitated hydrargillite are removed in the 'coarse' cyclone, washed with incoming water, and then transferred to a storage vessel. Small crystals are separated in the 'fines' cyclones, and returned to the crystallizer.

Each of the unit processes in the LLNL aluminium/air battery is known to be technically feasible and has been demonstrated on an appropriate scale. The various complex functions have not, however, been run simultaneously in the required continuous and automatic manner. Since the practicality of a total integrated system has still to be established, reported operational parameters are based on a conceptual battery. For typical EV applications, LLNL considered [76] that make-up water would be added, and hydrargillite product removed, after every 400 km of service. Fresh aluminium electrodes would be inserted every 1600 km. Taking this to be the driving range, rather than that between electrolyte replacements, the specific energy would be 268 Wh kg^{-1} [81] provided the optimum performance of experimental cells using electrodes of 25-cm^2 area is realized in the total system. A value of 325 Wh kg^{-1} was projected for a battery that incorporated anticipated future improvements [76]. The specific power from these two battery designs was 158 and 160 W kg^{-1}, respectively. It should be noted that much lower values for the specific energy (viz., 173 Wh kg^{-1} for cruising at 90 km h^{-1}) and peak specific power (viz., 102 W kg^{-1}) were derived by Will [69] on the basis of the initial, non-degraded performance reported by LLNL for 1000-cm^2 cells and the estimated total weight of the battery components and auxiliaries.

Because of the restricted power output of the aluminium/air battery, the idea has been advanced of a battery–battery hybrid EV in which a lead/acid battery or a nickel/cadmium battery provides peak power output, while an aluminium/air battery acts as a range extender [82]. Alupower in Canada built such an all-electric hybrid mini-van based on a 7.8-kW (gross) aluminium/air fuel battery (consisting of 56 series-connected cells) and a lead/acid battery to boost the power when required. The aluminium/air battery gave a steady output which was used to recharge the lead/acid battery when not required for propulsion. Refuelling was necessary after 75 kWh of discharge (approximately 10 h) and required the addition of 35 kg of aluminium and 180 l of electrolyte. This operation took about 10 min to be completed [78].

Among the candidate batteries for EVs, the aluminium/air system has the lowest energy efficiency. In this respect, it should be noted that only about half of the electricity input into the smelter is converted to aluminium metal (the charge cycle), and only about half of the energy stored in the aluminium is recovered as electricity during discharge. This poor efficiency is an important factor when comparing the use of energy in aluminium/air-powered EVs with that achieved in other transport options. With coal as the primary fuel, LLNL calculations [72] have placed the aluminium/air option ahead of ICEVs that are powered by synthetic fuels. By contrast, Will [69] presented energy data which reverse this order for state-of-the-art technology and emphasized that the LLNL conclusion is based on optimistic scenarios.

6.9. Zinc / Air Batteries

The theoretical specific energy of zinc/air is greater than that of iron/air and is more than five times that of lead/acid. The electrode and cell reactions are given in Table 6.1 and in Table 3.3, Chapter 3, respectively. The open-circuit cell voltage is 1.6 V, but the operating value is in the region of 1.1 to 1.3 V due to overpotential losses at the air electrode. During the 1960s and 1970s, there was considerable activity directed towards the development of the zinc/air system for EV applications. The principal investigators were: Alsthom, Compagnie Générale d'Electricité (CGE) and Citroën in France; Sony and Yuasa in Japan; General Motors and Gulf General Atomic in the USA.

Some development of *electrically rechargeable* zinc/air cells has been pursued in the 1990s. The leaders in this area included Zinc Air Power Corporation (previously, Dreisbach Electromotive, Inc. (DEMI)) and Westinghouse. As has been noted above (Section 6.4), however, zinc electrodes suffer from shape change and dendritic growth during charge–discharge cycling in alkaline media. Furthermore, the life of air electrodes is reduced considerably when the electrodes are operated in a bifunctional mode (see Section 6.7). In order to avoid these life-limiting problems in zinc/air batteries, other developers have chosen to regenerate the zinc

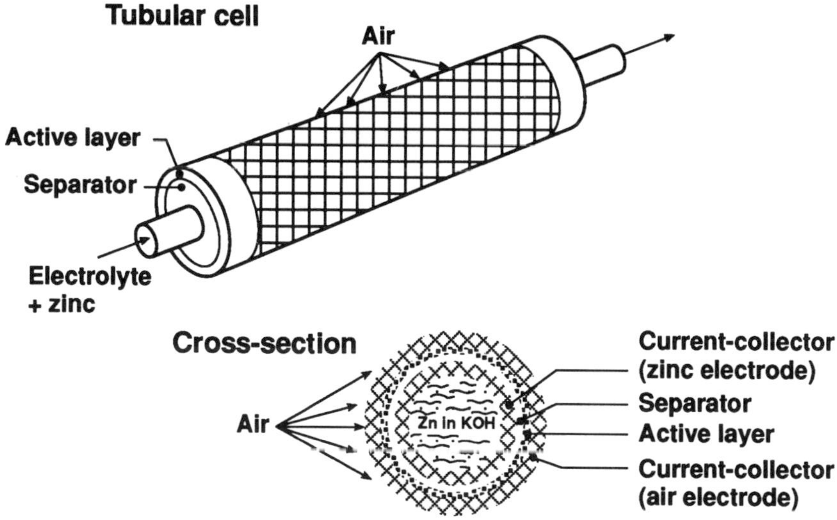

Fig. 6.14. Zinc/air battery developed by Compagnie Générale d'Electricité: schematic view of a tubular cell and cross-section. (Taken from [83] with permission of Elsevier Science Limited, UK.)

metal external to the battery. As with aluminium/air, the cells are recharged by adding fresh quantities of both the metal and the electrolyte. Such *'mechanical recharging'* is attractive for metal/air batteries because the positive material — atmospheric oxygen — is available in a continuous supply and it is not necessary to develop a bi-functional oxygen electrode.

An interesting approach to designing zinc/air batteries for EV applications was that pursued by CGE in France [83]. In this concept, a slurry of zinc powder and electrolyte was pumped through tubular cells of the type shown in Fig. 6.14. During discharge, the zinc was converted to both zincate and zinc hydroxide at a tubular current-collector made from mild-steel screen. The zinc hydroxide was stabilized in a colloidal form by the addition of silicate to the electrolyte. A thin asbestos separator was located between the inner negative and the outer (concentric) air electrode. The latter consisted of a layer of PTFE-bonded active carbon on a mild-steel screen; no metal catalyst was included. The air entered through an outer layer of porous PTFE which, because of its non-wetting properties, retained the electrolyte within the cells. The battery was operated at 50–55 °C. The spent slurry was regenerated in a separate electrolysis unit. Such units were designed to be either carried on board the vehicle or sited at a service station. Cells used for recharging were also of the tubular type; the zinc was deposited as fine dendrites on magnesium tips which were located along the tube length. By reversing the flow of electrolyte, the fine zinc deposit was removed to form a fresh slurry. The CGE battery achieved a specific energy of 110 Wh kg^{-1} and operated successfully for 1800 h, i.e., equivalent to 600 cycles under typical EV operation. Only 40% of the electrical energy required to generate the zinc was recovered in battery operation. This low energy efficiency is due to overpotential losses not only at the air electrode in the battery, but also at both the oxygen and the zinc electrodes in the recharging cells.

Calculations [84] indicated that the overall resource-to-wheels energy conversion efficiency of the CGE zinc/air system was only 9%, which is less than that of ICEVs. For example, an ICEV burning oil derived from coal has an overall efficiency of 11%. Therefore, this particular version of the zinc/air battery had serious disadvantages in terms of resource utilization and environmental impact.

Greater consideration was given to the aluminium/air system than to other metal/air counterparts in the 1980s because of the higher theoretical specific energy offered by that system. Later, the growing demand for reductions in greenhouse-gas emissions placed greater emphasis on achieving good efficiency in the conversion–storage of energy. Given that the overall fuel cycle of aluminium/air has a poor energy efficiency, attention was directed again towards zinc/air.

Electric Fuel Limited (EFL) is a leading developer of zinc/air batteries for EVs and was incorporated in 1990 to advance and market technology effected by Luz Industries (Israel) Limited. The EFL system is recharged mechanically and combines a high specific energy (200 Wh kg^{-1}), a modest power density (100 W kg^{-1} at 80% DoD), and a fast refuelling capability (a few minutes) [85–87]. The system is based on a discharge-only zinc/air battery pack which is made up of modules of 44 or 66 cells connected in

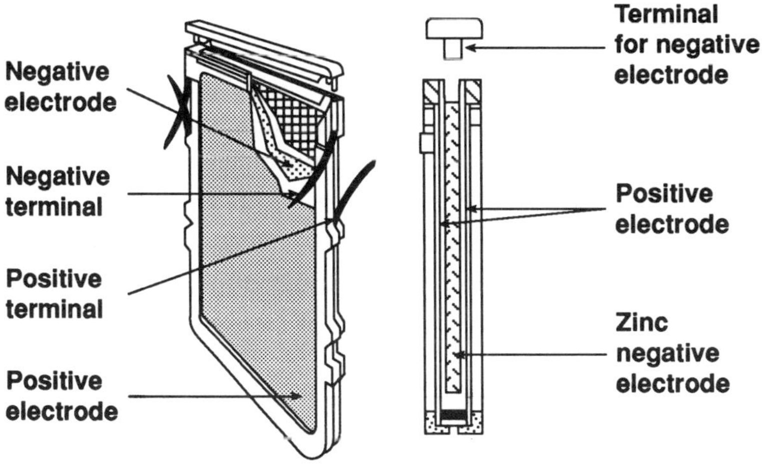

Fig. 6.15. Zinc/air cell and cassette system. (Taken from [88] with permission of *Batteries International*.)

series. The modules can be arranged in any combination of series and parallel connections according to the requirements of the vehicle, motor and controller. Each cell contains a central static bed of zinc particles slurried in potassium hydroxide solution. The bed is compacted on to a current-collection frame and inserted in a separator envelope, flanked on both sides with air electrodes (Fig. 6.15). The current-collector frame serves as the mechanical means of supporting, inserting and replacing the zinc electrode cassette. The capacity of each cell is 246 Ah when discharge is taken to 80% zinc utilization. A 66-cell module weighs 89 kg. A 150-kWh battery which has been built for the 4.6-t (GVW) Mercedes-Benz 410 electric postal van consists of two parallel strings of four modules in series (total 528 cells). The open-circuit voltage is 375 V and the operating voltage is 306 V. The battery has a specific energy of 215 Wh kg^{-1}, an energy density of 252 Wh l^{-1}, a constant specific power (C_5/5 rate) of 40 W kg^{-1}, a peak specific power at 80% DoD of 98 W kg^{-1}, a constant power density (C_5/5 rate) of 46 W l^{-1}, and a peak power density at 80% DoD of 115 W l^{-1}. With this power source, the vehicle has a range of ~300 km between mechanical recharges, although with only a modest power performance.

The on-board battery is refuelled by exchanging spent zinc electrode cassettes (i.e., the current-collector frame and separator envelope) with recharged units. In order to do this, it is necessary first to remove the modules from the vehicle, by sliding them out on rollers. Refuelling is accomplished by an automated machine that allows a turn-around time at the service station similar to that required by ICEVs. A development refuelling machine is shown in Fig. 6.16. In the concept under consideration for commercial implementation, the depleted cassettes are recharged electrochemically at centralized facilities which serve networks of refuelling stations.

The recharging process involves the following sequence: disassembly of the cassettes; removal of the zinc oxide and its dissolution in potassium hydroxide; electrowinning to produce particulate zinc; reassembly of the cassettes. The three integrally-linked elements of the proposed fuel system are shown in Fig. 6.17. The overall electrical efficiency of the discharge–charge cycle is 60–65%, and the projected cost (US$ 100 per kWh) is well below that of lead/acid.

The EFL zinc/air technology is considered by Deutsche Post AG to be the most promising battery for zero-emission, fleet-vehicle operation [89]. Assessment of the battery is being carried out in vehicles, both by Deutsche Post AG and by Edison Termoelecttrica SpA in Italy. Trials have also been carried out with a Mercedes-Benz 410 electric van, and has included both on-road driving in the Munich area and tests on a dynamometer. In 1993, the range achieved was 208 km for a city route and 300 km for suburban driving. Those distances compared very favourably with the corresponding values (40 and 50 km, respectively) that were accomplished by the same vehicle when powered by a lead/acid battery which was 1.4 times heavier.

Other developers, particularly in the USA, are also targeting the mechanically-rechargeable zinc/air battery [90–95]. For example, Zinc Air Transportation (ZAT), a US and German Consortium, is manufacturing traction batteries and evaluating them in road vehicles. A world distance

Fig. 6.16. Refuelling machine developed by EFL. This removes and replaces all the zinc cassettes from a 44-cell module at once. (Taken from [88] with permission of *Batteries International* and Electric Fuel Limited, Israel.)

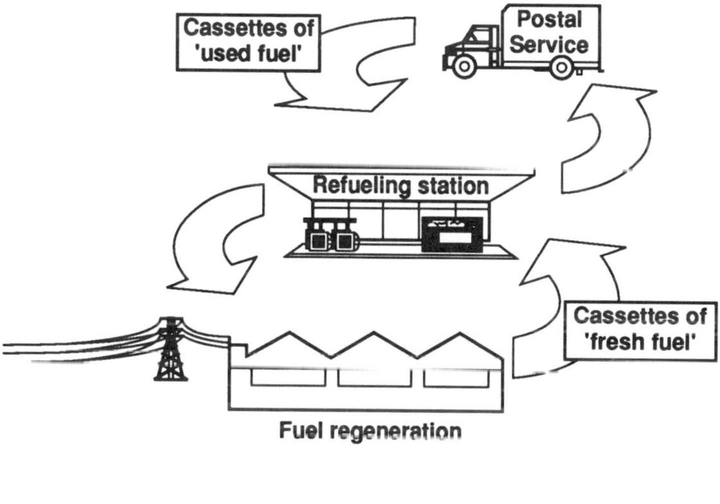

Fig. 6.17. Schematic representation of the operation of the EFL system for EVs powered by zinc/air batteries. (Taken from [87] with permission of Electric Fuel Limited, Israel.)

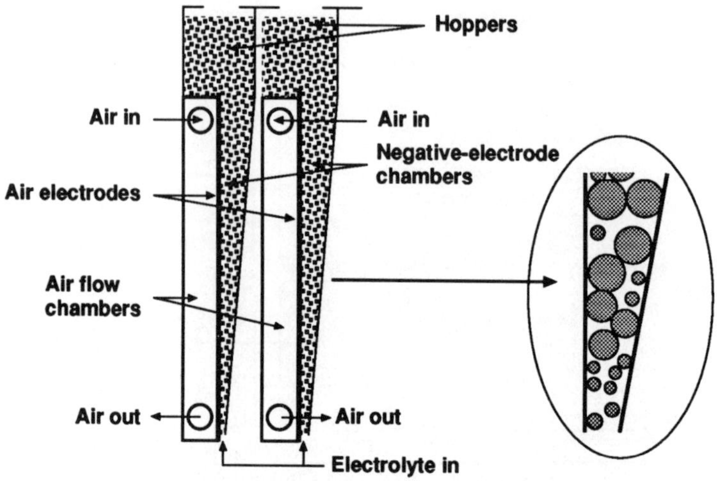

Fig. 6.18. LLNL zinc/air cells (bipolar configuration) in which zinc particles are gravity fed. (Based on diagram given in [91] with permission of the University of California, Lawrence Livermore National Laboratory, and the US Department of Energy.)

record of 765 km between recharges was set for a minivan in February 1997 and was later extended to 1669 km. A perceived short-coming of the EFL concept is the need to establish a new infrastructure to recharge mechanically and re-build the batteries. A more attractive approach would be to refuel batteries hydraulically with a zinc slurry. Both the Lawrence Livermore National Laboratory (LLNL) [90–93] and the Lawrence Berkeley Laboratories (LBL) [94–98] are developing systems which are designed to meet this goal.

In the LLNL battery, zinc particles (size: 1mm) are stored in hoppers, from which they fall continuously through a narrow (<3 mm) opening at the top of each individual cell, see Fig. 6.18. Since the cell gap is only a few times larger than the particles, close packing does not occur, even under strong vibration. Rather, particle bridging causes voids to develop and a particle packing-density of about 40% is established. Electrolyte is pumped upward through the cell and the hopper to remove heat and reaction products. The wedge-shaped cell allows dissolving zinc particles

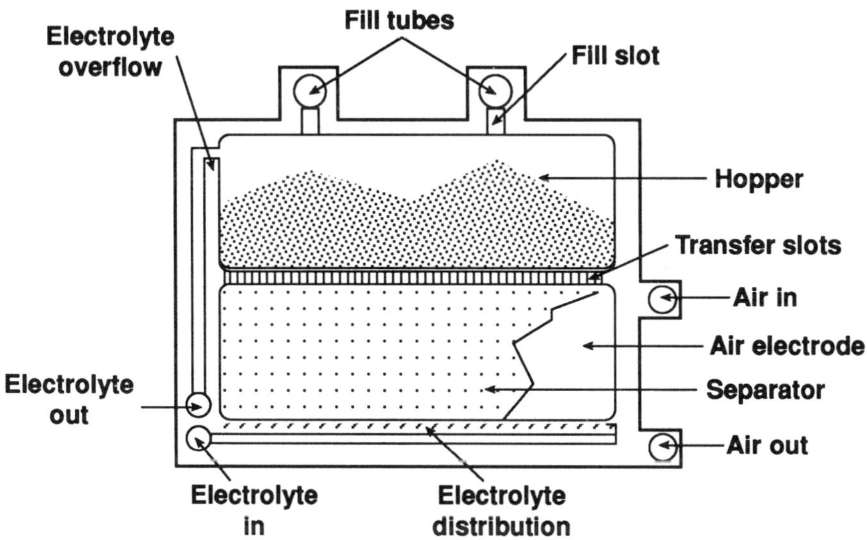

Fig. 6.19. LLNL bipolar zinc/air cell, schematic view at right angles to that shown in Fig. 6.18. (Taken from [91] with permission of the University of California, Lawrence Livermore National Laboratory, and the US Department of Energy.)

to fall into the increasingly narrow gap. Air is pumped into the side of the cell, whence it diffuses into the air electrode. Refuelling is accomplished by draining spent electrolyte and flowing fresh electrolyte sufficiently fast to drive the new zinc particles along the bottom of a horizontal fill-tube which connects the cell hoppers. A view of the cell from the air-electrode side is shown in Fig. 6.19 (note, fill-tubes are perpendicular to the page).

Two methods for recovering metallic zinc from the spent electrolyte of the LLNL cell have been proposed [92]. The first involves a combined electrochemical and mechanical process in which zinc is electrodeposited on to a non-adhering metal substrate under conditions which produce 'mossy zinc' that has 15–20% of full density. The zinc deposit is then pressed into cylindrical pellets (length: 1 mm) of 60% density by using a pin-and-die press. In the second method, zinc spheres are produced by electrodeposition in a fluidized bed which is seeded with particles (size: 0.1–0.2 mm). The seeds gradually accumulate layers of zinc as they flow in

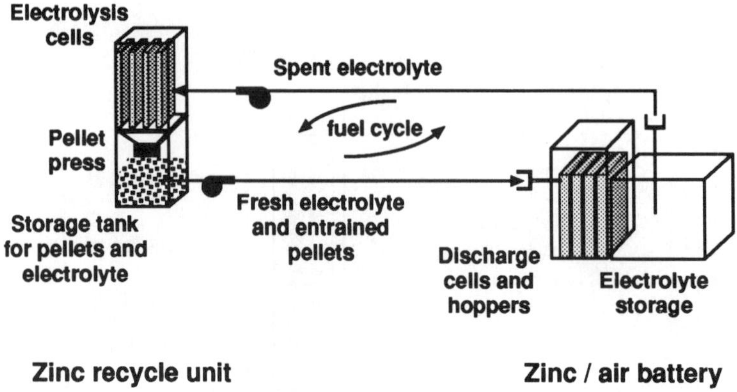

Fig. 6.20. Schematic representation of the fuel cycle of the LLNL zinc/air battery system. (Taken from [91] with permission of the University of California, Lawrence Livermore National Laboratory, and the US Department of Energy.)

contact with a negative electrode, and grow to a mature size of 1 mm before being harvested. Fresh electrolyte which contains entrained pellets or spheres is used to recharge the cell; it is considered that electrolyte replacement can be achieved in less than 10 min. The complete zinc/air fuel cycle is shown schematically in Fig. 6.20. The zinc recycle unit uses the existing electric power distribution infrastructure. The battery has a specific energy and an energy density of 150 Wh kg^{-1} and 225 Wh l^{-1}, respectively [92]. A conceptual vehicle powered by the system is predicted to have a range of 400 km over a simulated urban driving cycle.

The LLNL zinc/air battery has been successfully trailed [93] in a hybrid combination with lead/acid batteries in the 'Downtown Waterfront Electric Shuttle', an electric bus on loan from the Santa Barbara Metropolitan Transit District (see Fig. 1.14, Chapter 1). It was concluded that, although the zinc/air battery could power large EVs by itself, combining it with another power source (lead/acid battery, ultracapacitor, or flywheel) is preferred since such a hybrid power plant provides better acceleration and greater flexibility in selecting routes with varying grades. It has been further reported [93] that Sydney Electricity (now Energy Australia), a

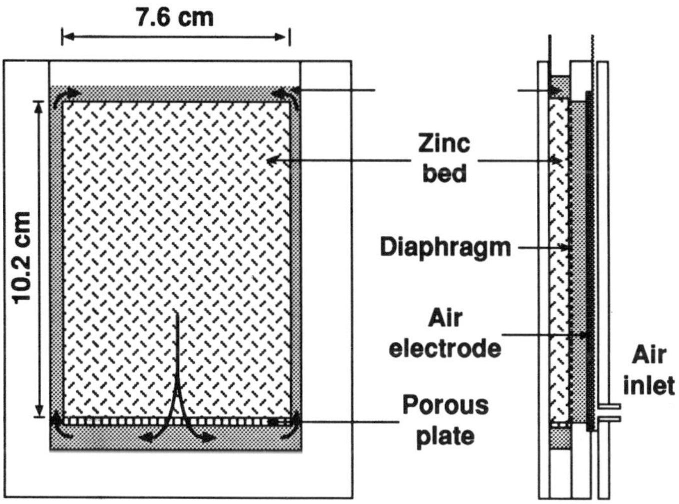

Fig. 6.21. Schematic of LBL zinc/air cell. (Taken from [97] with permission of Chapman & Hall, UK.)

member of the organizing committee for the 2000 Olympic Games in Sydney, Australia, plans to order ZEV people movers (buses, vans, and other vehicles) and is considering LLNL zinc/air technology as a power source.

The cell in the LBL version of a hydraulically-recharged zinc/air battery contains a stationary, 5–6 mm thick, packed bed of zinc particles (size: ~0.5 mm), Fig. 6.21 [97]. On each side of the bed is a channel (3 × 6 mm) which serves as a negative electrode reservoir and a conduit for electrolyte. As a result of the anodic dissolution of zinc on discharge, the electrolyte between the particles becomes denser than that in the conduits. The natural convection which results from this concentration gradient provides the electrolyte flow necessary for mass transport to and from the particle surfaces. Projections of the performance of laboratory cells at ambient temperature to that anticipated from a 55-kW battery have yielded a specific energy of 202 Wh kg^{-1} and a specific power of 88 W kg^{-1} over the SFUDS driving cycle. When the cell is operated at 55 °C, these values increase to 290 Wh kg^{-1} and 100 W kg^{-1}, respectively.

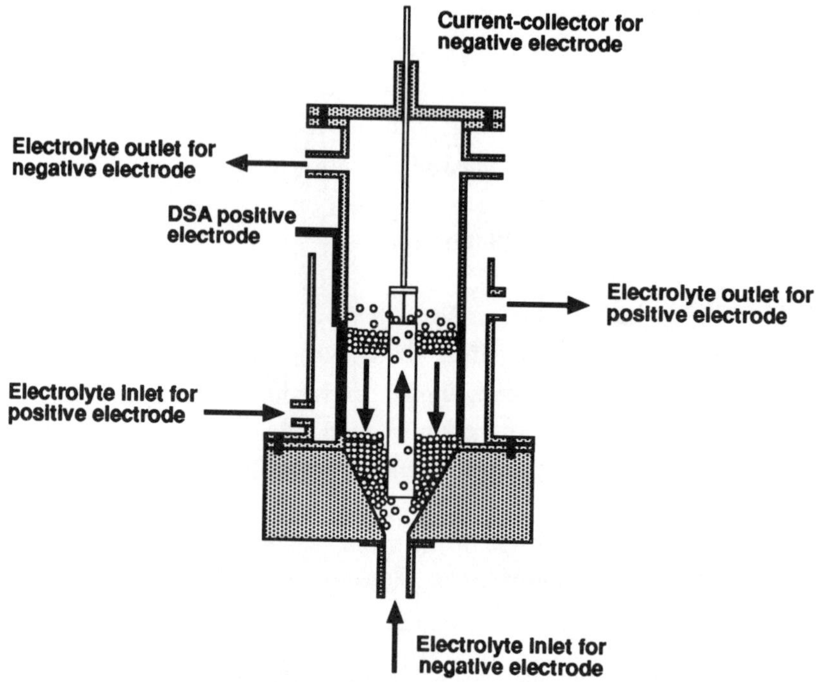

Fig. 6.22. 'Spouted-bed' electrowinning cell used to produce particulate zinc from spent electrolyte that could act as a 'charger' for mechanically-rechargeable zinc/air batteries. (Taken from [99] with permission of the Minerals, Metals & Materials Society, USA.)

Fluidized-bed electrodeposition has been shown by LBL [96] to be an efficient way of regenerating zinc particles from alkaline electrolyte; this process constitutes the 'charger' for the mechanically-rechargeable zinc/air system. A 'round-trip' electrical efficiency of ~50% is obtained from a system which comprises the LBL cell for power generation and fluidized-bed electrodeposition for recharging. In a further development, Evans and coworkers [98–100] have devised a 'spouted bed' cell in which zinc is electro-deposited on to seed particles. In this system, a jet of electrolyte is forced rapidly up the centre of the bed and carries zinc particles with it (Fig. 6.22). On reaching the top of the bed, the particles fall out of the electrolyte stream and then descend slowly as a packed bed until they are recaptured by the electrolyte spout. The motion of the particles is sufficient to prevent their adherence during metal deposition. The energy efficiency of the spouted-bed cell is greater than that of the fluidized-bed equivalent because the former system requires much less pumping power. The power requirements for 'charging' with the spouted bed should result in a round-trip electrical efficiency for the zinc/air system of over 60%.

6.10. Silver Oxide Batteries

The nickel oxide electrode in alkaline batteries may be replaced not only by an air electrode, but also by a silver oxide electrode. Silver/zinc and silver/cadmium batteries are well known and commercially available, principally for applications in the defence field. They are unsuited for commercial use in EVs on account of their high cost and short cycle life. Silver/zinc has a particularly good specific energy (90–120 Wh kg^{-1}), however, and it is noteworthy that in the 1996 World Solar Challenge for solar-powered cars (see Section 1.4, Chapter 1) six out of the first seven competitors to complete the course were equipped with packs of silver/zinc batteries.

6.11. Future Prospects

In France, there has been a firm national commitment to nickel/cadmium technology and a battery factory has been commissioned. To what extent the use of nickel/cadmium will spread elsewhere in the world will depend on the level of success achieved by the French, both technically and also commercially in implementing a viable battery-leasing plan. Looking further ahead, the nickel/zinc battery is more attractive on the grounds of its higher specific energy and lower cost, as well as the better environmental acceptability and increased availability of zinc compared with cadmium. Should the fundamental problem of short cycle life of the zinc electrode in alkaline media be overcome — and some progress has been made in this direction — nickel/cadmium battery manufacturers would be able to draw upon their technology and production experience and rapidly start the manufacture of nickel/zinc traction batteries.

Nickel/metal hydride batteries are receiving increasing attention and multi-kWh units are now being field-tested. Certainly, cost is a major problem to be addressed. It should be noted that nickel/metal hydride batteries can provide very high specific power and, therefore, are well suited to hybrid applications (see Section 10.4, Chapter 10).

The future for metal/air batteries is tied closely to the future of the air electrode. Interest in this electrode continues through its application in alkaline fuel cells. The prospects for an electrically rechargeable metal/air battery are not good, but the mechanically rechargeable zinc/air system may be a possibility. Certainly, the zinc/air system which is being developed in Israel is considered to be a serious industrial project. In addition, an Australian–USA consortium, Power Air Tech Pty Limited, has been formed (September 1997) to consolidate and develop further the LLNL and LBL technologies. In the long term, the aluminium/air battery is highly attractive as it would give an EV almost unlimited range. There are, however, both power and electrical efficiency limitations, as well as cell design problems, to be overcome before aluminium/air becomes a real option.

References

1. MYATT, J. *Secondary Battery Requirements for Space Use in the Late 1980's–1990's, ESTEC Contract No. 4404/80/NL/JS(SC). Final Report* (UK Atomic Energy Authority, AERE, Harwell, Oxfordshire, England, April 1982).

2. ZAGRODNIK, J.P., JONES, K.R., and BEAUCHAMP, R.L. *Proc. 34th International Power Sources Symposium,* 25–28 June 1990, Cherry Hill, New Jersey, USA (The Institute of Electrical and Electronics Engineers, Inc., New York, USA, 1990), pp. 243–245.

3. COATES, D., and MILLER, L. *Proc. 11th International Electric Vehicle Symposium (EVS-11),* Florence, Italy, 27–30 September 1992, **Vol. 2**, Paper 14.03.

4. PATIL, P., and ZEGERS, P. *J. Power Sources,* 1994, **49**, pp. 169–184.

5. Von STURM, F. *In* BOCKRIS, J.O'M., CONWAY, B.E., YEAGER, E., and WHITE, R.E. (Eds.) *Comprehensive Treatise of Electrochemistry, Vol. 3: Electrochemical Energy Conversion and Storage* (Plenum Press, New York, USA, 1981), pp. 407–419.

6. JUNGNER, W. *Swedish Patent 10, 177* (issued 1899).

7. EDISON, T.A. *British Patent 20, 960* (issued 1900).

8. FALK, S.U., and SALKIND, A.J. *Alkaline Storage Batteries* (John Wiley & Sons, New York, USA, 1969), p. 89.

9. ROBERTS, R. *Status of the DOE Battery and Electrochemical Technology Program IV* (Mitre Corp., McLean, USA, May 1983).

10. YAO, N.P. *J. Power Sources*, 1982, **7**, pp. 317–329.

11. *Electric and Hybrid Vehicles Program, 9th Annual Report to Congress for Fiscal Year 1985.* (U.S. Department of Energy, Washington, DC, USA, March 1986).

12. LABAT, J. *Prog. Batteries Sol. Cells*, 1987, **6**, pp. 236–238.

13. IWAKI, T., MITSUMATA, T., and OGAWA, H. *Prog. Batteries Sol. Cells*, 1987, **6**, pp. 225–257.

14. PATIL, P.G., WALSH, W.J., and MILLER, J.F. *Proc. 8th International Electric Vehicle Symposium (EVS-8),* Washington, DC, USA, 20–23 October 1986, pp. 48–52.

15. *The Future of Urban Transport: the Nickel-Iron Battery* (SAFT Department Accumulateurs, Romainville, France, 1987).

16. CORNU, J.-P. *NiCad94: Report on Conference, Geneva, Switzerland, September 1994* (International Cadmium Association, London, UK, January 1995), pp. 39–54.

17. MAYER, M.G. *Proc. Battery Council International 1985 Convention*, 5–8 May 1985, Toronto, Canada (Battery Council International, Chicago, USA, 1985), p. 24.

18. HASCHKA, F. *Batteries International*, 1994, **21**, pp. 80–81.

19. BUGNET, B., ROUGET, R., and DONIAT, D. *In* KEILY, T., and BAXTER, B.W. (Eds.) *Power Sources 12, Research and Development in Non-Mechanical Power Sources* (International Power Sources Symposium Committee, Leatherhead, England, 1988), pp. 157–163.

20. MORROW, H. *Electric & Hybrid Vehicle Technology '96* (UK & International Press, Dorking, UK, 1996), pp. 87–91.

21. MORROW, H. *Electric & Hybrid Vehicle Technology '95* (UK & International Press, Dorking, UK, 1995), pp. 91–94.

22. McBREEN, J. *In* McNICOL, B.D., and RAND, D.A.J. (Eds.) *Power Sources for Electric Vehicles* (Elsevier, Amsterdam, The Netherlands, 1984), pp. 541–572.

23. WALSH, W.J., and SYMONS, P.C. *Transfer of Battery Technology Developed by the US Department of Energy* (Argonne National Laboratory, Illinois, USA, April 1984).

24. McLARNON, F.R., and CAIRNS, E.J. *J. Electrochem. Soc.,* 1991, **138**, pp. 645–664.

25. DEUTSCHER, R.L., FLETCHER, S., FRAZER, E.J., PHANG, S., and WOODS, R. *Advanced Batteries for Traction; NERDDC Project 78/2612; Progress Report No. 3* (Australian Government Department of Resources and Energy, Canberra, Australia, June 1981).

26. BIEGLER, C., DEUTSCHER, R.L., FLETCHER, S., HUA, S., and WOODS, R. *J. Electrochem. Soc.*, 1983, **130**, pp. 2303–2309.

27. GAGNON, E.G. *Extended Abstracts, Proc. Vol. 85-2, Fall Meeting of the Electrochemical Society,* 13–18 October 1985, Las Vegas, Nevada, USA (The Electrochemical Society, Pennington, New Jersey, USA, 1985), p. 137.

28. WITHERSPOON, R.R. *US Patent No. 4,294,099* (issued 1981).

29. HIETBRINK, E.H., BOAK, R.W., CORBIN, R.L., JONES, R.A., and ATKINS, L.P. *Performance Characteristics of Nickel-Zinc Cells and Battery Packs. General Motors Research Publication: GMR-4218* (General Motors Corporation, Warren, Michigan, USA, October 1982).

30. BAKER, K. *Electric & Hybrid Vehicle Technology '95* (UK & International Press, Dorking, UK, 1995), pp. 38–40.

31. PYUN, Y.B., CHANG, H., YOO, Y.C., JUNG, B.H., and PARK, J.M. *Proc. 13th International Seminar on Primary and Secondary Battery Technology,* 4–7 March 1996, Boca Raton, Florida, USA.

32. REISNER, D.E., and EISENBERG, M. *Proc. 24th Intersoc. Energy Conv. Eng. Conf.,* 6–11 August 1989, Washington, DC, USA (The Institute of Electrical and Electronics Engineers, Inc., New York, USA, 1989), pp. 1677–1682.

33. BRONOEL, G., MILLOT, A., and TASSIN, N. *J. Power Sources,* 1991, **34**, pp. 243–255.

34. WAGNER, O.C. *High Cycle Life, High Energy Density Nickel/Zinc Batteries, Report No. 4. ERADCOM Report No. PSD-4C* (Electronics Research and Development Command, Fort Monmouth, USA, January 1981).

35. DEUTSCHER, R.L., FLETCHER, S., GALEA, J., HAMILTON, J.A., MARKOUTSAS, E., and WOODS, R. *Advanced Batteries for Traction; NERDDC Project 78/2612; Final Report* (Australian Government Department of Energy and Natural Resources, Canberra, Australia, January 1985).

36. *Annual Report for 1981 on Research, Development and Demonstration of Nickel-Zinc Batteries for Electric Vehicle Propulsion, Argonne National Laboratory Report ANL-OEPM, 81-82* (Exide Management and Technology Co., USA, March 1982).

37. BRONOEL, G., ROUGET, R., TASSIN, N., and MILLOT, A. *Extended Abstracts, Proc. Vol. 89-2, Fall Meeting of the Electrochemical Society,* 15–20 October 1989, Hollywood, Florida, USA (The Electrochemical Society, Pennington, New Jersey, USA, 1989), p. 1.

38. *Development of Aqueous Batteries for Electric Vehicles* (Argonne National Laboratory, Illinois, USA, April 1984).

39. McBREEN, J. *J. Power Sources,* 1994, **51**, pp. 37–44.

40. JINDRA, J. *J. Power Sources,* 1997, **66**, pp. 15–25.

41. DAGGET, J.S. *Batteries International,* 1995, **23,** pp. 64–65.

42. FLETCHER, S., GALEA, J., HAMILTON, J.A., TRAN, T., and WOODS, R. *J. Electrochem. Soc.*, 1986, **133**, pp. 1277–1281.

43. DZIECIUCH, M., GUPTA, N., and WROBLOWA, H.S. *Extended Abstracts, Proc. Vol. 86-2, Fall Meeting of the Electrochemical Society,* 19–24 October 1986, San Diego, California, USA (The Electrochemical Society, Pennington, New Jersey, USA, 1986), pp. 129–130.

44. KORDESCH, K.V., and WEISSENBACHER, M. *J. Power Sources,* 1994, **51,** 61–78.

45. BERNDT, D. *Maintenance-Free Batteries.* Power Sources Technology Series No. 1, BAGSHAW, N.E. (Ed.) (Research Studies Press Ltd., Taunton, UK, 1993), p. 247.

46. ISHIWA, K., AMIMOTO, K., and KOCHIWA, K. *Proc. 13th International Seminar on Primary and Secondary Battery Technology,* 4–7 March 1996, Boca Raton, Florida, USA.

47. OVSHINSKY, S.R., DHAR, S.K., VENKATESAN, S., FETCENKO, M.M., GIFFORD, P.R., and CORRIGAN, D.A., *Proc. 11th International Electric Vehicle Symposium (EVS-11),* Florence, Italy, 27–30 September 1992, **Vol. 2**, Paper 14.02.

48. CORRIGAN, D.A., VENKATESAN, S., GIFFORD, P.R., FETCENKO, M.A., DHAR, S.K., and OVSHINSKY, S.R. *Proc. 12th International Electric Vehicle Symposium (EVS-12)* Anaheim, California, USA, 5–7 December 1994, **Vol. 2**, pp. 208–217.

49. HARTKOPF, T., and BIRNBREIER, H. *Proc. 11th International Electric Vehicle Symposium (EVS-11),* Florence, Italy, 27–30 September 1992, **Vol. 2**, Paper 14.04.

50. CORNU, J.-P., MADERY, C., and LEONARDI, J. *Proc. 11th International Electric Vehicle Symposium (EVS-11),* Florence, Italy, 27–30 September 1992, **Vol. 2**, Paper 14.07.

51. CORNU, J.-P. *Proc. 12th International Electric Vehicle Symposium (EVS-12),* Anaheim, California, USA, 5–7 December 1994, **Vol. 2**, pp. 218–225.

52. MORISHITA, N., YOKOTA, A., KITAGAWA, K., IKOMA, M., TAGADA, K., YANAGIHARA, N., YAMASAKI, H., MATSUMOTO, I., and TOYOGUCHI, Y. *Proc. 11th International Electric Vehicle Symposium (EVS-11)*, Florence, Italy, 27–30 September 1992, **Vol. 2**, Papers 14.08 and 14.09.

53. IKOMA, M., HAMADA, S., MORISHITA, N., HOSHINA, Y., MATSUDA, H., OHTA, K., and KIMURA, T. *Proc. 12th International Electric Vehicle Symposium (EVS-12)*, Anaheim, California, USA, 5–7 December 1994, **Vol. 2**, pp. 547–556.

54. BRANDT, K., KÖHLER, U., NIGGEMANN, E., BÖTTCHER, F., and KRUGER, F.J. *Proc. 13th International Seminar on Primary and Secondary Battery Technology,* 4–7 March 1996, Boca Raton, Florida, USA.

55. SAKAI, T., ISHIKAWA, H., OGURO, K., IWAKURA, C., and YONEYAMA, H. *J. Electrochem. Soc.,* 1987, **134,** pp. 558–562.

56. BOONSTRA, A.H., and BERNARDS, T.M.N. *J. Less-Common Met.,* 1990, **161**, pp. 245–255.

57. SAKAI, T. *J. Less-Common Met.,* 1990, **161**, pp. 193–202.

58. SAKAI, T., MIYAMURA, H., KURIYAMA, N., KATO, A., OGURO, K., and ISHIKAWA, H. *J. Electrochem. Soc.,* 1990, **137,** pp. 795–799.

59. ADZIC, G.D., JOHNSON, J.R., MUKERJEE, S., McBREEN, J., and REILLY, J.J. *Extended Abstracts, Proc. Vol. 94-2, The Fall Meeting of the Electrochemical Society,* 9–14 October 1994, Miami Beach, Florida, USA (The Electrochemical Society, Pennington, New Jersey, USA, 1994), pp. 50–51.

60. RÜETSCHI, P., MELI, F., and DESILVESTRO, J. *J. Power Sources,* 1995, **57**, pp. 85–91.

61. FETCENKO, M.A., VENKATESAN, S., and OVSHINSKY, S.R. *Extended Abstracts, Proc. Vol. 91-2, The Fall Meeting of the Electrochemical Society,* 13–17 October 1991, Phoenix, Arizona, USA (The Electrochemical Society, Pennington, New Jersey, USA, 1991), pp. 166–167.

62. OVSHINSKY, S.R., FETCENKO, M.A., and ROSS, J. *Science,* 1993, **260**, pp. 176–181.

63. DHAR, S.K., OVSHINSKY, S.R., GIFFORD, P.R., CORRIGAN, D.A., FETCENKO, M.A., and VENKATESAN, S. *J. Power Sources,* 1997, **65**, pp. 1–7.

64. DHAR, S.K., OVSHINSKY, S.R., VENKATESAN, S., GIFFORD, P.R., and CORRIGAN, D.A. *Proc. 7th Asian Battery Conference,* 2–5 September 1997, Penang, Malaysia.

65. REISNER, D.E., and KLEIN, M. *Proc. 12th International Electric Vehicle Symposium (EVS-12),* Anaheim, California, USA, 5–7 December 1994, **Vol. 1,** pp. 340–349.

66. WARTHMANN, W., BENCZUR-URMOSSY, G., HASCHKA, F., and OHMS, D. *Proc. 13th International Seminar on Primary and Secondary Battery Technology,* 4–7 March 1996, Boca Raton, Florida, USA.

67. NAGANO, M., OYAMA, K., SATO, F., NOMURA, T., and YABUMOTO, T. *Electric & Hybrid Vehicle Technology '96* (UK & International Press, Dorking, UK, 1996), pp. 93–95.

68. CNOBLOCH, H., GROPPEL, D., KUHL, D., NIPPE, W., and SIEMEN, G. *In* COLLINS, D.H. (Ed.) *Power Sources 5, Research and Development in Non-Mechanical Electrical Power Sources* (Academic Press, London, UK, 1975), pp. 261–282.

69. WILL, F.G. *In* McNICOL, B.D., and RAND, D.A.J. (Eds.) *Power Sources for Electric Vehicles* (Elsevier, Amsterdam, The Netherlands, 1984), pp. 573–619.

70. LINDSTROM, O. *In* COLLINS, D.H. (Ed.) *Power Sources 5, Research and Development in Non-Mechanical Electrical Power Sources* (Academic Press, London, UK, 1975), pp. 283–301.

71. OJEFORS, L., and CARLSSON, L. *J. Power Sources*, 1978, **2**, pp. 287–296.

72. MOYES, H., BERK, I., ZUCKERBOD, D., and BUZZELLI, E. *Final Report of Westinghouse R & D Centre to Lawrence Berkeley Laboratory, Contract DE-ACO3-76SF00098*, USA, 1984.

73. LINDSTROM, O. *Extended Abstracts, Proc. Vol. 85-2, Fall Meeting of the Electrochemical Society,* 13–18 October 1985, Las Vegas, Nevada, USA (The Electrochemical Society, Pennington, New Jersey, USA, 1985), p. 25.

74. BUZZELLI, E.S., MOYES, H., ZUCKERBROD, D., and LIU, C.T. *Extended Abstracts, Proc. Vol. 85-2, Fall Meeting of the Electrochemical Society,* 13–18 October 1985, Las Vegas, Nevada, USA (The Electrochemical Society, Pennington, New Jersey, USA, 1985), pp. 26–27.

75. ZUCKERBROD, D., BUZZELLI, E.S., MOYES, H., and LIU, C.T. *Extended Abstracts, Proc. Vol. 85-2, Fall Meeting of the Electrochemical Society,* 13–18 October 1985, Las Vegas, Nevada, USA (The Electrochemical Society, Pennington, New Jersey, USA, 1985), p. 28.

76. BEHRIN, E., WOOD, R.L., SALISBURY, J.D., WHISLER, D.J., and HUDSON, C.L. *Design Analysis of an Aluminium-Air Battery for Vehicle Operations; Report No. UCRL-53382* (Lawrence Livermore National Laboratory, University of California, Livermore, USA, March 1983).

77. MAIMONI, A. *Proc. 20th Intersoc. Energy Conv. Eng. Conf.*, 18–23 August 1985, Miami Beach, Florida, USA (Society of Automotive Engineers, Inc., Warrendale, Pennsylvania, USA, 1985), pp. 2.14–2.20.

78. LAPP, S., and DAWSON, J. *Proc. 11th International Electric Vehicle Symposium (EVS-11),* Florence, Italy, 27–30 September 1992, **Vol. 2,** Paper 18.03.

79. KNERR, L.A., SCHUE, T.J., KRAUSE, C.L., and WHEELER, L.J. *Extended Abstracts, Proc. Vol. 85-2, Fall Meeting of the Electrochemical Society,* 13–18 October, Las Vegas, Nevada, USA (The Electrochemical Society, Pennington, New Jersey, USA, 1985), pp. 20–21.

80. GUPTA, S.L., TRYK, D., SCHERSON, D., ALFRED, W., and YEAGER, E. *Extended Abstracts, Proc. Vol. 85-2, Fall Meeting of the Electrochemical Society,* 13–18 October, Las Vegas, Nevada, USA (The Electrochemical Society, Pennington, New Jersey, USA, 1985), pp. 5–6.

81. COOPER, J.F., HOMSY, R.V., and LANDROM, J.H. *Proc. 15th Intersoc. Energy Conv. Eng. Conf.*, 18–22 August 1985, Seattle, Washington, USA (The American Institute of Aeronautics and Astronautics, New York, USA, 1985), pp. 1487–1495.

82. ADAMS, W.A., FELL, W., SONG, G.S., and WALKER, M. *Proc. 11th International Electric Vehicle Symposium (EVS-11)*, Florence, Italy, 27–30 September 1992, **Vol. 2,** Paper 18.02.

83. APPLEBY, A.J., and JACQUIER, M. *J. Power Sources*, 1976, **1**, pp. 17–34.

84. FOLLER, P.C. *J. Appl. Electrochem.*, 1986, **16**, pp. 527–543.

85. GOLDSTEIN, J.R., and KORETZ, B. *Proc. 28th Intersoc. Energy Conv. Eng. Conf.,* 8–13 August 1993, Atlanta, Georgia, USA (American Chemical Society, Washington, DC, USA), **Vol. 2**, pp. 2.279–2.284.

86. Electric Fuel Limited. *Electric & Hybrid Vehicle Technology '95* (UK & International Press, Dorking, UK, 1995), pp. 117–120.

87. GOLDSTEIN, J.R., and KORETZ, B. *11th International Seminar on Primary and Secondary Battery Technology and Application,* 28 February – 3 March 1994, Deerfield, Florida, USA.

88. *Batteries International,* 1994, **21,** pp. 10–13.

89. BOEHM, G. *Electric & Hybrid Vehicle Technology '95* (UK & International Press, Dorking, UK, 1995), pp. 266–267.

90. COOPER, J.F., KEENE, L.E., NORING, J., MAIMONI, A., and PETERMAN, K. *Proc. 36th International Power Sources Conference,* 6–9 June 1994, Cherry Hill, New Jersey, USA (The Institute of Electrical and Electronics Engineers, Inc., New York, USA, 1994).

91. TOKARZ, F., SMITH, J.R., COOPER, J., BENDER, D., and ACEVES, S. *Electric & Hybrid Vehicle Technology '95* (UK & International Press, Dorking, UK, 1995), pp. 126–130.

92. COOPER, J.F., and WOODROW, W.C., Jr. *Int. J. Environmentally Conscious Design & Manufacturing,* 1996, **5**, pp. 49–54.

93. *Science & Technology Review/Industrial Ecology Collection 1996,* pp. 12–15.

94. EVANS, J.W., and SAVASKAN, G. *J. Appl. Electrochem.*, 1991, **21**, pp. 105–110.

95. SAVASKAN, G., HUH, T., and EVANS, J.W. *J. Appl. Electrochem.*, 1992, **22**, pp. 909–915.

96. HUH, T., SAVASKAN, G., and EVANS, J.W. *J. Appl. Electrochem.*, 1992, **22**, pp. 916–921.

97. SALAS-MORALES, J.C., and EVANS, J.W. *J. Appl. Electrochem.*, 1994, **24**, pp. 858–862.

98. SIU, S., EVANS, J.W., SALAS, J., and NEWMAN, M. *Proc. 125th Annual Meetings of the Minerals, Metals & Materials Society (TMS) and the American Institute of Mining, Metallurgical and Petroleum Engineers, Inc. (AIME)*, 4–8 February 1996, Anaheim, California, USA, p. B10.

99. SALAS-MORALES, J.C., EVANS, J.W., NEWMAN, O.M.G., and ADCOCK, P.A. *Metall. Trans. B*, 1997, **28B**, pp. 59–68.

100. VERMA, A., SALAS-MORALES, J.C., and EVANS, J.W. *Metall. Trans. B*, 1997, **28B**, pp. 69–79.

CHAPTER 7

Flow Batteries: Zinc/Halogen & Redox

7.1. Basic Concepts

Zinc/halogen batteries operate through the discharge of elemental zinc and a halogen to form the zinc halide. Zinc is a most attractive material for battery electrodes because it has the highest available specific energy of any metal which can be effectively electrodeposited from aqueous solution. It combines a large metal/metal-ion potential (E^o = −0.763 V) with a high hydrogen overpotential. The latter allows zinc to be plated from aqueous solutions in the pH range 2–5 with a coulombic efficiency of 95–99%. Furthermore, zinc is relatively abundant in the earth's crust, is produced at a relatively low cost, and does not pose any handling problems. By virtue of these desirable properties, primary batteries based on zinc negative electrodes command a dominant position in the market-place.

The possibilities of utilizing zinc electrodes in secondary batteries with alkaline electrolytes have been discussed in Chapter 6. It is seen that such batteries suffer serious limitations through dendritic growth and 'shape change' of the zinc electrode on charge–discharge cycling. Although additives that diminish zinc solubility have improved battery performance, a fully acceptable remedy that allows a reasonable service life has yet to be found. In order to avoid such problems, an alternative approach is to adopt acid electrolytes in which the discharge product is soluble. Thus, zinc dissolution replaces the formation of zinc oxide as the discharge reaction at the negative electrode, i.e.,

$$Zn \rightarrow Zn^{2+} + 2e^- \tag{7.1}$$

Chlorine and bromine have been applied as positive electrodes in combination with zinc in acid electrolytes, and such systems have been developed, engineered, and road-tested in EVs. Ferricyanide has also been proposed and evaluated as a positive electrode [1,2]. It offers the advantage of being a benign material — it presents neither a toxicological nor a chemical hazard. On the debit side, the zinc/ferricyanide system has a much lower specific energy than the halogen analogues. The discharge reactions and corresponding standard electrode potentials (E^o) for the three types of positive electrode are:

$$Cl_2 + 2e^- \rightarrow 2Cl^- \qquad E^o = 1.360\ V \qquad (7.2)$$

$$Br_2 + 2e^- \rightarrow 2Br^- \qquad E^o = 1.065\ V \qquad (7.3)$$

$$Fe(CN)_6{}^{3-} + e^- \rightarrow Fe(CN)_6{}^{4-} \qquad E^o = 0.361\ V \qquad (7.4)$$

The zinc/ferricyanide battery was initially championed in the 1980s for load-levelling and solar-photovoltaic/wind energy-storage applications [1]. It failed, however, to emerge as a competitive power source for EVs and, accordingly, will not be discussed further.

Zinc/halogen batteries allow storage of active materials external to the battery; the reactants are circulated through the cell stack when required. Systems of this type are referred to as **flow batteries**. Not all the battery active materials are contained in the flowing electrolyte; zinc metal formed on charge is retained on the negative electrodes within the stack.

The flow-battery concept is taken further in **redox batteries** by employing two soluble redox couples to store the energy, i.e., both the charged and the discharged states of the two electrode systems are present in solution. The redox couples that have been promoted are iron/chromium and vanadium(II)/vanadium(V). The electrode reactions on discharge of the former system are:

$$\text{positive:} \quad Fe^{3+} + e^- \rightarrow Fe^{2+} \qquad (7.5)$$

$$\text{negative:} \quad Cr^{2+} \rightarrow Cr^{3+} + e^- \qquad (7.6)$$

The E^o values for the above two reactions are 0.77 and –0.41 V, respectively, and thus yield a standard cell voltage (V^o) of 1.18 V. The corresponding reactions for the vanadium redox system are:

$$\text{positive:} \quad VO_2^+ + 2H^+ + e^- \rightarrow VO^{2+} + H_2O \qquad (7.7)$$

$$\text{negative:} \quad V^{2+} \rightarrow V^{3+} + e^- \qquad (7.8)$$

with E^o values of 1.00 and –0.26 V, respectively, to give a V^o of 1.26 V.

Redox batteries have been advanced for stationary applications where the energy-storage capability per unit mass or volume is not critical. US and Japanese battery developers of the iron/chromium system first targeted prospective energy-storage markets in electrical utility operations and in remote-area power supplies [3]. Iron/chromium redox stacks of up to 60 kW were tested successfully by Mitsui in Japan as part of the Moonlight Project [4]. Further scale-up and evaluation was abandoned, however, due to the problem of cross-contamination of the electrolytes and to the very low specific energy and poor efficiency of this system compared with the sodium/sulfur and zinc/bromine batteries which were also under investigation at that time. In Australia, a research group led by Skyllas-Kazacos pioneered [5] the development of the vanadium redox battery to overcome the deficiencies of the iron/chromium as well as other energy-storage systems. In addition to stationary applications, the vanadium redox battery is being promoted as a power source for EVs. The perceived advantage of a redox system in EV duties arises from the external storage of both reactants and products. This offers the prospect of recharging of the vehicle battery at a rate equal to that for ICEVs through reagent exchange at a 'service station'. Regeneration of the spent vanadium solutions would then be carried out by either electrochemical or chemical means at a dedicated facility. The vanadium redox battery electrolyte can be treated as an alternative 'liquid fuel' which can be regenerated indefinitely. This feature is of great significance in EV applications where public acceptance will be more readily achieved if behaviour patterns do not need to be radically modified.

7.2. Zinc/Chlorine

7.2.1. Historical perspectives

A zinc/chlorine battery was employed by Charles Renard in 1884–85 to power his airship 'La France' [6]. The chlorine was fed to the battery from a chemical reactor that contained chromium trioxide and hydrochloric acid.

The zinc/chlorine battery remained merely a historical curiosity until the mid-1970s when Energy Development Associates (EDA, a Gulf and Western subsidiary) in the USA [7] and the Furukawa Electric Company in Japan [4] developed rechargeable versions to an advanced engineering stage for both energy storage and traction applications. Nevertheless, enthusiasm for zinc/chlorine waned in the late 1980s given the design and operational complexity of the battery, as well as a growing interest in the bromine analogue as a more practical power source for EVs.

7.2.2. Battery design and performance

A schematic representation of the EDA system is shown in Fig. 7.1. The battery stack is a bipolar design that consists of compact carbon negative and porous carbon positive plates. Zinc chloride solution is pumped through the stack and, during charge, zinc is plated on the negative electrodes and chlorine is generated at the positives. The solution containing the evolved chlorine is transferred to a cooled storage unit where the temperature is decreased to below 5 °C to separate out solid chlorine hydrate, $Cl_2{\cdot}xH_2O$, in which the value of x is 5.7–8 as dictated by the temperature conditions. The battery requires a refrigeration unit for the chlorine store, but this is operative only on charge and, hence, is powered from the mains electricity supply.

On discharge, hydrate is redissolved by circulating warm electrolyte from the cell stack to the store and the resulting chlorine solution is passed through the porous positive electrodes. In this design, therefore, chlorine is isolated from the zinc negative electrode on both charge and discharge.

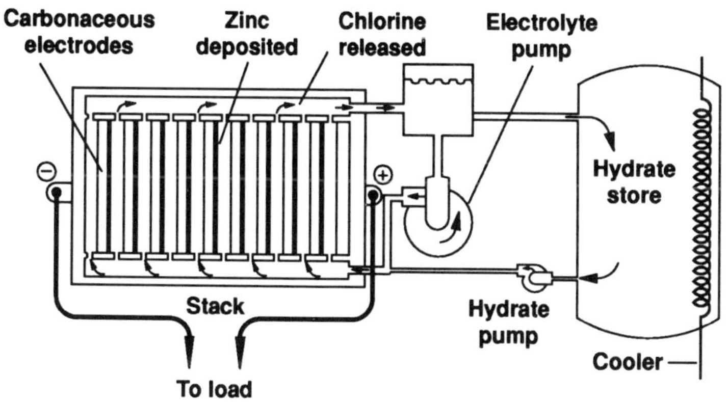

Fig. 7.1. Schematic representation of a zinc/chlorine battery during a charging period. (Reprinted with permission from SAE Paper No. 730248 ©1973, Society of Automotive Engineers, Inc. [8].)

Consequently, it is not necessary to include a separator and this feature, together with the use of a narrow cell gap of ~1 mm, minimizes the ohmic losses. Some hydrogen is inevitably generated at the negative electrode at the operative pH range of ~0.5 to 1. The hydrogen is combined with chlorine in a light-activated reactor and the hydrochloric acid produced is returned to the electrolyte.

The cell operates at between 30 and 50 °C. The fully-discharged electrolyte contains 3 M $ZnCl_2$, together with 1 M KCl and 2 M NaCl to enhance conductivity. Circulation of the electrolyte yields a zinc deposit with acceptable morphology; the latter can be improved further through the inclusion of additives such as thallium and bismuth [4]. The zinc electrode generally governs the charge capacity and the coulombic efficiency of the system, while the chlorine electrode determines the delivered power and the voltage efficiency. Batteries can operate with carbon alone as the positive electrode material, but the power capabilities are enhanced by the inclusion of catalysts such as ruthenium oxide [9].

It should be remembered that flow batteries consist of independent units (i.e., cell stack, electrolyte storage vessels, pumps) which can readily be replaced on breakdown to restore operation of the power source. The key component is the cell stack, and the period that expires before stack performance fails to meet the goals is the most appropriate definition of 'life'. Reaction of chlorine with the carbon positive plate is the most serious mode of degradation. Nevertheless, the rate of attack by chlorine is relatively slow and allows a practical cell life.

The battery developed by EDA had a specific energy of 111 Wh kg^{-1} and an energy density of 131 Wh l^{-1}, with a delivered power of 87 W kg^{-1} at the $C_1/1$ rate [10]. The performance of full-sized modules was evaluated in both a Renault Trafic van and a Grumman Kubvan [7]. An energy efficiency of 68% was achieved; the target was 70–75% and this would correspond to an overall a.c.-to-a.c. efficiency of 60–67%. A battery life of 500–5000 cycles was predicted by EDA.

7.2.3. Safety considerations

Chlorine hydrate is a stable compound that is not hazardous provided it is maintained at a temperature below 10 °C. Inevitably, the temperature of the hydrate store will rise on standing, but chlorine evolved in this process can be fed to the cell and allowed to react with the zinc negatives. This results in a modicum of self-discharge, but avoids the possibility of release of chlorine to the atmosphere. A safety audit of the zinc/chlorine battery was carried out by Factory Mutual [11]. Various EV crash scenarios were considered and it was concluded that released chlorine presents no greater hazards to drivers than does gasoline in ICEVs. Despite this favourable finding, it is still open to question whether the public at large will accept that a battery containing refrigerated chlorine is a totally safe proposition.

7.3. Zinc/Bromine

7.3.1. Bromine vs. chlorine electrodes

The application of bromine in place of chlorine in a zinc/halogen battery lowers the open-circuit voltage of the cell from 2.12 to 1.83 V, and the theoretical specific energy from 833 to 436 Wh kg^{-1}. This range of specific energy becomes 465–337 Wh kg^{-1} when the actual form of the stored halogens in the two battery systems (chlorine hydrate or morpholinium polybromide) is taken into account. Thus, the use of bromine in place of chlorine results in a 27% loss in theoretical specific energy.

On the other hand, bromine is much easier to contain than chlorine. In addition, bromine is more soluble in water and, therefore, higher concentrations can be supplied to the cell. These two properties result in a simpler and safer system for halogen storage, lower pumping rates, and more manageable plumbing. Given such advantages, zinc/bromine has replaced its chlorine counterpart as the major zinc/halogen candidate for energy-storage applications.

7.3.2. System development

The zinc/bromine battery was patented by Bradley in the 1880s [12]. It remained no more than an interesting concept for almost a century because of the high self-discharge rate (due to bromine reaching the zinc electrode), as well as the tendency of zinc to form dendritic deposits during charge (causing internal shorting of the cell). In the 1970s, a number of battery research groups revisited the zinc/bromine system and conducted research which was largely aimed at overcoming these two fundamental problems.

The self-discharge problem has been tackled through three approaches. The one which proved to be the most successful was pioneered by Exxon and Gould in the USA. It involved complexing bromine with quaternary ammonium bromide compounds such as N-methyl N-ethyl morpholinium bromide (MEMBr) or N-methyl N-ethyl pyrrolidium bromide (MEPBr). These compounds incorporate bromine to form polybromides which are oils that separate from the aqueous phase and can be stored external to the cell. The polybromide system has been exclusively adopted by companies which are presently engineering zinc/bromine systems, and batteries based on this bromine-storage concept are considered in detail in the latter parts of this Section.

The second approach to overcoming the self-discharge of zinc/bromine batteries was devised by the General Electric Company (GEC) in the USA. A cation-exchange membrane was incorporated as a cell separator to constrain bromine from reaching the zinc electrode [13]. Dupont 'Nafion', a perfluorosulfonic acid, proved to be an effective membrane and permitted the bromine concentration on the positive side to rise to 3 M without the transport of bromine becoming excessive. Provided sufficient bromide ions are present to form tribromide or pentabromide complexes, the electrolyte is a single phase of low viscosity. The GEC cells were of a bipolar design and employed electrodes made from vitreous carbon foam which was bonded to graphite plates; the electrode spacing was 2–6 mm [14]. The reported performance was comparable with that achieved by the polybromide systems, but the materials cost was much greater. In view of the latter, GEC discontinued development of this system in the 1980s.

The third approach to the attenuation of self-discharge of zinc/bromine cells was conceived by Parker and co-workers [15] at Murdoch University in Western Australia. A solution of bromine in propionitrile (PN) was used as the positive-side electrolyte, while the negative contained aqueous zinc bromide as in the above two systems. Propionitrile is immiscible with water and has a high partition coefficient for bromine so that transport of bromine to the zinc electrode is restrained. Due to the formation of charge-transfer complexes of PN and Br_2, and of tribromide ions during discharge, the conductivity of the non-aqueous solution is satisfactory, i.e., more than half that of the corresponding aqueous system. It is claimed that the electrolyte circulation is simpler in the PN system and requires less pumping power. These advantages are due to the fact that both the aqueous and PN phases have a low viscosity. The specific energy of such cells is significantly greater than that of the corresponding polybromide system. Moreover, the cell can deliver higher power.

ZBB (Australia) Limited was established in 1987 to research and commercialize the zinc/bromine concept which was proposed by Murdoch University. The Company established a research/marketing arrangement with the Energy Research Corporation (ERC) in the USA, a battery developer which had acquired the technology initiated by Gould. Together, ZBB and ERC concluded that a battery system based on aqueous electrolytes with polybromide storage, rather than the PN system, offered the best prospect for commercial success. In the early 1990s, ERC decided to discontinue its advancement of the technology. Subsequently, ZBB became allied with Multiplex Constructions Proprietary Limited and formulated a plan to bring zinc/bromine into the market-place.

Exxon curtailed its zinc/bromine battery activities in the 1980s and licensed the technology to Studiengesellschaft für Energiespeicher und Antriebssysteme (SEA) in Austria, to Johnson Controls, Inc. (JCI) in the USA, to Toyota Motor Corporation and Meidensha Corporation in Japan, and to Sherwood Overseas Limited in Australia. These companies continued to develop the battery. Units were engineered and field-tested in both EV and stationary-storage applications. Powercell Corporation was established in the USA and Austria in 1993 to refine the Exxon zinc/bromine technology as advanced by SEA. At present, the EV version of the battery is being commercialized under the trade name Wave™ and is to be installed in an EV which is being developed by the Daewoo Motor Corporation of South Korea (see Section 7.3.7). In 1992, JCI withdrew from the zinc/bromine research area following a bromine release incident which occurred during the Solar and Electric Racing Association races in Phoenix, Arizona. ZBB purchased the JCI technology and thus gained access to both ERC and Exxon expertise.

7.3.3. Polybromide storage

A schematic representation of a zinc/bromine battery with polybromide storage is presented in Fig. 7.2. The cell stack employs carbon bipolar electrodes and separate electrolyte streams are circulated through the positive and negative electrode compartments. In the fully-discharged state of the battery, the electrolyte in both streams is typically 3 M in zinc bromide and 1 M in a quaternary ammonium bromide. Potassium chloride or ammonium chloride at a concentration of 3–4 M is generally included to enhance conductivity and to diminish ohmic losses. The pH of the

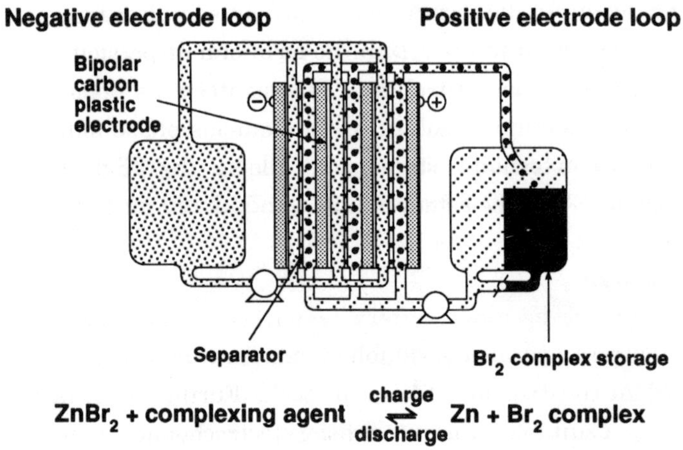

Fig. 7.2. Schematic representation of a zinc/bromine battery with polybromide storage. (Taken from [16] with permission of *Batteries International*.)

electrolyte is set at ~3. At higher pH values, the zinc deposited on charge has an unsatisfactory mossy morphology, and bromate can be formed as a side reaction at the positive. At lower pHs, corrosion of zinc metal can occur together with the evolution of hydrogen.

On charge, zinc is deposited on the negative electrode. As pointed out above, the use of a flowing electrolyte results in a satisfactory morphology for the zinc deposit. The quality of the deposit is improved by the presence of the quaternary ammonium compound [17]. The performance of the zinc electrode can also be enhanced by the addition of typical electroplating or electrowinning agents.

Bromine is liberated at the positive electrode during charge and is complexed with the quaternary ammonium compound, Q^+Br^-, to produce complexes of the type $Q^+Br_n^-$ (n = 3, 5 or 7 [18]) that form a separate liquid phase. The polybromide oil separates out in the electrolyte store on the positive-electrode side. The bromine concentration in the electrolyte in equilibrium with the polybromide phase is typically 0.1 M over a wide range of n values. Thus, most of the liberated bromine is stored external to the cell and, when the battery is on stand, only a fraction of the bromine can diffuse to the zinc electrode, with consequent low self-discharge. On discharge, bromine is returned to the cell in the form of a dispersion of the polybromide oil in the aqueous electrolyte. This strategy, as opposed to using the circulation of bromine dissolved in the aqueous phase following equilibration with the externally stored oil, allows the high power capabilities required for EVs to be achieved [19].

7.3.4. Cell stack design

The zinc/bromine battery operates at ambient temperatures and, hence, plastics can be used for construction of the cell stack. Furthermore, carbon contained in plastic is a suitable substrate for the electrochemical reactions which take place at the positive and the negative electrodes. Indeed, the ability to fabricate cell stacks from inexpensive plastic materials is a major advantage of the zinc/bromine battery. Figure 7.3 presents the basic building blocks of the bipolar, filter-press, cell design which was developed by Exxon [20]. The cell stack is built up from two components: a carbon-plastic, bipolar, electrode plate and an injection-moulded separator element. The bipolar electrode component is formed by co-extruding the central conductive carbon-plastic portions with two non-conductive side bonds. The injection-moulded separator component contains an insert of a microporous separator material (Daramic) on which posts have been embossed. The separator is housed in a non-conductive plastic frame which incorporates the required electrolyte flow channels.

A materials problem arises from the limited chemical stability of most plastics towards bromine. Polyvinylidene difluoride (PVDF) and polytetrafluoroethylene (PTFE) are the preferred plastic materials to use, but both are expensive. Cheaper, but less stable, alternatives are polyethylene, polypropylene, and polyvinylchloride (PVC). Most commercial plastics contain appreciable amounts of inorganic materials which are added as mechanical stabilizers (e.g., chopped glass-fibres), as pigments, or simply as fillers. Many of these are also subject to attack by bromine. Additives such as SnO_2 and Fe_2O_3 are undesirable and may give rise to hydrogen evolution. By contrast, PbO_2 is satisfactory and PVC loaded with this

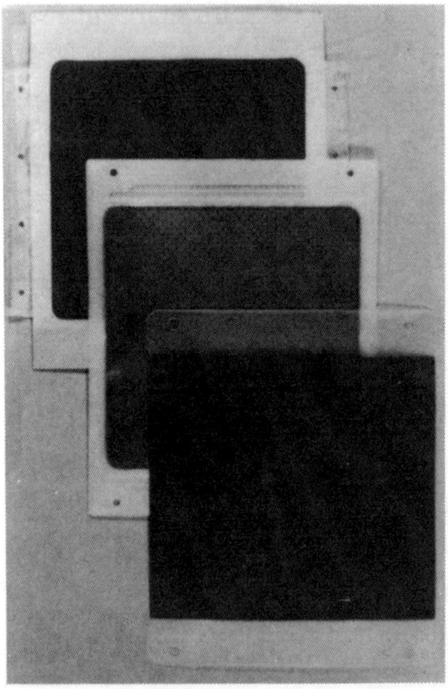

Fig. 7.3. Two-piece zinc/bromine cell with collector (end electrode). (Reprinted with permission from SAE Paper No. 830289 ©1983, Society of Automotive Engineers, Inc. [20].)

compound may be used for tubing, unions, etc. Gaskets and O-rings require elastomers which are stable to bromine and do not swell, deform or crack. VITON™ loaded with PbO_2 is acceptable for static seals, but the rotary shaft seals of the circulating pumps (Fig. 7.2) pose a more difficult problem and it may be necessary to use a magnetically coupled, seal-less pump.

The two-piece cell design was a significant advance on Exxon's earlier six-piece unit construction [21]. Further improvements in cell stack construction have been made by SEA [16]. The nut-and-bolt assembly has been replaced by a plastic-welding process which eliminates electrolyte leakage. SEA has also substituted polypropylene with polyethylene as the polymeric material for the carbon-plastic bipolar plates, while a more stable and highly active carbon layer is now used for the electrodes. These modifications reduce bromine adsorption by the carbon electrodes, a process which can lead to cell warping with consequent limitation in battery service life. In addition, the vertical arrangement of plates adopted by Exxon has been changed to a horizontal one; this permits more uniform flow characteristics and increases the design flexibility.

7.3.5. Shunt-current protection

The inclusion of a common electrolyte system to series-connected, bipolar cells results in an even supply of reactant, uniform deposition of metal, easy removal of product, and efficient thermal management through effective transfer of heat. On the debit side, energy loss through shunt currents (that is, self-discharge) occurs because electrodes at different voltages are connected via the electrolyte feed channels. The loss can be minimized by making the channels long and narrow, but this involves a compromise since a channel structure with high tortuosity demands higher levels of pumping power. Proper design can diminish shunt-current losses to ~3% of the discharged energy [19].

In addition to wasting energy, shunt currents flowing through the electrolyte manifold result in the transference of zinc from negative electrodes at higher voltages to those at lower voltages. This can result in zinc growing into the channels and blocking, or diverting, the flow of electrolyte. Exxon devised a control system to eliminate shunt currents in which several small electrodes were mounted in the manifold and a current of matching magnitude was applied in the opposite direction to the shunt current [22]. The siting of the electrodes in the shunt-current protection system is illustrated in Fig. 7.4. In this way, the deleterious effects of the shunt current on the zinc active-mass are eliminated, but the energy losses remain. An alternative approach researched by Exxon was to pass a bias current through a parallel tunnel port to match the voltage gradient in the active cells [22].

SEA has developed an automatic procedure to shut off shunt currents by isolating each cell from the manifold during stand-by periods [16]. This eliminates the need to include the above shunt-current protection in the battery design.

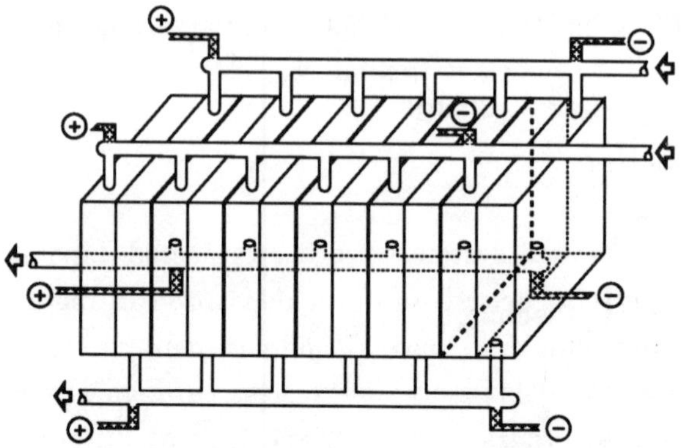

Fig. 7.4. Manifold shunt-current protection in a zinc/bromine system with a common electrolyte. (Taken from [22] with permission of Plenum Press Corporation, USA.)

7.3.6. Charge–discharge characteristics

The variation in average cell voltage reported by Exxon [23] for a 30-kWh (nominal) zinc/bromine battery subjected to constant-current charge and discharge is presented in Fig. 7.5. This battery consisted of 125 cells and the voltage was measured for banks of 25 cells. The charge profile was almost constant at a level just above 2.0 V. The voltage on discharge displayed an approximately linear decrease to about 1.7 V, at which point nearly all the capacity had been removed from the battery. The cells in the battery were closely matched and the banks of cells became discharged at the same time. The battery acted as its own 'fuel gauge'. The state-of-charge was determined by 'viewing' the level of the interface between the polybromide oil and the aqueous phase in the bromine storage reservoir.

Unlike non-flow batteries, such as lead/acid, the life of a zinc/bromine battery is not affected by totally discharging the battery. Indeed, zinc-based flow batteries perform much better when taken to 100% DoD as this produces a more satisfactory morphology for the zinc deposit on subsequent charging. Shorting the battery before charging so that the zinc is removed completely from the negative plates has been recommended to ensure good zinc electrode behaviour. This diminishes the effective energy efficiency since the capacity remaining in the battery before charge is lost.

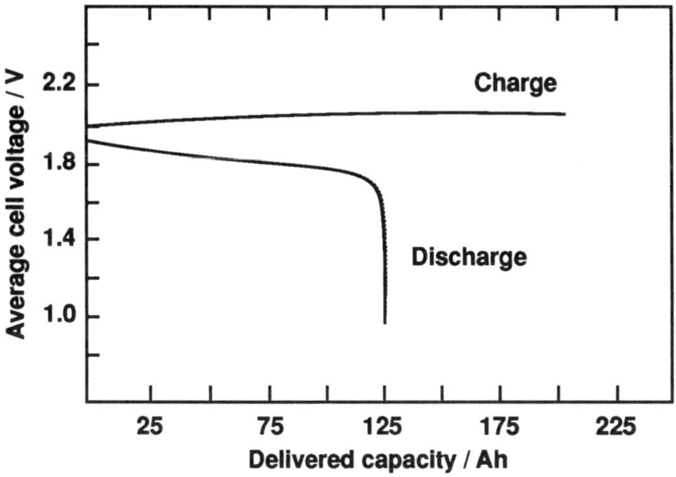

Fig. 7.5. Cell voltage behaviour for a 30-kWh zinc/bromine battery. Charge = 50 A; discharge = 40 A. (Reprinted with permission from SAE Paper No. 859186 ©1985, Society of Automotive Engineers, Inc. [23].)

The introduction of a shunt-current protection system decreased the number of such equilibration cycles which were required for maintaining efficient performance. Occasional equilibration cycles are still required in the latest zinc/bromine technology, but their application can be automatically initiated from an in-built computer [17].

Some hydrogen is evolved at the negative electrode during charge, although this is considerably less than that in the zinc/chlorine system because the pH of the electrolyte is significantly higher. Hydrogen evolution will result in a loss of water together with an increase in the pH of the electrolyte. The latter can, in turn, cause the precipitation of zinc hydroxide. Passive catalyst electrodes can be employed to combine hydrogen and bromine to form hydrogen bromide which is returned to the cell to maintain the pH at the initial value. An alternative approach for compensating for the effects of hydrogen evolution is to add either water or acid to the system. In general, however, developers of the battery consider hydrogen evolution to be sufficiently slow in normal application for no adjustments in electrolyte composition to be necessary.

7.3.7. Battery performance

The performance characteristics reported by SEA for its state-of-the-art EV battery pack in 1992 are presented in Table 7.1. The specific energy is double that of today's best lead/acid batteries, while the energy density and power density are similar. The energy efficiency is slightly less than that of lead/acid, but is still acceptable for EV applications and is significantly greater than that achieved in early zinc/bromine designs. For example, the energy efficiency which corresponds to the charge–discharge curves in Fig. 7.5 is only 53%.

Table 7.1. SEA zinc/bromine battery for EVs [16].

	Value
• Rated capacity (kWh)	22.5
• Nominal voltage (V)	108
• Battery weight (kg)	340
• Use regime	
– discharge time (h)	3
– charge time (h)	5
• Maximum power, fully charged (kW)	32
• Specific energy (Wh kg^{-1})	65–70
• Energy density (Wh l^{-1})	60–70
• Specific power	
– peak (W kg^{-1})	100
– sustained (W kg^{-1})	60–70
• Coulomic efficiency (%)	80–93
• Energy efficiency (%)	65–70
• Cycle life (cycles)	>500
• Maximum permissible temperature (°C)	42

Two of the above SEA packs were used to power an Elin bus (a modified Volkswagen Transporter Type 2) for the Austrian Post Office [16]. The

Fig. 7.6. Microvans, light-duty pickup trucks, sedans and purpose-built intra-city commuter vehicles powered by Wave™ zinc/bromine technology. (Taken from [24] with permission of Powercell GmbH, Austria.)

Elin bus operated in alpine areas and completed 15 000 km in 1990. It had a maximum driving range of 200 km and a top speed of 110 km h^{-1}. Five mountain passes with gradients of up to 14% were traversed; the total altitude climbed was more than 2100 m and the average speed was 41 km h^{-1}. This is just one of a number of trials in EVs made by SEA and other developers that have demonstrated that zinc/bromine battery packs can provide an effective power source. Figure 7.6 shows some of the vehicles which are being road-tested with zinc/bromine battery power.

A vehicle powered by Powercell zinc/bromine batteries finished in first place in the EV division of both the 1994 and the 1995 World Clean Air Vehicle Rallies in Los Angeles, California. The battery used in the 1994 rally is shown in Fig. 7.7. In a test of range in 1995, a Powercell vehicle took first place overall and exceeded the distance achieved by vehicles powered with natural gas. A similarly-powered vehicle won the 1995 Public Service EV race in Phoenix, Arizona. A standard General Motors GEO Prizm converted to Wave™ power demonstrated a cruising range of greater than 270 km per charge, highway speeds of 125 km h^{-1}, and an acceleration of zero to 100 km h^{-1} in 15 s.

Fig. 7.7. 35-kWh, 380-V zinc/bromine battery for EVs. (Taken from [25] with permission of Powercell GmbH, Austria.)

The promotional literature published by Powercell in 1997 claimed a specific energy of 80–85 Wh kg^{-1} and a battery life of >1000 full-discharge cycles. As discussed above for the zinc/chlorine system (see Section 7.2.2), battery life is determined principally by the cell stack since, for example, pump breakdowns can readily be repaired. One concept which has been considered [17] is to exchange battery modules as a routine maintenance procedure after a service of the order of 2000 cycles, rather than to repair defective parts or extend life at the penalty of increasing unit cost.

In 1996, Powercell announced that it had joined with Daewoo Motor Company, South Korea, in a development programme to integrate Wave™ zinc/bromine battery technology into the aluminium space-frame chassis of the new-generation Daewoo EV (Fig. 7.8). The battery fits snugly in the vehicle without raising unduly the centre of gravity or reducing the space available for components and passengers. A prototype vehicle was demonstrated in 1995 on a test track in Atlanta, Georgia, and gave a range of 175 km in sub-freezing (–8 °C) weather conditions. This is outstanding

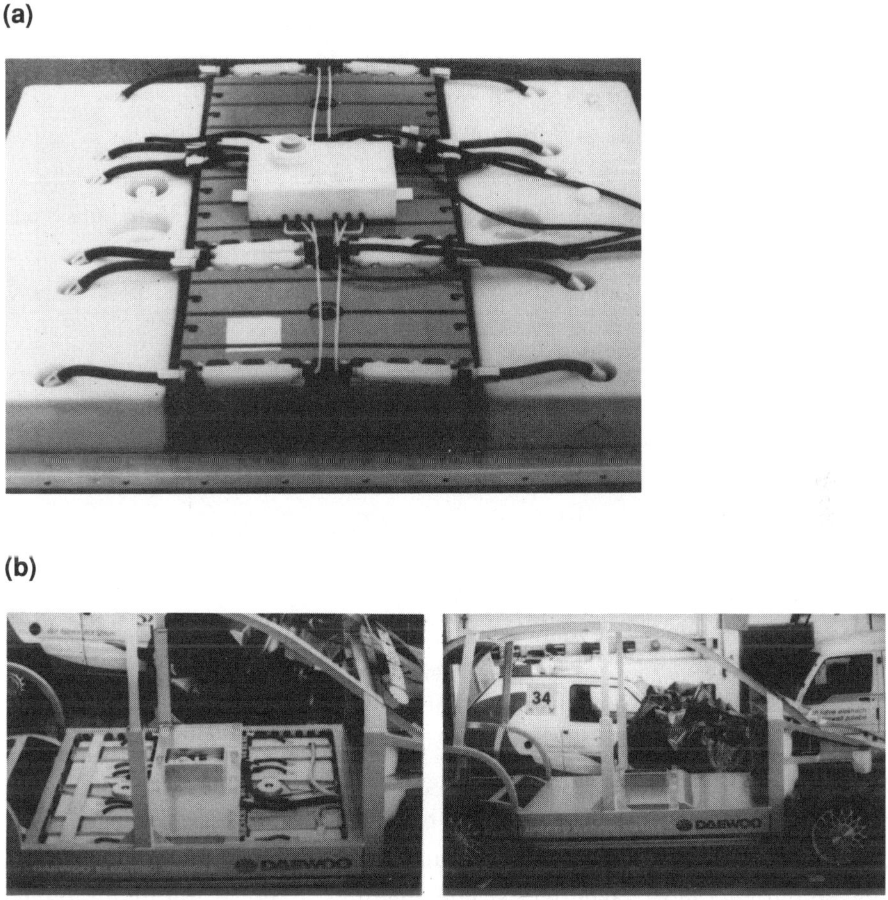

Fig. 7.8. (a) 33-kWh, 280-V zinc/bromine battery. (b) Battery assembled in space-frame of Daewoo EV prototype. (By courtesy of Powercell GmbH, Austria.)

in comparison with a range of <30 km which was obtained from lead/acid-powered vehicles under the same low-temperature conditions. Powercell's present research objective is to develop a 9-kWh / 81-V battery with a weight of 105 kg. A further aim is to lower the manufacturing cost of the batteries, even at the expense of both specific power and specific energy, because it is considered that cost reduction is the most effective path to the commercialization of EVs.

Toyota has also demonstrated zinc/bromine batteries in EVs. The EV-30 is a conceptual urban vehicle and has been designed to convey people in building complexes, shopping centres and small communities, as well as to transport passengers to and from railway stations.

The success of zinc/bromine batteries in powering EVs in both road trials and races indicates that this power source must be considered as a strong contender for a place in the future transportation sector. In summary, the advantages of the zinc/bromine battery are:

- high specific energy compared with present battery systems
- good specific power
- good cycle life
- full discharge capability
- maintenance-free operation
- relatively fast charge rate (3–5 h)
- facile state-of-charge monitoring
- low-cost active materials and materials of construction
- modular design
- flexible configuration
- all materials can be readily recycled.

The major disadvantages of the battery relate to the use of a toxic active material (namely, bromine) and to the temperature limitation of the plastic components (~60 °C).

7.3.8. Safety considerations

Bromine is a toxic gas, but it is considered that containment as a polybromide renders it safe to handle [16]. Factory Mutual carried out a safety audit analogous to that previously performed for zinc/chlorine batteries. It was concluded [26] that vapour propagation in the event of a spill would be much less than that from chlorine hydrate. Furthermore, potential hazards would be no greater than those encountered in accidents with ICEVs.

Following a bromine spillage from the JCI-built battery at the 1992 Solar and Electric Racing Association races in Phoenix, Arizona, Powercell Corporation reviewed safety aspects of the technology in collaboration with Exxon Research, Collmer Semiconductor, and the University of California at Davis [27]. The outcome of this study was that the Powercell battery exceeded the safety requirements set for the 1994 competition.

Powercell Corporation has incorporated a rugged solid-state electronic battery controller which manages battery operational modes (such as start-up and shut-down) and ensures reliable and safe operation of the battery. The controller automatically stops the electrolyte circulation pumps in the event of the detection of a leak, the failure of the flow sensors, or the battery reaching an excessive temperature. Following an impact (greater than a 2-*G* force), or other anomalies which the controller's sensors might interpret as indicative of a potentially hazardous situation, the controller reacts by shutting down the battery.

Crash tests were performed on a number of EV battery packs by Hochtief Abteilung Ingenieurbau und Konstrucktions Software in Frankfurt, Germany. These tests simulated head-on vehicle collisions at 50 km h^{-1}. A Powercell zinc/bromine battery, dropped from a height of 12.5 m, suffered little deformation and no leakage of bromine. The battery was found to function normally after being subjected to such an impact.

7.4. Vanadium Redox

7.4.1. Battery design

A secondary battery based on two vanadium redox couples was patented by the University of New South Wales (UNSW) in Australia in 1986 [5]. Since then, Skyllas-Kazacos and co-workers at the UNSW have developed the vanadium redox battery [28–32], and have built on previous studies of the iron/chromium system that had been conducted mainly at NASA-Lewis in the USA [3] and in the Moonlight Project in Japan [4].

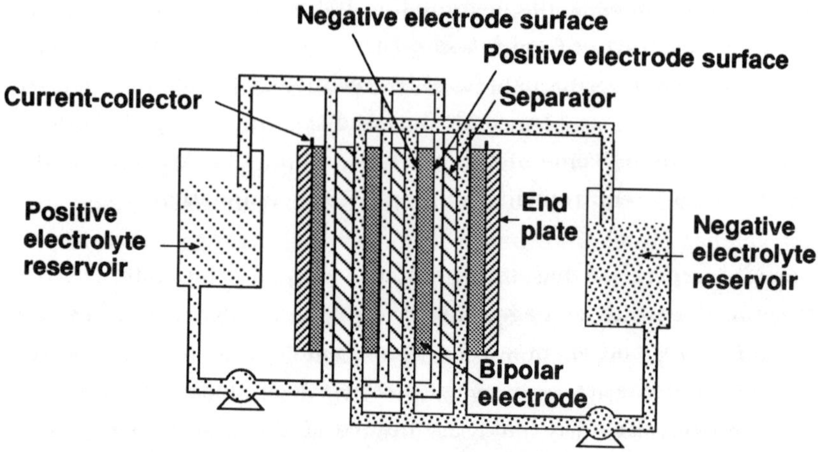

Fig. 7.9. Schematic representation of the vanadium redox battery. (Taken from [30] with permission of the University of New South Wales, Australia.)

The vanadium redox battery consists of a cell stack (similar to that used in zinc/halogen systems) through which the two reactants are passed. The design of the battery is shown in Fig. 7.9. The bipolar electrodes are composed of graphite-impregnated polyethylene plates of 0.7 mm thickness on to which a graphite felt (thickness: 3–6 mm) is heat bonded. The graphite felt acts as the electroactive surface for the electron-transfer reactions and as a flow-through electrode with a very large contact area. The end electrodes are similar to the bipolar electrodes, except that felt is bonded to one side and the other side is copper plated to minimize ohmic losses in the current-collectors. The development of these electrodes has reached the stage where polyethylene-based bipolar electrodes and end electrodes with resistivities as low as 0.6 and 0.8 Ω cm^2, respectively, have been obtained [31]. A polypropylene-based material has also been extruded and this gives resistivities in the range 0.5 to 0.6 Ω cm^2 for the bipolar electrodes [31]. The conductive polymer electrodes are extruded as sheets of uniform thickness. Large quantities of these sheets can be produced in a single production run at a cost which is as low as \$1 per square metre [33].

Selemion CMV cation-exchange membrane acted as the separator in early versions, but this material has now been replaced by New Selemion membrane (both the materials are produced by Asahi Glass Co., Japan). A microporous separator of much lower cost is also under development at UNSW. Transfer of electrolyte between the positive and negative electrode compartments is not a problem with respect to maintaining uniformity in electrolyte composition since both solutions contain 2 M vanadium in 2–3 M sulfuric acid. On the other hand, the two vanadium solutions are in different valence states and, hence, transfer of electrolyte will constitute an energy loss through self-discharge. No hydrogen is evolved in this system under normal battery operating conditions and, consequently, no procedures are required for maintaining charge balance between the two electrolytes. Occasional remixing of the two discharged electrolytes is, however, used to rebalance the system and to recover fully the capacity losses which can result from electrolyte transfer across the membrane.

7.4.2. Battery performance

The voltage profile of the vanadium redox cell is fairly flat during constant-current charge and discharge, see Fig. 7.10. The open-circuit voltage is 1.6 V at 100% state-of-charge and the operating cell voltage under typical EV duty will be 1.0–1.5 V, as determined by the current load. The energy efficiency is ~85%, but reduces to an overall value of 75% when the power requirements of the auxiliaries are taken into account. The battery is expected to meet the power requirements of EV operation, and a long service life is expected for the cell stack since the carbon is not attacked by the electrolyte.

By contrast, the practical specific energy and energy density are both low. For example, if the vanadium solutions alone are considered, and the total discharge at an average voltage is assumed to be 1.4 V, then only 37 Wh can be obtained for each litre, or 27 Wh for each kilogram. Obviously, when the necessary containers and cell stack are taken into account, the stored energy per unit weight or volume will be appreciably less. A practical vanadium redox battery at the present state of the art would be expected to have a specific energy of less than 20 Wh kg^{-1}. Recent studies suggest that it should be possible to increase the vanadium

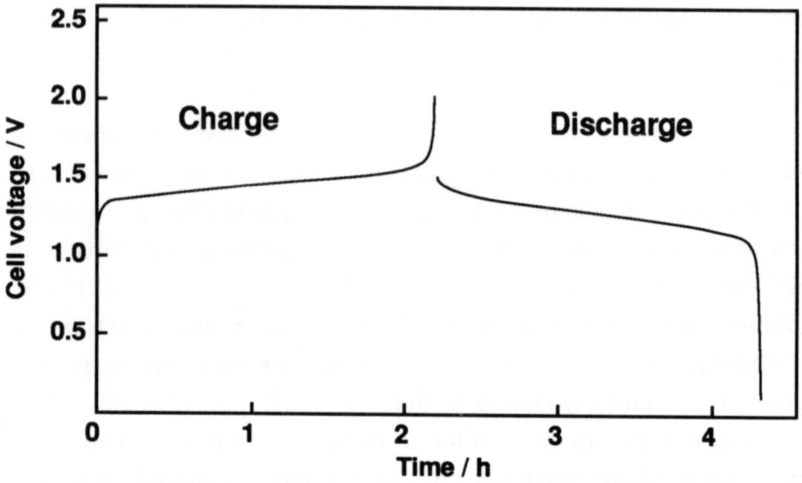

Fig. 7.10. Charge–discharge curve for vanadium cell. Charge and discharge current density = 30 mA cm^{-2}; temperature = 35 °C. (Taken from [30] with permission of the University of New South Wales, Australia.)

concentrations to 3 M, but this would only increase the practical specific energy for such a second-generation vanadium redox battery to, at best, 30 Wh kg^{-1}.

A third-generation battery concept [34] has been assessed by computer modelling of the battery when subjected to the GSFUDS (Section 4.3.1, Chapter 4). In this conceptual battery, the positive vanadium solution is regenerated *in situ* with oxygen from the atmosphere. This approach lowers the volume of the positive electrode solution that has to be carried in the vehicle, but the resulting specific energy will be less than double that of the 3 M redox battery (i.e., <60 Wh kg^{-1}). This would raise the specific energy to close to the value which is presently being achieved with zinc/bromine batteries (Table 7.1). The study concluded [34] that such a vanadium redox system would provide the General Motors 'Impact' with a performance superior to that obtained from its existing lead/acid battery pack.

In summary, the advantages of the vanadium redox battery are as follows [30]:

- the electrolyte solutions have an indefinite life and it is therefore necessary to replace only the mechanical components at the end of life
- fast recharge is possible by replacing the spent electrolytes
- the battery capacity can be readily increased by increasing the volume of the electrolyte solutions
- the battery can be fully discharged without any detrimental effects
- the cost per kWh decreases as the energy capacity increases; this renders EV-size batteries cost-effective
- the state-of-charge can be monitored on line by measuring the open-circuit voltage of the system with a small open-circuit cell which is inserted in the electrolyte lines
- the battery components are environmentally benign since no waste products are produced.

While the low specific energy and energy density are both sufficient for stationary applications of the battery, only the third-generation concept has values for these parameters that would make vanadium redox a possible candidate for EV applications.

References

1. ADAMS, G.B., HOLLANDSWORTH, R.P., and LITTAUER, E.L. *Proc. 16th Intersoc. Energy Conv. Eng. Conf.*, 9–14 August 1981, Atlanta, Georgia, USA (The American Society of Mechanical Engineers, New York, USA), pp. 812–816.

2. ASSINK, R.A., and ARNOLD, C., Jr. *Proc. 20th Intersoc. Energy Conv. Eng. Conf.*, 18–23 August 1985, Miami Beach, Florida, USA (Society of Automotive Engineers, Inc., Warrendale, Pennsylvania, USA, 1985), pp. 2.105–2.110.

3. BARTOLOZZI, M. *J. Power Sources*, 1989, **27**, pp. 219–234.

4. MISRAWA, Y., SUZUKI, A., SHIMIZU, S., SATO, H., ASHIZAWA, K., SUMII, T., and OHTOMO, M. *Proc. 24th Intersoc. Energy Conv. Eng. Conf.*, 6–11 August 1989, Washington, DC, USA (Institute of Electrical and Electronics Engineers, New York, USA, 1989), pp. 1325–1329.

5. SKYLLAS-KAZACOS, M., and ROBBINS, R.G. *US Patent 4,786,567* (issued 1986).

6. BAILLEAUX, C. *In* SALKIND, A.J. (Ed.) *Proc. Symposium on History of Battery Technology* (The Electrochemical Society, Pennington, New Jersey, USA, 1987), Proc. Vol. 87-14, pp. 268–278.

7. ROWAN, R.W., WARDE, C.J., and CARR, P. *In* PEARCE, L.J. (Ed.) *Power Sources 10, Research and Development in Non-Mechanical Electrical Power Sources* (The Paul Press Limited, London, UK, 1985), pp. 353–367.

8. AMATO, C.J. *Proc. International Automotive Engineers Congress*, 8–12 January 1973, Detroit, Michigan, USA (Society of Automotive Engineers, Inc., Warrendale, Pennsylvania, USA, 1973), Paper 730248.

9. HORIE, T., FUJIWARA, K., WATAKABE, Y., YABUMOTO, T., and MISAKI, H. *Proc. 20th Intersoc. Energy Conv. Eng. Conf.*, 18–23 August 1985, Miami Beach, Florida, USA (Society of Automotive Engineers, Inc., Warrendale, USA, 1985), pp. 2.85–2.90.

10. CHI, C.H., CARR, P., and SYMONS, P.C. *Proc. 14th Intersoc. Energy. Conv. Eng. Conf.*, 5–10 August 1979, Boston, Massachusetts, USA (American Chemical Society, Washington, DC, USA, 1979), pp. 692–697.

11. ZALOASH, R., BAJPAI, S., SHORT, T., and TSUI, E.R. *Chlorine Hazard Evaluation for the Zinc-Chlorine Electric Vehicle Battery*. DOE/ET/25405-1, April 1980. U.S. Department of Energy, Washington, DC, USA.

12. BRADLEY, C.S. *US Patent 312,802* (issued 1885); *US Patent 409,448* (issued 1889).

13. WILL, F.G. *In* THOMPSON, J. (Ed.) *Power Sources 7, Research and Development in Non-Mechanical Electrical Power Sources* (Academic Press, London, UK, 1979), pp. 313–328.

14. WILL, F.G., and SPACIL, H.S. *J. Power Sources,* 1980, **5**, pp. 173–188.

15. SINGH, P., WHITE, K., and PARKER, A.J. *J. Power Sources,* 1983, **10,** pp. 309–318.

16. FABJAN, Ch. *Batteries International*, 1992, **10**, pp. 80–81.

17. KORDESCH, K.V., WERTGARNER, A., FABJAN, Ch., and TOMAZIC, G. *In* LANDGREBE, A.R., RAND, D.A.J., VAN VOORHEES, S.L., and SEN, R.K. (Eds.) *Proc. Symposium on Stationary Energy Storage: Load Leveling and Remote Applications* (The Electrochemical Society, Pennington, New Jersey, USA, 1988), Proc. Vol. 88-11, pp. 293–304.

18. EUSTACE, D.J. *J. Electrochem. Soc.,* 1980, **127**, pp. 528–532.

19. ZAGRODNIK, J.P., and ESKRA, M.D. *Proc. 22nd Intersoc. Energy Conv. Eng. Conf.*, 10–14 August 1987, Philadelphia, Pennsylvania, USA (American Institute of Aeronautics and Astronautics, New York, USA, 1987), pp. 1039–1044.

20. MALACHESKY, P.A., BELLOWS, R.J., EINSTEIN, H.E., GRIMES, P.G., NEWBY, K.R., and YOUNG, A.R. *Proc. International Congress and Exposition,* 1983, Detroit, Michigan, USA (Society of Automotive Engineers, Inc., Warrendale, Pennsylvania, USA, 1983), SEA Technical Paper 830289, pp. 57–61.

21. BELLOWS, R.J., GRIMES, P.G., EINSTEIN, H.E., KANTNER, E., MALACHESKY, P.A., and NEWBY, K.R. *IEEE Transactions on Vehicular Technology,* 1983, **VT-32**, pp. 26–32.

22. GRIMES, P.G., BELLOWS, R.J., and ZAHN, M. *In* WHITE, R.E. (Ed.) *Electrochemical Cell Design* (Plenum Publishing Corporation, New York, USA, 1984), pp. 259–275.

23. BELLOWS, R.J., EINSTEIN, H.E., KANTNER, E., GRIMES, P.G., and MALACHESKY, P.A. *Proc. 20th Intersoc. Energy Conv. Eng. Conf.,* 18–23 August 1985, Miami Beach, Florida, USA (Society of Automotive Engineers, Inc., Warrendale, Pennsylvania, USA, 1985), pp. 2.70–2.78.

24. *Powerplants for New Generation Electric Vehicles,* Powercell Corporation, Cambridge, Massachusetts, USA.

25. TOMAZIC, G. *Zinc-Flow™ Battery Ready for Mass Production*, Powercell Tech. Paper (Powercell GmbH, Mürzzuschlag A-8680, Austria).

26. BAJPAI, S. *J. Chem. Eng. Data*, 1981, **26**, pp. 2–4.

27. MARINCIC, N. *Safety Aspects of Zinc/Bromine Batteries,* Powercell Tech. Paper (Powercell Corporation, Cambridge, Massachusetts, USA).

28. RYCHCIK, M., and SKYLLAS-KAZACOS, M. *J. Power Sources*, 1987, **19**, pp. 45–54.

29. SKYLLAS-KAZACOS, M., and GROSSMITH, F. *J. Electrochem. Soc.,* 1987, **134**, pp. 2950–2953.

30. SKYLLAS-KAZACOS, M., KASHERMAN, D., and KAZACOS, M. *In* WOOD, L., and JONES, A.J. (Eds.) *New Developments in Electrode Materials and their Applications* (Department of Industry, Technology and Commerce, Canberra, Australia, 1990), pp. 208–213.

31. MENICTAS, C., HONG, D.R., YAN, Z.H., WILSON, J., KAZACOS, M., and SKYLLAS-KAZACOS, M. *Status of the Vanadium Redox Battery Development Program,* Vanadium Battery Research Laboratory, School of Chemical Engineering and Industrial Chemistry, The University of New South Wales, Sydney, NSW 2052, Australia.

32. SKYLLAS-KAZACOS, M., and MENICTAS, C. *Proc. Nineteenth International Telecommunications Energy Conference, INTELEC '97*, 19–23 October 1997, Melbourne, Victoria, Australia (The Institute of Electrical and Electronics Engineers, Piscataway, New Jersey, USA), pp. 463–471.

33. HADDADI-ASL, V., KAZACOS, M., and SKYLLAS-KAZACOS, M. *J. Appl. Electrochem.*, 1995, **25**, pp. 29–33.

34. SKYLLAS-KAZACOS, M. *Vanadium Redox Battery in Vehicular Applications*, School of Chemical Engineering and Industrial Chemistry, The University of New South Wales, Sydney, NSW 2052, Australia.

CHAPTER 8

Sodium Batteries

8.1. Why Sodium Batteries?

At first sight, sodium would appear to be a strange choice of negative active-material to employ in a battery, given the highly reactive nature of this metal. An EV traction battery is a particularly large battery, which is likely to contain tens of kilograms of sodium and be exposed to the risk of damage and rupture through road accidents, followed possibly by exposure to water. Obviously, it is necessary to provide a strong justification for basing an EV battery on hot liquid sodium, together with demonstration of a high level of operational safety.

The battery systems reviewed in the previous Chapters have all used aqueous electrolytes. Zinc, with a reduction potential of –0.76 V, is the most electropositive metal that can be plated from aqueous solution. By contrast, the reduction potential of sodium is –2.71 V. This factor, together with the low atomic mass of sodium (23.0), offers the prospect of a battery with a high specific energy when a sodium negative electrode is coupled to an appropriate lightweight positive electrode. For example, the theoretical specific energy of the sodium/sulfur electrochemical couple (reactants only) is 760 Wh kg^{-1}, compared with 171 Wh kg^{-1} for lead/acid and 217 Wh kg^{-1} for nickel/cadmium. Only through the employment of batteries with high specific energies will it be possible to achieve the desired driving range for the all-purpose, battery-powered EV.

Other attractions of sodium as an electrode material are its natural abundance, its non-toxic nature, and its low cost. Also, sodium metal is a readily available item of commerce. These key advantages make sodium a prime candidate for large batteries, provided the developmental problems can be overcome to produce a design of battery that is reliable, safe and cost-effective. This, then, is the challenge which faces the international community engaged in the advancement of sodium batteries.

8.2. Early Development of Sodium Batteries

An alkali metal battery clearly requires a non-aqueous electrolyte. The available options are an organic liquid or solid polymer with dissolved salts (as employed in lithium batteries, see Chapter 9), a fused salt, or an ionically-conducting ceramic material. Salts mostly melt at elevated temperatures, while ceramic oxides which conduct sodium ions are too resistive at ambient temperatures to be of any practical value. No suitable organic or polymer electrolytes for use with sodium metal have been identified. Moreover, solid sodium (melting point: 98 °C) is considerably more difficult to fabricate than lithium. It follows that any sodium battery is likely to operate at raised temperatures with a negative electrode of liquid sodium and, therefore, attendant problems of thermal management and safety.

The first major technical breakthrough in sodium batteries came in 1966 with the discovery that sodium beta-alumina (a sodium aluminium oxide known since 1916) is highly conductive to sodium ions over temperatures of 200–400 °C as well as being a good electronic insulator. This discovery was made by scientists working in the Ford Motor Company's laboratory at Dearborn, Michigan [1] and encouraged them to demonstrate the concept of the sodium/sulfur battery [2] wherein a ceramic tube of beta-alumina (the electrolyte) served to separate a liquid sodium negative electrode from a liquid sulfur positive. The cell was heated in an oven at 300–400 °C and at such temperatures beta-alumina is sufficiently conducting to give a low-resistance cell. This was a revolutionary battery concept — instead of the conventional battery with solid electrodes and a liquid electrolyte, the new battery had liquid electrodes separated by a solid electrolyte.

A design for a sodium/sulfur cell is shown schematically in Fig. 8.1. Molten sodium is contained within a central tube of beta-alumina that is surrounded with molten sulfur. The sulfur is impregnated into a carbon or graphite felt. The latter provides the necessary electronic conduction since sulfur is an insulator.

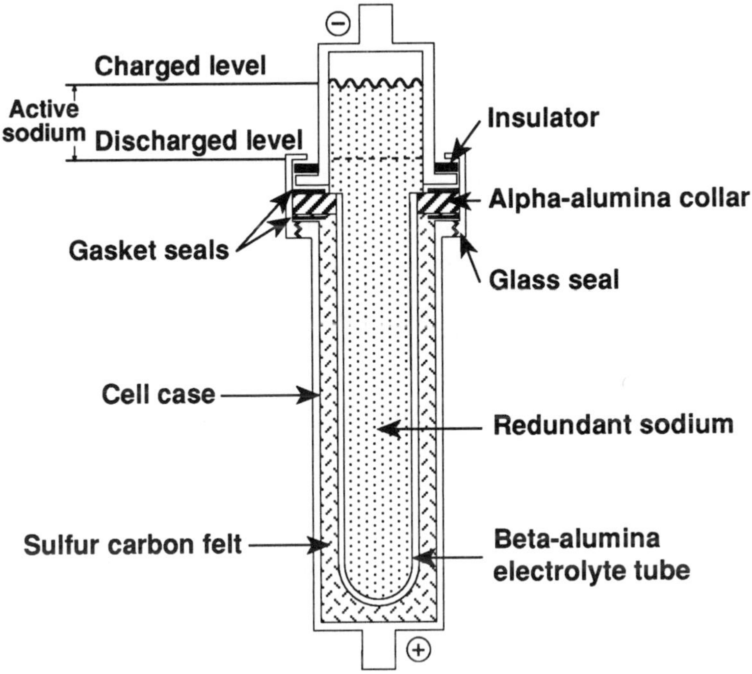

Fig. 8.1. Schematic of sodium/sulfur cell.

The cell discharges in two steps as Na^+ ions pass from the sodium negative electrode, through the beta-alumina, to the sulfur positive electrode, i.e.,

$$\text{step 1:}\quad 2Na + 5S \underset{(350\,°C)}{\rightarrow} Na_2S_5 \qquad E^o = 2.076\ V \tag{8.1}$$

$$\text{step 2:}\quad 2xNa + (5-x)Na_2S_5 \underset{(350\,°C)}{\rightarrow} 5Na_2S_{5-x} \quad (0 < x < 2) \qquad E^o = 2.076 \rightarrow 1.78\ V \tag{8.2}$$

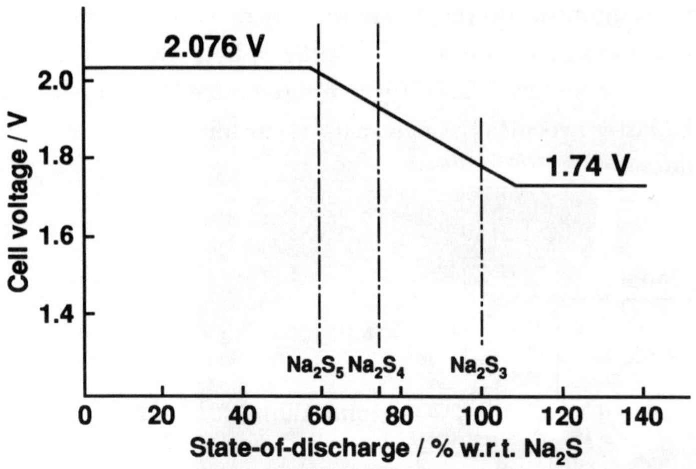

Fig. 8.2. Voltage of sodium/sulfur cell operating at 350 °C for increasing states-of-discharge. (Taken from [3] with permission of Chapman & Hall, UK.)

In step 1, liquid Na_2S_5 is produced in equilibrium with liquid sulfur; the two liquids are immiscible. When all the sulfur has been converted to Na_2S_5, further discharge results in the formation of lower polysulfides of sodium, Na_2S_{5-x}. There is now a single liquid phase and the open-circuit voltage falls progressively from 2.08 to 1.78 V at Na_2S_3 and to 1.74 V at $Na_2S_{2.7}$, as shown in Fig. 8.2.

The prospects offered by this new battery were soon recognized and numerous research and development programmes were set up in France, Germany, Japan, the UK, and the USA. Within a few years, upwards of 100 scientists were engaged in sodium/sulfur battery advancement. In the UK, there were two research teams, initially independent of each other, at the Electricity Council and at British Rail. Progress was rapid and by 1972 the Electricity Council had a 50-kWh experimental battery which operated at a nominal 350 °C. This battery was fitted in a medium-sized van (Fig. 8.3). The first battery consisted of 40 modules, each with 24 cells of 52-Wh capacity. The second experimental battery had 864 cells in 24 modules, each with 36 cells. Altogether, 2000 cells were manufactured and each cell had an specific energy of 160 Wh kg^{-1}. Figure 8.4 shows a

Fig. 8.3. Van powered by experimental sodium/sulfur battery. (By courtesy of Silent Power Limited, UK.)

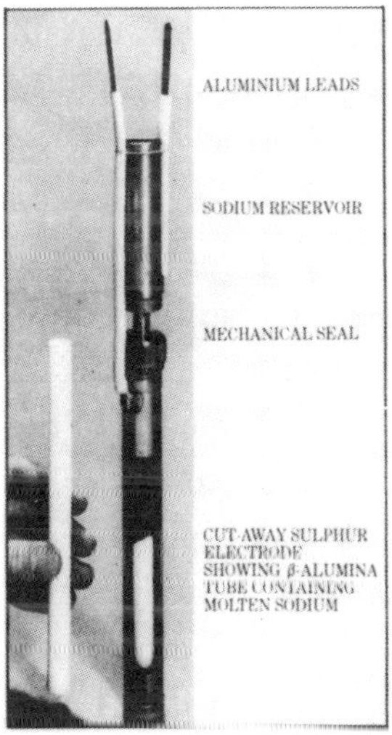

Fig. 8.4. Early (1972) UK central-sodium cell. (Taken from [4] with permission of Silent Power Limited, UK.)

photograph of one of these early cells with a cut-away sulfur electrode and an electrolyte tube alongside.

These batteries were constructed purely to demonstrate feasibility and were by no means optimized or engineered into the vehicle. Nevertheless, for one small group of scientists, starting from scratch, to have achieved this demonstration in three to four years was a remarkable feat, especially in the light of the difficulties encountered subsequently during the further development of sodium/sulfur technology. The van (Fig. 8.3) had a range of 125 km at 55 km h^{-1} on a level track. There is little doubt that this demonstration was the spur to greater investment and effort around the world. Throughout both the 1970s and the 1980s, sodium/sulfur work was undertaken by several hundred scientists, engineers and technicians who were scattered among industrial research laboratories in the above-mentioned five countries. Their efforts were complemented by a series of fundamental studies in universities.

Elsewhere, several variants of the sodium battery were being actively investigated. Because of the insulating nature of molten sulfur and its volatility, Mamantov and co-workers [5] attempted to dissolve the element in molten $NaAlCl_4$ to form a tetravalent sulfur positive electrode, namely, $SCl_3^+ AlCl_4^-$. This discharges in several steps, through S^{II}, S^0, S^{-II}. As an alternative approach, Werth et al. [6] used antimony trichloride dissolved in sodium chloroaluminate; the cell reaction is:

$$3Na + SbCl_3 \rightleftarrows 3NaCl + Sb \qquad (8.3)$$

This cell operated at the lower temperature of 210 °C. Both of these approaches encountered problems and neither was deemed sufficiently viable to warrant serious development.

Around 1980, research began in earnest in the UK and in South Africa on a new concept, the sodium/metal chloride battery, now known as the ZEBRA battery. This retained the molten sodium electrode and the beta-alumina electrolyte, but replaced the sulfur electrode with a solid metal chloride (normally, $NiCl_2$ or $FeCl_2$) which was saturated with molten $NaAlCl_4$ to make physical contact with the beta-alumina. The reversible cell reactions (charged state on the left) are:

$$NiCl_2 + Na \underset{(300\ ^\circ C)}{\rightleftarrows} Ni + 2NaCl \qquad V^o = 2.58\ V \tag{8.4}$$

$$FeCl_2 + 2Na \underset{(300\ ^\circ C)}{\rightleftarrows} Fe + 2NaCl \qquad V^o = 2.35\ V \tag{8.5}$$

The battery differs from the sodium/sulfur battery in many features. Some are improvements and some are not, but perhaps the key advantages are that the cell is assembled in the discharged state and that it is intrinsically safer in operational use. The latter property, in turn, allows the use of larger cells, and this is a significant economic gain. Building on the experience acquired previously with sodium/sulfur, progress was rapid and by 1984, the first EV powered by a sodium/nickel chloride battery was demonstrated on the streets of Derby, UK.

This Chapter deals first with those aspects which are common to both sodium/sulfur and ZEBRA batteries (namely, the ceramic electrolyte, the sodium electrode, the hermetic seals, and the thermal enclosure), and then discusses the individual batteries in greater detail. The aim is to provide an overview of the problems encountered and the solutions that have been adopted. A detailed account of sodium/sulfur development up to 1985 has been given by Sudworth and Tilley [3].

8.3. Beta-Alumina Electrolyte

Sodium beta-alumina, of general formula $Na_2O{\cdot}xAl_2O_3$, occurs in two crystalline forms which are designated: beta-alumina (β-Al_2O_3) and beta''-alumina (β''-Al_2O_3). Neither form normally exhibits exact stoichiometry and each may exist over a range of compositions and may also incorporate other ions as impurities. Both compounds are built up from a common structural unit. This is a spinel-like slab which is comprised of four layers

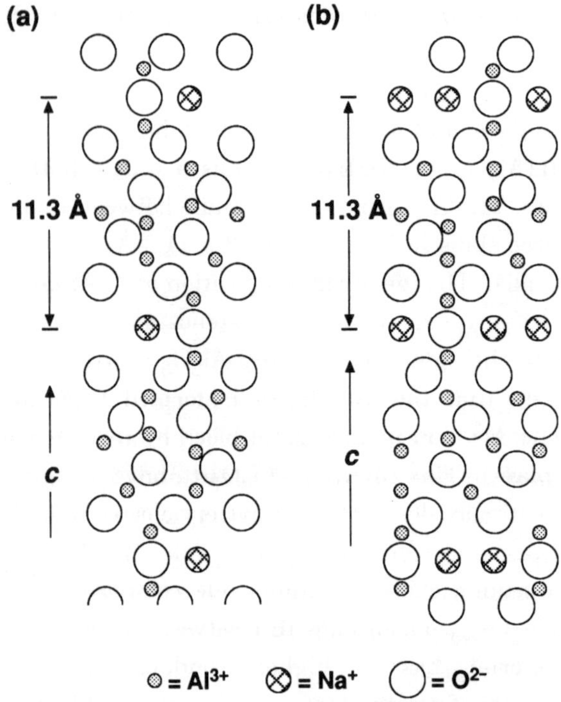

Fig. 8.5. Crystal structure of (a) β-Al_2O_3 and (b) β''-Al_2O_3; projections on $11\bar{2}0$ - plane. The sizes of the atoms are approximately to scale. (Taken from [7] with permission of Elsevier Science Limited, UK.)

of close-packed oxygen ions in a cubic stacking. The slabs are oriented perpendicular to the *c* axis. Between the slabs are relatively open planes which contain one-quarter as many oxygen ions as in the spinel structure. The sodium ions are located at these planes. The two forms of beta-alumina are shown in Fig. 8.5.

The sodium ions are constrained to move in the open planes which are perpendicular to the *c* axis. These planes are separated from each other by 11.3 Å, the thickness of the spinel blocks. The conductivity of Na^+ ions parallel to the *c* axis is effectively zero, while that in the perpendicular plane is so high that it is possible to draw an ionic current of several hundred mA cm^{-2} at 350 °C. This makes the material eminently suitable as a battery electrolyte.

The difference between β-Al_2O_3 and β''-Al_2O_3 is quite subtle. β-Al_2O_3, of idealized formula $Na_2O \cdot 11Al_2O_3$ (or $NaAl_{11}O_{17}$), has a unit cell that consists of two spinel slabs (2 × 11.3 Å) with a mirror plane between them. By contrast, β''-Al_2O_3 has three slabs in its unit cell (3 × 11.3 Å) that are related by a three-fold screw axis. The idealized composition of β''-Al_2O_3 is $Na_2O \cdot 5.33Al_2O_3$, or $Na_6Al_{32}[Vac]_{Al^{3+}}O_{51}$ with the existence of one cation vacancy, [Vac], per unit cell in the spinel blocks. If β-Al_2O_3 is written as $Na_3Al_{33}O_{51}$, it is clearly seen that the β'' variety is formed from the β counterpart by replacing one Al^{3+} ion in the spinel block by three extra Na^+ ions in the conduction plane. This causes β''-Al_2O_3 to have a higher conductivity and, thus, makes it more desirable as a battery electrolyte.

Beta-alumina normally contains excess sodium oxide compared with the idealized formula, viz., $(Na_2O)_{1+x} \cdot 11Al_2O_3$, with x between 0.2 and 0.3. The non-stoichiometric material has a higher conductivity than stoichiometric β-Al_2O_3. This stems from the large number of possible Na^+ sites between the Al-O-Al 'spacer units' that separate the spinel blocks. The detailed crystal chemistry of the conduction plane is a complex topic upon which a great deal of fundamental work has been undertaken [7].

Beta''-alumina is not easy to prepare in the pure form, but is stabilized by the addition of Mg^{2+} or Li^+ ions as dopants. These substitute for Al^{3+} ions in the spinel blocks. Charge neutrality is maintained by the filling of cation vacancies in the spinel blocks, and also by the elimination of interstitial oxygen from the conduction plane. This, in turn, removes a barrier to Na^+-ion diffusion. A second factor favouring Na^+-ion diffusion in the conduction plane of β''-Al_2O_3 is that more sites are crystallographically equivalent and, thereby, the activation energy for diffusion is reduced. The effect is quite significant and, typically, the ionic conductivity of β''-Al_2O_3 is several times that of β-Al_2O_3 at 300–350 °C. In practice, the polycrystalline ceramic material is often a mixture of the two phases and the conductivity varies approximately linearly with composition, provided that other factors such as grain size and density are held constant. The goal of the ceramic technologist is to prepare beta-alumina tubes of as high an electrical conductivity as possible, while maintaining mechanical strength and low susceptibility to atmospheric moisture. This translates to a fine-grained ceramic which is of high β''-Al_2O_3 content and sintered to near theoretical density.

8.3.1. Fabrication technology

The fabrication of beta-alumina electrolyte tubes involves three distinct steps: powder preparation; forming or shaping of the tubes; sintering. Having first settled on the chemical composition of the powder, there is more than one option available for each of the three steps. Beta-alumina is a highly sophisticated technical ceramic which must meet a demanding specification, as set out in Table 8.1. If success is to be attained, it is vital that tight process and quality control is exercised during each of the three preparative stages.

Table 8.1. Specifications for beta-alumina tubes.

• High ionic conductivity at 300 °C	• Good surface finish and closed end
• High density (impervious at 300–400 °C)	• Long life in cells at appropriate current densities
• High fracture toughness	• Resistant to electrical breakdown
• Stable to moisture	• Reproducible properties
• High strength, i.e., fine grain size, no flaws	• High manufacturing yield
• Straightness and close dimensional tolerances	• Rapid, low-cost production
• Low interfacial impedance/polarization, i.e., good wetting	

The first step (i.e., powder preparation) requires careful control of chemical purity and composition. The starting material is normally α-Al_2O_3 powder of a fine particle-size, with soda added as NaOH or Na_2CO_3, magnesia as $Mg(NO_3)_2$ or MgO, and lithia as LiOH or $LiAlO_2$. As an alternative to α-Al_2O_3, the monohydrate (boehmite, AlOOH) can be used. The reactants are generally vibro-milled together and then spray dried in order to improve homogeneity. Typical compositions to yield almost a pure β"-Al_2O_3 phase contain 4.0 wt.% MgO or 0.7 wt.% Li_2O as dopant. It is vital that these dopants are uniformly distributed.

The dried and blended powders are then usually pressed isostatically to form 'green' tubes which are handleable with care. The precise details of the mould filling and pressing operation are critical if the tubes are to be free of macro-defects and to have a good surface finish. The spray-dried powder should be soft so that the particles lose their identity in the pressing operation. An alternative approach used by some developers is to pre-react the mixed powder to produce beta-alumina without sintering, and then to form the tubes by electrophoretic deposition on metallic formers. Both approaches have been shown to work satisfactorily and the ultimate choice is a matter of production engineering and cost.

The final sintering stage may be carried out either by programmed sintering or by zone sintering. The former method involves loading a large furnace with several hundred tubes and following a predetermined sintering schedule. A complete cycle of heating, sintering and cooling takes 24 h. The peak temperature depends on the precise composition, but is near to 1600 °C. At this temperature, β"-Al_2O_3 has a significant vapour pressure of Na_2O and it is therefore necessary to take special steps to prevent loss of soda. Each tube is enclosed within an outer, sealed magnesia tube before loading into the furnace. The latter tubes can be reused many times. By contrast, zone sintering involves feeding tubes, one at a time, through an inductively heated furnace held at ~1700 °C. The feed rate is typically 40–50 mm per minute. The feed powder is converted to beta-alumina at ~1600 °C, before it reaches the final sintering zone. A Na_2O-rich atmosphere in the furnace represses the loss of soda and densification is rapid; it is completed within ~1 min. The tube must be supported and rotated to prevent sagging during sintering. A post-sintering annealling process at 1400 °C reduces quenched-in stresses and improves the conductivity.

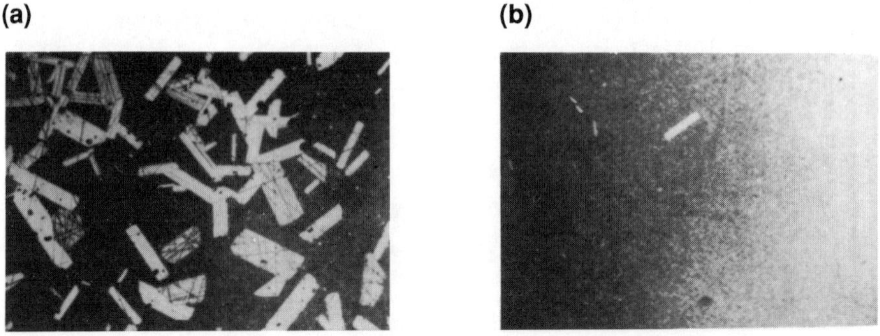

Fig. 8.6. Micrographs of: (a) duplex and (b) fine-grained beta-alumina. Etched specimens. (Taken from [8] with permission of Elsevier Science Limited, UK.)

Beta-alumina displays a marked tendency towards a characteristic duplex grain growth, as shown in Fig. 8.6(a), whereby a comparatively few grains grow at the expense of the rest. This results in a matrix of fine-grained material which contains some relatively large crystals. Such a structure leads to a serious weakening of mechanical strength and is therefore detrimental; fortunately, with careful control of the sintering conditions, the problem can be largely avoided, see Fig. 8.6(b) [8].

Electrolyte tubes are fabricated in many different diameters and lengths according to the desired application. Although tubes of up to 55 mm in diameter and 650 mm in length have been made successfully, small tubes are generally preferred for EV applications. After sintering, the tubes are subjected to optical examination as well as to a dye penetration test in order to identify any defects or cracks that may be present. The density of the sintered tube should exceed 3.22 g cm^{-3} (i.e., 99% of theoretical).

8.3.2. Electrical and mechanical properties

Measurements of a.c. impedance on single crystals of beta-alumina have shown [9] the activation energy for Na^+-ion conduction to be 0.13 eV. At 300 °C, the resistivity of a single crystal of β-Al_2O_3 is 4.7 Ω cm compared

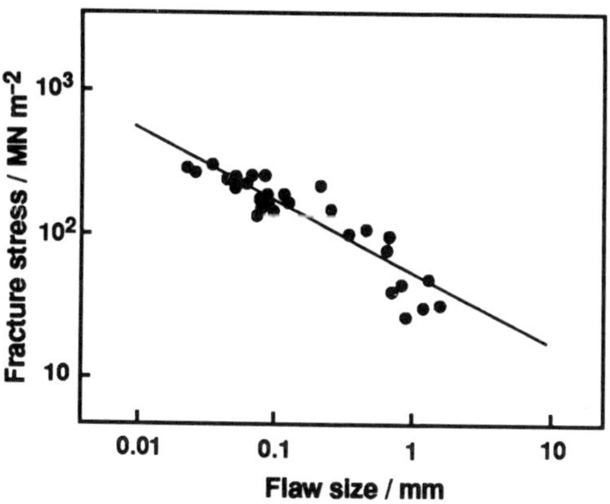

Fig. 8.7. Variation of fracture stress with flaw size of beta-alumina. (Taken from [10] with permission of Chapman & Hall, UK.)

with 1 Ω cm for β"-Al_2O_3. The resistivity of the polycrystalline material is 15–20 Ω cm at 350 °C for the pure β phase, and falls linearly to ~2 Ω cm for the pure β" phase. Unfortunately, the β" form is more susceptible to moisture than the β counterpart and this results in a deterioration in mechanical properties and the need to handle the material under dry-box conditions. A compromise is therefore necessary and, in practice, a two-phase mixture of resistivity ~5 Ω cm at 350 °C is regarded as acceptable for battery electrolyte tubes. It should be noted here that the attainment of a low overall resistance in an electrolytic cell depends not only upon having a low-resistance electrolyte, but also upon achieving a good contact between the electrolyte and the reactants, i.e., the tube should be well wetted by both the sodium (see Section 8.4) and the positive-electrode melt.

Good mechanical properties for the beta-alumina tube dictate the need for a uniform, fine-grain, ceramic structure. With such a structure, a specimen cut from an electrolyte tube will, typically, have a tensile strength of ~250 MN m^{-2}. By contrast, hydraulic rupture tests on whole tubes often report considerably lower strengths. This is explained by the presence of one or more defects (flaws) of considerable size in the tube. Such defects may be voids, cracks, porous regions or large crystals that arise from exaggerated grain growth. The data given in Fig. 8.7 show how the fracture stress declines logarithmically as the size of the defect increases logarithmically [10]. Clearly, the aim is to eliminate defects of size >30 μm.

8.3.3. Failure modes of beta-alumina

The problems of fabricating high quality, long life, beta-alumina electrolyte tubes have now largely been solved by commercial battery developers. Modern electrolyte tubes perform satisfactorily in both sodium/sulfur and sodium/metal chloride cells and have lives in excess of 1000 charge–discharge cycles under a regime pertinent to EV usage. Such performance has been the culmination of a substantial research and development effort during which much has been learnt about the various possible failure modes of beta-alumina. These failure modes are listed in Table 8.2 [8].

Table 8.2. Failure modes of beta-alumina tubes [8].

Mode of failure	Cause	Observation
Electrical breakdown	• Excessive current density on charge • Voltage breakdown	• Short-circuit by sodium dendrite • Gross physical damage
Mechanical shock	• Intrinsic defect in tube • Externally applied shock	• Cracking or fracture at weak spot
Thermal cycling of cell	• Expansion mismatch between tube and seal • Rapid heating and cooling	• Cracking near seal • Tube cracks below 150 °C as reactants melt

Electrical breakdown may occur by at least two different mechanisms; both take place during charging of the cell. High-quality electrolyte, when cycled under moderate current densities, has exceptionally long life; lives of 7000 Ah cm^{-2} at a current density of 1 A cm^{-2} have been reported. Under less favourable circumstances, it is a common experience that, after a variable period of cycling, a cell will fail to hold its charge. An internal short-circuit has developed and this rapidly worsens with further cycling.

Post-test examination reveals that a filament of sodium metal (or 'dendrite') has penetrated the wall of the electrolyte tube. These dendrites appear to be initiated at micro-cracks or imperfections at the exit (negative electrode) surface, and it seems that sodium is discharged preferentially in such fissures. The problem is minimized by having a smooth, blemish-free surface to the tube on the negative-electrode side and by not charging the cell at too high a rate.

The second class of electrical breakdown applies particularly to sodium/sulfur cells. As these cells have no mechanism for overcharge, they cannot pass current when fully charged. In a battery of series-connected cells, the cell which has the least capacity (the 'weakest' cell) will reach the top-of-charge first and then the entire battery voltage will fall across this cell. This can lead to voltage breakdown of the electrolyte at voltages above ~60 V and, consequently, to gross physical damage of the beta alumina. At first sight, it is surprising that such damage results at a few tens of volts. The explanation appears to be that, at the top-of-charge, a thin layer of electrolyte adjacent to the positive electrode is depleted in Na^+ ions and becomes highly resistive; the voltage is then dropped across this thin layer and yields a field strength which is sufficiently high to cause dielectric breakdown and localized melting. The problem is mitigated in sodium/sulfur batteries by using a series–parallel arrangement of cells with frequent cross-connections between the series chains (see Section 8.7.4).

Mechanical breakdown of the electrolyte tube, manifested as macro-cracking, can result from in-built stresses in the tube (especially at the glass seal), from stresses imposed during cell assembly, or from externally applied shock. Sodium, which at 300 °C has a viscosity lower than that of water at ambient temperature, readily penetrates cracks and reacts with the positive active material. The resultant rise in temperature is quite severe with liquid sulfur and leads to thermal shock and further cracking. This compounds the problem and, ultimately, the tube may disintegrate completely. In-built stresses can be avoided by exercising care in the manufacture of both the electrolyte and the glass-seal assembly, as well as in the welding of the cells. Problems of external shock are difficult to avoid for EV batteries, particularly under accident conditions, but much can be achieved by correct siting and mounting of the battery in the vehicle.

Thermal breakdown may result from primary electrical or mechanical breakdown, as just described, or from one of two other causes. The first is fracture of the seal assembly that results from a mismatch in the thermal expansion coefficients of the beta-alumina tube, the alpha-alumina collar, and the glass used to join them (see Section 8.5). The second is breakdown which arises from thermally cycling the cell between the ambient and the operating temperature. At ambient temperature, the cell is totally solid — electrodes as well as electrolyte. On warming, the sodium metal on the negative electrode side and the salt on the positive electrode side (S/NaS_x for sodium/sulfur, $NaAlCl_4$ for ZEBRA) both expand before melting. Furthermore, on melting, there is a significant increase in molar volume which, under unfavourable conditions, can give rise to unacceptable stress on the beta-alumina. In the early days of sodium battery research, the thermal cycling of cells posed a major problem. This has now largely been overcome by a combination of measures. These include arranging (through modified cell design) for the heating and melting to take place from the surface downwards so that the stress exerted on the tube is minimized, and carrying out the operation slowly, especially when passing through melting points. The ZEBRA cell has a certain advantage here as solid $NaAlCl_4$ is a very soft material which deforms readily. Modern designs of cell are perfectly capable of handling thermal cycling, and entire EV batteries have been subjected to such treatment without the failure of a single cell.

From the foregoing discussion, it will be appreciated that beta-alumina is a highly sophisticated technical ceramic, the manufacture of which poses demanding scientific and technological problems. Solid-state scientists and ceramic technologists have responded well to the challenge and there is now a wealth of scientific knowledge on the crystal structure and ionic conduction processes in beta-alumina, as well as established production routes for the manufacture of electrolyte of high quality.

8.4. Sodium Electrode

An essential pre-requisite for the molten sodium electrode is the use of pure sodium metal, uncontaminated by metallic impurities or by sodium oxide, hydroxide or carbonate. Contaminants tend to concentrate at the surface of the electrolyte. At best, this behaviour reduces the area of active electrolyte; at worst, it may lead to rapid failure of the tube. Calcium is particularly harmful in this regard. If present as an impurity in the electrolyte, calcium progressively diffuses to the sodium electrode surface where it reacts with oxide impurity in the sodium to form CaO, and this leads to a steady increase in cell resistance [3]. In order to ensure cleanliness, it is usual to fill laboratory sodium/sulfur cells with high-purity sodium in an argon-filled dry box. Larger-scale production techniques involve metering out the desired amount of molten sodium in a sealed, stainless-steel system and then transferring to the pre-evacuated cell using pressurized, inert gas. An alternative approach is to use a preformed sodium electrode assembly (see Section 8.7.2). The ZEBRA cell has a distinct advantage in that it is assembled in the discharged state and no handling of liquid sodium is necessary (see Section 8.8.2).

In order to minimize the internal resistance of the cell, it is necessary to ensure good wetting of the electrolyte surface by the molten sodium. This requires not only pure sodium, but also a smooth and clean surface. The concept of wetting a beta-alumina surface by molten sodium is not straightforward. The degree of wetting depends upon the smoothness of the surface, the temperature, and the presence of impurities adsorbed on the surface. Smooth surfaces are more readily wetted than rough surfaces. On heating a beta-alumina tube in molten sodium, wetting begins at >150 °C but is not complete until 300–350 °C. Once wetting has been achieved, it is maintained on cooling and there is then no interfacial polarization at the Na/β-Al_2O_3 surface.

When beta-alumina is exposed to moist air, it tends to form surface carbonate which, in extreme cases, can grow as crystalline needles. β''-Al_2O_3 is particularly susceptible to this effect, presumably on account of its higher Na^+-ion content. Such surface films are inclined to inhibit wetting and increase polarization. On the other hand, it is known that wetting is facilitated by treating the electrolyte surface with lead acetate solution [3]; this compound is reduced to a thin surface film of lead, which is wetted readily by sodium.

The design of the sodium electrode is intimately bound up with the design of the overall cell (see Section 8.7.2) and with considerations of safety (see Section 8.7.4). The general requirements are:

- to ensure that the entire electrolyte surface remains wetted with sodium throughout the discharge reaction without having excessive quantities of redundant sodium present
- to restrict the amount of sodium in proximity to the tube so that in the event of tube fracture the chemical reaction will be limited in extent.

These two requirements are best met by the use of a close-fitting wick around the electrolyte tube. Typically, a beta-alumina tube may be 20 cm tall in a vertically mounted cell, and it is necessary to wick the sodium up to this height. The wick consists of a cylindrical liner of aluminium or stainless steel that fits tightly around the inside or outside of the beta-alumina tube in the negative compartment. Stainless steel is preferred to aluminium on account of its higher melting point, a factor which becomes important in the event of tube fracture. Sometimes, a woven metal cloth of 300-mesh stainless steel forms the wick and is held in place by the cylindrical steel liner. The low viscosity of molten sodium at 300–400 °C facilitates penetration into the wick and, consequently, allows smooth operation of the electrode. Depending upon the wicking height and the ease of wetting, it may be necessary to use gas pressurization to assist in wicking the sodium. Here again, the ZEBRA cell has a definite advantage as the sodium metal is produced by diffusion through the electrolyte when the cell is first charged. This process yields the purest possible form of sodium. Furthermore, the sodium is deposited directly in the wick and simultaneously wets the beta-alumina surface.

The detailed design of the sodium electrode is also bound up with the procedures which are employed for cell manufacture and assembly. Two criteria are particularly important on the sodium side, namely: (i) to ensure that no gas bubbles are trapped in the electrode, as these will adversely affect both wicking and wetting; (ii) to effect a design such that the volume expansion of sodium on melting (during both initial heating and thermal cycling) will not exert undue stress on either the electrolyte tube or the seal assembly. These requirements are best accomplished by heating the cells slowly, and from above, so that the sodium metal melts from the surface downwards.

8.5. Seals

The design and manufacture of satisfactory seals for sodium cells is one of the more intractable problems which face the battery developer. In particular, there are difficult issues with respect to materials science and production engineering. Seals have to be made at the following three junctions: (i) the beta-alumina electrolyte tube and the insulating ceramic collar or 'header', the latter is usually fabricated from high purity alpha-alumina; (ii) the sodium (negative) compartment and the insulating collar; (iii) the positive compartment and the insulating collar.

The first of these seals serves to isolate the negative compartment from the positive compartment and must be resistant to chemical attack by both hot sodium metal and the constituents of the positive compartment. The alpha-alumina collar acts as an electrical insulator to prevent internal short-circuiting of the cell. The second and third seals are to prevent the ingress of air to the cell and the escape of toxic or corrosive fumes. These, too, must be resistant to chemical attack by sodium and by the positive active materials, respectively.

In addition to these already difficult specifications, the seals must meet four other criteria. First, they must be as lightweight as possible in order not to lower unduly the specific energy of the cell. Second, they must withstand repeated thermal cycling; this requires the expansion coefficients to be closely matched. Third, their design and manufacture must be seen as an integral part of the cell design and assembly operation. For instance, it is impractical to have a preformed seal assembly which does not allow the sodium and sulfur electrodes to be introduced subsequently into the cell, nor, for the alternative approach of filling the cell first, to have a seal which cannot be made in the presence of the reactants. Again, the cell design and the assembly operations must not impose unacceptable stresses or bending moments on the seals. Fourth, the seals must be capable of mass production at acceptable cost. When all these factors are taken into account, the problem changes from being a difficult task to being a formidable challenge. Fortunately, an enormous effort has been put into meeting this challenge by able materials scientists and engineers, and practical solutions have been achieved.

The first seal, that of the beta-alumina tube to the alpha-alumina collar, is universally made from glass. Generally, there is not too much of a chemical problem on the positive-electrode side as many glasses resist attack by sulfur/sodium polysulfide and by sodium chloraluminate. The negative-electrode side is a different matter as most glasses are susceptible to attack by sodium to some degree. The problem is to identify a glass composition which is both acceptably stable and has a thermal expansion coefficient which is well matched to that of alpha- and beta-alumina. Fortunately, these two ceramic materials have rather similar expansion coefficients (α-Al_2O_3: 8.8×10^{-6} K^{-1}; β-Al_2O_3: 7–8 $\times 10^{-6}$ K^{-1}) and, thus, it is necessary to identify a glass with an expansion coefficient in this range. The glass should contain neither too great a proportion of silica (which is readily attacked by sodium), nor leachable impurities (such as potassium and calcium) which are highly detrimental to beta-alumina. It has been found that glass compositions based on barium alumino-borate (BAB) are generally acceptable for use in sodium/sulfur cells, with dopants added to ensure that the seal remains in compression between room temperature and 350 °C. Since BAB dissolves slowly in $NaAlCl_4$, glass seals with other compositions are used in ZEBRA cells.

There are many possible geometric designs for the alpha-alumina header. Early cells had a relatively massive horizontal flange which was attached to the body of the cell and the sodium reservoir by compression seals with gaskets (as in Fig. 8.1). The electrolyte tube was glassed into a hole recessed in the flange. Later cell designs used a lightweight, 'slimline' collar of alpha-alumina, one example of which — an all-glass seal — is shown in Fig. 8.8. It was found convenient to fabricate the glass for the β-Al_2O_3/α-Al_2O_3 seal in the form of a spiral spring which fitted tightly around the electrolyte tube. Because of the high melting point of the glass (typically, 900–1000 °C), it is conventional to perform the glassing operation before other bonding and assembly activities.

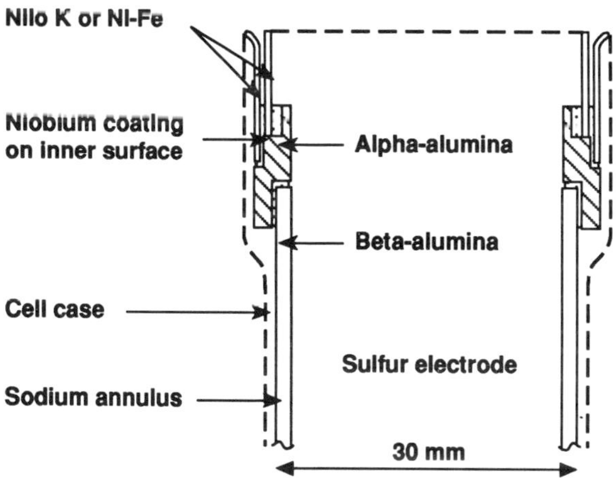

Fig. 8.8. Glass–metal seal arrangement for a central-sulfur cell. (By courtesy of AEA Technology, Harwell Laboratory, UK.)

There are numerous possibilities for positive and negative compartment seals. After the early gasket seals, with heavy bolts or disc springs to compress the members together, attention focused on truly hermetic seals. Three options have been evaluated extensively, viz., glazed seals, brazed seals, and thermo-compression bond (TCB) seals. Hermetic glass–metal seals are well known and are employed in the manufacture of incandescent and fluorescent lamps. Generally, a nickel–iron alloy, such as Nilo K, is employed because its thermal expansion coefficient is close to that of hard glass. In order to attain a good bond, it is necessary to pre-oxidize the metal surface. This procedure proved to be convenient and satisfactory for sealing the positive compartment. There was a problem, however, with the negative compartment. Sodium vapour gradually and progressively penetrated the interface between the alloy and the glass; it reduced the oxide film on the metal and, ultimately, this gave rise to seal failure. The problem was solved by pre-coating the surface of the nickel–iron alloy with a thin layer of niobium (as shown in Fig. 8.8), a metal whose oxide is not reduced by sodium; unfortunately, this is a relatively expensive solution. One attraction of using glass seals for the electrode compartments is that the entire glazing operation, for all three seals, can be carried out simultaneously in a single step. Because of the high temperature involved, it is necessary to use an inert-gas atmosphere to protect the metal components from further oxidation. Figure 8.9 shows the components for an all-glass seal assembly for a central-sulfur cell. In the central image, the alpha-alumina collar, the two glass circular preforms and the two Nilo K components are laid out in sequence.

Fig. 8.9. Components of glass–metal seal assembly for a central-sulfur cell. (Taken from [11] with permission of AEA Technology, Harwell Laboratory, UK.)

Brazed ceramic–metal joints are well known in industry and so it was natural to investigate brazing as a technique for making sodium/sulfur seals. Using alpha-alumina components of similar design to that shown in Fig. 8.8, the ceramic is first metallized in the regions where the seal is to be made. On the sodium side, a bronze alloy braze (Cu–Ni–B) was found to make a satisfactory seal. This was not achievable on the positive side because of the corrosive properties of sulfur. The only satisfactory joint proved to be one in which stainless steel was brazed to alumina through the use of a gold or gold–nickel alloy. This was deemed to be too expensive for EV batteries and, therefore, was not pursued further. Another feature of brazing is that because of the high temperature involved (1100 °C for the conventional braze), it is necessary to glass the preformed seal assembly to the beta-alumina tube *after* the brazing operation. This is perfectly feasible, but does dictate the order of cell assembly.

The third technique for making a hermetic seal, namely, thermo-compression bonding (or diffusion bonding), TCB, is the one which has been the most widely adopted. A thin foil of a soft metal, such as aluminium, is inserted between the ceramic and the metal member to be bonded, and is then subjected to heating under pressure. The aluminium diffuses atomically into the ceramic and forms an intermetallic compound with the metal. Under optimum conditions, the result is a truly hermetic seal which is highly durable. Aluminium is the ideal inter-layer material for use in sodium/sulfur seals: it is cheap, is readily available as foil, has a low melting point, is soft (which allows it to accommodate metal and ceramic components with substantially different expansion coefficients) and, most important of all, is passive to corrosion by both sodium and sulfur/polysulfide. Different geometric designs of TCB seal have been developed for use in either central-sodium or central-sulfur cells.

For ZEBRA cells, the situation is quite different on the positive-electrode side. The sodium chloraluminate electrolyte is highly corrosive towards aluminium under polarizing conditions; indeed, the corrosion of aluminium is the first step in the charging of newly manufactured cells (see Section 8.8). It is therefore not possible to use aluminium as the inter-layer material on the positive-electrode side. This situation has been overcome by the development of a satisfactory nickel diffusion bond. Although nickel is also corroded during the charging reaction, there is a sufficient amount of porous, reactive nickel present that the thin inter-layer, sandwiched between the ceramic and the metal components, is not in practice attacked at a significant rate.

In conclusion, it should be noted that a number of technically satisfactory solutions to the problem of hermetically sealing sodium/sulfur and ZEBRA cells have now been developed. Glass seals, brazed seals and TCB seals are all feasible solutions. The problem is now one of production engineering and cost. For applications where cost is a secondary consideration and reliability is paramount, for example in aerospace operations, the best technical solution will be adopted. For EVs, however, acceptable cost assumes far greater importance and it is necessary to minimize the overall cost of battery manufacture. The sealing operations then have to be optimized for cost within the overall framework of cell and battery assembly, while maintaining an acceptable level of performance, reliability and safety. One operation which is quite expensive is the manufacture, and particularly the surface grinding, of the alpha-alumina collar. Ideally, a slip-cast component would be employed that did not require machining. Attention is now being directed towards such cost-reduction measures.

8.6. The Battery Enclosure

Sodium batteries operate at elevated temperatures (250–450 °C) and this dictates the use of a thermal enclosure in order to maintain the operating temperature with minimum heat input. Electrical energy must be supplied in order to heat up the battery and to maintain it at the desired temperature when not in use. This heat is equivalent to an electrical inefficiency or self-discharge mechanism in a conventional battery and, therefore, must be kept to a minimum. For the heat loss to be a small percentage of the stored energy, it is necessary to consider only large batteries in which the surface-to-volume ratio is small. Accordingly, sodium batteries are normally considered only for multi-kWh applications, such as EV traction and load-levelling in the electricity-supply industry.

The minimum temperature at which sodium/sulfur batteries may be operated is 285 °C, as determined by the freezing point of Na_2S_4, and the maximum temperature is ~400 °C, above which the rapidly increasing vapour pressure of sulfur leads to enhanced corrosion. In the case of

ZEBRA batteries, the minimum temperature is ~175 °C, as determined by the increasing resistance of the beta-alumina electrolyte; the maximum temperature depends upon the positive active-material. With ferrous chloride positives, the maximum operating temperature is ~300 °C, above which complications in the electrode reactions are encountered. Nickel chloride batteries may, however, be operated safely up to 450 °C, or even higher. The normal range for the operating temperature of sodium/sulfur batteries is usually taken as 300–400 °C, and that for ZEBRA batteries as 250–300 °C ($FeCl_2$) or 250–400 °C ($NiCl_2$). The thermal challenge facing the battery designer is to maintain the batteries within these temperature limits under all conditions of charge, discharge and non-use. Moreover, this must be achieved without unduly degrading the gravimetric and volumetric energy densities, increasing the cost beyond an acceptable limit, or using excessive amounts of electricity.

The general principle adopted by all developers is to maintain the battery hot at all times; electrical heating is used when the battery is not in use. Only under exceptional circumstances would the battery be allowed to cool to ambient temperature, e.g., when the EV was not in use for an extended period. Overnight or over the weekend, the battery would be plugged into the mains for recharging. Nevertheless, it is accepted that the battery must maintain the required range of operating temperature for at least 24 h when unused if it is to be commercially acceptable. If necessary, this period could be achieved, or prolonged, by self-discharging the battery through the in-built electrical heaters. At high discharge rates, during EV duties (e.g., acceleration or hill-climbing), the rate of heat liberation is likely to exceed the rate of heat loss. The design engineer relies on the thermal mass of the battery to smooth out most of the temperature variations which arise from the driving profile, although under some conditions of use it may be necessary to have active cooling.

Early sodium/sulfur batteries, as used to power the van shown in Fig. 8.3, were contained in a double-skinned box which was insulated with mineral fibre material. This is not a particularly good insulant and heat losses were considerable. Since then, the technology has moved on and modern vehicle batteries, both sodium/sulfur and ZEBRA, employ double-skinned, stainless-steel, vacuum enclosures. These are of thin gauge steel and, in order to prevent collapse under the pressure differential, a load-bearing, glass-fibre thermal insulation board is fitted between the inner and outer skins. This provides excellent thermal insulation, particularly if an opacifier is added to reduce infrared radiation loss. Inevitably, however, there are small heat leaks where the high-current electrical connectors pass through the vacuum enclosure. Generally, these vehicle batteries have, at operating temperature, an outer surface skin temperature which is ~10 °C above ambient. The actual heat loss depends upon the battery size and temperature, but lies in the range 100–300 W.

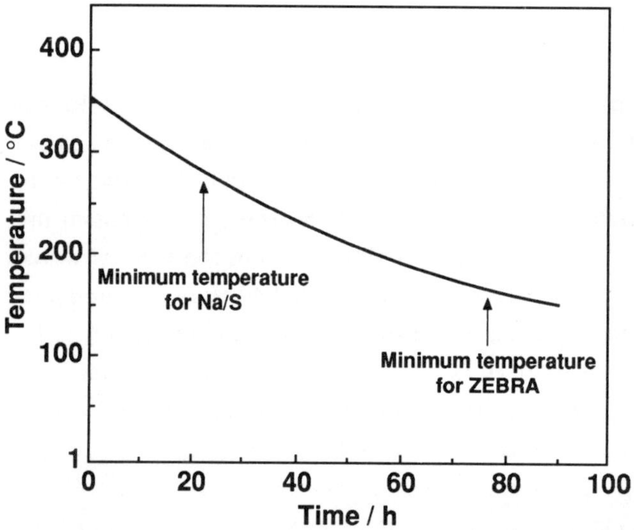

Fig. 8.10. Cooling curve for 40-kWh $NiCl_2$ ZEBRA battery. (Taken from [12] with permission of Beta Research & Development Limited, UK.)

A cooling curve for a 40-kWh ZEBRA battery is given in Fig. 8.10 [12]. Because the ZEBRA battery can operate to a lower temperature than sodium/sulfur (175 vs. 285 °C), it is able to stand unheated for much longer periods (viz., 3 days compared with 1 day). This is a significant advantage.

Continuous high-rate cycling as, for example, in an electric bus or taxi, is likely to cause a gradual increase in temperature to an unacceptable level. This would necessitate active cooling. Two systems have been considered: forced air cooling using a blower, and liquid cooling. Each has its drawbacks. Air cooling is not particularly efficient, is difficult to implement between the closely packed cells, and the blower is a further parasitic electrical load on the battery. Moreover, the low specific heat of gases necessitates the use of large volumes of air. One particular design of sodium/sulfur battery incorporates heat-conducting aluminium plates above and below the cells to help smooth temperature variations across the battery, and to conduct heat to channels where air cooling is effective [13].

Liquid cooling, on the other hand, is both efficient and compact. It employs an external radiator and pump to circulate high-temperature oil, as a cooling medium, through an internal heat exchanger. The ABB sodium/sulfur batteries mounted in the Ford Ecostar vans (Section 8.7.4) were cooled in this way and the batteries were held at 325 °C. ZEBRA batteries (Section 8.8.5) have successfully employed both air cooling and liquid cooling. The latter is preferred as it is more compact; it was found possible to remove 20 kW of heat continuously for an expenditure of only 250 W of pumping energy [14]. This surplus heat may then be used for heating the passenger cabin, although it is available only when the battery needs cooling. The integrated battery design must ensure that at other times the heat loss through the connecting pipes, battery terminals and current leads is held to a minimum.

The size and shape of the battery container depends primarily upon the volume, shape and distribution of the space available in the vehicle. Most sodium-powered EVs demonstrated to date have been commercial ICEVs adapted to electric drive. This is far from ideal as the battery has to be accommodated in a space which has not been designed specifically for such a purpose. In the case of a car, this is generally where the engine had been located, or in the luggage compartment. With vans, which are higher than cars, attempts are made to locate the battery under the floor although often it intrudes into the load-carrying space. These adaptations are by no means perfect and ideally the vehicle, the electric drive and controller, and the battery should all be designed together.

A characteristic of sodium batteries is that although they possess three or four times the energy of lead/acid batteries on a gravimetric basis, their volumetric energy density compares less favourably. For this reason, space is at a premium if the vehicle is to have a range several times that of a lead/acid-powered EV. Early designs of sodium/sulfur battery assumed that it would be necessary to gain access to the battery and to open the container in order to effect repairs and to exchange failed cells or modules of cells. As the difficulties of integrating the battery into the vehicle became apparent, and particularly the difficulty of gaining access to service the battery, a new philosophy developed. With the ever improving reliability of cells and the demonstration of longer cycle-lives without failure, the target changed to that of a battery which would be totally sealed and maintenance-free for its entire service life of three to five years. This simplifies the design of the battery box and where it may be located in the vehicle, although consideration has still to be given to factors such as the floor height and ground clearance of the vehicle and protection of the battery from serious impact damage.

From the foregoing, it becomes clear that the battery enclosure is the engineering interface between the sodium cells and the vehicle which they propel. The enclosure has to have the appropriate thermal insulation properties, to be lightweight so as not to degrade unduly the specific energy, and to be of low cost. But also it acts as the conduit whereby electricity passes to and from the battery cells and to the in-built electrical heaters, and whereby cooling is introduced if needed. Finally, the enclosure must be of a size which is adequate to accommodate the required level of energy storage, while its shape must be acceptable to both the cell and the vehicle design engineers. This is by no means a simple set of specifications.

8.7. Sodium / Sulfur Cells and Batteries

The original idea for a rechargeable sodium/sulfur cell was conceived by J.T. Kummer during his work at the Ford Motor Company in the mid-1960s. He believed that the concept was possible if a stable, ionically conducting membrane could be found [3]. There followed the search that led first to beta-alumina and then to beta"-alumina. Subsequently, the feasibility of the secondary sodium/sulfur cell was demonstrated by the Ford workers and this, in turn, resulted in the setting up of major research and development programmes in several countries. As part of these activities, a great deal of basic work was carried out on the physical chemistry and electrochemistry of the sulfur electrode.

8.7.1. The sulfur electrode

Sulfur melts at 113 to 120 °C (depending upon the allotrope) to form an amber coloured, mobile liquid. On heating further, it becomes exceedingly viscous above 160 °C (~20 000 cP) and darkens in colour. At still higher temperatures, the viscosity declines again (to ~500 cP at 350 °C) and the colour deepens to almost black. Liquid sulfur is an excellent insulator (it conducts neither electrons nor ions) and thus, at first sight, it is quite unpromising as a positive active-material. Fortunately, sodium polysulfides, the discharge products of the sodium/sulfur reaction, are good ionic conductors (though electronic insulators). In order to conduct electrons from the positive current-collector to the reaction site, the approach universally adopted is to fill the electrode compartment with a carbon or graphite felt into which liquid sulfur is impregnated. Sodium ions are conducted from the beta-alumina electrolyte to the reaction site via the liquid polysulfide phase (which is immiscible with liquid sulfur), while the carbon felt conducts the electrons which are required to reduce S to S_x^{2-}. This, then, is the basis of the sulfur electrode.

The sodium/sulfur phase diagram from S to Na_2S [15,16] is shown in Fig. 8.11; the region of interest for the battery is that between 300 and 400 °C. The first discharge product, Na_2S_5, has a melting point of around

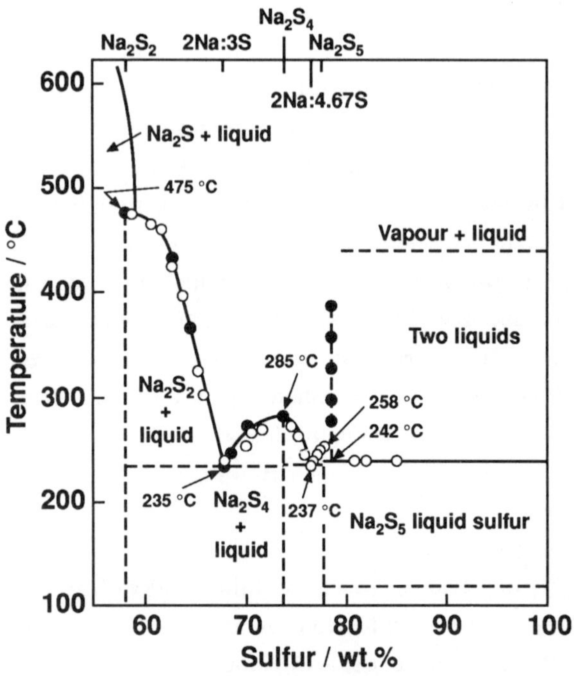

Fig. 8.11. Na_2S–S phase diagram. (Taken from [15] with permission of The Electrochemical Society, Inc.)

242 °C and the liquid is immiscible with liquid sulfur. This product is formed in the so-called 'two-phase' region of discharge over which the cell voltage is invariant (2.076 V, see Fig. 8.2). Initially, the discharge takes place at the beta-alumina surface and the electrons reach the reaction site via the carbon felt. Gradually, droplets of Na_2S_5 are formed and these serve to conduct Na^+ ions further into the sulfur melt. Thus, discharge progresses from the electrolyte towards the positive current-collector and always takes place at the three-phase contact of the polysulfide, the sulfur and the carbon felt. At the same time, it is likely that some Na_2S_5 droplets also break away and diffuse into the bulk of the liquid sulfur under the influence of convection currents. The extent of this diffusion is restricted by the fact that beta-alumina is wetted preferentially by polysulfide, while sulfur wets preferentially the carbon felt. Ultimately, all the sulfur is converted to Na_2S_5.

Beyond this point, further discharge occurs in the single-phase region where Na_2S_5 is progressively converted to lower polysulfides. Na_2S_4 has a melting point of 285 °C and it is necessary for the temperature to remain above this value during discharge of the cell. Within the single-phase region, the cell voltage falls linearly from 2.076 to 1.74 V, where solid Na_2S_2 separates out and is in equilibrium with the liquid polysulfide to yield a second constant-voltage plateau (Fig. 8.2). In practice, it is customary to terminate the discharge at 1.78 V, which corresponds to the formation of Na_2S_3. The specific energy of the cell reaction $2Na + 3S \rightarrow Na_2S_3$ is then 760 Wh kg^{-1}, the theoretical value normally cited for the sodium/sulfur cell. No difficulties are generally encountered in fully discharging the sulfur electrode at quite high current densities (i.e., several hundred mA cm^{-2}). This is attributed principally to the fact that polysulfides wet beta-alumina, while liquid sulfur wets carbon felt — factors which both favour the discharge reaction. It should be noted, however, that at high discharge rates there is a tendency for Na_2S_5 near the electrolyte tube to be reduced further before all of the sulfur has been converted to Na_2S_5; this is manifest by the point of inflection in the discharge curve moving to the left as the discharge rate increases.

Charging the sulfur electrode is more difficult. In particular, it is likely that the electrode will suffer severe polarization when it is charged at high rates. This is largely due to the wetting characteristics becoming unfavourable. The reaction product (sulfur) preferentially wets the carbon felt and it becomes increasingly difficult to abstract electrons from the S_x^{2-} ions. Pockets of unreacted Na_2S_5 are left and this causes the utilization of the electrode to be less than 100%, typically 80–90%. A worse situation obtains at high recharge rates. Regions of sulfur that do not conduct Na^+ ions may form adjacent to the electrolyte and, therefore, recharge is terminated prematurely in these regions. Ultimately, the beta-alumina becomes enveloped in a layer of sulfur and recharge ceases abruptly, i.e., the electrode polarizes. Charge-acceptance is thus a function of current density, and full recharge is most likely to be achieved at moderate current densities (<50 mA cm^{-2}). Fast charging of sodium/sulfur batteries is not a practical proposition.

It should also be noted that the common practice of charging at constant current can be highly misleading as there is a tendency to equate constant current with constant current density. While this may be true for many batteries, it is not true for sodium/sulfur. Towards the end of recharge, regions of the electrolyte surface become blanketed by sulfur, so that the effective electrolyte surface becomes less and, therefore, the *current density* becomes greater. The true current density may then be several times the nominal current density and this can lead to electrolyte failure. For this reason, it is preferable to use a composite charging regime viz., constant current (e.g., 50 mA cm^{-2}) for the first 60% of the recharge, followed by a constant-voltage taper charge of decreasing current for the remainder of the recharge.

Various expedients have been adopted to ameliorate these recharge problems. One successful approach is to use a structured electrode in which a non-conducting inorganic fibre that is preferentially wetted by polysulfides (such as alumina fibre) is incorporated into the carbon felt adjacent to the ceramic electrolyte. In this way, ionically conducting pathways to the beta-alumina surface are left open and polarization does not occur until much higher current densities. Another strategy which has been tried is the use of chemical additives to modify the viscosity of the sulfur and to promote its electromigration.

The materials and construction of the sulfur electrode have received considerable attention. A particular feature of tubular sodium batteries is that the thickness of the positive electrode lies in the range 6 to 15 mm, which is very much greater than for conventional flat-plate batteries. This is a potential source of both high resistance in the cell and polarization on rapid recharge. Early attempts to use metal foams were unsuccessful as corrosion was rapid. Attention was then directed towards carbon, and various types of carbon and graphite felts, cloths and rigid foams were investigated. The rigid foams were found not to be sufficiently porous and best results were obtained with low density (~0.1 g cm^{-3}) carbon and graphite felts that are manufactured commercially and have porosities of ~95%. A low-resistance sulfur electrode is prepared by compressing the felt slightly so that it grips both the beta-alumina tube and the current-collector in order to make good electrical contact. Liquid sulfur can then be impregnated into the felt once it is in position in the cell. In practice, it has proven easier to manufacture carbon-felt–sulfur preforms by hot injection-moulding. The felt is compressed and, after cooling, is held in place by the solid sulfur. The sulfur electrode is now a preformed component, ready for cell assembly. When the cell is first heated up, the sulfur melts and the carbon-felt springs back to fill the electrode compartment and make good electrical contact with both the beta-alumina and the current-collector. Due allowance must be made for the volume expansion of the sulfur on melting and, more especially, for the increase in molar volume during discharge as the sodium transfers from the negative to the positive compartment. To achieve this, only 60% of the volume of the positive electrode compartment is filled initially with sulfur.

Corrosion of the current-collector is the second major problem that has been encountered with the sulfur electrode. There are two basic designs of sodium/sulfur cell: (i) the 'central-sodium cell', in which the sodium is contained within the electrolyte tube; (ii) the 'central-sulfur cell', in which the sodium electrode forms a narrow annulus around the outside of the ceramic tube (see Section 8.7.2). In the central-sulfur cell, the positive current-collector is simply a metal rod which is located axially along the beta-alumina tube; it fulfils no structural purpose. In the central-sodium design, the positive current-collector serves also as the cell case and must therefore have suitable mechanical, as well as electrical, properties.

From many points of view, aluminium is the ideal material to use as the positive current-collector. It is lightweight, cheap, readily available and has a high electrical conductivity. Also, it is passivated rather than corroded in sulfur/polysulfide melts. Unfortunately, aluminium also has two serious drawbacks: the passivating film of aluminium sulfide formed on its surface is non-conducting, and the metal itself is structurally weak with a low melting point (660 °C). The poor strength of aluminium and its low melting point are not a disadvantage in normal battery operation, but become a safety hazard when the aluminium can of a central-sodium cell fails and there is a major temperature excursion. Loss of structural integrity of the cell, with the escape of fumes or even of liquid polysulfide, can cause serious damage to the remainder of the battery. Nevertheless, despite this drawback, the advantages of aluminium are such that at least one major developer has utilized aluminium cell containers for central-sodium cells. The other disadvantage of aluminium, namely, the high resistivity of the passivating film, is common to both designs of sodium/sulfur cell. In order to overcome this problem, it is necessary to precoat the aluminium with a conducting layer which protects it from contact with the positive-electrode melt and which, itself, is not attacked. The coating materials which seem to be most practicable are carbon, molybdenum and chromium. There is a variety of possible techniques for laying down thin and impermeable coatings on metals (e.g., chemical vapour deposition, plasma vapour deposition, ion sputtering, electroplating), although a difficulty encountered with aluminium is its high expansion coefficient (23×10^{-6} °C^{-1}). Most of the coating techniques are easier to execute on the surface of a rod than on the inside surface of a cylinder. It follows that both of the major disadvantages of aluminium as a current-collector for the positive electrode are easier to accept for the central-sulfur design than for the central-sodium design. Interestingly, though, the major developers of the former cell have still not seen fit to use aluminium as a current-collector.

Consideration has been given to the application of graphite as the positive current-collector (or 'pole') in a central-sulfur cell. While this may be feasible for a short cell, graphite is too resistive for employment in a cell of, say, 25–30 cm length. Most other bulk metals which show reasonable inertness to the polysulfide melt (e.g., chromium, molybdenum) are simply too expensive for use in EV batteries. Refractory metallic carbides and nitrides (SiC, TiC, TiN, etc.) are generally insufficiently conducting and are too difficult to fabricate. This leaves the possibility of using mild or stainless steel with suitable corrosion protection in the form of a surface layer or composite.

High alloy stainless steels containing ~20 wt.% chromium show better corrosion resistance than mild steel and are attacked only slowly. Nevertheless, the corrosion products lead to an increase in cell resistance, a progressive decline in capacity, partial blockage of the beta-alumina surface by corrosion products and, ultimately, electrolyte failure. The corrosion of stainless steel in sodium/sulfur cells has been studied by Bones et al. [17,18]. These workers showed that in the two-phase region ($S + Na_2S_5$) a protective sulfide film is formed on the surface, but that when the positive electrode is polarized negatively, such that its potential falls below 1.73 V, the protective sulfide film is stripped off to leave a shiny steel surface. Because of the resistance drop across the sulfur electrode, it is usual for the pole to reach this voltage towards the end of discharge. As the cell is charge–discharge cycled, the surface sulfide film will form and then be stripped, alternately. This results in progressive corrosion of the container and increasing accumulation of corrosion products in the melt. These studies show that stainless steel is unsatisfactory on its own; its surface must be coated or protected. Accordingly, mild steel might as well be employed as it is much cheaper.

A great deal of research has been conducted on coatings and claddings on mild and stainless steel; some of this is of limited value as it has consisted of simple immersion tests in S/Na_2S_5 melts without the application of an electrochemical potential. In-cell testing (including extended cycling tests) is the only really satisfactory guide to current-collector performance. For a while, it was thought that thin coatings of molybdenum metal, laid down by ion sputtering or plasma spraying, would be adequate, but these have not lived up to expectation. Chromium electroplating also failed to yield coatings which were free of pin-holes or porosity. Such defects caused rapid sub-surface corrosion and spalling of the chromium coat. In general, it is now agreed that the preferred approach is that of chromizing the steel current-collector. This is accomplished by using a modification of the industrial pack-chromizing technique. The process involves a high-temperature (viz., 800–1100 °C) treatment in an inert atmosphere with exposure to chromium halide vapour for up to 24 h. Chromium diffuses into the surface of the steel to enhance corrosion protection. The precise metallographic structure of the resulting steel depends on the carbon content of the mild-steel substrate, as well as on the conditions and duration of chromizing. Therefore, the degree of protection afforded is dependent upon the details of the process. This is by no means a simple process for automation and mass production, nor is it especially cheap, particularly for the central-sodium design. Nevertheless, it seems to be the most cost-effective compromise available at the present time.

8.7.2. Cell design

The design of sodium/sulfur cells is directed towards producing batteries of high specific energy and power, low manufacturing cost, high electrical efficiency, long life and reliability, and complete safety in use. These features are closely interrelated and, in a practical design, a compromise must be reached that takes account of constraints imposed by the basic properties of materials, the practicalities of production engineering, and the battery specification laid down by the user. There is unlikely to be one universally preferred design. Rather, sodium/sulfur is an electrochemical couple from which many engineered cell and battery designs may stem, optimized for different applications.

The early investigations of sodium/sulfur cells were entirely concerned with the central-sodium design. Failure to produce corrosion-free cell cases, coupled with the availability of larger diameter (33 mm) beta-alumina tubes, led to a proposal in 1976 [19] to adopt the central-sulfur design. At first, this was thought to be a neat way to avoid corrosion of the cell case, with the attendant hazard of rupture and liberation of the corrosive sulfur/polysulfide melt. Despite such an advantage, it soon became apparent that corrosion of the central current-collector still posed a technical problem, albeit one that was easier to solve by coating technology than by coating the inside surface of a cylinder. Moreover, the area to be coated was much reduced. Following this proposal, the first central-sulfur cells were constructed and tested [4]. From the mid-1970s to the mid-1980s, development work in the UK was focused entirely on the central-sulfur design, while efforts in the USA and Germany continued on the central-sodium design. As a result of these extensive studies, the technical attributes of both cell types are now well established, as are their limitations. Nevertheless, there is still no general agreement on the preferred cell design for EV batteries. This is because several other factors have to be taken into account, e.g., ease and cost of mass producing cells, the configuration of cells into batteries, battery performance, reliability, and safety. Examination of these factors, as well as the need for higher specific power, led the principal UK developer in the mid-1980s to revert to a central-sodium cell of smaller size than was previously considered to be viable [20]. Here, both cell types are briefly reviewed.

Central-sodium cells

Early cells of this type, as used in the first sodium-powered EV (Fig. 8.3), had a 'top-hat' reservoir of sodium with gravity feed (Fig. 8.1). While this simple design works well enough, it requires the electrolyte tube to be filled with sodium even when discharged, and thus an excess of sodium is carried. This imposes weight, volume and cost penalties, and also has safety implications. A safer alternative, while retaining gravity feed, is to fill the electrolyte tube partially with an inert material, so as to reduce the sodium inventory. One advantage of the gravity-fed design is that it requires only a small-diameter electrolyte tube, typically 15 mm.

Soon the idea took hold of eliminating the top-hat reservoir and using just the stoichiometric amount of sodium that corresponds to conversion of all the sulfur to Na_2S_3 (plus a margin for operational reasons). This may be achieved by means of a wick, as described in Section 8.4. All the sodium is then contained within the beta-alumina tube and yields a more compact cell design of higher specific energy. The volume of sodium in the wick is then minimal and, moreover, a flow restrictor can be fitted to regulate the rate at which sodium enters the wick. Unfortunately, the wicking height of solid-liner wicks is strictly limited. The most satisfactory wick design for a central-sodium cell was found to be a composite structure in which a layer of cloth woven from 300-mesh stainless steel is held in place against

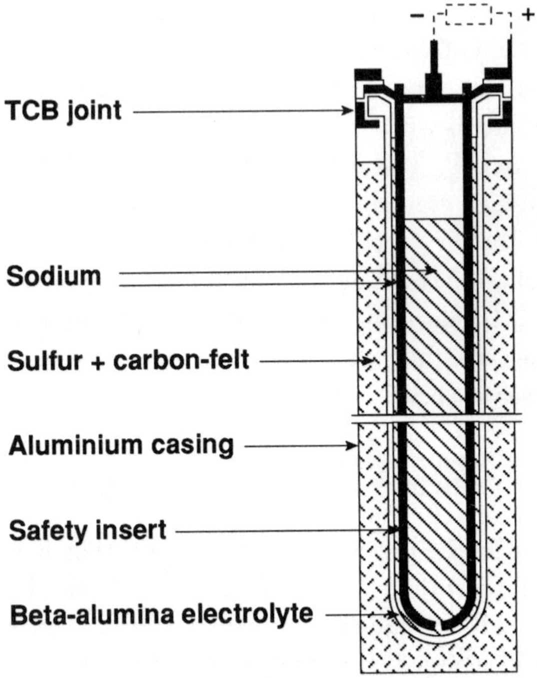

Fig. 8.12. Schematic of cross-section of a central-sodium cell (Type A04) for EV propulsion. Weight 420 g, capacity 40 Ah, resistance 9 mΩ. (Taken from [21] with permission of Asea Brown Boveri GmbH, Germany.)

the internal beta-alumina surface by a stainless-steel liner. This provides adequate wicking height (at least 200 mm) at 300 °C, with only a small amount of sodium present in the wick, but with a replenishment rate which is sufficient to cope with the maximum discharge rate of the cell. Figure 8.12 provides a schematic of a 40-Ah central-sodium cell with an internal sodium reservoir, as developed by the Brown Boveri Company in Germany [21]; notice the combined safety insert and wick, as well as the sodium flow restrictor at the bottom of the tube. The components of this cell, including the pre-fabricated carbon-felt–sulfur composite electrodes in the form of half cylinders, are shown in Fig. 8.13.

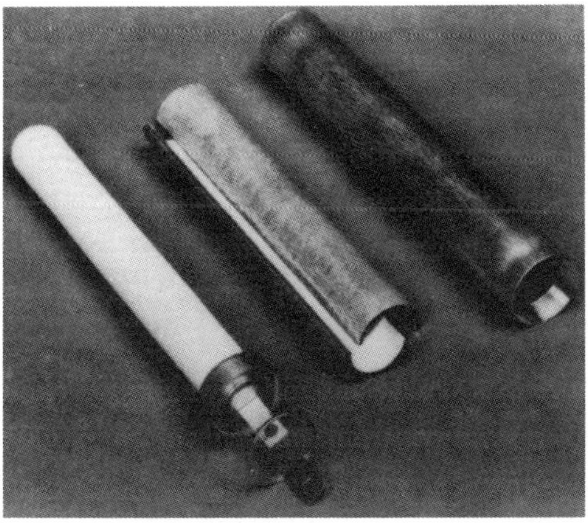

Fig. 8.13. Components of a Type A04 central-sodium cell (see Fig. 8.12). From left to right: solid electrolyte tube filled with sodium and sealed; prefabricated sulfur electrode half-cells; cell casing. (Taken from [21] with permission of Asea Brown Boveri GmbH, Germany.)

In the mid-1980s, the UK developers of the sodium/sulfur battery, Chloride Silent Power Limited (CSPL), elected to focus on a new design of internal reservoir, central-sodium cell. This was a relatively small (10 Ah, 20 Wh) cell of diameter and height both 45 mm, and weight ~100 g. The specific energy approached 200 Wh kg^{-1} at the 3-h rate. Figure 8.14 shows a cross-section of this cell, and Fig. 8.15 is a photograph of the actual cells.

A number of factors were responsible for this choice of small capacity cell. The principal ones were: (i) the decision not to replace failed cells in a battery; (ii) the fact that cells tend to fail to open-circuit and therefore lead to a requirement for a large number of parallel strings of cells (see Section 8.7.4); (iii) the greater safety of small cells. Also, wicking is easier to achieve in small cells and the freeze–thaw characteristics are superior. Against these factors must be set the higher manufacturing cost of small cells. A 60-kWh battery, typical of that used to power a medium-sized EV delivery van, would consist of 3000 cells. A pilot plant to manufacture 250 000 cells per year was built in the early 1990s. In 1992, CSPL was taken over by RWE (a German electricity utility) and given the new name, Silent Power Limited.

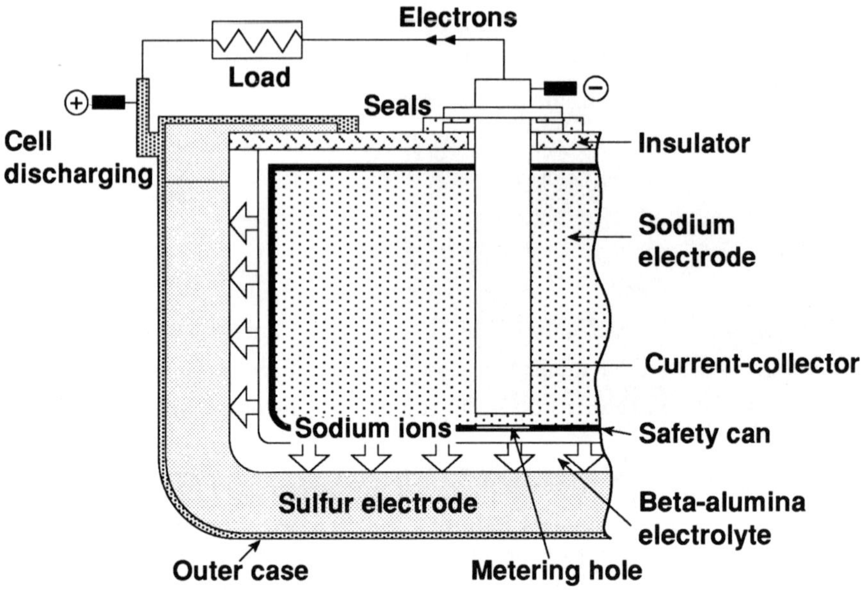

Fig. 8.14. Schematic of cross-section of central-sodium cell. (By courtesy of Silent Power Limited, UK.)

Fig. 8.15. Central-sodium cells (10 Ah). (By courtesy of Silent Power Limited, UK.)

Central-sulfur cells

The concept of a central-sulfur cell was first proposed by the Harwell Laboratory (UK), principally to circumvent the problem of cell case corrosion by polysulfide. There were soon found [22] to be a number of other advantages to this design, as follows.

- The sodium is readily accommodated in an external reservoir without adding to the overall volume of the cell. This longitudinal separation of sodium and sulfur facilitates flow restriction and enhances safety.
- It is possible to preform the sodium electrode sub-assembly in a simple fashion.
- As sodium does not attack mild steel, it is possible to use a very cheap cell casing, e.g., automotive exhaust pipe has proved suitable.

These favourable aspects, together with the lesser corrosion problem on the positive-electrode side and the greater ease of chromizing the pole, added up to a persuasive argument for the central-sulfur design. Accordingly, for about ten years, research in the UK was focused on this cell. Two negative aspects of the design are a reduced specific power, on account of the thicker and more resistive sulfur electrode, and a much increased current density on the positive current-collector.

The hermetic seal arrangement at the top of the cell does not allow gravity feed from an external sodium reservoir. One possible solution is to invert the cell (as in Fig. 8.16(a)) when gravity feeding is possible. This poses problems in cell inter-connection, however, and also results in the seal being immersed under liquid sodium which attacks it more readily than sodium vapour. The alternative arrangement (Fig. 8.16(b)) is to locate the sodium reservoir beneath the cell and use gas pressure to drive the liquid sodium up around the electrolyte tube. This is the design which was adopted. The sodium is contained in a cylindrical aluminium 'pressure can' which forms a sliding fit with the extended mild-steel cell case. At the base of the pressure can is a source of inert gas to act as a pressurizing agent. Originally, some graphite felt was placed at the bottom of the can.

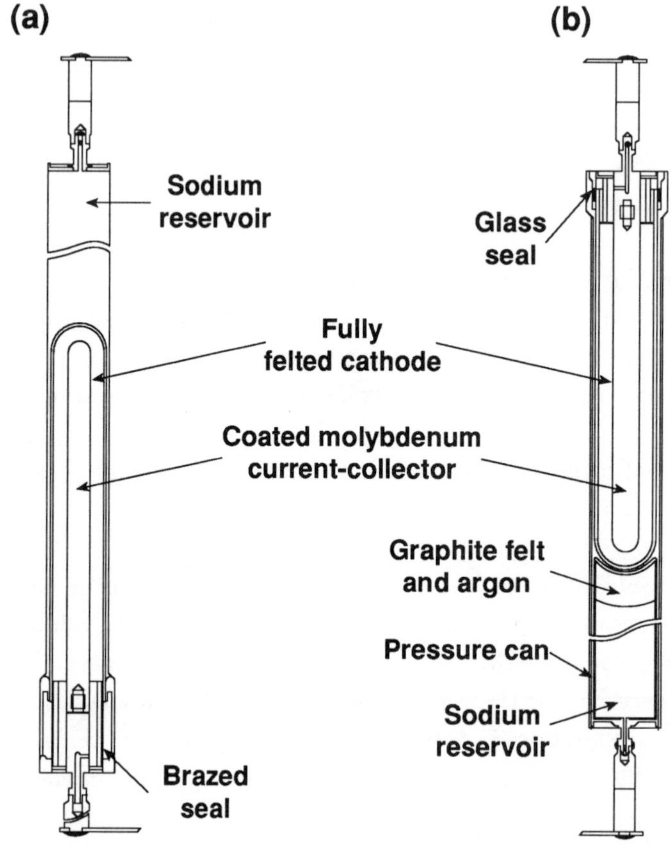

Fig. 8.16. Two alternative configurations of central-sulfur cell. (Taken from [22] with permission of Elsevier Science Limited, UK.)

The latter was filled with molten sodium in an argon-filled glove box and, thereby, argon was trapped in the felt. Later, it was found more beneficial to incorporate a pellet of sodium azide (NaN_3) at the bottom of the pressure can. The sodium metal was allowed to cool and solidify and the filled pressure can then constituted the sodium electrode sub-assembly. This was inverted and slid into the open (bottom) end of the cell casing. The casing was then sealed.

When the cell is first heated to 350 °C, the NaN_3 decomposes to liberate nitrogen gas which drives the sodium up the small annular gap around the pressure can, through an orifice in the restrictor plate situated below the beta-alumina tube (not shown in Fig. 8.16), and then up around the electrolyte tube. Because the sodium is pressure-fed, there is now no need for a wick. The restrictor orifice is of a size to allow cell discharge at the maximum rate desired, whilst acting as a throttle in the event of electrolyte fracture.

Central-sulfur cells have been based upon beta-alumina tubes with an internal diameter of 30 mm as standard. Larger tubes would result in too high resistance and polarization in the sulfur electrode. The overall cell diameter is 40 mm. Cells with electrolyte tubes of up to 650 mm in length and 175 Ah in capacity have been built and tested successfully. As these cells have an overall length of almost 1 m, they are unsuitable for use in electric road vehicles unless mounted horizontally. Rather, most central-sulfur cells have been based upon 160- and 300-mm long electrolyte tubes with capacities of 37 Ah and 70–100 Ah, respectively, as determined by the sulfur electrode thickness [23]. Experimental batteries using the 37-Ah cells have been assembled and tested successfully.

8.7.3. Cell performance

The specification for sodium/sulfur cells and batteries depends critically upon the application for which they are intended. Even within the traction battery field, there are major differences in specification for batteries to be used to power railway locomotives, buses, goods vehicles, and private cars. A key design parameter is the ratio, E : P, of stored energy (Wh) to peak power output (W). This ratio has the dimension of time (h) and defines the maximum discharge rate. A bus operating a 16-h or longer day must have an E : P ratio of at least 3, whereas a light-goods vehicle has a need for less range but better acceleration that leads to an E : P ratio of around 1, i.e., the battery should be capable of discharge at the 1-h peak rate. These requirements have to be translated into design parameters for the cells and batteries. In particular, the higher the desired power output, the lower the cell resistance should be and, therefore, the thinner the sulfur electrode.

The resistance of a cell is best expressed in Ω cm^2, i.e., the product of the measured cell resistance and the active area of the ceramic electrolyte. This allows a direct comparison between cells with electrolyte tubes of different diameter or length. Cell resistances are normally measured towards the start of discharge, but do vary somewhat over the course of the discharge–charge cycle. There are only two significant components to the overall cell resistance, namely, the electrolyte and the sulfur electrode. The contribution from a well-constructed sodium electrode and from the current-collectors is negligible. The resistance of the electrolyte tube depends upon its composition (i.e., β-Al_2O_3 : β''-Al_2O_3 ratio) and upon its wall thickness. Tubes with an internal diameter of 30 mm generally have a wall thickness of ~1.5 mm, although with the very short cells developed by CSPL for its 10-Ah central-sodium design it was possible to use tubes with a wall thickness of only 0.8 mm. The resistance of the beta-alumina electrolyte lies in the range 0.8 to 1.3 Ω cm^2. The resistance of the sulfur electrode depends upon the component thickness: typical values are 0.65 Ω cm^2 for a 6.5-mm thick electrode and 1.0 Ω cm^2 for a 10-mm thick electrode [23], although values as low as 0.4 Ω cm^2 have been reported [3]. For a given capacity, central-sodium cells have a thinner sulfur electrode than central-sulfur counterparts (this arises from purely geometric considerations) and, therefore, a lower resistance. Typical values of overall cell resistance are 1.5–1.7 and ~2 Ω cm^2 for central-sodium and central-sulfur cells, respectively. Thus, the central-sodium cell has the higher power output.

The EV designer is interested in expressing the stored energy and the peak power output of the battery in terms of battery mass and volume. The required levels reflect the energy and power of the cells that make up the battery, although other factors such as the closeness of packing of the cells and the mass of the battery box and heaters are equally important. As outlined in Chapter 2, the following four performance parameters are critical to cell design: stored energy per unit mass and volume (Wh kg^{-1}

and Wh l^{-1}); peak power per unit mass and volume (W kg^{-1} and W l^{-1}). The peak power is usually defined as the power available at 80% DoD.

Values quoted by Sudworth and Tilley [3] for the specific energy and energy density of both central-sulfur and central-sodium cells at three different discharge rates are given in Table 8.3. These data relate to rather large (100-Ah) cells that were based on beta-alumina tubes with dimensions: 300 mm (length) × 30 mm (diameter) × 2 mm (wall thickness).

Table 8.3. Specific energy and energy density of sodium/sulfur cells [3].

Discharge time	Central-sulfur cell		Central-sodium cell	
(h)	Wh kg^{-1}	Wh l^{-1}	Wh kg^{-1}	Wh l^{-1}
1	172	276	159	310
2	179	331	209	407
5	227	365	239	466

The peak power output of central-sodium cells can be as high as 225 W kg^{-1}, or over 400 W l^{-1}. The corresponding performances of central-sulfur designs are lower on account of the heavier cell and the thicker sulfur electrode.

The manufacturing cost of sodium/sulfur batteries is a critical consideration and, to be economically viable, the materials selection must meet strict cost targets. Features which introduce cost penalties include: stainless-steel cans to improve corrosion resistance; molybdenum as a corrosion protection barrier; niobium to protect glass–metal seals from attack by sodium.

A second cost consideration is that of cell size. Sodium and sulfur are intrinsically cheap commodities, whereas the manufacture of cells involves many operations and is comparatively expensive. It therefore follows that cells should be as large as practicably possible, both to reduce the cost per kWh of stored energy and to increase the specific energy (Wh kg^{-1}). Unfortunately, this runs counter to most other considerations. Large cells pose safety hazards, have reduced specific power, and are not suitable for use in a series–parallel network, except in a correspondingly large battery.

These cost considerations present a serious dilemma to the battery engineer as outlined below.

8.7.4. Battery construction

There are three salient features of the sodium/sulfur cell that the battery engineer has to consider when designing EV traction batteries. These are as follows:

- Cells do not pass current when in the fully-charged state but go to infinite resistance. This results in the entire charge voltage of a series chain falling across the cell which comes to top-of-charge first. Generally, this cell is the weakest.
- When cells fail, they usually fail to open-circuit — if not initially, then certainly after a while when the chemical reaction has taken place.
- Molten sodium and molten sulfur are intrinsically reactive materials which ignite when brought into contact. Once a fire starts, it may propagate from cell-to-cell and cannot be extinguished by conventional means (dry powder or carbon dioxide extinguishers) since the two reactant species are packaged together and the fire is independent of oxygen ingress.

The above three features dictate the design of sodium/sulfur batteries.

Clearly, it is impossible to guarantee that all cells will have identically the same capacity when manufactured and will deteriorate at exactly the same rate. The designer has therefore to circumvent the problem of infinite resistance at top-of-charge, which is a fundamental property of the sodium/sulfur cell. This is achieved by having a series–parallel matrix of cells with frequent cross-connections between the series chains. If, for example, there is a cross-connection every three cells, then the maximum voltage which can fall across any one cell is 6–8 V, and this should not result in breakdown of the electrolyte.

A systematic study of series–parallel battery performance was made by Bones and Teagle [24] who inter-connected six 20-Ah central-sulfur cells in 3 × 2 and 2 × 3 series–parallel arrays and cycled them for 1325 cycles over 1.5 years. It was shown that the presence of one weak cell, which resulted from gradual deterioration, caused a marked reduction in available capacity; the extent depended upon the imposed operating conditions. Since then, much practical experience has been gained of the behaviour of large series–parallel networks of sodium/sulfur cells in vehicle-sized batteries.

External factors which impact on the matrix design are the desired size of the battery (kWh) *vis-à-vis* the size of the cells employed (Wh) and the desired battery voltage. The voltage simply determines the number of series-connected cells that are required. Having fixed this, the number of parallel chains is dictated by the cell capacity and the desired battery capacity. Obviously, a rather small battery (say, 10 kWh) of high voltage (200 V) and comprised of large cells (100 Wh) will require just one series chain of cells and is quite impracticable. At the opposite extreme, a 100-kWh, 100-V battery using 20-Wh cells would require 5000 cells arranged in 100 parallel chains of 50 series-connected cells.

The EV battery which CSPL was developing in the UK was based on small 20-Wh cells. Four, series-connected cells (8 V) formed a unit string, and 30–50 strings in parallel comprised a module of capacity 300–500 Ah (i.e., stored energy 2.4–4.0 kWh). Three of the 4-kWh modules when assembled together provided a 12-kWh battery which was contained in an insulated battery box. The whole assembly weighed 100 kg. The specific energy of such a battery was 120 Wh kg^{-1} and the energy density was 150 Wh l^{-1}, both at the 3-h rate. Comparable figures for lead/acid traction batteries are 40 Wh kg^{-1} and 90 Wh l^{-1}. Later, CSPL built a number of 34-kWh EV batteries and subjected them to road trials in Ford Ecostar vans in both the USA and Europe, with mixed success. The batteries had a peak specific power of 110 W kg^{-1}, but following a re-design, this was raised to 230 W kg^{-1} while retaining a specific energy of 110 Wh kg^{-1}.

The cross-connection between parallel chains every few series-connected cells also has the merit of limiting the number of cells that are lost (isolated) when one cell fails to open-circuit. Even if cross-connections are as frequent as every third cell, there is still a loss of 6 V and this is likely to be significant. A chain with a failed cell will reach top-of-charge first, as a result of circulating currents, and it may be necessary to limit the extent of charge of the parallel chains. For this reason, the overall loss of capacity will be greater than that expected for the loss of just three cells. If the cell failure is initially to short-circuit, then the cells in parallel discharge themselves through the failed cell and are in danger of overdischarging. The uncertainty which arises if cells sometimes fail to open-circuit, and sometimes to short-circuit, creates difficult operational problems for the battery design engineer.

Traction batteries have also been constructed and tested in EVs by the Brown Boveri Company in both Germany and Canada [21]. The trials were conducted on central-sodium cells (40 Ah) that had internal reservoirs and aluminium cases, as shown in Fig. 8.12. An illustration of the Type B06 battery that contains 224 cells is given in Fig. 8.17. This battery was constructed in the mid-1980s and had a theoretical energy content of 18 kWh and a practical value of 14 kWh at the 2-h discharge rate. The power output at two-thirds open-circuit voltage was 21 kW. Two of these batteries were fitted into a Chrysler electric minivan which was road tested around Toronto, Canada. Later, the Company (now, Asea Brown Boveri (ABB)) developed 'standard' 10- and 20-kWh battery units, also based on 40-Ah cells (Fig. 8.18, with cut-away to show the cells). These standard Type B17 batteries gave specific energies of 93 and 110 Wh kg^{-1}, respectively.

Fig. 8.17. Central-sodium Type B06 battery with 224 × Type A04 cells. (Taken from [21] with permission of Asea Brown Boveri GmbH, Germany.)

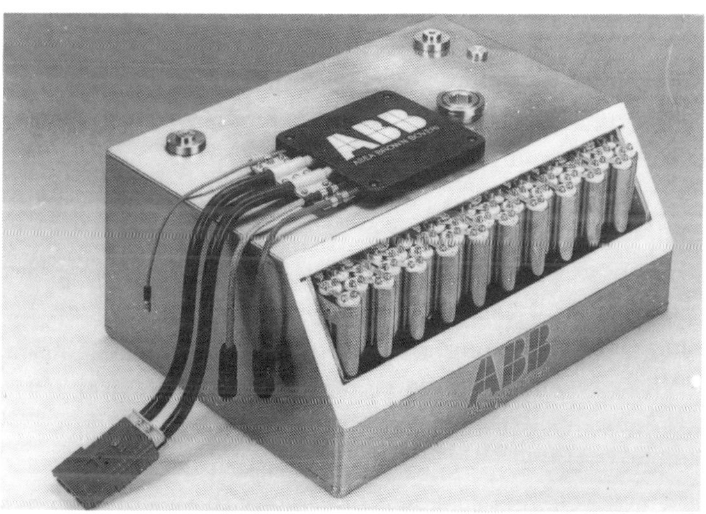

Fig. 8.18. Sodium/sulfur 'standard' Type B17 battery constructed by Asea Brown Boveri and fitted with main power leads, voltage monitor leads, heater lead, and earth lead. (Taken from [25] with permission of *Batteries International* and Asea Brown Boveri GmbH, Germany.)

Early in 1993, it was reported [25] that the Company had produced almost 100 000 cells in the previous year and had assembled them into 10- and 20-kWh batteries which were supplied for EV testing. By November 1992, cars fitted with ABB sodium/sulfur traction batteries had covered a total of 600 000 km without any kind of accident attributable to the batteries. A bus fitted with an ABB battery performed faultlessly for six months in the Bavarian National Park on a regular daily service of up to 200 km. These and similar vehicle trials provide reassurance that sodium/sulfur batteries can be manufactured reproducibly and that they perform reliably in vehicle traction applications.

Later in 1993, a 38-kWh battery (Type B19) of 480 cells was developed and used to power the Ford Ecostar van. Over 160 of these B19 batteries were delivered to the Ecostar programme. The ABB batteries were designed around groups of six cells (arranged two in series per string, with three parallel strings). Each group had a 'looping element' which was normally open, but which closed in the event of cell failure and thus short-circuited the group. The 38-kWh battery (480 cells) consisted of 80 groups wired in series (nominal voltage = 332 V). As cells failed and looping elements were activated, the terminal voltage dropped in increments of 4.15 V. Provided the reliability of cells is high, this provides a satisfactory solution to the design of sodium/sulfur.

Thermal control

There are several aspects of thermal control, namely: heating up the battery from cold; maintaining uniform temperature during operation; minimizing heat loss when standing on open-circuit. Heating from ambient to 350 °C must be executed slowly, i.e., over a period of many hours, in order to avoid the development of thermal stresses. These are particularly likely in the sodium compartment unless the metal is melted from the surface downwards. For this reason, it is preferable to situate the electric heaters above the cells at the top of the box.

Maintaining uniform temperature in operation is difficult because some of the cells are completely surrounded by hot neighbours while others, adjacent to the walls of the battery box, are in an asymmetrical situation. To help in smoothing out temperature fluctuations, horizontal aluminium plates have been built into some battery packs. When the battery is discharging at a high rate (i.e., during vehicle acceleration or hill-climbing), the rate of heat liberation considerably exceeds the rate of dissipation. The temperature is allowed to rise within the range 300 to 400 °C and the thermal mass of the battery is usually sufficient to accommodate heat liberation on the urban drive cycle. For continuous high-rate discharge, however, it will be necessary to have some form of active cooling, as discussed earlier.

The problem of safety

Sodium/sulfur cells and batteries are intrinsically hazardous. A 50-kWh EV battery, for example, will contain around 40 kg of molten sodium and 60 kg of molten sulfur at 300–400 °C. These two reactive materials are packaged in many small units, and are separated in each by a brittle ceramic tube. In the event of fracture (or even cracking) of either the tube or the header seal, the highly mobile sodium metal will penetrate the crack and react chemically with the sulfur. This is an exothermic reaction and the heat liberated can exacerbate the crack. In fortuitous circumstances, solid Na_2S_2 will form and plug the crack. Alternatively, the tube will fracture completely and, in the absence of in-built safety features, the reaction between sodium and sulfur will be unrestrained. The latter creates a very rapid build-up of temperature as ignition takes place. Sulfur boils at 444 °C and, above this temperature, its vapour pressure rises exponentially. This can lead to an explosion. More often, the course of events is such that the highly corrosive sodium polysulfide product burns its way through the cell case and a flame emerges under pressure. Cinematograph film has shown that this flame has all the characteristics of an oxy-acetylene torch, to which the adjacent cell is highly vulnerable. The fire can then propagate from cell-to-cell and, within a short time, the entire battery will be set ablaze. This was demonstrated in single-cell tests conducted at the Harwell Laboratory, in the early 1970s, and has since been confirmed by the number of fires that have been experienced in both laboratory and vehicle batteries. A prime task of the cell/battery designer is to tame this reaction by controlling the rate at which the reactants can come into contact. Even when catastrophic fire propagation does not occur, the rupture of a single cell is still likely to be highly detrimental to the battery because the sulfurous gases released will be corrosive towards the heaters, thermocouples, electrical connectors, etc.

As a result of work in the 1970s, the principles involved in the design of safe cells are well established [3,22]. Three criteria must be met.

Criterion 1: The quantity of sodium and/or sulfur immediately available for reaction after electrolyte failure must be minimized. This is normally achieved for the central-sodium design by the use of a solid shim or cartridge within the electrolyte tube that separates the bulk of the sodium from the beta-alumina (Figs. 8.12, 8.14). A wick, when employed, will lie in the narrow gap between the shim and the electrolyte. For the central-sulfur design, the bulk of the sodium is contained in a reservoir at the bottom of the cell, i.e., below the ceramic tube and well separated from it, and the small quantity of sodium adjacent to the electrolyte is contained in a narrow annular gap between the tube and the cell case (Fig. 8.16(b)).

Criterion 2: The flow of bulk sodium to the reaction site must be restricted. This requires the use of a flow restrictor, normally a pinhole orifice. For central-sodium cells, the use of a sodium cartridge, rather than a liner, allows a small hole to be drilled in the base of the can. In central-sulfur cells, the sodium can is upside down in the cell (i.e., open bottom) and, thus, it is common practice to incorporate the sodium-flow restrictor plate immediately below the bottom of the electrolyte tube.

Criterion 3: The outer cell case (mild steel) must be protected against corrosion by sodium polysulfides at elevated temperatures. Such protection is normally accomplished by wrapping a layer of 'Grafoil' around the inner surface of the cell casing. This is a suitable material because it is attacked neither by sodium nor by sodium polysulfides and is lightweight. With aluminium cans, as used by ABB, the problem is different. Aluminium is not corroded rapidly by polysulfides, but has very little strength and melts at a low temperature. As a consequence, the design is totally dependent upon the first two criteria to ensure safety.

When these three criteria are properly implemented, it has been demonstrated in the laboratory that sodium/sulfur cells are safe, and that fracture of the electrolyte tube leads to temperature excursions of only a few tens of degrees. Even so, the incidence of cell rupture and resulting fires in EV-size batteries is still worrying; it is not clear whether this is in some way associated with the battery configuration or with an inability to manufacture properly all the many cells in the battery. There seems to be some evidence that these fires occur overnight, when the battery is on charge, and therefore may be associated with electrolyte breakdown at the top-of-charge. One way to reduce the risk of propagation (and also to ensure that cells do not short out by touching each other) is to place thin layers of a refractory material between the cells; a mica-like substance, lightweight and cheap, has been used for this purpose. Other designs use a porous insulating material which is capable of absorbing reactants in the event of cell leakage.

The above discussion has been concerned with battery failure which is spontaneous. There is also the possibility of damage that is externally induced to the battery as a result of operational abuse or vehicle accident. Operational abuse might include failure to charge overnight, short-circuiting of the terminals, etc. Accidental cooling of the battery can be accommodated (provided it is reheated carefully), while suitable fuses will allow for short-circuiting. Violent damage in a vehicle accident is best mitigated by appropriate design of the vehicle and the location of the battery well inboard. Work on crash simulation by Silent Power Limited has provided encouraging results for sodium/sulfur batteries. Impact tests and drop tests resulted in no leakage of reactants, emission of gases, or fire. Two separate hazards are recognized as a result of cell and battery rupture, viz., fires between sodium and sulfur, and reaction of released sodium metal with water. If sodium batteries are to come into widespread use, the public will have to be made aware that water cannot be employed to extinguish sodium fires.

From this discussion, it is clear that more needs to be done to establish unequivocally the safety of sodium/sulfur batteries. Nevertheless, some encouragement may be drawn from the successful vehicle trials that have been carried out to date, and also from an incident in which a BMW electric vehicle fitted with an ABB battery was engulfed in a fire which was externally generated. The car was reported to have been completely gutted and yet, afterwards, the battery was still functioning satisfactorily [25].

Finally, further consideration should be given to the dismantling and disposal (or recycling) of used batteries on a large scale.

8.8. Sodium / Metal Chloride (ZEBRA) Batteries

8.8.1. A brief history of the ZEBRA project

The ZEBRA battery project had its origin in the mid-1970s at the Council for Scientific and Industrial Research (CSIR) laboratories, Pretoria, South Africa. A small team under J. Coetzer studied the basic electrochemistry of a sulfur electrode in which the liquid sulfur was incorporated into a zeolite structure. Lacking the facilities to investigate the behaviour of this electrode in an engineered sodium/sulfur cell, the team approached the Harwell Laboratory, in 1979 for help. Harwell had been working on sodium/sulfur batteries for some years and had available the appropriate technology. By that time, CSIR were in partnership with the Anglo-American Corporation of South Africa who sponsored the subsequent collaborative work at Harwell. Because of the requirement for commercial secrecy, Harwell devised the project code name ZEBRA. This stood for **ZE**olites applied to **B**attery **R**esearch **A**frica.

Unfortunately, in-cell experiments soon showed that this design of sulfur electrode was incapable of sustaining a realistic current density and the idea was dropped. J. Coetzer and colleagues had, however, devised an alternative electrode, based on chlorinated transition metal carbides and borides, specifically iron carbide [26]. This was tested for the first time in tube cells at Harwell with more encouraging results; molten sodium chloraluminate was used to make ionic contact between the beta-alumina electrolyte and the solid negative electrode. Subsequent collaborative research between the two laboratories revealed that the elemental metals were just as effective as the hard metal interstitial compounds. Thus, the latter were relinquished in favour of chlorinated iron and nickel negative electrodes [27]. Although this work was carried out in the early 1980s, the first scientific publications had to await the issue of patents and did not appear until 1986 [28,29].

Chlorination of the positive electrode emerged as a key problem during the course of the above work. The use of chlorine gas was cumbersome and in the case of iron gave $FeCl_3$ (rather than $FeCl_2$) which dissolved in the molten $NaAlCl_4$. Hydrogen chloride gas yielded $FeCl_2$, as desired, but was almost equally difficult to handle. Furthermore, the molar volume expansion in going from the metal to the metal chloride restricted the progress of the reaction and the resulting chlorinated electrode was non-porous and could only be discharged slowly. It was soon realized that the way to avoid these problems was to assemble the cells in the *discharged* state via the use of a highly porous metal electrode prepared by sintering the metal powder with sodium chloride in hydrogen [30,31]. After impregnating the porous electrode with molten sodium chloraluminate, the cell could then be charged to generate metal chloride and elemental sodium according to the reaction:

$$M + 2NaCl \rightarrow MCl_2 + 2Na \qquad (8.6)$$
$$(M = \text{Fe or Ni})$$

This had two important advantages: (i) the loss of sodium from the positive electrode gave rise to added porosity which facilitated the subsequent discharge reaction; (ii) there was no need to handle sodium metal under inert gas conditions and, moreover, the sodium generated by

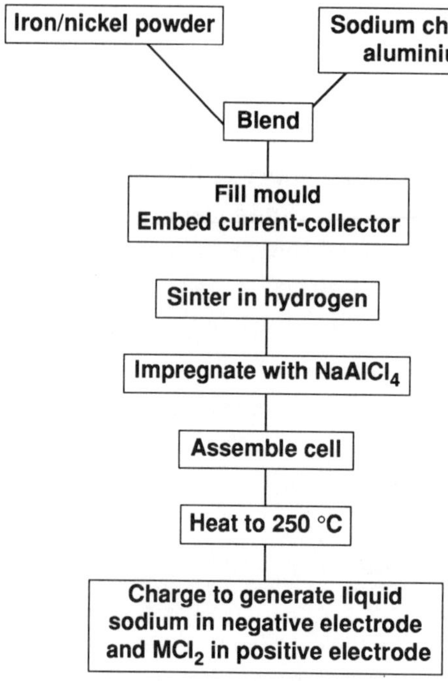

Fig. 8.19. Fabrication process of solid positive electrodes for ZEBRA batteries. (By courtesy of Beta Research & Development Limited, UK.)

diffusion through beta-alumina was of high purity. The procedure for preparing the positive electrode was quite simple in principle, as outlined in Fig. 8.19.

The beta-alumina tubes used in this early work were supplied by a group working on the sodium/sulfur battery at British Rail (Derby). In 1981, however, British Rail decided to terminate this research programme and the battery group left to form its own company, Beta Research & Development Limited. Their first contract was with the Anglo-American Corporation to develop the ZEBRA battery in partnership with the Harwell Laboratory. Subsequently, Beta Research & Development became a subsidiary of the Anglo-American Corporation. Based on the experience of the two groups with sodium/sulfur batteries, progress was rapid and after just two years (1984) the first ZEBRA-powered EV was in use at Derby. In

Fig. 8.20. Left to right: Ford Escort van and Suzuki Microbus fitted with ZEBRA traction batteries, photographed alongside a Bedford EV fitted with lead/acid batteries (1986). (By courtesy of Beta Research & Development Limited, UK.)

that year, Zebra Power Systems Pty (ZPS) was formed in Pretoria, South Africa, and J. Coetzer and his team left CSIR to join the new company. At this juncture, a three-party collaborative project existed and involved Beta Research & Development at Derby, ZPS in Pretoria, and Harwell.

The three laboratories worked closely together to develop viable EV traction batteries. By 1986, both a Ford Escort van and a Suzuki Microbus had been fitted with ZEBRA traction batteries (Fig. 8.20). These vehicles were evaluated thoroughly, both in city traffic and on test tracks, and demonstrated encouraging performance as regards reliability and driving range. It is to be noted, however, that each of the vehicles was somewhat underpowered.

At this point, the Anglo-American Group decided to seek a partner in the venture with complementary industrial skills (in particular, the manufacture of electrical equipment) and with connections to the vehicle manufacturing business. After extensive enquiries, the German company

AEG, now part of the Daimler-Benz consortium, was selected and an agreement was reached in 1989 to set up a joint company: AEG-Anglo Battery Holdings GmbH. This brought into the project a team of German scientists and engineers and there started a three-nation collaboration which was directed towards the commercialization of the battery. Since then, much of the endeavour has concentrated on perfecting the battery technically, improving its performance, reducing production costs, and setting up a pilot plant to manufacture traction batteries for EV field trials. The battery components (beta tubes, seal assemblies) are made in Derby, the cells are put together in Berlin, and the batteries are assembled in Ulm.

A somewhat more detailed account of the history of the project, as seen from a South African perspective, has been published in *South African Transport* [32].

8.8.2. ZEBRA cell chemistry

The sodium electrode of the ZEBRA cell is identical to that used in the sodium/sulfur system, apart from the fact that the sodium is produced *in situ* on charge. A metal wick, inserted in the steel can before the seal sub-assembly is welded in place, acts as the substrate for sodium deposition when the cell is charged.

Although the ferrous chloride electrode was investigated first, more work has been carried out on the nickel chloride counterpart. The latter electrode is intrinsically simpler and operates over a much wider range of temperature (viz., 175–350 °C). In addition, the nickel cell has a higher open-circuit voltage than the iron cell, namely, 2.59 vs. 2.35 V for $Na/FeCl_2$ at 250 °C [30,33,34]. Figure 8.21 provides a schematic of a sodium/nickel chloride cell. These cells are always of the central-positive design, since if they were of the reverse configuration, the steel cell case would be chlorinated preferentially to the nickel positive during charging. In principle, a central-negative cell would be possible with a pure-nickel cell case, given that the chlorination of the porous positive electrode is much

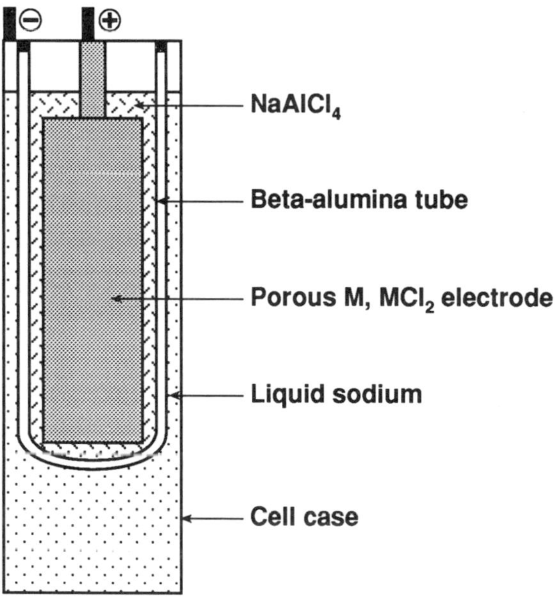

Fig. 8.21. Schematic of sodium/metal chloride cell. (Taken from [33] with permission of The Electrochemical Society, Inc.)

faster than that of the cell housing; in practice, however, the cost of a nickel cell case precludes it from being of interest for EV applications. The design is therefore limited to the central-positive electrode configuration with its attendant strengths and weaknesses.

The equilibrium open-circuit voltage of the $Na/NiCl_2$ cell as a function of state-of-charge is shown in Fig. 8.22. The theoretical specific energy of the cell (790 Wh kg^{-1}) is slightly higher than sodium/sulfur (760 Wh kg^{-1}), but the practical value will be lower because of the need to have both a solid electrolyte (beta-alumina) and a liquid electrolyte ($NaAlCl_4$). A key feature of these cells is that both the $NiCl_2$ and the excess NaCl are essentially insoluble in the basic $NaAlCl_4$ molten salt electrolyte. It is important that the alumino-chloride melt remains basic, i.e., an excess of NaCl should be present. If the melt becomes acidic, the free $AlCl_3$ abstracts Na^+ ions from the beta-alumina electrolyte and causes a rise in the resistance of the cell.

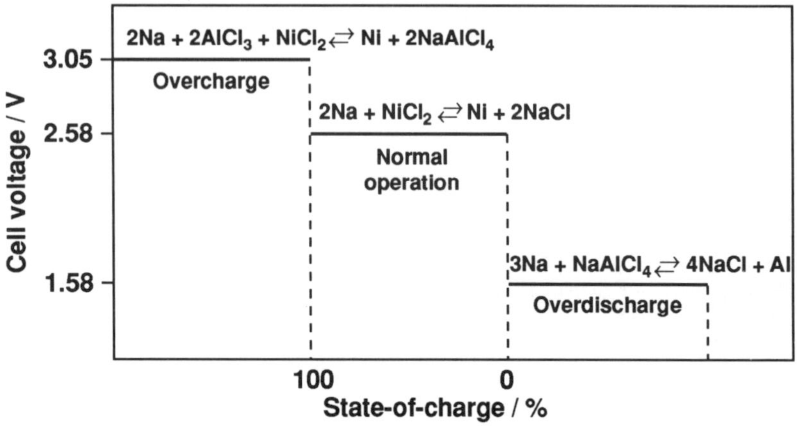

Fig. 8.22. Charge–discharge equilibrium voltages of sodium/nickel chloride cell at 250 °C. (By courtesy of Beta Research & Development Limited, UK.)

Both overcharge and overdischarge reactions are available. At the top-of-charge, when all the solid NaCl has been decomposed, overcharge results in depletion of Na^+ ions in the melt and further chlorination of the porous metal matrix, i.e.,

$$2NaAlCl_4 + Ni \rightleftarrows 2Na + 2AlCl_3 + NiCl_2 \tag{8.7}$$

Excessive overcharge is undesirable as it results in an acidic melt. Overdischarge involves the reversible reaction, i.e.,

$$3Na + NaAlCl_4 \rightleftarrows Al + 4NaCl \tag{8.8}$$

whereby sodium chloraluminate is reduced to aluminium. In practice, it is found to be convenient to assemble the cells in the *overdischarged* state by incorporating some fine aluminium powder and extra NaCl in the positive-electrode mix. When the cell is charged, this aluminium reacts first (according to the reverse of eqn. (8.8)) to yield a stoichiometric excess of sodium metal beyond that required for the discharge of the $NiCl_2$. A small sodium excess is always necessary in order to ensure that the negative-electrode compartment is not completely empty at the end of discharge.

This technique of overdischarge assembly avoids altogether the need to handle sodium metal during cell construction.

The theoretical capacity of the cell is determined by the quantity of NaCl in the positive-electrode mix. As charging takes place, the NaCl is consumed and thus leaves voids to accommodate the volume expansion associated with the conversion of nickel to $NiCl_2$. Around 30% of the nickel is converted to the chloride. Typical capacities lie in the range 0.2 to 0.4 Ah cm^{-3}.

Early charge–discharge cycling experiments with nickel electrodes revealed a rapid decline in capacity over just a few cycles. Post-test examination by scanning electron microscopy showed that the nickel metal matrix was undergoing recrystallization with considerable loss of surface area and porosity [33]. Individual particles grew from ~2 to ~10 μm in size and had an angular appearance, consistent with grain growth from the melt. Clearly, this behaviour is unacceptable for a long-life cell. Experiments were therefore performed to determine the effect of additives on the recrystallization process. It was found that small additions of sulfur to the positive-electrode melt almost completely repressed the phenomenon. Indeed, the nickel recovered from cycled sulfur-containing cells consisted of agglomerates of very finely-divided particles. The sulfurated positives retained their capacity well on cycling; after 2000 cycles at the fast 1-h discharge rate, the capacity retention was >75% and there was no increase in resistance. On dismantling the cell, the inner surface of the beta-alumina, adjacent to the positive electrode, was pure white and exhibited no evidence of chemical attack [33]. This success in repressing the recrystallization of the nickel formed the basis for future development of the nickel chloride electrode.

The attraction of using iron in place of nickel lies in its lower cost and the possibility of using a central-negative design if so desired. These attractions are offset to some degree by the lower cell voltage. Early work with the ferrous chloride matrix, however, soon revealed complications and limitations that were not present with nickel. A particular problem was associated with overcharge. The reactions for the Na/$FeCl_2$ cell are given in Fig. 8.23 [29]. The normal cell reaction and the overdischarge reaction are analogous to those for $NiCl_2$ but when the cell is at the top-of-charge, the voltage rises abruptly from 2.35 to 2.75 V and the overcharge reaction is then the oxidation of $FeCl_2$ to $FeCl_3$. Unfortunately, $FeCl_3$ — unlike $FeCl_2$ — is very soluble in $NaAlCl_4$ and the Fe^{3+} ion is highly detrimental to the beta-alumina electrolyte. In order to circumvent this problem, some nickel powder was mixed in with the iron powder during electrode preparation. As the Ni/Ni^{2+} voltage plateau is at 2.59 V, i.e., below that of Fe^{2+}/Fe^{3+}, the nickel will be chlorinated before iron is converted to the trivalent state (Fig. 8.23). This proved to be highly successful when ~20% of nickel was incorporated with the iron, and such electrodes were shown to have long cycle-lives [29]. Figure 8.24 shows charge–discharge curves for a small (10.8 Ah) Na/$FeCl_2$ cell that contains 20 wt.% nickel. The discharge curve on cycle 701 was virtually identical to that on cycle 72 (or even better!), while the charge curve included the step on to the nickel plateau at 2.59 V. The cycling of this cell was continuous for >700 cycles over a period of 1.5 years. Post-cycle examination of a specimen of a discharged iron–nickel electrode after washing out the NaCl showed that, unlike pure iron, a marked transformation in structure had occurred during the first few cycles to give a matrix made up of large agglomerates of very fine crystallites. The metals had alloyed fully.

Detailed crystallographic investigations of positive electrodes taken from Na/$FeCl_2$ cells (no nickel present) at various stages of charge and discharge revealed that the reaction was in fact more complex than that represented by eqn. (8.6). In particular, an intermediate compound, viz., Na_6FeCl_8 (or $6NaCl{\cdot}FeCl_2$), is produced during the course of the electrochemical processes. This compound has been well characterized by both electrochemical and powder X-ray diffraction studies; it has an ordered defect rocksalt structure. Although of scientific interest, the compound does not appear to affect cell operation in the temperature range 250–300 °C [31].

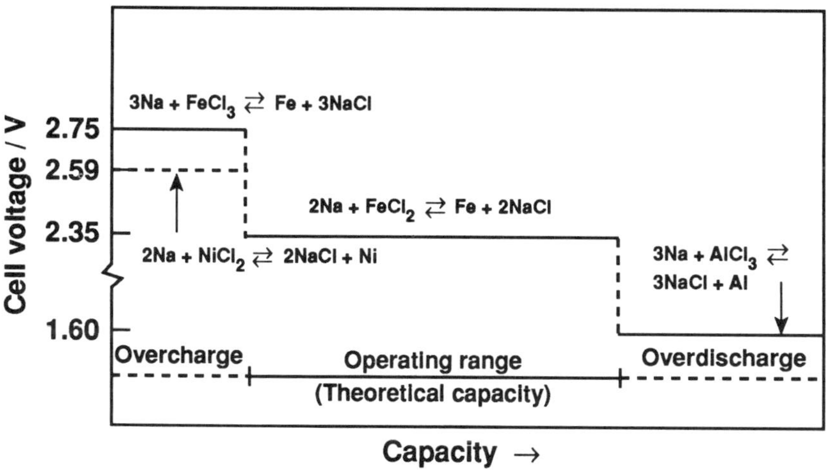

Fig. 8.23. Cell reactions of nickel-doped $FeCl_2$ electrode at 250 °C. (Taken from [29] with permission of The Electrochemical Society, Inc.)

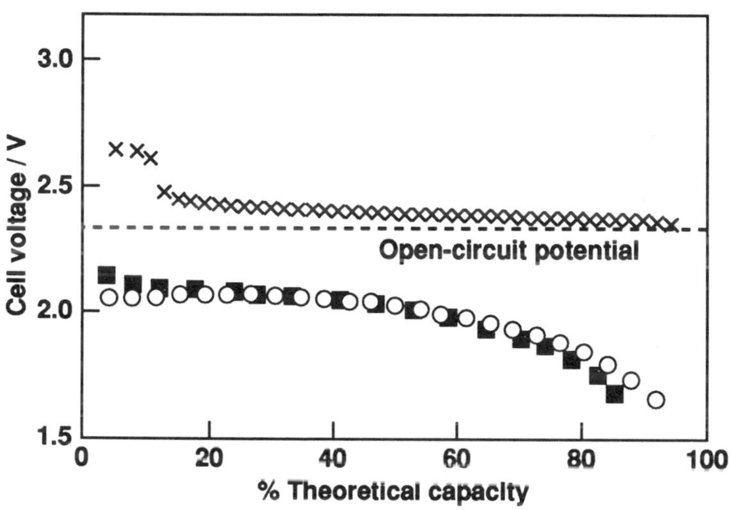

Fig. 8.24. Continuous cycling of a ZEBRA cell with a nickel-doped $FeCl_2$ electrode at 250 °C. (■) 5 A discharge, cycle 72; (○) 5 A discharge, cycle 701; (×) 0.7 A recharge, cycle 701. 5 A = 90 mA cm^{-2}. (Taken from [31] with permission of The Electrochemical Society, Inc.)

In addition to the problem associated with overcharge, the ferrous chloride electrode also suffers from a limited temperature range of cell operation. Despite the fact that ZEBRA cells will pass current down to 175 °C, they are preferably not operated in a steady state below 250 °C as the resistance of the beta-alumina becomes excessive. Ferrous chloride cells operate well between 250 and 300 °C. Above 300 °C, complications set in and these result in a rise in cell resistance. The latter is due to Fe^{2+} or Fe^{3+} entering the beta-alumina under conditions of current flow [35]. Regions of beta-alumina that contain iron are found to have very high resistivity. The addition of NaF as a dopant to the melt controls this problem to a certain extent, presumably by a lowering of the concentration of soluble iron cations. Cells operated at temperatures above 350 °C contain two other phases in the $FeCl_2$–NaCl system, viz., Na_2FeCl_4 and $Na_2Fe_3Cl_8$. These phases have been previously unreported and are isostructural with the known manganese analogues [36]. Because of these complications, it is not considered prudent to operate Na/$FeCl_2$ cells much above 300 °C. This is a serious limitation.

8.8.3. Fabrication of positive electrodes

From an economic viewpoint, the idea of making a battery essentially out of cheap raw materials (alumina, caustic soda, common salt and iron filings) is most attractive. It was soon discovered, however, that cheap, coarse iron powders, when sintered with salt in hydrogen, did not give a sufficiently reactive positive electrode. Moreover, finely divided, reactive iron powder was surprisingly expensive. This dilemma was neatly solved by a process for activating the cheap powder [37]. The required quantities of coarse (20 μm) iron powder, nickel, aluminium powder and sodium chloride are mixed and poured into a mould around a mild-steel current-collector. This compact is then sintered *in air* at high temperature to cause oxidation of the outer layers of the metal particles and the steel pole. The volume expansion associated with oxidation causes the particles to sinter both together and to the current-collector. The compact (now a hard, solid artefact) is removed from the mould and sintered in hydrogen to reduce the oxide back to metal. During this oxidation–reduction cycle, there is a large

Fig. 8.25. ZEBRA positive electrodes with a nominal capacity of 10, 60 and 100 Ah. (Taken from [29] with permission of The Electrochemical Society, Inc.)

increase in both the surface area and the reactivity of the sintered metal, together with the development of a much improved pore structure. Electrodes of various size (10-, 60- and 100-Ah capacity) prepared in this way are shown in Fig. 8.25. These electrodes were found to perform perfectly satisfactorily.

The pre-oxidation procedure was not needed for nickel chloride cells as nickel powder is available commercially in a finely divided, reactive form (ex nickel carbonyl). It is necessary to use a nickel current-collector as this is not corroded in normal operation of the cell. The electrode reaction occurs preferentially on the high surface area, reactive nickel. Nickel positives may therefore be fabricated following the procedure shown in Fig. 8.19 and using a nickel pole. Later work revealed that it was unnecessary to pre-fabricate the electrodes and that cells operated satisfactorily when a mixture of nickel powder, salt and aluminium power was simply poured into the beta-alumina tube around the central current-collector. This has a number of advantages: it saves the fabrication step; it is easier to incorporate the aluminium powder uniformly; there are no problems of dimensional tolerances between the electrode and the electrolyte tube. After tamping down the powder mixture, the molten sodium chloraluminate is added and the cell is sealed by welding on the positive terminal. During the first charge–discharge cycle, it seems that the electrochemical reactions cause the particles to fuse together and form the positive-electrode matrix.

8.8.4. Cell design and performance

The design of a ZEBRA cell for a given application is a specialized activity. There are several factors which permit a greater degree of freedom than for sodium/sulfur cells, namely: the demonstrated safety of the ZEBRA system allows larger cells to be constructed; (ii) the existence of an overcharge reaction enables the use of longer series chains of cells; (iii) the failure to a short-circuit state does away with the need for parallel connections. These factors permit the possibility of few, if any, parallel chains of cells in a battery and, therefore, larger individual cells can be used.

The variables in cell design, assuming a cell with a central-positive electrode which is aligned vertically, are: nickel or iron chloride; length and diameter of the electrolyte tube; loading of the electrode (i.e., percent conversion of metal matrix to chloride when charged); porosity of the electrode.

For a central-positive cell, the diameter of the beta-alumina tube fixes the diameter of the positive electrode. Deducting the diameter of the central current-collector gives the annular thickness of the electrode. Standard cells using a beta-alumina tube with an internal diameter of 30 mm have a positive-electrode thickness of ~10 mm (which is much greater than that of most conventional battery plates), while large cells may have corresponding electrode thicknesses of up to 25 mm. The larger the cell diameter, the higher its specific energy and energy density (Wh kg^{-1} and Wh l^{-1}) will be, but also the higher its resistance and the lower its power (W kg^{-1}, W l^{-1}). For large vehicles such as buses or trains, where energy and cost are more important than specific power, large cells would be employed. For electric cars, where power is essential, it is necessary to find a compromise between all the desired features.

The loading of the positive electrode can be raised, up to a point, by incorporating more sodium chloride and converting a greater proportion of the metal matrix to chloride. This will provide a higher specific energy. Higher power is achieved at the expense of energy by employing a positive-electrode of higher porosity. The limit to each of these approaches is determined by the need to maintain the structural integrity of the electrode and keep its resistance low. Cell design is therefore specific to the application and is an exercise in compromise and optimization.

The effect of positive-electrode thickness on cell performance can be illustrated by the following comparison of data for Na/$FeCl_2$ cells [29]. Figure 8.26(a) shows charge–discharge curves for a small (8 Ah) cell with an electrode thickness of 13 mm, quite thick for a battery plate. Thirty percent of the iron matrix was converted to chloride in this high porosity electrode which yielded a capacity density of 0.25 Ah cm^{-3}. Note the flat discharge curves at all rates up to 10 A (the 0.8-h rate) and the 100% utilization of capacity. Figure 8.26(b) relates to a much larger (73 Ah) cell where the radial electrode thickness was >20 mm and the matrix conversion was 37%. Clearly, the shape of the discharge curve at the 3.6-h rate is inferior to that of the small cell at the 0.8-h rate and the utilization is only ~80%. Nevertheless, these performances are still remarkable for such a thick electrode, and it is clear that large Na/$FeCl_2$ cells are very promising for low-rate applications.

For high-rate applications, the nickel system has advantages over the iron system since maximum power is proportional to the square of the open-circuit voltage and so, for the same cell resistance, the power of the nickel cell is 20% greater than that of the iron counterpart. More than any other factor, it is this demand for specific power that determined the choice of nickel for EV batteries. For the same reason, the design has standardized on a beta-alumina tube with an internal diameter of 30 mm, as larger diameters are not compatible with high rates. The length of the cell is restricted by the height which is available in a vehicle. Taken together, these two factors have defined a 40-Ah cell as the current basis of the AEG-Anglo Battery Holdings GmbH battery programme. At an earlier

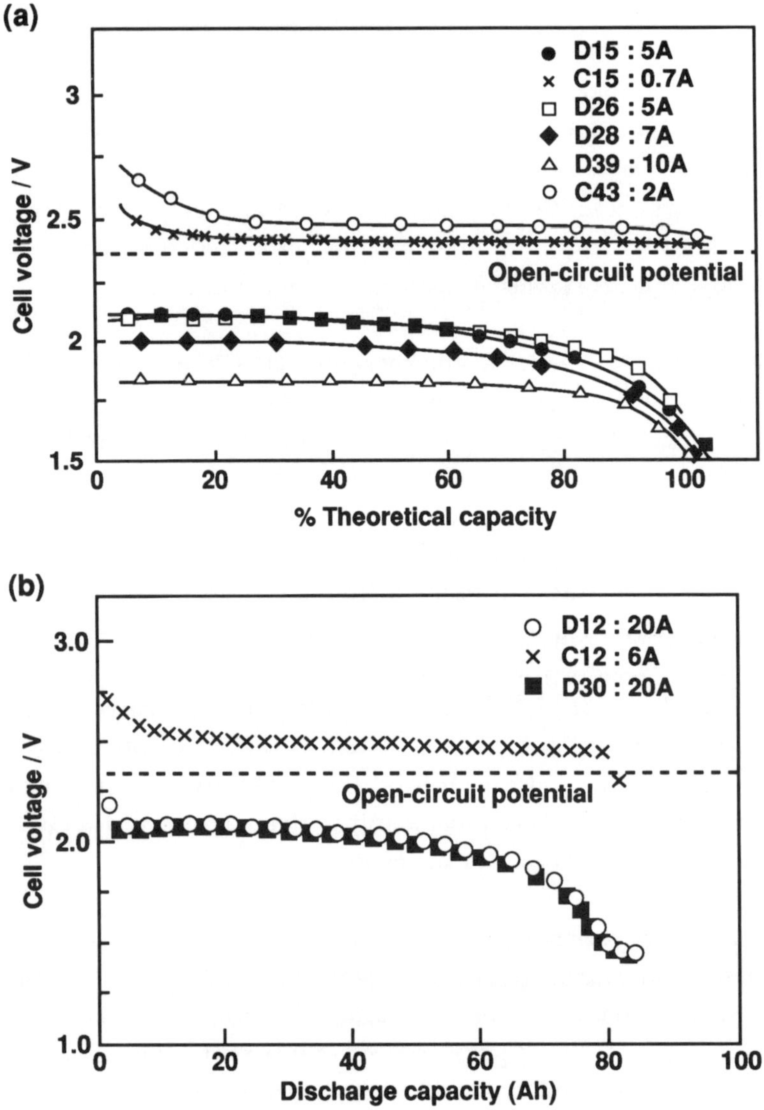

Fig. 8.26. Selected half-cycles (C = charge, D = discharge) for Na/$FeCl_2$ ZEBRA cells. (a) 8-Ah cell at 250 °C, 30% Fe converted to $FeCl_2$ in charged state, 5 A = 90 mA cm^{-2}; (b) 73-Ah cell at 275 °C, 37% Fe converted to $FeCl_2$ in charged state, 10 A = 33 mA cm^{-2}. (Taken from [29] with permission of The Electrochemical Society, Inc.)

Table 8.4. Performance data for ZEBRA cells [14].

	$Na/NiCl_2$ (Central positive)	$Na/FeCl_2$ (Central negative)
Cell capacity (Ah)	40	42
Energy, 3-h rate (Wh)	96	90
Power* (W)	118	88
Specific energy, 3-h rate (Wh kg^{-1})	124	142
Energy density, 3-h rate (Wh l^{-1})	310	260
Specific power* (W kg^{-1})	153	139
Power density* (W l^{-1})	380	250

* Power measured at two-thirds open-circuit voltage and 70% DoD.

stage, experimental vehicles were fitted with larger cells and with iron cells for trial purposes (see Section 8.8.5).

A comparison of performance data for a 40-Ah $Na/NiCl_2$ cell (central-positive electrode) and a comparable $Na/FeCl_2$ cell (central-negative electrode) is given in Table 8.4. The superior power performance of the $NiCl_2$ cell is immediately apparent, particularly on a volumetric basis, despite the fact that this cell has a thicker positive electrode than the $FeCl_2$ cell. On a gravimetric basis, the $NiCl_2$ cell is penalized by having a higher overall density (2.5 g cm^{-3}) than the iron cell (1.8 g cm^{-3}); this is on account of the larger diameter (and, therefore, heavier) electrolyte tube in a central-positive cell. Other factors being equal, a central-negative cell will always exhibit higher specific energy and higher power than a central-positive cell. A comparison of the specific energies in Table 8.4 with those in Table 8.3 indicates most clearly that the sodium/sulfur cell has the advantage; this is primarily because the ZEBRA cell has to carry around a second electrolyte ($NaAlCl_4$) which is a heavy component. As will be seen in Section 8.8.5, however, this disadvantage at the cell level is offset when it comes to battery construction.

8.8.5. Battery construction

Several ZEBRA batteries in the size range 15 to 40 kWh were built in 1984–86 for experimental trials in EVs. The characteristics and performance of these batteries were reported in 1987 [12,14]. The first batteries were based upon cells with 30-mm diameter electrolyte tubes, but as a result of extensive safety testing, confidence was developed that it was safe to build and handle large ZEBRA cells. Therefore, some batteries were built and were based upon cells with beta-alumina tubes of diameter 55 mm. At that early stage, the emphasis was on demonstrating specific energy rather than specific power. The characteristics of three of these batteries are summarized in Table 8.5.

Table 8.5. Experimental EV traction batteries [14].

Positive electrode:	**Central $FeCl_2$**	**Central $NiCl_2$**	**External $FeCl_2$**
Cell orientation	Horizontal	Vertical	Horizontal
β''-Al_2O_3 diameter (mm)	55	55	55
Capacity (Ah)	140	100	200
No. of cells	152	66	76
Array	8 × 19	6 × 11	4 × 19
No. of cells in series	38	33	38
Voltage (V)	89	85	89
Stored energy (kWh)	40	15	30
Specific energy (Wh kg^{-1})	88	88	89
Energy density (Wh l^{-1})	103	95	97

Note: two of the $NiCl_2$ batteries were used to power a vehicle.

Cell-to-battery degradation factors of 0.81 and 0.61 were achieved for the gravimetric and volumetric energy densities, respectively. These data were regarded as very encouraging at that stage of cell and battery development. By contrast, the power capabilities of the batteries were less than satisfactory. Two of the 15-kWh $NiCl_2$ batteries gave a combined power output of only 22 kW, thanks to the comparatively high resistance of cells constructed with 55-mm diameter electrolyte tubes. Nevertheless, these two batteries, when installed in the Suzuki microbus (Fig. 8.20), gave a range of 115 km to 60% DoD at an average speed of 60 km h^{-1}. The top speed, limited by the characteristics of the motor used, was 71 km h^{-1}. The 40-kWh $FeCl_2$ battery provided the Ford Escort with a range of 230 km on a test track, while in a similar trial in South Africa, the Suzuki microbus fitted with a 50-kWh $FeCl_2$ battery achieved 252 km [12]. These first vehicle batteries and trials are now primarily of historical interest as there have been substantial further developments since the mid-1980s. Even so, they served to demonstrate technical feasibility and safety, and to point the way in which developments needed to proceed.

Since that time, the emphasis has been on smaller, 40-Ah cells (based on beta-alumina tubes of 30-mm internal diameter) in order to improve the power rating. One problem encountered is that the resistance of the cells tends to rise during discharge and thus causes a degradation in power output. This behaviour results from the operational nature of the positive electrode — a reaction front moves through the electrode outwards from the beta-alumina interface on both charge and discharge [38]. The ionic path through the melt therefore becomes progressively longer and the cell resistance rises. For this reason, the power output is normally stated at two-thirds of the open-circuit voltage and 80% DoD; the EV operator is just as interested in the power output at the end of the day as when the battery is freshly charged. For similar reasons, the 'end of battery life' is defined as the stage at which the power output has dropped to 80% of its initial value.

Since 1988, when the collaboration between AEG and the Daimler-Benz Group first began, the emphasis of the ZEBRA programme has been on the development of traction batteries for use in conventional Mercedes cars. The Mercedes 190E was selected as a prime test vehicle (Fig. 8.27). By early 1992, four of these cars powered by ZEBRA batteries were undergoing trials in Stuttgart. The cars have been driven on public roads by a variety of drivers [39]; the performance of the cars is presented in Table 8.6.

Fig. 8.27. ZEBRA-powered Mercedes-Benz 190E test vehicle. (By courtesy of AEG-Anglo Battery Holdings GmbH.)

Table 8.6. Performance data for Mercedes-Benz 190E cars powered by ZEBRA batteries (February 1992) [39].

Vehicle gross weight (kg)	1565
Battery weight (kg)	330
Maximum speed (km h^{-1})	130
Range at 50 km h^{-1} (km)	175
Rating of motor (kW)	
– at 3 min	31
– at 1 h	19

The vehicles were fitted with 152-V batteries of total energy 26 kWh, made up of six parallel chains of cells. The operational temperature range was 250 to 350 °C. Even though these were still immature designs of the $Na/NiCl_2$ cell and battery, the performance compared with that of other batteries in the same car was impressive (Fig. 8.28). Also, the scope for reducing the vehicle mass by custom designing an EV is considerable. Altogether, the four cars completed more than 200 000 km of driving on public roads with few cell failures. The cars were fully traffic-compatible for city and suburban driving. One car has completed 100 000 km over a four-year period, with no cell failures and no battery maintenance. Examination of cells retrieved from the batteries displayed no evidence of positive-electrode corrosion, nor of beta-alumina degradation. In fact, no wear-out mechanisms for the nickel ZEBRA cell have been identified [40].

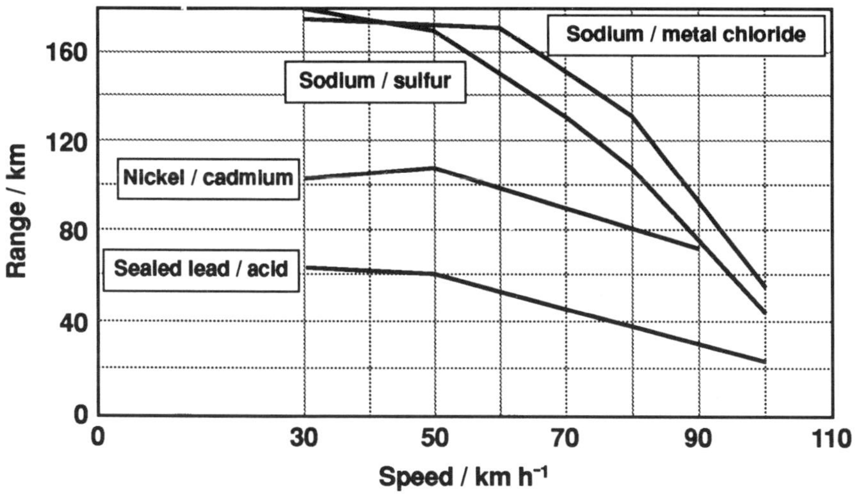

Fig. 8.28. Performance of ZEBRA-powered Mercedes-Benz 190E cars. (Taken from [41] with permission of *Batteries International.*)

Because of space constraints in a car, it is important to minimize the battery volume, as well as to maximize the power output. This is easier with the ZEBRA battery than with the sodium/sulfur counterpart because the safety of the cells allows them to be packed more closely together. Incidentally, this also improves the specific energy of the battery as its housing becomes smaller. Furthermore, the use of a central-positive design permits the possibility of a cell case with a square cross-section [42,43,44]. This gives maximum packing efficiency and battery sizes can be changed in increments of 37 mm (i.e., cell width plus a thin layer of mica between cells to provide electrical insulation) to suit the space available in the vehicle. By 1992, 40-Ah cells with a square cross-section design had been adopted for battery construction (Fig. 8.29), and later vehicles were fitted with these cells. This resulted in the expected improvements in battery energy with values approaching 100 Wh kg^{-1} and 150 Wh l^{-1} at the 3-h rate. These values are adequate to meet the required daily range of a city car. The specific power of a fully-charged battery is ~150 W kg^{-1}, but this falls off to about 80 W kg^{-1} at 80% DoD as the cell resistance increases.

Fig. 8.29. Two views of ZEBRA cells in square mild-steel cans that are separated only by mica. The square connector and tab of the positive electrode allow electrical connections in any direction. (Taken from [44] with permission of *Batteries International* and AEG-Anglo Battery Holdings, GmbH.)

The problem of resistance rise during discharge is a fundamental one and stems from the cell chemistry. In order to decrease the resistance of the positive electrode (the main contributor to cell resistance), it is necessary to reduce the electrode thickness. Attempts to produce beta-alumina in the form of flat-plate, ceramic envelopes proved successful, but gave rise to both sealing and cost problems [45]. Accordingly, the strategy adopted was to fabricate the ceramic as a tube of non-circular cross-section with four lobes (Fig. 8.30). This shaped tube, termed a 'monolith', can be used directly in the square cross-section can with a modified header seal. As a consequence, the positive electrode is now only 10 mm in thickness, and the reaction front at 80% DoD is reduced to 5 mm (compared with 8 mm in the former cell). This increases the specific power to 115 W kg^{-1} at 80% DoD.

Further improvements have been obtained through minor changes to the chemical composition of the positive electrode with the result that the cell resistance has fallen further and the specific power of the battery

(a) (b)

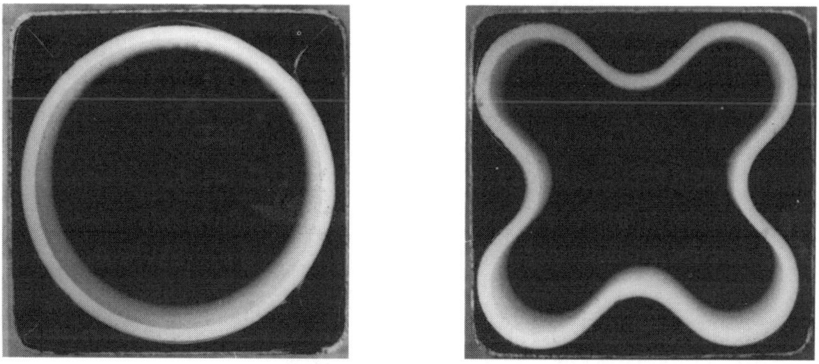

Fig. 8.30. Cross-section of: (a) conventional SL09 cell; (b) shaped beta-alumina electrolyte tube designed for high-power ML1 cell. (Taken from [45] with permission of AEG-Anglo Battery Holdings, GmbH.)

at 80% DoD now approaches the USABC target of 150 W kg^{-1} [45,46]. Moreover, the specific energy of the monolith cell has been raised to 108 Wh kg^{-1} (compared with 98 Wh kg^{-1} for the cell with a simple tubular electrolyte). This, in turn, leads to a 29-kWh battery module with a specific energy of 94 Wh kg^{-1}. Because the strength of the shaped ceramic is less than that of the simple tube, it has been necessary to repeat all the safety tests. These have been completed satisfactorily and batteries based on the monolith cell are now being supplied for vehicle trials.

A key question in battery design is the voltage selected for the EV and, accordingly, the number of parallel chains of cells to give a battery of stated size. For the Mercedes-Benz 190E car, it is convenient to split the battery into two, with one half accommodated under the bonnet (hood) in place of the engine, and the other half under the floor of the boot (trunk). Each module is contained in its own thermal enclosure. Using square cross-section cells, each of these batteries had a 'nameplate energy' of 17 kWh (see Fig. 8.31 [43]), although the latest batteries which use the convoluted electrolyte tubes have an energy of 29 kWh [45]. The 40-Ah cells have a designated 'nameplate capacity' of 30 Ah. Thus, each 17-kWh

Fig. 8.31. 17-kWh, $Na/NiCl_2$ battery module, Type Z5. (By courtesy of AEG-Anglo Battery Holdings GmbH.)

battery contains 220 cells. Two possible examples of how to connect these together are: (i) five parallel strings, each of 44 cells (113.5 V), or (ii) a single string of 220 cells (567.7 V). A single string of series-connected cells is perfectly feasible as ZEBRA cells always pass current, even if they fail and suffer voltage loss. Moreover, the existence of an overcharge and overdischarge capability means that it is not necessary to have parallel strings. From a purely battery standpoint, a single string of cells is indeed preferable as it avoids the possibility of internal circulating currents which arise when cells fail in parallel strings. Also, thinner electrical cables may be used with a high-voltage system, which reduces mass, cost and heat loss from the battery. On the other hand, safety engineers may not welcome a 500-V system.

When parallel chains are employed in battery assembly, it is important that the cells are well matched in capacity. Every cell is tested by charging

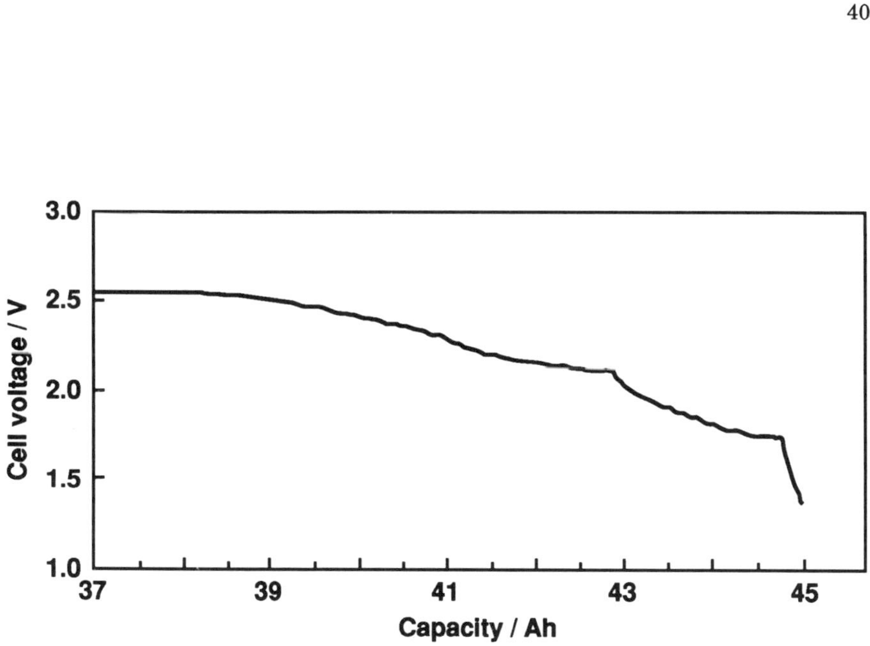

Fig. 8.32. Shape of end of discharge curve of ZEBRA battery; very slow discharge rate (45 h) at 1 A. (Taken from [40] with permission of Elsevier Science Limited, UK.)

and discharging before assembly into a battery. This procedure is done coulometrically and the cells must meet strict acceptance criteria which include constant capacity within ± 0.5%. In this way, circulating currents which result from cell imbalance are kept to a minimum. Cells are also subjected to four freeze–thaw cycles before they become operational.

When one or more cells fail in a battery, there is a danger that circulating currents will cause cells in parallel strings to be driven negative. Fortunately, the battery discharge curve exhibits a lower voltage plateau of width ~2 Ah at the end of discharge (Fig. 8.32) [40]. This is attributed to discharge of the added sulfur and serves to protect the cells. When a series-connected chain continues to discharge under the action of circulating currents, the cells will drop on to the lower voltage plateau and the current will fall to zero. This prevents the cells from being driven negative and subsequently failing. This combination of cells well matched in capacity and the presence of the short low-voltage plateau towards the end of discharge is responsible for the success which has been achieved in vehicle trials with batteries of up to six parallel strings of cells.

Control of the charging regime (in particular, the charge rate and the top-of-charge voltage) is important for all sodium batteries. This is best achieved by first applying a constant-current charge and then a constant-voltage taper charge. Such a procedure has resulted in the development of a microprocessor-based battery controller with the following functions: (i) battery charging, with integration of current for state-of-charge display; (ii) measurement and control of the battery temperature; (iii) monitoring of the electrical insulation resistance; (iv) voltage control; (v) fault detection and recording. As charge control is an important issue, it is likely that the charger and the battery controller will be the responsibility of the battery manufacturer rather than the vehicle manufacturer.

In 1992, a four-year project to test EVs was set up on the German island of Rügen in the Baltic Sea. A total of 60 electric cars, trucks and buses were to be used by a variety of local authorities and departments. The project was sponsored by the German government. Initially, Mercedes supplied 20 vehicles, of which 10 were 190E models. Seven of the 190E cars were fitted with ZEBRA batteries. Because of the comparative success of the ZEBRA batteries in the trials, they have now been fitted to almost half of the vehicles. Meanwhile, Mercedes 190E cars continue to be driven around Stuttgart over routes of up to 150 km per day.

The pilot plant which has been built in Derby and in Berlin has the design capacity to manufacture 160 000 cells per year. These are shipped to Ulm for assembly into 30-kWh batteries. The Ulm facility is able to produce 450 batteries per year [44]. Initial teething problems have been overcome and batteries are becoming available in increasing numbers for EV trials. Plans to test a fleet of 16 vehicles (cars and buses) in Germany and 10 buses in Italy have been announced.

As more batteries are manufactured, the demand for battery recycling and disposal facilities grows. Here again, the ZEBRA battery has an distinct advantage as it has no toxic components such as lead or cadmium. A recycling process is under development. This involves first heating the cells to operational temperature with the terminals short-circuited to ensure that the sodium and nickel chloride are fully discharged. The cell is then cut open, the beta-alumina tube removed, and the case recycled as steel scrap. Next, the ceramic tube is broken, the positive electrode removed and leached with water to dissolve NaCl, $NaAlCl_4$ and NaS_x, and the nickel is recovered as a sludge which contains some aluminium. This nickel may possibly be re-usable to make batteries, either directly or after further processing. Either way, the recovery of nickel may be the key to the economics of the ZEBRA battery.

8.8.6. Safety evaluation

The safety of ZEBRA cells/batteries in both manufacture and use, as well as under accident or abuse conditions, is a vital issue to be addressed. Considerable effort has been devoted to this subject.

ZEBRA cells

Much has already been said or implied about the greater safety of ZEBRA cells and batteries. A principal reason for this safety is the use of a solid positive electrode which is separated from the molten sodium metal by both a solid ceramic and a liquid electrolyte. What happens when the ceramic cracks? The mobile liquid sodium passes through the crack to meet the molten sodium chloraluminate where it reacts chemically according to the overdischarge reaction, see eqn. (8.8). Because both reaction products (sodium chloride and aluminium) are solid, they seal the crack and further reaction is prevented. Furthermore, the fact that aluminium is a metal ensures that the failure is to short-circuit rather than to open-circuit. What happens if the fracture is more than just a crack and significant quantities of sodium enter the beta-alumina tube? Again very little, as the bulk of the sodium chloraluminate is localized and immobilized within the structure of the positive electrode, as also is the metal chloride. Here, the electrode structure acts as an anchor for the solid reaction products and causes the reaction rate to be very sluggish. To illustrate this behaviour, a fully-charged 10-Ah electrode was removed from a working cell and plunged hot into a bath of liquid sodium held at

250 °C. The reaction was difficult to observe visually, but resulted in a temperature rise of less than 100 °C [47].

Other key safety features of the ZEBRA cell, in contrast to sodium/sulfur, are: (i) the reaction products are solid, non-toxic and non-corrosive, unlike sodium polysulfide which is liquid and highly corrosive towards steel and, therefore, can bring about failure of the cell case through the absence of a protective layer; (ii) there is no volatile reactant or product present that could give rise to cell pressurization and possible rupture in the event of high temperatures which result from unrestrained reaction (cf., sulfur, which boils at 444 °C and has a vapour pressure of 20 atm at 720 °C). These features ensure the integrity of the cell case in the event of internal fracture of the electrolyte tube. Should the glass seal or compression bond seal at the top of the cell fail, there is no volatile product to corrode either the internal battery wiring or the resistance heaters.

In further out-of-cell safety tests at the Harwell Laboratory, 20 g of molten sodium was poured into a similar quantity of molten $NaAlCl_4$ at 250 °C [47]. This resulted in the formation of two liquid layers, separated by a crust of NaCl and aluminium. No violent reaction was observed until this crust was broken by vigorous agitation. Then, there was a fire and the measured temperature rose instantly to 1000 °C. On repeating the experiment with the electrolyte absorbed in graphite felt, the temperature rise was much slower (minutes) and limited to 150 °C. This demonstrates clearly the effectiveness of a porous solid matrix in immobilizing the chloraluminate and slowing the reaction, as described above.

Many other tests have been conducted in which the electrolyte tube in operating cells has been deliberately crushed to ensure thorough mixing of the sodium with the positive electrode reactants. Some of these tests were on large (150 Ah) iron and nickel cells in which no safety features were included. In none of the tests was any cell breaching or deformation observed. Moreover, the cells remained conducting after failure. The maximum temperature excursion recorded was ~300 °C with cells which incorporated sodium pressure cans. The safest cells were those with wicked sodium electrodes where the temperature rise was only a few tens of degrees. Other tests on cells were as follows.

Short-circuit behaviour

The short-circuiting of a ZEBRA cell often does not lead to failure. Some designs may be short-circuited for limited periods without any loss of performance. In no instance was any safety hazard observed.

Overheating and freeze–thaw

ZEBRA cells have been deliberately overheated to 800 °C. In these tests, no cell breaching has been observed. This is in accord with expectation as there are no volatile constituents. Cells are regularly overheated to 600 °C with no adverse effects. Thermally-induced cell failure results first from melting of the glass seal above this temperature.

Freeze–thaw tests are now routinely conducted on all production cells and the failure rate is very low indeed. Some cells have been subjected to 100 freeze–thaw cycles with no sign of deterioration. This has been attributed to the soft, friable nature of sodium chloraluminate [47], and also to the use of the central-positive design [44]. The external sodium is soft and exerts no great stress on the beta-alumina, while the chloraluminate contracts on cooling and shrinks away from the inside of the tube.

Overcharging and overdischarging

As the matrix of the positive electrode is only 30% chlorinated, there is more than a two-fold excess of metal to be chlorinated in overcharging before electrolysis of chloraluminate leads to a build-up in the pressure of gaseous chlorine. In practice, all the sodium ions in the sodium chloroaluminate are consumed before this stage is reached. Experiments on a specially designed cell with a large negative-electrode compartment and an abundant excess of sodium confirmed that 200% overcharge was possible at 6 V, with no significant build-up of chlorine pressure.

Overdischarge of cells is possible to at least 20%. Beyond this point, either the sodium will be depleted (thus reducing the active area of the beta-alumina and promoting high current densities), or the positive-electrode compartment will be overfilled. Both of these effects can result in electrolyte failure.

Vibration testing

Vibration testing of cells has been carried out using a computer-controlled programme to produce a random amplitude and acceleration test which simulates driving over 1600 km of cobble stones (pave). This is equivalent to 160 000 km of normal driving. All the cells tested passed this test.

ZEBRA batteries

With the growing interest in the future use of EVs on public roads, the regulatory authorities and vehicle manufacturers began to interest themselves in the safe operation of traction batteries, especially in the case of sodium batteries as these were perceived to pose a significant threat. Although the cell tests were acknowledged to be promising, it was deemed vital to carry out safety evaluations on traction batteries of full size. Accordingly, a consortium of German motor manufacturers compiled a number of tests [48] of which the most noteworthy are as follows.

- To drop a fully-charged battery at working temperature on to the end of a section of corrugated-steel motorway barrier to simulate an impact at 50 km h^{-1}.
- To perform the same drop test on a 150-mm diameter steel cylinder which is used to simulate a telegraph pole.
- To drive a conical steel spike into the battery, and pump aqueous foam into the hole soon afterwards.
- To hold the battery for 30 min over a petrol fire without protective vacuum insulation.
- To forcibly charge the battery at 50% above the open-circuit voltage for 12 h.

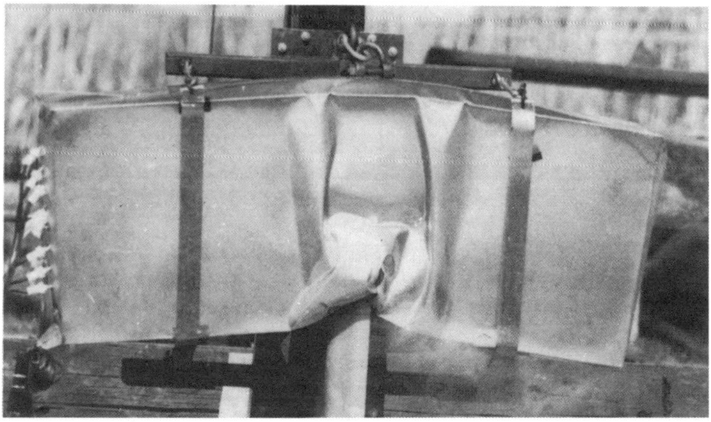

Fig. 8.33. Impact of a hot $Na/NiCl_2$ battery on a corrugated-steel motorway barrier. (Taken from [40] with permission of Elsevier Science Limited, UK.)

A 30-kWh $Na/NiCl_2$ battery was subjected to the first of these tests in 1992 at the proving ground of the Motor Industry Research Association (UK). The steel section penetrated the battery case and intruded about two-thirds of the battery width (Fig. 8.33). There was no fire or obvious hazard. Some white smoke was emitted and the external temperature of the battery increased locally.

In the second test, using a semi-cylindrical steel pole, there was considerable distortion of the battery box, although again there was no evidence that either the vehicle occupants or passers-by would have been exposed to any hazard. There was no leakage from the battery in either of the tests. The remaining three tests were all completed satisfactorily with no perceived safety problems. In particular, the fire test demonstrated spectacularly that the hazard with burning petroleum was much greater than that with an overheated ZEBRA battery.

In all the vehicle trials conducted to date just one spontaneous battery failure has been reported [49]; this occurred in one of the 30 vehicles in Rügen that was fitted with ZEBRA batteries. Apparently, an internal short-circuit resulted in high currents and overheating to >700 °C. The interior of the battery was completely destroyed, but the vacuum insulation was undamaged and the outside of the battery case was only slightly deformed. Non-toxic alumina smoke (resulting from hydrolysis of chloraluminate) and some hydrogen chloride gas were emitted from the battery, but only slowly over a period of two hours. It seems that the battery 'self-destruct' is a very slow process; such battery failure is highly undesirable, but hardly dangerous. Further electrical design work should reduce the risk of the development of internal short-circuits.

8.8.7. Favourable features of the ZEBRA battery

The favourable features of the ZEBRA battery, as seen from the standpoint of the manufacturer and the user, can be summarized as follows.

Manufacturer:

- low cost materials, apart from nickel which may be recycled
- assembled in the discharged state — no sodium handling
- no safety features needed
- no corrosion protection necessary
- flexible cell design
- large cells (up to 500 Ah) possible
- cells may be close-packed in the battery
- failed cells conduct current; this allows long series strings with no cross-connections.

User:

- wide operating temperature range for $NiCl_2$ cells
- high cell voltage
- high specific energy (4–5 times that of lead/acid)
- 100% coulombic efficiency
- overcharge and overdischarge mechanisms
- long life (>1000 cycles)
- no volatile components

- safety in use fully demonstrated
- sealed and maintenance-free
- very low cell failure rate
- independent of ambient temperature.

8.9. Future Prospects

As will be evident from the foregoing discussion, sodium batteries, though simple in concept, have been fraught with technical complications, both in development and in the subsequent transfer to commercial practice. Only by understanding these difficulties is it possible to comprehend why sodium/sulfur batteries have taken 30 years to reach their present state of refinement. Even ZEBRA batteries required almost 15 years of research before EV fleet trials were possible — and they had the advantage of building upon a pool of knowledge and experience derived from the work on sodium/sulfur systems. Moreover, the development teams for both batteries, notably in Germany, South Africa and the UK, were substantial in size and well supported financially. It is clear that the industrial backers of these projects have high expectations for the future of the EV, together with high hopes of sodium batteries as the chosen power source.

Stepping back from the technical detail, what broad statements can be made concerning why sodium battery development is so difficult?

First, it should be said that all batteries, especially secondary batteries, are complex entities. When it is recalled that the lead/acid battery is still not fully understood and is still being developed after almost 150 years since its discovery, the above time-scale for sodium batteries may not be so unusual. A secondary battery is expected to give ~1000 charge–discharge cycles with minimal loss of capacity. In chemical terms, such a life performance relates to a reversible reaction (usually one that involves solid-state transformations) which may be cycled 1000 times with no loss of reactants through side reactions, no loss of reactivity through recrystallization, etc. In no other branch of applied solid-state chemistry is so severe a specification sought.

Second, sodium batteries are novel in three important respects, namely, the use of a solid ceramic electrolyte, operation at elevated temperatures, and the use of molten sodium as the negative electrode. This brings into the battery development team not only electrochemists and materials scientists — the usual players — but also ceramic scientists and technologists, corrosion specialists, and thermal engineers.

Third, unlike a motor vehicle, a battery is not a collection of independent component parts, where the performance of one is largely independent of the rest, where each component can be developed in isolation and assembled on a production line, and where the failure of one broadly has little effect on the rest. Rather, each component in a battery is highly interactive with the others in terms of design, assembly and operation. Moreover, degeneration of any one component may lead to cell failure. When the electrochemistry of the cell starts to decline this is usually a downhill spiral, often with impact on the performance and life of other cells in the battery. While this situation is difficult enough for any type of secondary battery, it is many times worse with the complications of ceramic electrolytes and operation at high temperature. Essentially, all the professions have to work closely together to find a unique solution which is optimal for a particular application.

Fourth, there is the safety aspect associated with the use of molten sodium and, in the case of sodium/sulfur batteries, molten sulfur and polysulfides.

Another problem stems from the fact that sodium batteries are only viable in sizes which are sufficiently large to be thermally sustainable (i.e., multi-kWh). Even when a new cell design has been developed, and hopefully perfected, there is no opportunity to test it in realistic batteries until a small production facility has been set up that is capable of making hundreds of cells reproducibly. Since this takes time and costs money, sponsors are reluctant to enter into this commitment until they have sufficient confidence. Once that has been achieved and a first battery built, it is likely that some new aspect associated with safety, thermal engineering, or the behaviour of cells in series–parallel circuits will be discovered and will necessitate redesign of the cell and repetition of the total development process. When finally a laboratory-based production facility has produced a few experimental vehicle-size batteries, and these have performed well in bench tests, there comes the problem of modifying the battery (and possibly the cells) to suit the space available in a particular vehicle and the performance expected from that vehicle. In the case of monobloc lead/acid or alkaline batteries, moving from one vehicle to another usually involves no more than making a new battery carrier, installing fewer or more batteries, and wiring them to give the desired voltage and current. With sodium batteries, however, it may involve a redesign of the cell, the battery, or both.

When a successful demonstration has been achieved in one vehicle, there is next a call for fleet trials which involve tens or hundreds of batteries. This necessitates the building of a pilot plant, which is even more time consuming and costly. Moreover, unlike lead/acid or alkaline batteries, there is no non-traction market to which one can revert or against which one can set costs. The confidence level has to be even higher before making this large financial commitment, and this takes time to establish and consolidate. As with any other industrial product, the ultimate decision is that of building a full-scale manufacturing plant — the largest investment of all. This is likely to be a market-driven decision and will not take place until sales for sodium batteries are assured. It is interesting to reflect that if sodium batteries had been easier and faster to develop, they would have been technically ready before the major EV market existed. The project might then have been shelved, the teams disbanded, and the expertise lost. (In this regard, the Nuclear Fast Breeder Reactor provides a salutary lesson.) As it is, the Californian initiative for zero-emission vehicles provides a positive goal for sodium battery developers and EV manufacturers, and it appears that the product and the market may be coming together at about the same time.

What are the prospects for the two competing sodium batteries? In principle, sodium/sulfur should have the edge over ZEBRA in terms of both performance and cost. This is because the central-sodium cell has a higher specific energy and power than the central-$NiCl_2$ cell, and because sulfur is cheaper than nickel. The situation is not that simple, however, and other issues such as corrosion, safety and inter-cell connection play an important role in the design of both cells and batteries. Because the ZEBRA cell is intrinsically safe, the packing of cells in a battery may be closer than with sodium/sulfur cells and this largely cancels out the difference in energy at the battery level. Also, the ability to utilize large cells to reduce production cost, and the prospect of recycling nickel, are favourable economic factors for ZEBRA.

In late 1995, it was announced that two of the major sodium/sulfur developers, Asea Brown Boveri in Germany and Silent Power Limited in the UK, had decided to terminate their activities. This is a serious setback for the sodium/sulfur system, although development continues in Japan for load-levelling, if not for traction.

At present, the international ZEBRA programme continues unabated, with encouragement derived from the successful vehicle trials, the safety demonstrations, and interest from motor manufacturers. The remaining obstacles to be overcome have been summarized by J.L. Sudworth, the UK manager of the ZEBRA project, as follows: (i) to increase the specific power of the battery towards the end of discharge; (ii) to demonstrate in the pilot plant that the processes which have been developed for cell and battery production may be fully automated and are capable of meeting the cost targets [50]. If these targets are attained and the market for EVs develops, then it seems probable that a full-scale production plant will be built. Beyond that lies the tantalizing prospect of developing a fully commercial battery based upon the cheaper $FeCl_2$ cells.

References

1. YAO, Y.F.Y., and KUMMER, J.T. *J. Inorg. Nucl. Chem.*, 1967, **29**, pp. 2453–2475.

2. WEBER, N., and KUMMER, J.T. *Adv. Energy Conv. Eng., ASME Conference,* Florida, USA, 1967, pp. 913–916.

3. SUDWORTH, J.L., and TILLEY, A.R. (Eds.) *The Sodium Sulfur Battery* (Chapman and Hall, London, UK, 1985).

4. DELL, R.M., SUDWORTH, J.L., and JONES, I.W. *Proc. 11th Intersoc. Energy Conv. Eng. Conf.*, 12–17 September 1976, State Line, Nevada, USA (American Institute of Chemical Engineers, New York, USA, 1976), pp. 503–509.

5. MAMANTOV, G., MARASSI, R., MATSUNAGA, M., OGATA, Y., WIAUX, J.P., and FRAZER, E.J. *J. Electrochem. Soc.*, 1980, **127**, pp. 2319–2325.

6. WERTH, J., KLEIN, I., and WYLIE, R. *Extended Abstracts Proc. Vol. 75-2, Fall Meeting of the Electrochemical Society,* 5–10 October 1975, Dallas, USA (The Electrochemical Society, Pennington, New Jersey, USA, 1985), pp. 611–613.

7. DELL, R.M., and MOSELEY, P.T. *J. Power Sources,* 1981, **6**, pp. 143–160.

8. DELL, R.M., and MOSELEY, P.T. *J. Power Sources,* 1981/2, **7**, pp. 45–63.

9. HOOPER, A. *J. Phys.(D) Appl. Phys.*, 1977, **10**, p. 1487.

10. TAN, S.R., and MAY, G.J. *J. Mater. Sci.,* 1977, **12**, pp. 1058–1061.

11. BAKER, D.J., BONES, R.J., and PARTRIDGE, B.A. *Deutscher Verlag für Schweisstechnik Conference*, 1980, Baden-Baden, West Germany.

12. TILLEY, A.R., and WEDLAKE, R.J. *Electric Vehicle Developments,* 1987, **6**, pp. 119–122; 140.

13. GRIBBLE, D. *ETV World,* 1986, **1**, pp. 8–10.

14. TILLEY, A.R., and BULL, R.N. *Proc. 22nd. Intersoc. Energy Conv. Eng. Conf.,* 10–14 August 1987, Philadelphia, Pennsylvania, USA (American Institute of Aeronautics and Astronautics, New York, USA, 1987), pp. 1078–1084.

15. GUPTA, N.K., and TISCHER, R.P. *J. Electrochem. Soc.,* 1972, **119**, pp. 1033–1037.

16. OEI, D.-G. *Inorg. Chem.,* 1973, **12**, pp. 435–441.

17. BONES, R.J., BROOK, R.J., and MARKIN, T.L. *In* COLLINS, D.H. (Ed.) *Power Sources 5, Research and Development in Non-Mechanical Electrical Power Sources* (Academic Press, London, UK, 1975), pp. 539–557.

18. BONES, R.J., and MARKIN, T.L. *J. Electrochem. Soc.,* 1978, **125**, pp. 1587–1591.

19. PARTRIDGE, B.A., JENKINS, T.R., and McGUIRE, M. *US Patent 3, 982,959* (issued 1976).

20. HAMER, M. *New Scientist,* 17th July 1986, pp. 37–39.

21. FISCHER, W., BIRNBREIER, H., and BENNINGER, G.N. *Proc. 21st Intersoc. Energy Conv. Eng. Conf.*, 25–29 August 1986, San Diego, California, USA (American Chemical Society, Washington, DC, USA, 1986), pp. 1041–1047.

22. DELL, R.M., and BONES, R.J. *In* VASHISHTA, P., MUNDY, J.N., and SHENDY, G.K. (Eds.) *Fast Ion Transport in Solids* (Elsevier/North Holland, Amsterdam, The Netherlands, 1979).

23. BIRD, J.M., TILLEY, A.R., and SUDWORTH, J.L. *Proc. 13th Intersoc. Energy Conv. Eng. Conf.*, 20–25 August 1978, San Diego, California, USA (Society of Automotive Engineers, Inc., Warrendale, Pennsylvania, USA, 1978), pp. 685–689.

24. BONES, R.J., and TEAGLE, D.A. *J. Power Sources,* 1978, **3**, pp. 45–58.

25. *Batteries International,* 1993, **14**, pp. 11–13.

26. COETZER, J., and THACKERAY, M.M. *US Patent 4,288,506*.

27. COETZER, J., GALLOWAY, R.C., BONES, R.J., TEAGLE, D.A., and MOSELEY, P.T. *US Patent 4,546,005* (issued 1985).

28. COETZER, J. *J. Power Sources,* 1986, **18**, pp. 377–380.

29. BONES, R.J., COETZER, J., GALLOWAY, R.C., and TEAGLE, D.A. *J. Electrochem. Soc.,* 1987, **134**, pp. 2379–2382.

30. GALLOWAY, R.C. *J. Electrochem. Soc.,* 1987, **134**, pp. 256–257.

31. BONES, R.J., TEAGLE, D.A., BROOKER, S.D., CULLEN, F.L., and LUMSDON, J. *Proc. 2nd Symposium on Electrode Materials and Processes for Energy Conversion and Storage* (The Electrochemical Society, Pennington, New Jersey, USA, 1987), Proc. Vol. 87-12, p. 537.

32. McDOWELL, M. *South African Transport,* August 1995, pp. 33–40.

33. BONES, R.J., TEAGLE, D.A., BROOKER, S.D., and CULLEN, F.L. *J. Electrochem. Soc.,* 1989, **136**, pp. 1274–1277.

34. GALLOWAY, R.C. *Extended Abstracts, Proc. Vol. 87-2, Fall Meeting of the Electrochemical Society,* 18–23 October 1987, Honolulu, Hawaii, USA (The Electrochemical Society, Pennington, New Jersey, USA, 1987), p. 232.

35. MOSELEY, P.T., BONES, R.J., TEAGLE, D.A., BELLAMY, B.A., and HAWES, R.W.M. *J. Electrochem. Soc.,* 1989, **136**, pp. 1361–1368.

36. ADENDORFF, K.T., and THACKERAY, M.M. *Proc. 2nd Symposium on Electrode Materials and Processes for Energy Conversion and Storage* (The Electrochemical Society, Pennington, New Jersey, USA, 1987), Proc. Vol. 87-12, p. 552.

37. BONES, R.J., TEAGLE, D.A., and BROOKER, S.D. *In* KEILY, T., and BAXTER, B.W. (Eds.) *Power Sources 12, Research and Development in Non-Mechanical Electrical Power Sources* (International Power Sources Symposium Committee, Leatherhead, England, 1989), pp. 537–547.

38. NICHOLSON, N.D., DEMOTT, D.S., and HUTCHINGS, R. *In* KEILY, T., and BAXTER, B.W. (Eds.) *Power Sources 12, Research and Development in Non-Mechanical Electrical Power Sources* (International Power Sources Symposium Committee, Leatherhead, England, 1989), pp. 549–561.

39. TILLEY, A.R. *Motor Industry Research Association (UK) Conference,* February 1992.

40. SUDWORTH, J.L. *J. Power Sources*, 1994, **51**, pp. 105–114.

41. *Batteries International,* 1991, **7**, pp. 28–30.

42. DUSTMANN, C.-H., and SUDWORTH, J.L. *Proc. 11th International Electric Vehicle Symposium (EVS-11),* Florence, Italy, 27–30 September 1992, Vol. 2, paper 14.05.

43. Brochure published by AEG-Anglo Battery Holdings GmbH.

44. *Batteries International,* 1994, **20**, pp. 72–73; 76–77.

45. COETZER, J., and SUDWORTH, J.L. *Proc. 13th International Electric Vehicle Symposium (EVS-13),* Osaka, Japan, 1996.

46. HAMMERLING, H., BOEHM, H., DUSTMANN, C.-H., TILLEY, A.R., and VAN ZYL, A. *Proc. 13th International Seminar on Primary and Secondary Battery Technology and Applications,* 4–7 March 1996, Boca Raton, Florida, USA.

47. WEDLAKE, R.J., TILLEY, A.R., and TEAGLE, D.A. *Bull. Electrochem.*, 1988, **4**, p. 41.

48. *Batteries International,* 1992, **13**, pp. 92–93.

49. *Batteries International,* 1995, **23**, pp. 76–77.

50. SUDWORTH, J.L. *Phil. Trans. R. Soc. Lond.* A, 1996, **354**, pp. 1595–1612.

CHAPTER 9

Lithium Batteries

9.1. Advantages of Lithium Batteries

At the beginning of Chapter 8, we asked: "Why *sodium* batteries?" The corresponding question for lithium is easier to answer. Lithium is the lightest of all metals with an atomic mass of only 6.94 (cf., sodium 23.0). It has a specific capacity (3.86 Ah g^{-1}) which is over three times that of sodium (1.165 Ah g^{-1}), as well as a higher electrochemical reduction potential, viz., –3.045 vs. –2.714 V. Lithium has also a higher melting point (180.5 vs. 97.8 °C) and, thus, it is more amenable for use as a solid electrode at ambient temperatures. Finally, lithium is far easier than sodium to fabricate as thin sheets or foil, and is less prone to bulk oxidation because a semi-protective film forms on its surface in air. All these factors have made lithium attractive as a potential negative electrode for use in ambient-temperature batteries. Accordingly, lithium batteries have been the subject of intensive research for several decades.

It is of interest to compare the properties of lithium with those of the other common negative electrodes, viz., cadmium and zinc (see Table 9.1). Because lithium discharge involves a single electron, its capacity density is inferior to that of cadmium or zinc, both of which involve two-electron transfer. In gravimetric terms, this is more than compensated by the low density of lithium metal (0.53 g cm^{-3}) and results in a substantially higher specific capacity for lithium electrodes. When the higher electrode potential is also taken into account, the stored specific energy of lithium is much greater than that of zinc or cadmium. It follows that lithium will not have such an energy advantage over conventional batteries in volumetric terms as it has in gravimetric terms.

Table 9.1. Comparison of the performance parameters of lithium, cadmium and zinc negative electrodes.

Parameter	Lithium	Cadmium	Zinc
Capacity density (Ah cm^{-3})	2.05	4.13	5.85
Specific capacity (Ah g^{-1})	3.86	0.48	0.82
Standard potential (V)	–3.04	–0.40	–0.76
Specific energy (Wh g^{-1})	11.7	0.19	0.62

The overall scenario for lithium batteries stands in sharp contrast to that for sodium batteries. As pointed out in Chapter 8, the latter are viable only as 'hot' batteries with a ceramic electrolyte. This restricts the use of sodium batteries to applications which require multi-kWh and larger sizes. The major potential markets are traction batteries and load-levelling batteries. Given the size and the exceedingly difficult technology of the system, only a few, large international teams have been involved with the development of sodium batteries. By contrast, lithium batteries are based upon a much wider range of possible electrolytes and electrode materials, and may be made in any size from milli-Wh units upwards. Also, both ambient-temperature and high-temperature lithium batteries are possible. The low-temperature technology has some features in common with conventional batteries, particularly in the design of small button and coin cells. This factor, together with the strong interest in new batteries for electronic applications and for portable power sources, has encouraged many companies around the world to become involved with lithium batteries. At the same time, many university researchers have been attracted by the scientific challenges of such batteries. The initial entry cost into the field of lithium batteries is low; the requirements are little more than a dry-box, basic electrochemical equipment for small-scale or fundamental studies, and a dry-room for development work. The result is intense international interest and activity on lithium batteries, as reflected in the extensive scientific literature and the many international conferences which have been held.

Until very recently, it was only the high-temperature, rechargeable lithium batteries which had been developed in kWh sizes; few rechargeable ambient-temperature systems had progressed beyond C- or D-size cells. Only now is the possibility of producing larger ambient-temperature lithium batteries receiving serious attention. The technology of EV-size lithium batteries is still in its infancy compared with sodium technology, although lately the Sony Corporation has reported the assembly of the first lithium EV battery. This Chapter highlights some of the developments in lithium batteries over the past two decades, together with the problems encountered. The initial discussion deals with primary lithium batteries since a working knowledge of these systems provides a sound basis for understanding secondary analogues and the scale-up to EV-size units.

9.2. Primary Lithium Batteries

The most commonly used electrolyte in small lithium cells is a solution of a lithium salt in an organic liquid. Typical aprotic solvents are: cyclic esters e.g., ethylene carbonate (EC), propylene carbonate (PC), γ-butyrolactone (BL); linear esters, e.g., methyl formate and acetate, diethyl carbonate; cyclic ethers, e.g., dioxolane; acetonitrile; tetrahydrofuran; sulfoxides. The soluble lithium salts which have been used include: $LiClO_4$, $LiAlCl_4$, $LiBF_4$, $LiPF_6$, $LiAsF_6$, $LiCF_3SO_3$, and $LiN(CF_3SO_2)_2$. These solutions provide reasonable conductivities, although the values are never as high as those exhibited by aqueous electrolytes. The first lithium cells of this type to be developed were primary systems which employed a variety of positive electrode materials, e.g., CuO, CuS, CF_x, MnO_2, MoO_3, Ag_2CrO_4, V_2O_5. The cells have been manufactured in small sizes (button cells, coin cells, cylindrical cells up to D-size) and marketed successfully for a variety of applications that include watch batteries, cameras, heart pacemakers, and hearing aids. The most widely adopted positive-electrode materials are CF_x and MnO_2.

An all-solid-state lithium cell can be used for very low drain applications as, for instance, in watches, pocket calculators, and heart pacemakers. The cell is comprised of a positive electrode which is a complex formed between iodine and poly-2-vinylpyridine and a negative made from lithium foil. When these two solids are brought into contact, a thin film of lithium iodide is formed at the interface and this serves as a solid electrolyte. Although the film has a low conductivity for Li^+ ions, it is adequate for low-drain service. Obviously, leakage problems that are a potential hazard with cells which employ a liquid component are not relevant to cells made from all-solid materials.

High-rate lithium primary cells have also been developed, particularly in large sizes for military applications (naval sonar buoys, army field communications, space power, etc.). There are two types, namely, the 'dissolved positive-electrode cell' and the 'liquid positive-electrode cell'. In both designs, the positive active-material is in direct contact with the lithium-metal negative, with which it reacts superficially to form a passivating layer. The dissolved positive cell consists of a solution of sulfur dioxide in an electrolyte of lithium bromide dissolved in acetonitrile. The discharge reaction is:

$$2Li + 2SO_{2\,(dissolved)} \rightarrow Li_2S_2O_4 \qquad V^o = 3.0\ V \qquad (9.1)$$

The partial pressure of the dissolved sulfur dioxide is ~300 kPa at room temperature and rises to 3 MPa at 100 °C. Thus, the cell must be enclosed in a pressure housing. Alternatively, a liquid SO_2 solvate of $LiAlCl_4$, such as $LiAlCl_4{\cdot}xSO_2$, may be used as the electrolyte.

The liquid positive cell employs thionyl chloride ($SOCl_2$) or sulfuryl chloride (SO_2Cl_2) as the positive active-material, with dissolved $LiAlCl_4$ salt to provide the conductivity. Thionyl chloride is the more common choice and the discharge reaction is:

$$4Li + 2SOCl_2 \rightarrow 4LiCl + SO_{2\,(dissolved)} + S \qquad V^o = 3.6\ V \quad (9.2)$$

This cell has a high specific energy (330 Wh kg^{-1}), a high energy density (>700 Wh l^{-1}), a good low-temperature performance, and a long shelf-life. Thus, the cell is very suitable for military use in cold climates, or for space power applications.

Although none of these primary cells are directly applicable as EV power sources, the basic science behind their mode of operation may be significant. The development of primary systems is well documented and the reader is referred to other textbooks for further information, e.g., [1,2].

9.3. Early Work on Rechargeable Lithium Cells

With the advent of successful primary lithium cells in the 1970s, interest turned towards the investigation of rechargeable alternatives. Several large companies became involved, e.g., Exxon and Bell Telephone in the USA. A significant step forward was the discovery by Whittingham [3], while working at Exxon, that certain inorganic compounds (e.g., TiS_2) react with lithium ions by an 'intercalation' mechanism. This involves the incorporation, electrochemically, of Li^+ ions into the crystalline lattice of the host with no significant change in crystal structure, merely a change in lattice parameter, i.e.,

$$x\mathrm{Li}^+ + \mathrm{TiS}_2 + xe^- \underset{\text{charge}}{\overset{\text{discharge}}{\rightleftharpoons}} \mathrm{Li}_x\mathrm{TiS}_2 \quad (0 > x < 1) \tag{9.3}$$

The intercalation reaction is both reversible and rapid [3,4]. It provides the basis for a positive electrode which can be discharged and recharged ('de-intercalated') at rates in excess of 10 mA cm^{-2}. Because only a single solid phase is involved, thermodynamics decrees a sloping discharge voltage curve (from 2.5 V down to 1.9 V, Fig. 9.1).

Over 400 discharge–charge cycles were demonstrated to > 50% capacity. The essential features of TiS_2 are that it has a high capacity for lithium, rapid intercalation and de-intercalation, no significant change in crystal structure, and a good electrical conductivity. Subsequently, other reversible lithium intercalation electrodes were discovered, e.g., MoS_3 [6], V_6O_{13} and $V_xFe_{1-x}S_2$ [7]. Amorphous MoS_3 was found to be capable of reacting electrochemically with up to 3.8 Li per MoS_3 formula unit, and gave a cell with a high theoretical specific energy.

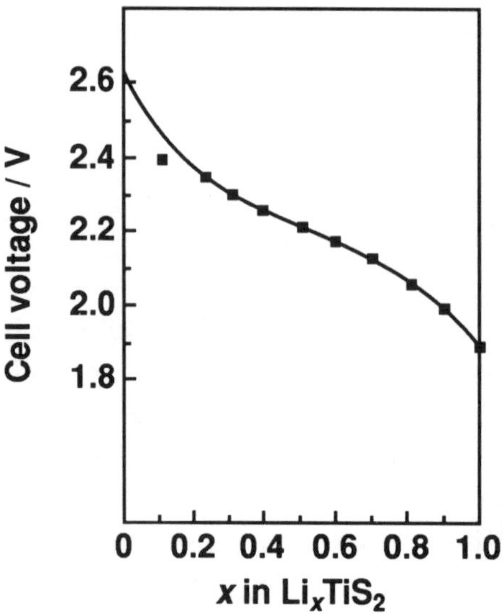

Fig. 9.1. Discharge curve for Li/TiS_2 cell. (Taken from [5] with permission of The Electrochemical Society, Inc.)

Whittingham and colleagues at Exxon used a solution of lithium perchlorate ($LiClO_4$) in dioxolane as the electrolyte for Li/TiS_2 cells. This proved to be very good for cycling both the TiS_2 and the lithium; the latter was replated very smoothly on recharge. On the basis of the results obtained with a $Li–Al/TiS_2$ button cell, it was predicted that an EV battery could be designed with a specific energy of 100 Wh kg^{-1} and a specific power of 100 W kg^{-1} [8]. Unfortunately, it was found that perchlorate was unstable in dioxolane and liable to detonate. All attempts to find a stable electrolyte combination which gave a smooth lithium electroplate on recharge proved to be unsuccessful. Eventually, Exxon terminated the programme and withdrew their button cell from the market in 1979. Whittingham concluded that there remained a critical need for basic research to understand the electroplating of lithium from organic solutions at reasonable rates, i.e., at several mA cm^{-2}. Later, the Jet Propulsion Laboratory developed 5-Ah Li/TiS_2 cells for possible space applications [9]. These were cycled 50–100 times, but with serious loss in capacity. This performance added weight to Whittingham's earlier conclusion.

At AT&T Bell Laboratories, Broadhead [10] researched a different type of intercalation material. The starting point was the high theoretical specific energies of the lithium/iodine and lithium/sulfur electrochemical couples (697 and 2689 Wh kg^{-1}, respectively). Broadhead showed that certain transition metal dichalcogenides could act as 'host' structures for iodine and sulfur, these included: $MoSe_2$, $MoTe_2$, NbS_2, $NbSe_2$, $NbTe_2$, TaS_2, $TaTe_2$, TiS_2, $TiTe_2$, WS_2, WSe_2, and WTe_2. All these materials have open-circuit potentials which approach 3 V when intercalated with iodine and, when used in cells with lithium negative electrodes, provide working voltages which fall from 3.0 to 1.5 V. $NbSe_2$ was selected for investigation and many hundreds of cycles were demonstrated with small pellet cells, at rather low depths-of-discharge. Later, Broadhead focused on niobium triselenide (without intercalated iodine or sulfur) as a positive electrode; this undergoes the following reversible reaction:

$$3Li + NbSe_3 \underset{\text{charge}}{\overset{\text{discharge}}{\rightleftarrows}} Li_3NbSe_3 \qquad (9.4)$$

The cell has an open-circuit voltage of 2.2 V and can be discharged down to 1.4 V. AT&T set up a pilot line for the manufacture of these cells (the 'Faraday™ cell') in the AA-size. The cells give almost twice the capacity of AA-size nickel/cadmium cells as well as a higher voltage, so that the stored specific energy is two-to-three times greater (Fig. 9.2). Cycle lives in excess of 200 cycles were demonstrated and the self-discharge rate of the cell was very low (shelf-life >5 years). Despite this encouraging performance and the investment in a pilot manufacturing facility, the project was ultimately abandoned. It seems unlikely that this type of cell is relevant to EV traction batteries, not least because of the cost and mass of niobium. Nevertheless, the Exxon and AT&T industrial research projects in the 1970s did much to awaken world-wide interest in lithium intercalation batteries.

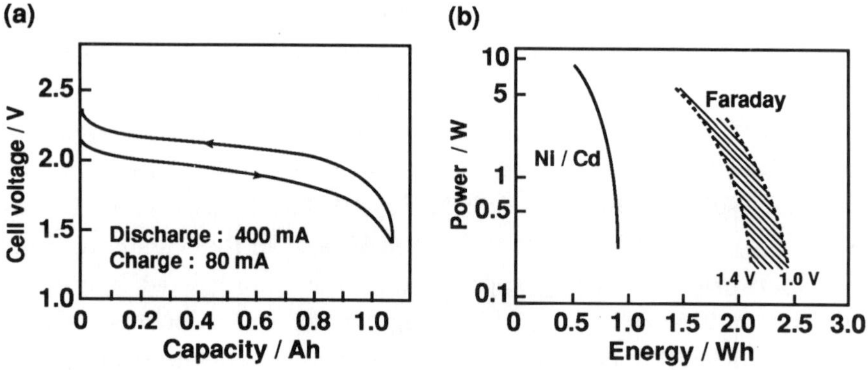

Fig. 9.2. (a) Discharge curve and (b) electrical performance characteristics of the Faraday™ cell at 20 °C. (By courtesy of AT&T, USA.)

9.4. Rechargeable Liquid-Electrolyte Cells

The first rechargeable lithium cell to be commercialized, in the late 1980s, was the Molicel™ which was manufactured by Moli Energy Limited in British Columbia, Canada. This was available in both AA- and C-sizes (spirally wound) with a lithium-metal negative, an organic electrolyte, and an intercalation positive of naturally occurring molybdenum disulfide (molybdenite) [11,12]. The nominal cell voltage was 1.8 V and the cell had a specific energy which was 2–3 times higher than either lead/acid or nickel/cadmium. When operating at over 80% of its theoretical capacity, the cell provided 200–300 cycles. The shelf-life was many years. Some problems were experienced with electroplating of the lithium on recharge, as is the common experience with cells which employ negative electrodes of lithium metal. Larger, 65-Ah cells were also constructed and tested experimentally; a battery of six series-connected cells (700 Wh) was operated for more than 100 cycles with only a modest loss of capacity [13]. The specific energy of the battery was nominally 94 Wh kg^{-1} in an unoptimized configuration, although the average value over 100 cycles when discharged at the $C_3/3$ rate was only 49 Wh kg^{-1}.

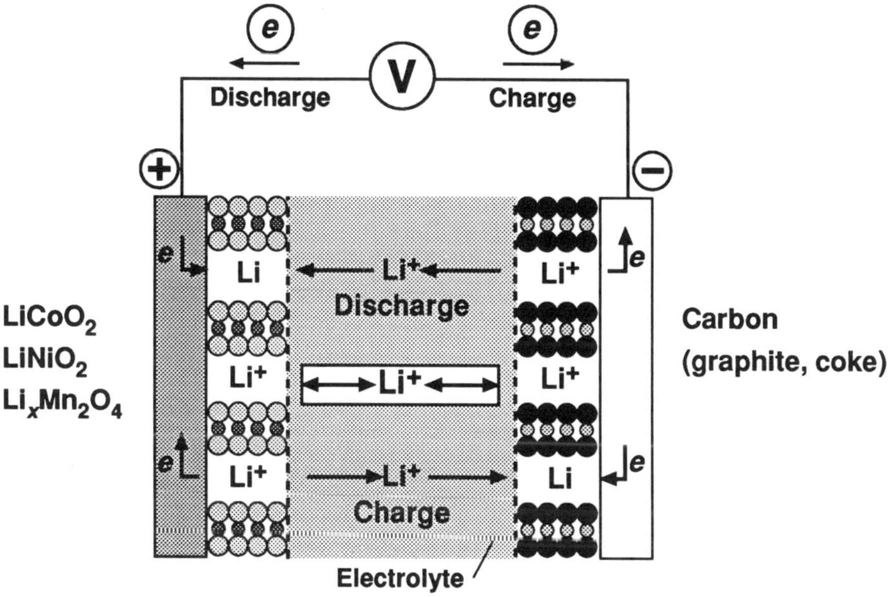

Fig. 9.3. Schematic representation of the charge–discharge operation of a lithium-ion battery.

Because of the problems with the electroplating of lithium metal from organic electrolytes, attention has concentrated recently on a new type of secondary lithium battery, the 'lithium-ion battery' (Fig. 9.3). This battery employs not one, but two, lithium intercalation electrodes — a negative as well as a positive. Provided that the difference in free energy of the lithium in the two electrodes is sufficiently large, a useful voltage may be obtained. This avoids the problems of electroplating lithium metal since the lithium ions are intercalated into the structure of the negative electrode. The usual negative material is carbon (graphite or coke) and this intercalates lithium up to a composition of C_6Li at a voltage of zero to 1 V with respect to a lithium reference electrode. The favoured positive material is a mixed oxide. The lithium-ion battery is known by a number of other names, including the 'swing' battery and the 'rocking-chair' battery, so called because the lithium ions 'swing' or 'rock' from side-to-side as the battery charges and discharges. Inevitably, the specific energy of the lithium-ion battery is lower than that of the corresponding lithium-metal battery because of the reduced cell voltage and the extra mass of carbon in the cell.

9.4.1. Electrolytes

Numerous different solvents and dissolved salts have been employed in rechargeable liquid-electrolyte batteries. The majority of the solvents are organic liquids, although some work has been carried out by Duracell Incorporated on a rechargeable Li/SO_2 cell [14]. A prototype C-size cell was developed based on the electrolyte $LiAlCl_4 \cdot 6SO_2$ with a lithium-metal negative and carbon-black positive. This had an open-circuit voltage of 3.2 V, a nominal capacity of 1.8 Ah, and was capable of 50 cycles at 1 A to a cut-off voltage of 2.0 V. The specific energy and energy density of the cell were 135 Wh kg^{-1} and 197 Wh l^{-1}, respectively. This was an encouraging development, although it was noted that discharge below 1.0 V was hazardous. Other authors [15] have emphasized the dangers that are inherent in attempting to recharge primary Li/SO_2 cells.

In selecting an organic-solvent/lithium-salt combination as electrolyte, consideration must be given to a number of physical and chemical factors. Foremost among these is the requirement for the electrolyte to be compatible with the negative electrode. This essentially excludes aqueous systems; in non-aqueous systems, the formation of a stable passivating layer on the negative electrode (a so-called 'Solid Electrolyte Interface', SEI) is necessary. A second important consideration is ionic conductivity, since most of the electrolytes have appreciably lower conductivities than aqueous or fused-salt electrolytes. Other important factors are the solubility of the lithium salt in the solvent (that largely determines the conductivity), the melting point and boiling point of the solvent (that limit the temperature range over which the battery can operate), the solvent density and viscosity, its chemical compatibility with the positive electrode, and its electrochemical stability window. A wide stability range is essential if the electrolyte is to be compatible with a positive electrode that may be charged at more than 4 V. To avoid excessive polarization, the electrolyte should ideally have a Li^+ ion transport number which approaches unity. Finally, for batteries which have lithium-metal negatives, a particularly important consideration is the role of the electrolyte in determining the nature and smoothness of the electroplate formed on recharging (see Section 9.4.2).

The solvents and solutes considered for use in rechargeable lithium batteries are mainly the same as those listed in Section 9.2 for primary systems, although $LiClO_4$ is no longer favoured as the salt because it constitutes an explosive hazard in an organic solvent. The conductivities of the electrolytes generally lie in the range 0.1–3.0 S m^{-1}; the value goes through a maximum as the concentration of solute is increased. A solution of 2 M $LiAsF_6$ in methyl formate is one of the best electrolytes; it has a conductivity of 4.5 S m^{-1} at 25 °C, and is suitable for high-rate cells.

Researchers at the University of Waterloo, Canada [16] used Raman spectroscopy to demonstrate the presence of the solvated ion $Li(MA)_4^+$ in a solution of 1 M $LiAsF_6$ in pure methyl acetate (MA). In addition, through measurements of corrosion potential–time transients on freshly cut surfaces of lithium metal, it was possible to make deductions concerning the interaction between the electrolyte and the surface. The solvents tetrahydrofuran, 2-methyltetrahydrofuran and PC were found to be stable with respect to the AsF_6^- ion, and the films formed on the surface of the lithium metal were shown to be derived from electrochemical reduction of the anion rather than the solvent. Other salts, i.e., $LiBF_4$, $LiClO_4$ and $Li(CF_3SO_2)_2N$, displayed much less reactivity towards the bare metal than $LiAsF_6$ and the surface film tended to be formed by decomposition of the solvent. Corrosion potentials were much enhanced in the presence of water as an impurity in the electrolyte. Basic physicochemical measurements such as these are valuable in revealing the nature of the species present in the electrolyte and the complexities of the solvent/solute/lithium interaction. In the same study [16], Raman spectroscopy was also used to investigate the crystallographic changes which take place during the intercalation of lithium into various forms of pyrolytic carbon.

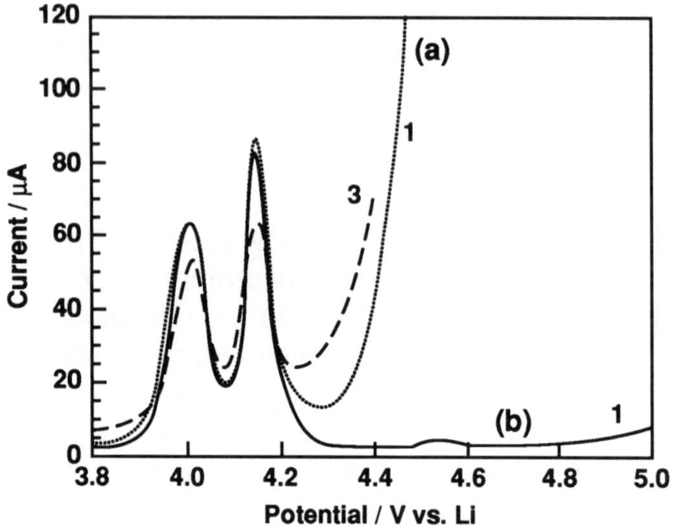

Fig. 9.4. Voltammograms for $Li_{1+x}Mn_2O_4$ at 55 °C using: (a) EC + DEE (1 : 1) + 1 M $LiClO_4$ electrolyte, the numbers 1 and 3 refer to the first and third charge, respectively; (b) EC + DMC (2 : 1) + 1 M $LiPF_6$ electrolyte. DEE=diethoxyethane. (Taken from [17] with permission of Elsevier Science Limited, UK.)

When a high-voltage intercalation positive electrode such as $LiCoO_2$ or $Li_{1+x}Mn_2O_4$ is employed, the electrolyte needs to be kinetically stable towards oxidation at potentials of up to almost 5 V vs. lithium and at temperatures of ~50 °C. Workers at the Bell Corporation (Bellcore) investigated [17] the high-voltage stability of various liquid electrolytes against $Li_{1+x}Mn_2O_4$/carbon positives. The findings highlighted the considerable complexity of the system — the voltage stability of the electrolyte depends both upon the solvent mixture employed and the dissolved lithium salt. Bearing in mind the need for adequate ionic conductivity (which ruled out one of the stable salts, $LiBF_4$) and safety (which ruled out another, $LiClO_4$), a comprehensive study eventually identified a solution of 1–2 M $LiPF_6$ in a mixed solvent of ethylene carbonate and dimethyl carbonate (EC + DMC) as being optimum for the $Li_{1+x}Mn_2O_4$ positive electrode, with a breakdown voltage that exceeds 5 V. In Fig. 9.4, a voltammogram for this new electrolyte composition (curve (b)) is compared with that for the electrolyte used previously (curve (a)). The breakdown voltage of the electrolyte has increased from 4.3 to over 5.0 V. In practical cells, the EC + DMC electrolyte is suitable for voltages up to 4.9 V at room temperature, and up to 4.8 V at 55 °C. It is liquid over a wide temperature range and exhibits excellent ionic conductivity between –20 and 65 °C.

These and similar fundamental studies illustrate the rapid progress which is now being made in understanding the chemistry of complex liquid organic electrolytes and their interaction with lithium-metal negatives and intercalation positives.

9.4.2. Negative electrodes

Lithium metal

The principal reason for the slow evolution of rechargeable lithium batteries has been the inability to cycle the lithium electrode with high efficiency over long periods of time. A freshly formed surface of lithium metal is thermodynamically unstable with respect to virtually all organic solutions and, accordingly, will rapidly decompose such solutions until the metal is covered by a passivating (protective) surface film of decomposition product. A typical reaction, with PC, proceeds as follows:

$$\begin{array}{l} CH_3{-}\underset{\substack{| \\ O \\ \diagdown}}{CH}{-}\underset{\substack{| \\ O \\ \diagup}}{CH_2} + 2Li \rightarrow Li_2CO_3 + CH_3{-}CH{=}CH_2 \\ \qquad\quad CO \end{array} \qquad (9.5)$$

to give a surface film of lithium carbonate and the liberation of propylene gas. It is a common experience that lithium cells tend to liberate gases on recharging. The protective surface film often serves as a solid electrolyte interface (SEI) which is permeable to Li^+ ions and thus allows the charge and discharge reactions to continue. It is responsible for the long shelf-life of lithium cells, but may also result in poor cycleability of the lithium electrode, particularly if it is non-uniform in nature. The passivating film may also determine the rate capability of the cell and the characteristics of its voltage delay.

The morphology of the lithium electroplate formed on recharging a lithium battery is found to depend critically upon the choice of solvent and dissolved lithium salt, any additives or impurities which may be present, and the rate of recharge. As cycling proceeds, the cell capacity generally declines because the stripping efficiency of the electrodeposited lithium is often poor. If a long cycle life is to be achieved, it is necessary to employ a large excess of lithium (typically four- or five-fold) over that required to react with the positive electrode. This is normally attributed to individual particles of lithium metal becoming encapsulated by a surface film of reaction product, and thus becoming isolated electrically from the bulk of the deposited electroplate and no longer available to take part in the subsequent discharge reaction. This large quantity of finely-divided lithium does, of course, pose a safety hazard in the cell, especially if there are lithium dendrites present that can lead to an internal short-circuit. A short-circuit can, in turn, result in local overheating and melting of the lithium metal (melting point: 180.5 °C) and disruption of the SEI. Without a protective film, the molten lithium reacts exothermically with the electrolyte and results in cell venting, often with flames. Possible means of controlling the formation of electrically isolated lithium, or its subsequent removal, have been discussed in detail by Koch [18].

More recently, the lithium-metal/electrolyte interface has been the subject of considerable research as battery scientists have begun to realize the complexity of the situation and the importance of prolonging the cycle life of cells. The charge–discharge behaviour of the lithium electrode is intensely affected by the structure of the interface which, in turn, is determined by the composition of the electrolyte. The interface has been studied by electrochemical cyclic voltammetry and by a.c. impedance analysis, while the morphology of the deposit has been investigated by scanning electron microscopy. Fourier-transform infrared spectroscopy has been used to analyse the reaction products between lithium and the electrolyte solution. These studies have served to highlight the specificity of the lithium/electrolyte interface and how it may be modified by changing the composition of the electrolyte, or even by adding minor amounts of inorganic cations or certain organic additives [19]. These modifications result in changes in the coulombic efficiency for lithium cycling.

A thorough study of the lithium/electrolyte interface has been made by Aurbach et al. [20] with a number of different experimental techniques. Four conclusions were reached. First, the chemical structure of the surface films formed on lithium is determined by a delicate balance between the reduction processes of the solvents, salts, and common contaminants. The greater the heterogeneity in the films, the more they induce non-uniform lithium deposition and dendrite formation. Second, ageing processes gradually change the structure and properties of the surface films. For instance, CO_2 may be a desirable additive or impurity since it reacts to form Li_2CO_3, which is highly stable and protective. Again, traces of water react with organic carbonates ($ROCO_2Li$) to form Li_2CO_3. Third, stable salts such as $LiAsF_6$ are more efficacious than salts which are more readily reduced ($LiPF_6$, $LiBF_4$). Fourth, the cycling efficiency is high in systems where the lithium deposition is smooth and/or the deposited lithium is efficiently passivated by the instantaneously formed surface species. What is beyond dispute is that the electrochemical performance of lithium in batteries is determined by the stability, uniformity, composition and morphology of the solid electrolyte interface.

Microscopic observation of the morphological changes in a lithium electrode during cycling has produced an explanation for the known fact that a high-rate discharge can enhance cycle life [21]. Two forms of electrodeposited lithium were observed, namely, acicular (needle-like) and granular. The nature of the deposit depends not only on the charge rate, but also on the rate of the previous discharge. The latter defines the sites and area available to receive the fresh deposit. The acicular or 'mossy' lithium is inactive ('dead') while the granular lithium is active. A fast discharge serves to smooth the surface and to recombine isolated lithium. This change in structure results in an increase in cycle life. Another possible approach is to use chemical 'levelling agents', as employed in conventional electroplating procedures.

Studies such as these are progressively throwing light on the complex series of phenomena which are involved in the electrodeposition of lithium from organic media, and they are helping towards a practical solution to the problem of achieving a lithium electrode with long cycle life. It may be noted that there are certain similarities between the organic electrolyte, lithium-metal battery and the aqueous electrolyte, nickel/zinc battery (see Section 6.4, Chapter 6). Both are ambient-temperature systems and both have high specific energies; thus, they are attractive candidates for EV power sources. Both suffer a marked decline in capacity on cycling, which stems from replating of the metal on the negative electrode, and both may experience internal short-circuits via the formation of dendrites. In each battery, a several-fold excess of metal is employed in the negative electrode and, in spite of this and considerable research and development on the problem, life performance is still measured in hundreds, rather than in thousands, of cycles. In order to enhance battery life beyond 1000 cycles, it is necessary to improve both the electroplating of lithium from organic electrolytes and that of zinc from aqueous electrolytes — each represents a great challenge to the research community.

Composite negative electrodes

A possible solution to the problem of electroplating lithium is to employ an alloy of lithium such as Li–Al, Li–Si, or Li–B that can be decomposed and reformed electrochemically. Lithium–aluminium alloy has been used extensively in the high-temperature Li/FeS_x battery (see Section 9.6.2) and seems to be the best option. The phase diagram of the lithium–aluminium system (Fig. 9.5) shows a two-phase region ($\alpha+\beta$) from 9 to 47 at.% lithium. This is the region where the lithium–aluminium electrode is cycled and where the potential vs. lithium is 0.3 V, i.e., the cell voltage is reduced by 0.3 V compared with a lithium-metal cell. Unfortunately, the diffusion coefficient of lithium in the alloy is low at ambient temperature and this restricts the thickness of the alloy to a thin foil and also limits the rate at

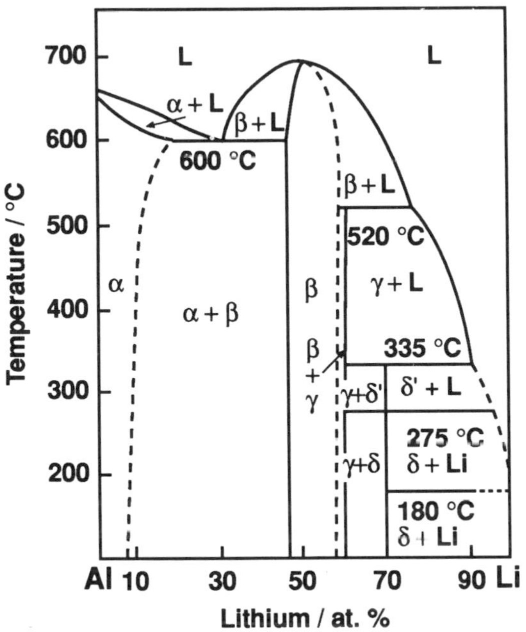

Fig. 9.5. Lithium–aluminium phase diagram. (Taken from [22] with permission of The Electrochemical Society, Inc.)

which the cell may be cycled. Although lithium–aluminium alloy may be an option for use in a low-rate cell, experience has shown that cycling invariably causes severe roughening and ultimate disintegration of the electrode.

The second type of composite electrode, under extensive study at present, is the carbon–lithium intercalation electrode, as used in the lithium-ion battery [23]. Carbon electrodes intercalate lithium ions according to the basic scheme:

$$6C + xLi^+ + xe^- \underset{\text{discharge}}{\overset{\text{charge}}{\rightleftarrows}} Li_xC_6 \quad \text{where } 0 < x < 1 \qquad (9.6)$$

Carbon can be obtained from many different sources (graphite, petroleum coke, carbon blacks, pyrolyzed polymers, etc.), and with varying degrees of both structure and order. It is important to determine: (i) how lithium intercalation depends upon the structure and surface area of the carbon; (ii) the extent to which the kinetics of intercalation and de-intercalation, which define charge–discharge rates, vary with structure; (iii) the extent to which the potential of the electrode vs. lithium varies with structure. Early work established that the ability of disordered carbons, such as petroleum coke, to take up lithium (0.186 Ah g^{-1}, corresponding to $Li_{0.5}C_6$) was only half that of graphite (0.372 Ah g^{-1}, corresponding to LiC_6). Moreover, even the graphite value is an order of magnitude smaller than the specific capacity of lithium metal itself (3.86 Ah g^{-1}). This comparison is not strictly fair, however, as it is necessary to use four-to-five times the theoretical quantity of lithium if an acceptable cycle life is to be achieved in batteries with lithium-metal negatives. Even so, the specific energy of lithium-ion batteries is likely to be considerably smaller than that of lithium-metal batteries.

Carbon intercalation electrodes are prepared by slurrying the carbon powder and a binder in a suitable solvent, casting into the desired shape, and allowing the solvent to evaporate. In the first lithium-intercalation cycle of graphite, the measured capacity is both greater than 'theoretical' and greater than that obtained in subsequent cycles. This arises through partial decomposition of the electrolyte caused by the expansion and exfoliation of the graphite during intercalation; the decomposition products are adsorbed on the graphite to form a passivating surface layer which inhibits further decomposition. A similar phenomenon is observed with coke electrodes. Qualitatively, such passivation may be considered to be similar to that exhibited by lithium-metal negative electrodes. Indeed, it serves much the same purpose, namely, to stabilize the surface against further electrolyte decomposition while permitting Li^+-ion transport. Both the amount and the nature of the surface species depend on the type of carbon, the reactivity of the solution, and the passivating properties of the reaction products.

Although coke electrodes have a lower specific capacity than graphite, they have the advantage of being less sensitive to the nature of the electrolyte employed. This permits the use of high conductivity electrolytes, such as $LiClO_4$–PC, that are unsuited to graphite. On the other hand, graphite electrodes have a flatter voltage curve on charge–discharge cycling. Coke electrodes ($Li_{0.5}C_6$) show a linear voltage decline as high as 1.2 V on removing all the lithium. Clearly, neither graphite nor coke is ideal, but since this is such an important component of the lithium-ion battery, further research is required to optimize the properties of the electrode. A useful approach has been made by Dahn et al. [24] who have studied the dependence of the intercalation process on the crystal structure for a wide variety of carbons. It seems that the optimum choice of carbon for a negative electrode will lie between the highly-ordered graphite and the less-ordered carbon black [25]. Recent work has demonstrated [26] that lightly pyrolyzed carbons (sugar or pitch prepared at 500–700 °C) which retain a high hydrogen-to-carbon ratio may have a lithium capacity as high as 0.5–0.7 Ah g^{-1}, but at the expense of a significantly higher potential vs. lithium and considerable hysteresis in the charge–discharge curves. Whether or not this is a useful finding is not yet clear. It should be possible to identify a carbon which intercalates lithium reversibly to near the optimum degree (LiC_6), but which is less susceptible than graphite to the nature of the organic electrolyte.

Another possible negative material is a low-voltage intercalation spinel such as $Li_4Ti_5O_{12}$, in spinel notation Li $[Ti_{1.67}Li_{0.33}]O_4$. This has a stable operating voltage of about 1.5 V vs. Li. When combined with a 4-V positive electrode, such as $LiCoO_2$, it is still possible to have a cell with a voltage of 2.5 V, i.e., approximately twice the voltage of nickel/cadmium or nickel/metal hydride.

9.4.3. Positive electrodes

Discussion here centres on solid positive electrodes that intercalate lithium into their structure. The major advantage of intercalation materials is that no major recrystallization process is involved in charge–discharge cycling and there is minimal reorganization of the electrode structure. The downside of this is that the cell voltage falls as discharge proceeds. Thus, the electrode material should possess a structure through which Li^+ ions may diffuse readily, and it should also be electronically conducting so that compensating electrons may be taken up from the external circuit.

Broadly speaking, the intercalation materials fall into two categories: (i) di- and tri-chalcogenides of transition metals; (ii) the mixed-oxides: $LiCoO_2$, $LiNiO_2$ and $Li_xMn_2O_4$. A key difference between these two categories is that cells based on the chalcogenides have normally been assembled in the charged state with negative electrodes fabricated from lithium foil, whereas those based on the oxides are lithium-ion cells which are assembled in the discharged state with carbon-intercalation negative electrodes.

The development of cells based on chalcogenide positives was carried out principally in the 1970s and early 1980s, as discussed in Section 9.3. With the single exception of the MoS_2-based Molicel™, little serious development has continued on this type of cell. The disinterest is due to the fact that cells using chalcogenide positives and lithium-metal negatives suffer from a limited cycle life, require a large stoichiometric excess of lithium, and are prone to safety problems.

The mixed-oxide positive electrodes $LiCoO_2$ and $LiNiO_2$ and the concept of cell assembly in the discharged state were conceived by Goodenough and co-workers [27]. The possibility of high-potential positive electrodes for use in a 'rocking-chair' type of cell was explored. During the course of the study, it was recognized that the voltage of the cell would be determined by the difference between the chemical potential of the lithium present in the two electrode polarities.

There are two types of oxide or sulfide framework structure that can act as a 'host' to lithium 'guest' atoms [28]. The first type of structure includes the oxides $LiMO_2$ (M = V, Cr, Co or Ni) and $LiTiS_2$. These compounds have a layered structure in which O–M–O layers (or S–Ti–S layers) are held together electrostatically by Li^+ ions which are intercalated between these layers. The Li^+ ions may diffuse readily in two dimensions between

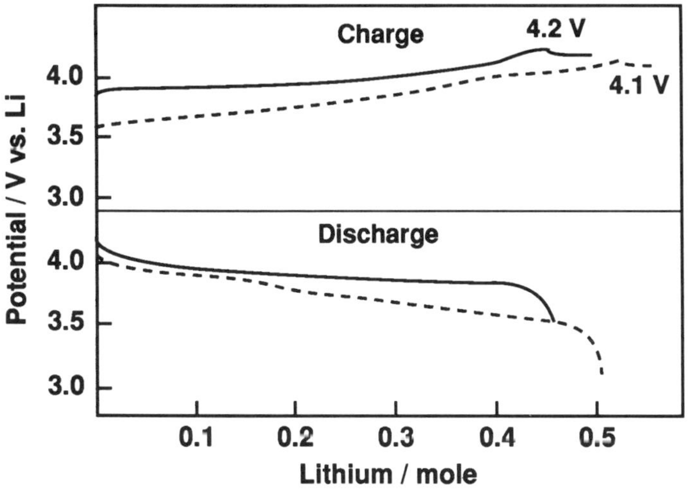

Fig. 9.6. Charge–discharge curves for: (--------) Li_xNiO_2 and (——) Li_xCoO_2 after a few cycles. (Taken from [29] with permission of Elsevier Science Limited, UK.)

the layers. Thus, the reversible charge–discharge reaction for the lithium-ion cell is:

$$LiMO_2 + 6C \underset{\text{discharge}}{\overset{\text{charge}}{\rightleftarrows}} Li_{(1-x)}MO_2 + Li_xC_6. \quad (9.7)$$

The maximum value of x that may be achieved in practice varies from compound to compound and the value determines the specific capacity of the cell.

Charge discharge curves for $LiNiO_2$ and $LiCoO_2$ measured at low current densities are shown in Fig. 9.6 [29]. The cobalt curves are less sloping than the nickel curves and $LiCoO_2$ has a slightly higher cut-off voltage on charging (4.2 V) than $LiNiO_2$ (4.1 V). After a few initial cycles, about one-half of the intercalated lithium may be removed reversibly. The lithium intercalation–de-intercalation processes are by no means simple and involve subtle structural rearrangements, especially in the case of the nickel compound. On continued cycling, the reversible reactions are:

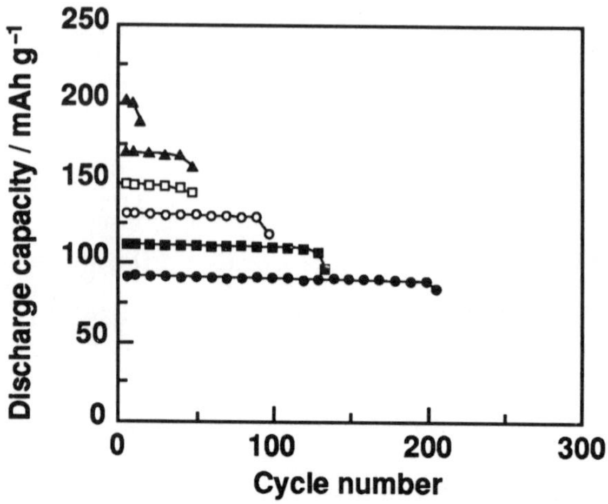

Fig. 9.7. Cycle characteristics of $LiNiO_2$ charged at 1.0 mA cm^{-2} in 1 M $LiClO_4$ in a 1 : 1 mixture of PC and 1,2-dimethoxyethane. (Taken from [30] with permission of Elsevier Science Limited, UK.)

$$Li_{0.35}NiO_2 + 0.5Li^+ + 0.5e^- \underset{\text{charge}}{\overset{\text{discharge}}{\rightleftarrows}} Li_{0.85}NiO_2 \text{ (135 mAh g}^{-1}) \qquad (9.8)$$

and

$$Li_{0.55}CoO_2 + 0.45Li^+ + 0.45e^- \underset{\text{charge}}{\overset{\text{discharge}}{\rightleftarrows}} LiCoO_2 \text{ (123 mAh g}^{-1}) \qquad (9.9)$$

It seems possible that the difference between the two electrodes stems from a partial cation mixing in $Li_{1-x}NiO_2$ that does not occur in $LiCoO_2$.

Yamada et al. [30] have recently investigated the properties of the $LiNiO_2$ positive as a function of the conditions used for its preparation. Under optimum conditions, discharge capacities as high as 200 mAh g^{-1} were obtained. Unfortunately, the cycle life of the electrode was a strong function of its capacity and ranged from 200 cycles at 80 mAh g^{-1} to less than 20 cycles at 200 mAh g^{-1} (Fig. 9.7). Other authors have reported 1200 cycles for this electrode in coin cells [31]. On overcharging or on heating above ~100 °C, the charged positive electrodes (low values of Li_x) both decompose with liberation of oxygen gas: Li_xNiO_2 gives $LiNi_2O_4$ (spinel), while Li_xCoO_2 yields Co_3O_4. For this reason, careful control of the voltage at top-of-charge is vital for lithium-ion cells.

The TiS_2 compound exists in the lithium-free state and, accordingly, all the lithium may be de-intercalated from $LiTiS_2$. The S–Ti–S layers are then held together by van der Waal's forces.

The second type of framework structure is the spinel structure $A[B_2]X_4$. Examples are $Li[Mn_2]O_4$ and $Li[Ti_2]S_4$. The spinel structure is based upon a cubic-close-packed array of oxygen ions with octahedral and tetrahedral sites. In the ideal spinel, one-half of the octahedral sites and one-eighth of the tetrahedral sites are occupied by cations and the remainder are vacant. This is a rigid structure and lithium can move freely into and out of the interstitial sites of the lithium spinel $Li[B_2]O_4$ to form $Li_x[B_2]O_4$ ($0 < x < 2$) without disrupting the $[B_2]O_4$ sub-lattice.

Manganese oxide is particularly attractive as a positive-electrode material on account of its wide availability, low cost, and non-toxicity. The manganese spinels have therefore been extensively researched [32–36]. The optimized $LiMn_2O_4$ electrode has a reversible capacity of 130 mAh g^{-1}, which falls to 115 mAh g^{-1} after 100 cycles. The $Li/Li_x[Mn_2]O_4$ cell has an open-circuit voltage of >4 V over the range $0 < x < 1$. This voltage is rather high for many electrolytes. Nevertheless, 750-mAh cells of this type have been shown to perform well up to at least 100 cycles with less than 10% capacity loss [37]. At $x = 1$, the open-circuit voltage drops abruptly to 2.9 V and remains at this value until $x = 2$, after which it falls again to ~1.2 V (Fig. 9.8). The theoretical specific energy of the $Li/Li_x[Mn_2]O_4$ cell over the composition range $1 < x < 2$ is still an attractive 400+ Wh kg^{-1}. Moreover, if the electrode is overcharged, it does not liberate oxygen (in contrast to $LiCoO_2$ and $LiNiO_2$), but merely moves on to the upper voltage plateau ($x < 1$). This is a safety feature which favours sealed cells.

The constant voltage during cell discharge, also an attractive feature, implies the occurrence of a two-phase reaction rather than an intercalation process, although the ready reversibility at ambient temperature suggests that the structural change is not profound. In fact, the insertion of lithium ions increases the Mn^{3+} ion concentration and this, in turn, induces a Jahn-Teller distortion and reduces the crystal symmetry from a cubic to a

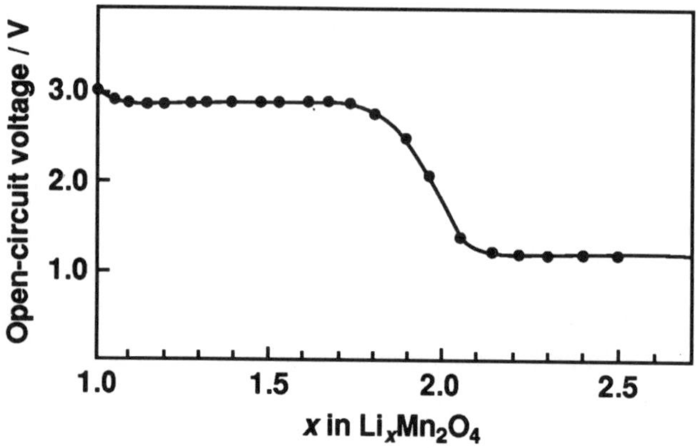

Fig. 9.8. Open-circuit voltage of a Li/Li$_x$Mn$_2$O$_4$ ($1 < x < 2.5$) cell as a function of positive-electrode composition. (Taken from [36] with permission of Elsevier Science Limited, UK.)

tetragonal form. This is ideal behaviour for an electrode, that is, a two-phase reaction to give a flat voltage profile with minor crystallographic transformation so as not to impede diffusion nor affect the reversibility of the cell. Unfortunately, there is evidence that even this minor structural change does result in progressive disintegration of the electrode on cycling. Also, it has been suggested [26] that, with increasing values of x in an electrode of composition Li_{1+x} $Mn_{2-x}O_4$, the cycling behaviour improves as the capacity declines; this is a commonly observed trade-off.

Other materials which have been investigated as prospective positive electrodes are the spinels: Fe_3O_4, Co_3O_4 and Mn_3O_4 [36]. These have very high capacities for lithium (~900 mAh g^{-1}) but at low voltages (0.8, 1.0 and 0.3 V, respectively). The first lithium ion per unit formula, at which the voltage is substantially higher, is held irreversibly. The vanadium oxides V_2O_5, V_6O_{13} and V_3O_8 show considerable promise and have the ability to intercalate 4, 8 and 4 lithium ions per unit formula, respectively. The corresponding specific capacities are 250, 290 and 210 mAh g^{-1}. Extensive studies have also been conducted on V_6O_{13} in the context of polymer-electrolyte batteries (see Section 9.5.2). All of the above-described oxides have to be mixed with carbon black and a binder in order to form an effective positive electrode.

9.4.4. Cell design and performance

The rechargeable lithium cells which have been commercialized or semi-commercialized to date have all been in small sizes, namely, button or coin cells, or AA-, C- or D-size cylindrical cells. Mention has already been made of the Faraday™ cell ($Li/NbSe_3$) in Section 9.3, and the Molicel™ (Li/MoS_2) in Section 9.4.

Table 9.2 summarizes the characteristics of $LiNiO_2$ and $LiCoO_2$ cells with lithium-metal foil negatives as reported by SAFT [29]. These cells had three-to-five times excess capacity of lithium over that of the positive electrode, and the electrolyte was a solution of $LiAsF_6$ in an unspecified organic solvent. The nickel cells displayed substantially better charge retention on storage than the cobalt equivalents. For both systems, the specific energies and energy densities are excellent. Clearly, therefore, the cells are highly attractive candidates for scale-up to EV traction batteries, were it not for considerations of cycle life and safety associated with the use of lithium-metal negative electrodes, as well as for the cost of cobalt. $LiCoO_2$ costs approximately four times as much as $LiNiO_2$. Tadiran in Israel is also marketing a small lithium-metal cell with a positive electrode made from a manganese oxide. The cycle life of this cell is comparable with that of the SAFT cells in Table 9.2.

Because of the problems associated with lithium-metal electrodes, recent development work has focused largely on the lithium-ion battery, first introduced by the Sony Corporation in the early 1990s for use in camcorders and other portable electronic equipment. Since then, several other companies have become involved in the manufacture of small (~1 Ah) lithium-ion cells, particularly A-T Batteries (a consortium of Asahi and Toshiba), Matsushita, Sanyo, and Moli-Energy (Canada). The market for such cells in powering camcorders and many other consumer electronics applications has escalated, particularly in Japan. By 1995, tens of millions of these cells were being produced annually. Only a few larger cells have been constructed, either for experimental purposes or for specialized applications. All the cylindrical cells produced to date have had a spiral

Table 9.2. Characteristics of $LiNiO_2$ and $LiCoO_2$ spirally-wound cells (lithium–metal negatives) [29].

System	$LiNiO_2$	$LiCoO_2$	$LiCoO_2$
Cell size	C	C	D
Voltage limits (V)	4.1–3.0	4.2–3.5	4.2–3.0
Charge rate	*C*/20	*C*/20	*C*/10
Discharge rate	*C*/2	*C*/10	*C*/2.8
Capacity (Ah)	1.8	2.0	3.6
Average discharge voltage (V)	3.6	3.92	3.85
Specific energy (Wh kg^{-1})	145	174	145
Energy density (Wh l^{-1})	246	300	260
Cycle life (100% DoD)	150	100	110
Self-discharge (% per month at 45 °C)			
– 1st month	1–3	10–12	
– 2nd month	3–6	20–25	
– 3rd month	3–10	short-circuit	

construction. The Sony battery employs $LiCoO_2$ as the positive electrode, although other developers have argued in favour of either $LiNiO_2$ on the grounds of better electrolyte stability at the slightly lower top-of-charge voltage and better shelf-life, or $LiMn_2O_4$ on the grounds of lower cost. Nevertheless, $LiCoO_2$ is the industry standard, except in the case of SAFT where the preference is for $LiNiO_2$. It appears, however, that all three positives are contenders for use in lithium-ion cells, as dictated by the electrolyte employed and the application envisaged.

The performance of the carbon/$LiCoO_2$ production cell and of the carbon/ $Li_{1+x}Mn_2O_4$ prototype cell, both of AA-size, has been reported by Tarascon and Guyomard [38], see Table 9.3. Data for nickel/cadmium and nickel/ metal hydride are included for the purposes of comparison. It is at once apparent that although the energy-storage capability of lithium-ion cells is substantially better than that of nickel/cadmium, the cells show little improvement on nickel/metal hydride and have only about half the specific energy of their lithium-metal counterparts (cf., data in Table 9.2). Thus, the outstanding performance of the lithium-metal cells has been traded for

Table 9.3. Stored energy in AA-size lithium-ion cells [38]: a comparison with nickel/cadmium and nickel/metal hydride cells.

Performance parameter	C/$LiCoO_2$	C/$Li_{1+x}Mn_2O_4$	Ni/Cd	Ni/MH_x
Specific energy (Wh kg^{-1})	78*	75	30	70–80
Energy density (Wh l^{-1})	192*	180	87	170

* Higher values for have been reported by Sony, namely, 90–110 Wh kg^{-1} and 216–261 Wh l^{-1} [39].

improved safety and cycle life. When cycled at the $C_2/2$ rate, laboratory lithium-ion cells with $Li_{1+x}Mn_2O_4$ positives reached 2300 cycles at room temperature and more than 400 cycles at 55 °C, in both cases with a capacity still higher than 60% of the initial value [38]. Commercial cells with $LiCoO_2$ positives are reported to have service lives of ~600 cycles.

The performance of rechargeable carbon/$LiNiO_2$ coin cells (petroleum coke negatives) has been investigated by Dahn et al. [40]. $LiNiO_2$ was chosen as the positive material because its top-of-charge voltage is lower than that of $LiCoO_2$ or $LiMn_2O_4$ and, therefore, oxidation of the organic electrolyte is less likely to occur on continuous cycling. These small cells gave long life (hundreds of cycles), excellent high-temperature performance (up to 85 °C), low self-discharge, and the possibility to employ electrodes of at least 1 mm in thickness. It was concluded that this system is well-suited for use in coin, prismatic, spiral-wound cylindrical, and small bobbin cells.

A feature of lithium-ion cells is the requirement for accurate control of the charging voltage; the required tolerance is 1.0–1.5%. Higher voltage reduces cycle life dramatically, while a reduced voltage of 100 mV leads to a significant loss in capacity. The lower voltage limit on discharge is also important. Lithium-batteries cannot be employed safely and efficiently without electronic voltage control.

Of particular interest for EV traction is the recent work by the Sony Corporation in Japan on the scale-up of their lithium-ion technology (using $LiCoO_2$ positives) to produce prototype 100-Ah cells [41]. Eight cells were series-connected to form a battery module which weighed 29 kg and had dimensions of: 290 mm (length) × 150 mm (height) × 400 mm (width), see

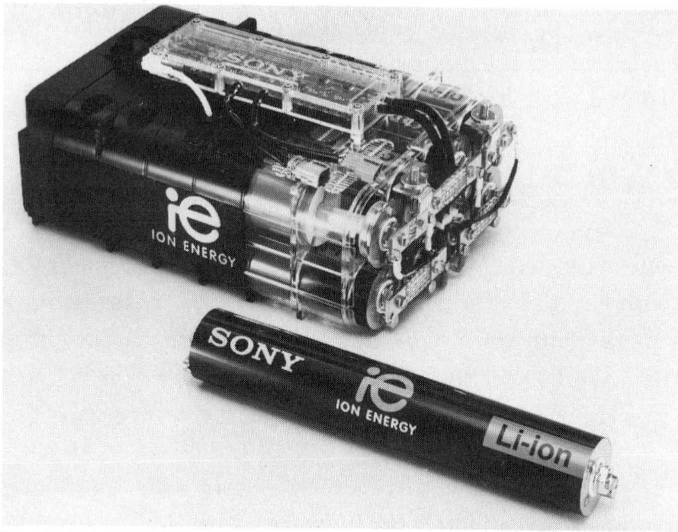

Fig. 9.9. Sony lithium-ion EV traction battery module (2.9 kWh). (By courtesy of Sony, Japan.)

Fig. 9.9. The module had a specific energy of 100 Wh kg^{-1} and an energy density of 160 Wh l^{-1}. The series-connection of modules gave a 35-kWh EV battery with a specific energy of 91 Wh kg^{-1}. This accomplishment is remarkable for the energy achieved, for the cycle life reported (1200 cycles with 78% capacity retention) and, especially, for the peak power output (900 W kg^{-1} initially, with 220 W kg^{-1} at 80% DoD). The Sony EV battery meets, or exceeds, the USABC mid-term targets (see Table 3.4, Chapter 3) in all respects save cost. The cost target is only likely to be achieved by moving away from cobalt-containing positive electrodes. The scale-up from the manufacture of small, C-size cells to the production of 100-Ah units in quantities of hundreds over a short period of time is also a remarkable feat. These are early days for this development. It remains to be seen what technical problems still exist and whether batteries of similar performance can be made from the cheaper $LiNiO_2$ or $LiMn_2O_4$ material. Also, safety will need to be carefully evaluated. Such valid questions in no way detract from Sony's remarkable achievement to date. Other companies such as Duracell, Japan Storage Battery Company, SAFT and VARTA are also actively interested in scaling-up lithium-ion batteries for use in EVs. For example, in December 1996, Japan Storage announced a battery with a specific energy of 133 Wh kg^{-1} that could power an EV over 300 km. SAFT has developed prismatic cells of the type C–Li/$LiNiO_2$ with a capacity of 110 Ah and with specific power values similar to those of the Sony cylindrical cells based on $LiCoO_2$ (viz., 900 W kg^{-1} fully charged; 200 W kg^{-1} at 90% DoD). These cells performed well, even at −10 °C, and survived more than 500 cycles.

In summary, rechargeable lithium cells with liquid electrolytes have a number of performance advantages over competing technologies, namely, higher voltage, higher specific energy, less self-discharge, and longer shelf-lives. The cells also have some serious disadvantages. For example, cells with lithium-metal negative electrodes suffer from an inability to attain a lithium cycling efficiency of 100% and, consequently, this necessitates a substantial stoichiometric excess of lithium compared with that required to intercalate the positive electrode fully. This, in turn, leads to the hazards of dendritic short-circuits. A further problem associated with the excess lithium is that of overdischarge. When this process takes place, many intercalation positive electrodes decompose irreversibly, with liberation of oxygen. For these reasons, the emphasis is now on lithium-ion cells which are negative-limited. With these cells, however, improved cycle life and safety is purchased at the expense of specific energy. Also, there is the problem of restricted charging rate and accurate electronic control of charging voltage. For many applications, including EV traction, this may not pose too serious a limitation. Certainly, the number of Japanese and other companies that are following Sony into the market with their own lithium-ion batteries for portable electronics and communications applications demonstrates a widely held view that these are indeed the batteries of the future for such markets. Whether or not this future extends to multi-kWh batteries for EV traction is largely a matter of cost and safety, as well as possible competition from other battery systems. Already, Nissan has announced its intention to introduce small fleets of EVs powered by Sony's lithium-ion battery into the Japanese market in 1997, and into California in 1998. The vehicle, known as the Altra-EV, is equipped with an inductive charging facility (see Section 10.5.1, Chapter 10). High-powered versions of the batteries are also being designed specifically for use in the Nissan S-HEV, a series hybrid which will deliver a range of 50 km in the pure battery–electric mode. SAFT modules (100 Wh kg^{-1}) have been tested in a Peugeot 106 and a range of 200 km is expected.

9.5. Solid-Electrolyte Cells

Many of the features which have been discussed for liquid-electrolyte cells apply also to solid-electrolyte cells, but the two classes of cells are sufficiently different to merit separate consideration. Solid electrolytes may be either crystalline inorganic compounds, amorphous glasses, or polymers. One obvious candidate material is lithium beta-alumina, but this has proved more difficult to prepare than sodium beta-alumina. Moreover, lithium beta-alumina has a lower conductivity — the lithium ions display a tendency to enter the spinel slabs rather than to remain in the conduction planes between the slabs. No serious attention has therefore been given to a lithium analogue of either the sodium/sulfur or the sodium/metal-chloride batteries.

9.5.1. Inorganic electrolytes

The search for new solid electrolytes was directed towards compounds with framework structures and with inter-connected channels through which cations could diffuse, or with a layer structure with diffusion in the inter-layer planes. Many different inorganic compounds have subsequently been considered as lithium-ion conductors [42]. The conductivities of some of these compounds are compared with the value for sodium beta-alumina in Table 9.4. The data show that Li_3N has quite a good conductivity but has a low stability window with respect to lithium and, hence, readily undergoes voltage breakdown. Pure lithium iodide (without added alumina) has a low conductivity (~10^{-5} S m^{-1} at room temperature) which is just adequate for low-rate primary cells. Commercial primary cells based on lithium iodide electrolyte are being manufactured for use in heart pacemakers. These cells are of the form: $Li_{(solid)} | LiI_{(solid)} | P2VP.nI_{2(solid)}$, where $P2VP.nI_2$ is a solid complex which is formed between iodine and poly-2-vinylpyridine (see Section 9.2).

Table 9.4. Properties of solid electrolytes.

	Conductivity ($S\ m^{-1}$)		Activation energy
	25 °C	**300 °C**	**(eV)**
Li–β-Al_2O_3	10^{-2}	0.6	0.37
Li_3N	6×10^{-2}	3	0.26
LiI/Al_2O_3	10^{-3}	10	0.4
Li glasses	$10^{-5} - 10^{-4}$	2	0.8
Na–β-Al_2O_3	10^{-1}	20	0.24

At ambient temperature, solid inorganic electrolytes generally (with the exception of some Ag^+- and Cu^+-ion conductors) have ionic conductivities which are several orders of magnitude lower than those of aqueous electrolytes. To obtain acceptable cell internal resistance, it is necessary either to use very thin electrolyte membranes, which may be difficult to fabricate, or to operate at elevated temperatures, or both. One area in which some success is being obtained with ambient temperature solid-state batteries is that of micro-batteries for use with integrated circuits. In this technology, the M | MX | X laminate is built up by using conventional semiconductor fabrication technology which allows the laying down of thin layers.

Attempts to prepare secondary cells based on solid inorganic or glass electrolytes soon ran into other problems. As Li^+ ions move across the electrolyte from one electrode to the other, a volume contraction takes place in the exit electrode and an expansion occurs in the entry electrode. This results in disruption at the respective electrode/electrolyte interfaces that leads to polarization and an increase in interface resistance. Also, there is a tendency for the discharge reaction product to accumulate at the interface between the electrolyte and the positive electrode. To overcome these problems, researchers recognized that it would be necessary to have a flexible or plastic electrolyte which could deform to accommodate the volume changes of the electrodes and maintain the interfaces intact. In addition, the diffusion rate of lithium ions in the positive should be at least equal to that in the electrolyte. Thus, the idea of the solid-polymer electrolyte was conceived.

9.5.2. Solid-polymer electrolytes

Early work on solid-polymer electrolytes was carried out in the 1970s by Armand [43,44] at the University of Grenoble. He realized that ion-which contain polymers, in particular those based on poly(ethylene oxide) PEO, could be utilized as electrolytes in all-solid-state batteries. Polymers which contain a hetero-atom, such as oxygen or sulfur, tend to have a relatively high dielectric constant and to dissolve lithium salts in reasonable concentration. Among the favoured lithium salts are lithium perchlorate $LiClO_4$ and lithium trifluoromethane sulfonate $LiCF_3SO_3$ ('triflate'). The polymer electrolyte is prepared, typically, by dissolving PEO and the chosen lithium salt in anhydrous acetonitrile and then casting a film of thickness 25–50 μm from the viscous solution. A continuous-casting, solvent-evaporation technique ('doctor blade' casting) is employed and yields large-area, uniform membranes which are pinhole free.

The target thickness for the electrolyte film is determined by the ionic conductivity and by the acceptable voltage drop ($I\mathcal{R}$) across the film. Log–log plots of the voltage drop versus the electrolyte film thickness for a constant current density of 10 mA cm^{-2} are presented in Fig. 9.10 for electrolytes of various conductivity. If a voltage drop of 10 mV is acceptable, then an electrolyte conductivity of 0.1–1.0 S m^{-1} is required to give a reasonable film thickness of 10–100 μm. If the current density is reduced by an order of magnitude to 1 mA cm^{-2}, then the corresponding conductivities for the same film thickness are 0.01–0.1 S m^{-1}. Ionic conductivity data as a function of temperature for electrolytes formed from PEO with dissolved $LiClO_4$ and $LiCF_3SO_3$ are given in Fig. 9.11. Both electrolytes exhibit a sharp fall in conductivity below 50–60 °C, and it is only above this temperature that the conductivity comes into the desired range. Research has shown that this sharp drop in conductivity is associated with crystallization of the polymer [46]. Below the transition temperature, the polymer is a mixture of crystalline and amorphous material, and it is only the amorphous phase which has appreciable ionic conductivity. In order to keep the thickness of the electrolyte film within practical limits, it is necessary to operate cells based on these PEO

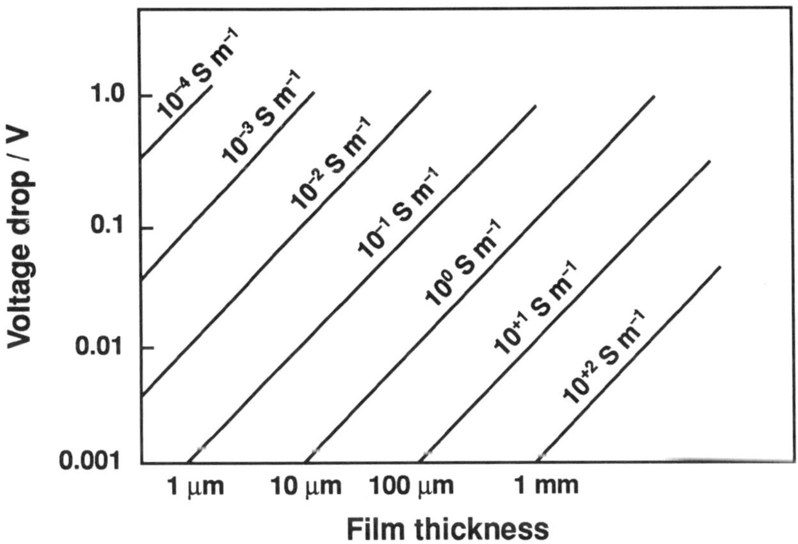

Fig. 9.10. Voltage drop ($I\mathcal{R}$) vs. electrolyte thickness for a constant current density of 10 mA cm^{-2}. Plots for films of different conductivity (S m^{-1}).

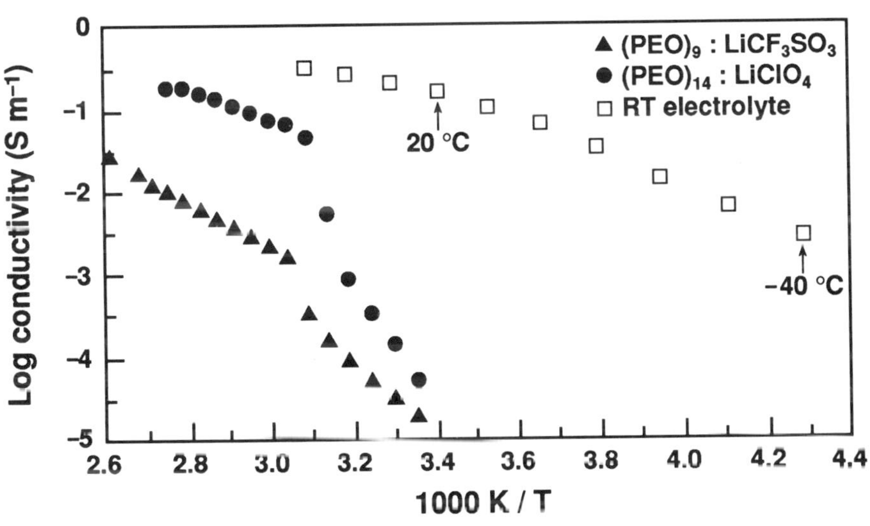

Fig. 9.11. Ionic conductivity vs. temperature plots for PEO electrolytes. Note, RT is a proprietary electrolyte developed by AEA Technology, Harwell Laboratory, UK. (Taken from [45] with permission of AEA Technology, Harwell Laboratory, UK.)

electrolytes at temperatures of 80–140 °C. Although such a temperature restriction is inappropriate for many battery consumer applications, it is not a serious problem for EVs. Indeed, in many ways, it is an ideal temperature range, i.e., low enough not to present materials science problems (sealing, etc.), but high enough to require insulation and battery temperature control and, therefore, to be independent of the prevailing ambient temperature. Such a traction battery could be used in extremes of climate without much difficulty. Nevertheless, in attempts to reduce the operating temperature to ambient or below (i.e., for non-EV applications), various organic liquids were dissolved in the polymer with a view to obtaining maximum inhibition of the phase transition (crystallization). This strategy was successful and produced an electrolyte which could be used below room temperature (Fig. 9.11). At 20 °C, the conductivity was almost 10^4 times higher than that for the simple solution of salts in PEO. The electrolyte may be regarded as a quasi-solid, or gel type, and is more difficult to prepare and handle than a true solid-polymer electrolyte.

The electrical and mechanical properties of a solid-polymer electrolyte depend, naturally, on the concentration of dissolved salt. The conductivity data for $(PEO)_xLiCF_3SO_3$, where x is the oxygen-to-lithium ratio, as a function of reciprocal temperature are presented in Fig. 9.12 [47]. An appropriate combination of electrical and mechanical properties is achieved at a composition of $(PEO)_9LiCF_3SO_3$, and this electrolyte has been utilized in the temperature range 100–140 °C.

A positive material which has been used extensively in solid polymer cells is V_6O_{13}. This intercalates up to 8 lithium atoms per formula unit:

$$Li_x{}^+ + V_6O_{13} + xe^- \underset{\text{charge}}{\overset{\text{discharge}}{\rightleftharpoons}} Li_xV_6O_{13} \qquad (9.10)$$

$$(0 < x < 8; \quad V^{4+}/V^{5+} \rightarrow V^{3+})$$

The electronic and ionic conductivities of V_6O_{13} are not adequate for use as a planar electrode. It is therefore necessary to mix this oxide with other materials to provide conduction pathways to the individual particles. The

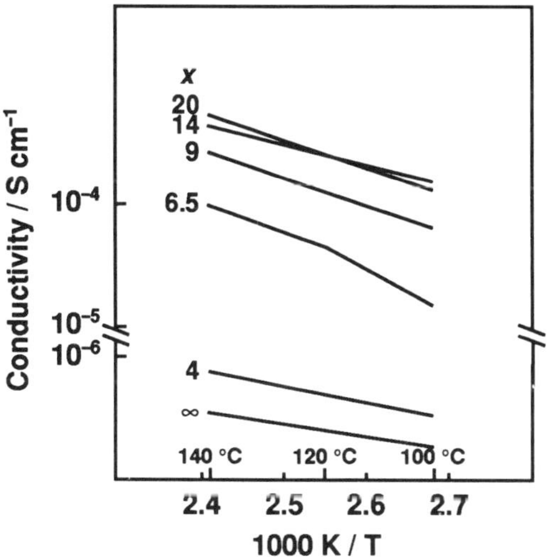

Fig. 9.12. Conductivity of $(PEO)_xLiCF_3SO_3$ as a function of x. Numbers indicate value of x. (Taken from [47] with permission of AEA Technology, Harwell Laboratory, UK.)

procedure adopted is to use the solid polymer electrolyte itself as a component of the electrode to provide the ionic conduction path to the V_6O_{13} particles, with added carbon black to enhance the electronic conduction. Composite positive electrodes are fabricated by adding V_6O_{13} and carbon to the electrolyte mix of PEO, $LiCF_3SO_3$ (or $LiClO_4$) and acetonitrile, and casting directly on to a nickel current-collector. A typical positive electrode is 50 μm thick and has a composition of 45 vol.% V_6O_{13}, 5 vol.% carbon and 50 vol.% PEO–$LiClO_4$. The porosity present in the V_6O_{13} accommodates the volume expansion associated with lithium intercalation so that there are no dimensional changes during charge and discharge.

The all-solid-state cell is fabricated simply by laminating together a lithium-foil negative electrode (~50 μm), an electrolyte film (~50 μm), and a composite positive electrode (~50 μm) cast on a thin nickel-foil current-collector (10 μm). The cell has an overall thickness of 150–200 μm. Lamination is conducted in a dry room and is accomplished by using a combination of heat and pressure. The capacity of such cells is ~2 mAh cm^{-2}

and the area of the cell can be as large as desired, but within the ability to cast and handle the films. Most research has been performed on small cells (~40 cm^2 area), although facilities are now available for scaling-up the operation to produce square metres of battery laminate. The general technology of manufacturing thin polymer films and laminates is, of course, well established for low-cost applications such as packaging.

An interesting aspect of these solid-polymer cells is that they can be rolled into cylinders, folded, or otherwise shaped to suit the given application. With care, it is even possible to cut a cell into half with scissors so that both halves are still functional. Because there is no liquid electrolyte to leak, and because of their high surface area, these cells are ideal for assembly into batteries which use the bipolar configuration. It has even been suggested that for EV traction applications, they might be accommodated in the roof lining or door panels of the car.

Early work on the fabrication and performance of solid-polymer cells was reported by Hooper and North [48] working at the Harwell Laboratory. Small cells (1–2 mAh) which employed a $(PEO)_9LiCF_3SO_3$ electrolyte and V_6O_{13} positives were cycled at 120–140 °C. Later development work in the same laboratory [45] found that the preferred electrolyte at temperatures >100 °C is $PEO_{14}LiClO_4$. The first discharge curve of a 80-mAh (40-cm^2 area) test cell of this type at low rate ($C_{10}/10$) is shown in Fig. 9.13(a) [49]. The distinctive plateaux in the curve are characteristic of V_6O_{13} and are indicative of structural transformations which occur during the discharge. After a few cycles, the discharge curve becomes smoother (Fig. 9.13(b)), and the vanadium oxide is found by X-ray diffraction analysis to have lost its crystallinity and to be largely amorphous. The capacity falls to ~65% of theoretical over the first 20–30 cycles. Thereafter, it declines only slowly and several hundred cycles are possible at the 60% capacity level. This reduced, but stable, capacity is associated with the amorphous form of V_6O_{13}. The capacity measured as a function of discharge rate is given in Fig. 9.14; more than 50% of the theoretical capacity is available at the $C_1/1$ rate, 80% at the $C_2/2$ rate, and over 90% at the $C_4/4$ rate. This is a very satisfactory rate performance.

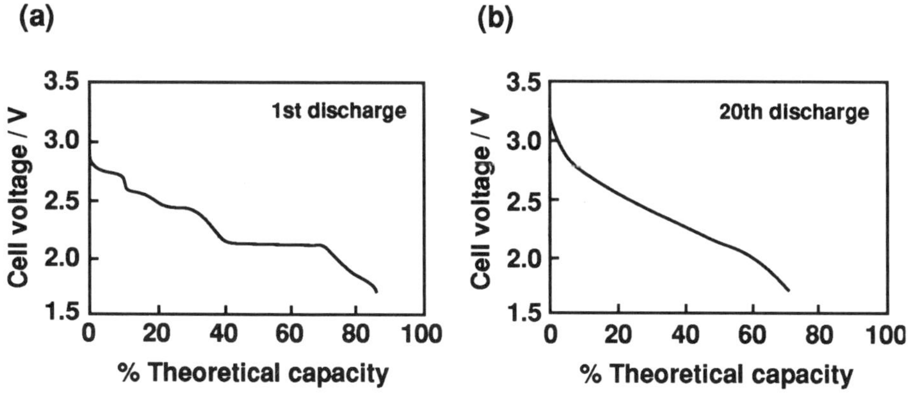

Fig. 9.13 Discharge at $C_{10}/10$ rate of a Li/V_6O_{13} polymer cell at 120 °C: (a) 1st cycle; (b) 20th cycle. (Taken from [49] with permission of AEA Technology, Harwell Laboratory, UK.)

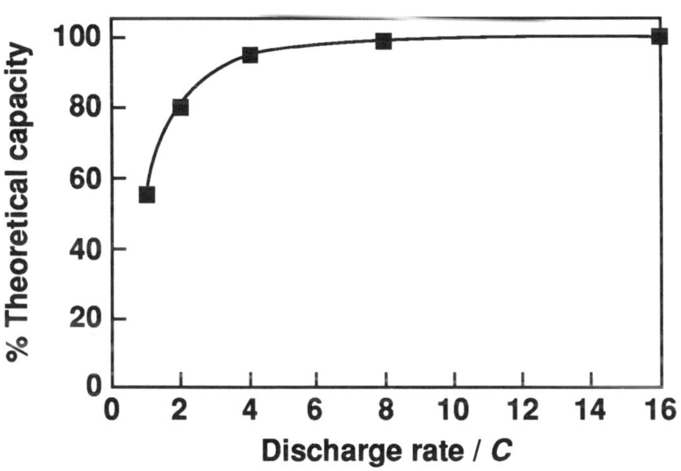

Fig. 9.14. Rate performance of a Li/V_6O_{13} polymer cell at 120 °C. (Taken from [45] with permission of AEA Technology, Harwell Laboratory, UK.)

In view of the complexity of the composite positive electrode based on V_6O_{13} and the well-documented problems of the lithium-metal negative electrode, it is perhaps not surprising that the capacity declines with cycling. Gel-type cells with organic liquids immobilized in PEO exhibit much worse cycling behaviour at 20 °C; the available capacity rapidly fades away.

Some research has been carried out on the behaviour of solid-polymer cells with positive-electrode materials other than V_6O_{13}. It has been found that V_2O_5 behaves in a similar fashion when cycled down to 1.7 V and undergoes a transition from the crystalline to the amorphous phase [49]. On the other hand, when discharged in the range 3.4 to 2.5 V ($V_2O_5 \rightarrow LiV_2O_5$) in liquid-electrolyte cells, the V_2O_5 electrode is capable of several hundred cycles, limited only by the rechargeability of lithium metal [50]. Lithium ions may also be intercalated into TiO_2 (rutile phase) to form $Li_{0.5}TiO_2$; this cycles at 120 °C for at least 30 cycles with an average cell voltage of 1.73 V [51]. The behaviour of $LiMn_2O_4$ towards the insertion of further lithium at 120 °C is similar to that shown in Fig. 9.8; the best capacity is obtained for the spinel synthesized at the lowest temperature (450 °C) [52]. At least 100 cycles can be performed at >80% capacity retention. These preliminary research results are encouraging, especially as vanadium, titanium and manganese are all cheap and non-toxic oxides. Research is now in progress on the performance of various lithium-ion cells with solid-polymer electrolytes.

Scrosati and co-workers [23,53] have investigated the performance of some new gel-electrolyte membranes which are based on immobilizing a solution of a lithium salt in an organic solvent in a poly-acrylonitrile (PAN) or poly-methylmethacrylate (PMMA) matrix. Such membranes are termed 'gelionic' electrolytes and have ionic conductivities which approach those of organic liquid electrolytes, namely, $\sim 10^{-1}$ S m^{-1} at 20 °C. The membranes are described by the general designation: 'polymer–organic liquid–salt', for example, PAN–PC/EC–$LiClO_4$. The electrochemical properties of some of these gelionic electrolytes are given in Table 9.5. All the electrolytes listed have conductivities and transference numbers comparable with those of the parent liquid electrolytes. Also, the excellent anodic stabilities make the electrolytes very suitable for use in lithium-ion cells. The earlier gelionic electrolytes, based on PEO polymers, have ionic conductivities of the same order of magnitude at 25 °C (Fig. 9.11). The gelionic electrolytes appear to be stable at the high voltages required for recharging intercalation positive electrodes. One of the best of these is claimed [53] to be a solution of $LiN(CF_3SO_2)_2$ in an EC + DMC mixture immobilized in PMMA.

Table 9.5. Electrochemical properties at 25 °C of gelionic electrolyte membranes based on PAN and PMMA [23].

Electrolyte	Conductivity (10^{-1} S m^{-1})	Lithium transfer number	Anodic stability (V)
PAN–PC/EC–$LiClO_4$	4.5	0.5–0.6	4.8
PAN–PC/EC–$LiAsF_6$	4.6	0.6–0.7	4.5
PAN–PC/EC–$LiN(CF_3SO_2)_2$	2.0	0.7–0.8	4.6
PAN–BL–$LiClO_4$	4.4	0.5–0.6	4.9
PAN–BL–$LiAsF_6$	6.0	0.6–0.7	4.6
PAN–BL–$LiN(CF_3SO_2)_2$	4.0	0.7–0.8	4.7
PMMA–PC/EC–$LiAsF_6$	0.8	0.6	4.8
PMMA–PC/EC–$LiN(CF_3SO_2)_2$	0.7	—	4.9

BL = γ-butyrolactone

Many potential advantages have been cited for solid-polymer batteries, particularly in the context of EV traction. These include:

- high specific energy; this is dependant upon which positive electrode material is used and whether the negative is lithium metal or carbon
- safety — dendrites are less likely to penetrate a polymer electrolyte than a liquid electrolyte; whether this is true also for gelionics is less certain
- flexibility of geometry
- ease of bipolar assembly
- ruggedness — no liquid components to leak
- low self-discharge rates
- wide temperature range of operation (80–140 °C)
- long shelf-life.

While these are highly attractive features, it must be emphasized that research is still at the laboratory stage. The technology for manufacturing lithium | polymer | metal-oxide laminates as large, thin sheets — the first step towards constructing a pilot plant — is being developed in the UK (at the Harwell Laboratory) and in the USA (3M Corporation and others).

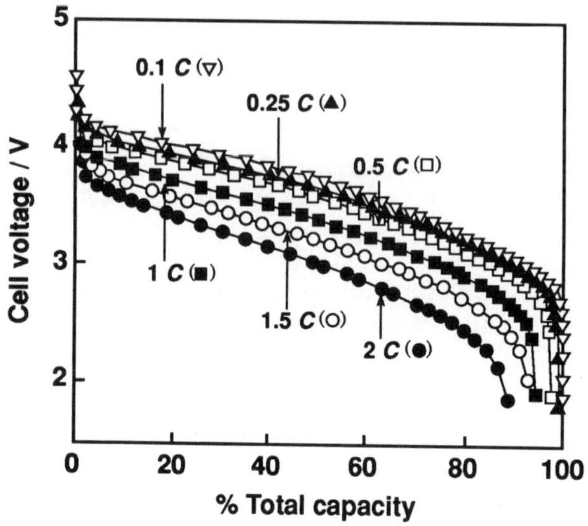

Fig. 9.15. Discharge of Bellcore cell at 25 °C. (By courtesy of Bellcore, USA.)

Obvious applications for these batteries lie in situations where a liquid electrolyte cell is unacceptable, or where an ultra-thin, flexible cell has positive advantages. These would include small batteries on printed-circuit boards or on 'smart' credit cards. It is doubtful whether any batteries of more than tens of Ah capacity have yet been constructed and demonstrated to have specific energies, cycle lives and capacity retention appropriate to EV traction applications. Some of the problems encountered in the past may be attributed to the use of lithium-metal negatives and the obvious next step is to investigate in some detail the performance of lithium-ion batteries with solid polymer electrolytes.

A start in this direction has been made by Bellcore which has recently announced a lithium-ion cell based on a plastic gelionic electrolyte [54,55]. The liquid electrolyte is immobilized in a fluoropolymer matrix which is electrochemically inert. The C–Li negative electrode and the $LiMn_2O_4$ positive electrode are polymer-bonded to a copper mesh and an aluminium current-collector, respectively. The flexible cell has an overall thickness of 0.5 mm. The nominal voltage is 3.8 V and the cell has a sloping discharge characteristic (Fig. 9.15). The loss of capacity in early cycles is shown in

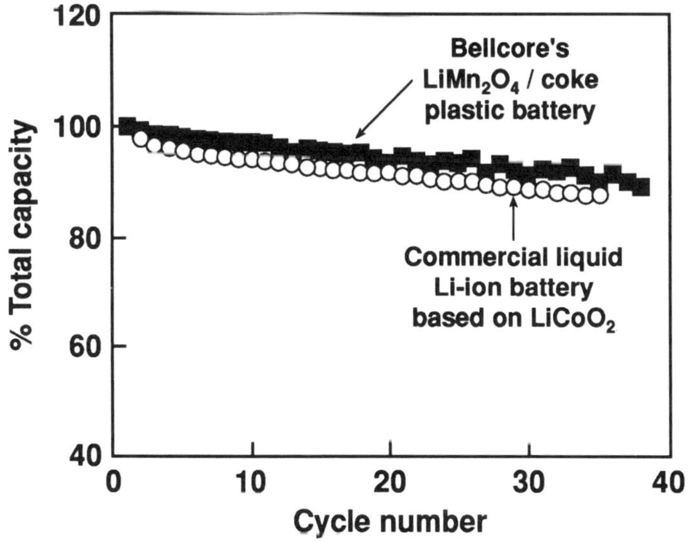

Fig. 9.16. Performance of Bellcore cell at 55 °C. (By courtesy of Bellcore, USA.)

Fig. 9.16. With further service, ~80% and ~60% of the capacity is retained after 1200 and 2500 recharges, respectively. The cell has a specific energy of 100 Wh kg^{-1} and can be used over a wide range of temperatures (–20 to +60 °C). At present, these cells are limited to small sizes for consumer applications and scale-up for EV applications has not been reported.

9.6. High-Temperature Lithium Batteries

As discussed in Chapter 8, it is perfectly feasible to design EV batteries based on sodium and operating at temperatures up to 400–450°C. In the absence of a suitable solid-state electrolyte, high-temperature lithium batteries have been based exclusively on fused-salt electrolytes, generally halides on account of their stability. Fused salts have several advantages, notably: high ionic conductivity (between 100 and 500 S m^{-1}) which allows operation at high current densities, and high exchange-currents which result in rapid and reversible electrode reactions. Against these advantages must be set the materials-science problems which are encountered at high temperatures, i.e., corrosion, compatibility and cell sealing.

9.6.1. Early work

The alkali metals are compatible with molten halide electrolytes, but as sodium shows a significant solubility, which results in self-discharge, attention has been directed towards lithium. This metal, too, has a small solubility in the alkali halides, but not such as to preclude its use in a battery. Because of the high melting points of the simple lithium halides, most researchers have chosen to use eutectic mixtures of lower melting points, e.g., LiCl–KCl which melts at 352 °C.

The first couple to be investigated systematically, by General Motors Corporation in the USA (mid-1960s), was lithium/chlorine — first, in molten LiCl at 650 °C [56], and later in a LiCl–KCl eutectic at a lower temperature [57], i.e.,

$$2Li_{(liquid)} + Cl_{2(gas)} \underset{\text{charge}}{\overset{\text{discharge}}{\rightleftarrows}} 2LiCl_{(liquid)} \quad V^{o} = 3.47\text{ V} \quad (9.11)$$

This cell has a theoretical specific energy (reactants only) of 2190 Wh kg^{-1}, but in a practical system this would be much degraded by the requirement for chlorine storage cylinders and pumps, as well as for the more usual items of electrolyte, separators, current-collectors and straps, and battery housing and insulation. In the experimental work, the negative electrode consisted of liquid lithium held by capillary forces in a metal mesh current-collector (stainless steel or nickel). Chlorine gas was fed into the cell through a porous graphite electrode. In a practical EV traction battery, the chlorine liberated during recharge would have to be pressurized and pumped back into the cylinder. This external storage of chlorine adds to the complexity and cost, as well as to the overall weight of the system. Later designs by the Sohio Company [58] avoided this complexity by adsorbing the chlorine on a carbon positive electrode with a high surface area; this allowed the cell to be sealed. A solid Li–Al negative electrode was employed in a LiCl–KCl eutectic at 400 °C. The specific energy attainable was limited to ~70 Wh kg^{-1} by the adsorptive capacity of the carbon, but it was possible to attain a very high power output, namely, ~40 W cm^{-2}.

Table 9.6. Performance characteristics of lithium/chalcogen cells [59].

Parameter	Li/S	Li/Se	Li/Te
Electrolyte	LiCl–LiI–KI	LiF–LiCl–LiI	LiF–LiCl–LiI
Melting point of chalcogen (°C)	118	220	450
Operating temperature (°C)	347	365	470
Voltage (V)	2.3	2.1	1.7
Theoretical energy (Wh kg^{-1})	2680	1210	612
Expected energy (Wh kg^{-1})	250–360	220–330	185–265
Expected power (W kg^{-1})	550–800	600–1000	550–800

Nevertheless, the severe materials problems encountered, coupled with the concern over using chlorine in EVs, led to the abandonment of the project.

The next class of fused-salt battery to be extensively studied, also in USA but at the Argonne National Laboratory, was that based on the lithium/chalcogens [59]. The characteristics of these cells are listed in Table 9.6. On moving from Li/S through Li/Se to Li/Te, there is a more than four-fold drop in theoretical specific energy as a result of both the lower cell voltage and the increased atomic weight of the chalcogen. Experiments at the Argonne National Laboratory were performed on very small laboratory cells using an inert, porous ceramic filler ($LiAlO_2$, Y_2O_3, or MgO) to immobilize the fused halide electrolyte as a semi-rigid paste. In view of the small scale of these tests, the values forecasted for both the specific energy and the specific power of developed batteries must be seen as optimistic. Compared with Li/Cl_2, the Li/S cell had the advantage of a somewhat lower operating temperature, as well as a liquid oxidant which is easier to store than a gas. On the other hand, in a three-liquid system, i.e., $Li_{(liquid)} | LiX_{(liquid)} | S_{(liquid)}$, it is exceedingly difficult to control the interfaces and there is a major corrosion problem with liquid chalcogens. Sulfur tended to be lost through volatilization and dissolution in the electrolyte, while liquid lithium corroded the ceramic separators and bridged to the positive electrode to cause internal short-circuiting. For these reasons, the work with chalcogen batteries was terminated in the early 1970s in favour of using solid electrodes.

Several different alloys of lithium may be considered as possible solid negative electrodes (see Section 9.6.2), but most work has focused on lithium–aluminium alloys. Similarly, a variety of metal sulfides were evaluated as possible positive-electrode materials; these included copper, iron, nickel, chromium and cobalt. Considerations of cost led to the selection of iron sulfides as most suitable for EV traction batteries, and almost all studies have centred on the Li–Al/fused-salt/FeS_x system. Although most of the research and development effort has been in the USA — by the Argonne National Laboratory and its industrial sub-contractors (viz., Eagle-Picher, Gould, Rockwell International), and independently by General Motors — there have been separate European programmes at VARTA in Germany and at the Admiralty in the UK.

9.6.2. Negative electrodes

The two best known alloys for use as the negative electrode are Li–Al and Li–Si. The voltages and specific energies of the corresponding cells are listed in Table 9.7. It is clear that FeS_2 is preferable to FeS on specific energy grounds, although there are other complications (see Section 9.6.3). Of the negative electrodes, Li–Si is more attractive than Li–Al, although the former does have a higher resistance, especially in the later stages of discharge. A whole series of voltage plateaux exist which correspond to two-phase regions of the compounds: Li_5Si, $Li_{22}Si_5$, Li_4Si, $Li_{15}Si_4$, $Li_{14}Si_5$, Li_2Si [60]. Although such behaviour poses a complication, the voltage spread over the entire range is only ~350 mV vs. lithium, and this does not seem to be a valid reason for rejecting Li–Si alloy as a negative electrode when the specific energy is so favourable. Nevertheless, research concentrated on the Li–Al/FeS cell, which has the least favourable voltage but was seen to be the most practical of the systems and to promise the longest cycle life. Throughout most of the 1970s and 1980s, development efforts were directed towards this cell in a monopolar configuration.

The lithium–aluminium phase diagram is shown in Fig. 9.5. During normal cell operation, the alloy is confined to the two-phase $\alpha + \beta$ region in which the composition varies between ~9 and 47 at.% Li. The potential of

Table 9.7. Properties of lithium alloy/iron sulfide cells.

Cell	Average voltage (V)	Theoretical specific energy (Wh kg^{-1})
Li/FeS_2	2.1 → 1.6	1320
Li/FeS	1.6	850
LiAl/FeS_2	1.8 → 1.3	640
LiAl/FeS	1.3	460
Li_4Si/FeS_2	2.0 → 1.4	950
Li_4Si/FeS	1.5	640

the electrode in this region is approximately 300 mV vs. lithium at 400 °C. The chemical diffusion coefficient of lithium in β-Li–Al is at least four orders of magnitude higher than in α-Li–Al at the same temperature [61]. Consequently, on charging an aluminium electrode, it is expected that the β-Li–Al phase will form at the electrolyte/electrode interface and grow into the bulk at the α/β phase boundary. Conversely, on discharge, the electrode is expected to polarize severely as lithium is removed from the β phase to leave a surface layer of the α phase. In practice, the situation is complicated by the large volume increase which occurs in going from aluminium to lithium–aluminium and which, in turn, gives rise to structural change, disintegration, and a substantial increase in electrode surface area.

Lithium–aluminium alloy is a brittle material which is not readily fabricated into sheet form. Three methods have been employed to prepare the electrode [62], namely: (i) lithium is electrochemically charged from a molten salt into compressed aluminium fibre; (ii) powdered mixtures of alloy and electrolyte are hot pressed on to a stainless-steel screen; (iii) a powdered Li–Al/electrolyte mixture is loaded vibratorily into a porous stainless-steel structure. Each of these techniques is feasible, but more needs to be done to optimize the electrode as regards structure (thickness, porosity, etc.) for particular cell operating regimes, and to determine precisely which of the electrode manufacturing routes is preferable in terms of both commercial feasibility and cost.

In the 1990s, work at the Argonne National Laboratory turned towards bipolar battery designs (see Section 9.6.6) [63]. In a bipolar battery, it is essential that the component cells have a tolerance to overcharge so that the state-of-charge of all the cells in the stack can be equalized. Under normal cell operating conditions, there is a tendency for the cells to become unbalanced and, occasionally, it is necessary to overcharge some cells if the others are to be brought to top-of-charge. Tolerance to overcharge may be achieved by providing an alternative charge reaction (cf., gassing of aqueous electrolyte batteries, or reaction with the $NaAlCl_4$ electrolyte in $Na/NiCl_2$ cells), or by a gas-recombination mechanism (as, for example, in valve-regulated lead/acid cells and nickel/hydrogen cells). With bipolar Li–Al/FeS, the latter procedure is adopted; the lithium activity in the negative electrode is allowed to increase to the point where the metal is much more soluble in the halide electrolyte and back-diffuses to the positive electrode where it is re-oxidized. In order to avoid the Li–Al alloy becoming liquid in the electrode (at ~50 mV vs. lithium), it is necessary to add a further alloying constituent to control the lithium activity. The Argonne workers added approximately 10 mol% Al_5Fe_2. This reacts to form $Li_5Al_5Fe_2$ which has a sufficiently high lithium activity and releases lithium metal into the electrolyte to diffuse to the positive electrode and be reoxidized. In this way, it was found possible to provide the overcharge capability required by a bipolar battery. Typically, the lithium 'shuttle reaction' corresponds to an overcharge current density of a few mA cm^{-2}.

9.6.3. Positive electrodes

Two sulfides of iron, namely, FeS and FeS_2, have been employed as positive electrodes in lithium–aluminium/iron sulfide cells. They may be compared as follows:

- FeS_2 has a higher sulfur activity; this gives a higher cell voltage than FeS, but also poses more severe corrosion problems
- FeS_2 reduces via FeS to iron, so that discharge and charge take place over two distinct voltage levels

- FeS_2 is a better electronic conductor than FeS and this allows the use of thicker electrodes for an equivalent voltage drop.

Early work with the FeS_2 electrode proved disappointing. Corrosion was a severe problem and necessitated the use of expensive molybdenum cell components, while capacity loss on cycling was rapid. For these reasons, the FeS electrode became the system of choice.

The phase diagram of the Li_2S–FeS binary system [64], which is part of the Li–Fe–S ternary, reveals that below ~800 °C there is one intermediate phase, viz., Li_2FeS_2; at less than 50 mol% FeS, this compound co-exists with Li_2S, and above 50 mol% with FeS. The cell discharge reactions are therefore quite simple:

$$2Li + 2FeS \underset{\text{charge}}{\overset{\text{discharge}}{\rightleftarrows}} Li_2FeS_2 + Fe \qquad V^o = 1.64\ V \quad (9.12)$$

$$2Li + Li_2FeS_2 \underset{\text{charge}}{\overset{\text{discharge}}{\rightleftarrows}} 2Li_2S + Fe \qquad V^o = 1.62\ V \quad (9.13)$$

(potentials measured vs. lithium at 450 °C)

By contrast, the FeS_2 electrode discharges in quite a complex manner; at least three stages are involved. The overall reaction is a four-electron process, i.e.,

$$4Li + FeS_2 \underset{\text{charge}}{\overset{\text{discharge}}{\rightleftarrows}} 2Li_2S + Fe \qquad (9.14)$$

The intermediate steps involve: (i) a phase variously described as $Li_4Fe_2S_5$ or $Li_3Fe_2S_4$; (ii) Li_2FeS_2, as above. The latest reported discharge phase progression is the following [65]:

$$FeS_2 \underset{\text{charge}}{\overset{\text{discharge}}{\underset{(1)}{\rightleftharpoons}}} Li_3Fe_2S_4 \underset{\text{charge}}{\overset{\text{discharge}}{\underset{(2)}{\rightleftharpoons}}} Li_2Fe_{1-x}S_2 + Fe_{1-x}S \quad (x \sim 0.2)$$

$$\underset{\text{charge}}{\overset{\text{discharge}}{\underset{(3)}{\rightleftharpoons}}} Li_2FeS_2 \underset{\text{charge}}{\overset{\text{discharge}}{\underset{(4)}{\rightleftharpoons}}} Li_2S + Fe \qquad (9.15)$$

Discharge on the upper voltage plateau involves the first three phase transitions to produce Li_2FeS_2. Stopping the discharge reaction at this stage enhances the reversibility of the FeS_2 electrode and prolongs its cycle life. For this reason, the FeS_2 electrode is now normally operated only on its upper voltage plateau (1.75 V vs. Li–Al) and, accordingly, is termed the 'UP FeS_2 electrode'. The fully-discharged FeS_2 electrode (1.3 V vs. Li–Al) exhibits a marked overpotential (viz., 95 mV) for nucleation of FeS_2 on recharge. Only above 2.5 V, will FeS_2 be completely reformed. At this high voltage, soluble $FeCl_2$ and free sulfur are produced; these migrate to the separator where solid species are precipitated. This behaviour causes an increase in internal resistance and a rapid decline in capacity. By contrast, if the discharge is confined to the upper voltage plateau, the overpotential for recharge back to FeS_2 is only 20 mV, soluble species are not formed, and the cycle life of the cell is much longer.

A complication which arises in potassium-containing electrolytes (such as LiCl–KCl) is the formation of a complex solid phase of composition $LiK_6Fe_{24}S_{26}Cl$. This phase, popularly known as the 'J-phase', hinders the kinetics of the FeS electrode in potassium-containing melts. Accordingly, there has been a shift towards all-lithium melts or lithium-rich melts (see Section 9.6.4).

Positive electrodes may be fabricated in a similar fashion to negatives, i.e., compressing a mixture of iron sulfide and electrolyte on to a metal current-collector, or vibratory loading of powdered material into a honeycomb metal matrix. Graphite powder may be included to improve the electronic conductivity and, sometimes, 10–15 mol% of CoS_2 or NiS_2 is added as well. Optimization of the electrode is quite complicated — it involves optimization not only of the composition, but also of both the density (i.e., porosity and pore-size distribution) and the thickness. In the case of the bipolar battery, where 10 mol% Al_5Fe_2 is added to the negative to cater for overcharge (Section 9.6.2), it is necessary to add LiS_2 to the charged positive electrode to balance the Li_5Fe_2 which is formed in the negative.

When the cell is discharged and lithium is transferred from the negative to the positive electrode, the latter undergoes substantial swelling if it has inadequate porosity. Such behaviour can have undesirable side-effects. One way to avoid the problem is to assemble the cell in the discharged state. This involves using Li_2FeS_2 powder for the UP FeS_2 electrode, while finely-divided iron powder and Li_2S is used for the FeS electrode. These active materials are mixed with electrolyte salt (and optionally with graphite) and then hot pressed into a plaque on a metal current-collector. After charging, the porosity necessary to accommodate the lithium on subsequent discharge is then present in the positive electrode. This has the added advantage in that the negative electrode is fabricated from aluminium powder which is easier to prepare and handle than Li–Al alloy powder. Cells made in this fashion at the Argonne National Laboratory yielded much improved cycling properties over cells which were assembled in the charged state [66].

9.6.4. Electrolytes and separators

The most widely used electrolyte throughout the 1970s and 1980s was the binary eutectic LiCl–KCl of composition 58.5 mol% LiCl, and 41.5 mol% KCl. The melting point of the eutectic is 352 °C. As mentioned above, a problem with this electrolyte is the formation of the J-phase which changes the composition of the liquid. As the eutectic composition has a narrow liquidus region at the melting point, it is necessary to operate at a significantly raised temperature in order to ensure that the electrolyte remains liquid as potassium is lost. Thus, cells which are based on this electrolyte are generally run at 450–500 °C.

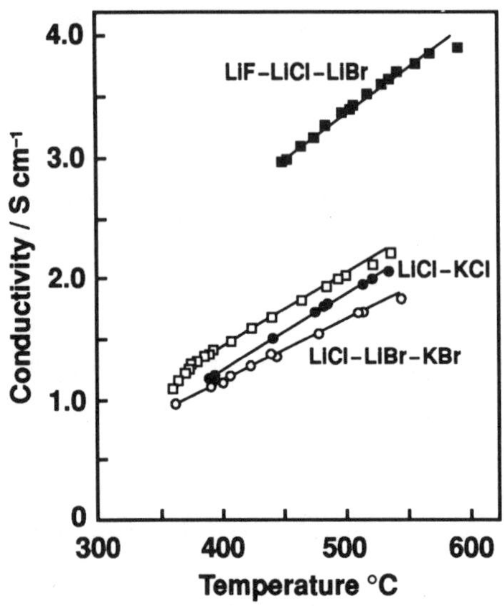

Fig. 9.17. Ionic conductivities of molten-salt electrolytes for rechargeable lithium batteries. (❐) Conductivity for LiCl-rich (34 mol%) LiCl–LiBr–KBr. (Taken from [63] with permission of Elsevier Science Limited, UK.)

In order to avoid battery performance problems which arise from the development of the J-phase, ternary lithium electrolytes have been employed, e.g., LiF–LiCl–LiBr and LiF–LiCl–LiI. The first of these has a particularly high ionic conductivity, viz., ~300 S m^{-1} at 450 °C, Fig. 9.17. Unfortunately, the eutectic temperature is such that the electrolyte must be operated at even higher temperature than LiCl–KCl. For this reason, studies at the Argonne National Laboratory have concentrated on the LiCl–LiBr–KBr system. The eutectic composition has a conductivity only slightly below that of LiCl–KCl (Fig. 9.17). Moreover, the system has a low melting point (320 °C), a broad liquidus region (thus allowing large Li/K ion deviations from the eutectic), and may be used safely at 400–425 °C. This lower operating temperature has significant advantages, in terms of both thermal management and reduced corrosion. Kaun et al. [63] have introduced a LiCl-rich version of this ternary electrolyte with a composition of 34 mol% LiCl, 32.5 mol% LiBr, and 33.5 mol% KBr. This composition represses the formation of the J-phase and also has an improved ionic conductivity (Fig. 9.17). The higher conductivity allows 'electrolyte-starved' cells to operate with an area-specific-impedance (ASI) comparable with that of flooded cells based on the eutectic electrolyte.

Early US cells employed a boron nitride fabric separator, but because this was not available in a suitable weave, a zirconia cloth was used in addition as a particle retainer. Later designs switched to boron nitride felt which is cheaper than woven cloth. Technically, these separators proved satisfactory, but they were far too expensive for EV battery applications. Attention both in the USA and in Europe turned to the use of inert ceramic powders as 'immobilizers'. The preferred powder was calcined magnesia. This material has the advantage of being far cheaper than boron nitride, as well as being more stable in contact with lithium at high activity.

Cells with a magnesia powder separator must be operated in the electrolyte-starved regime in order to avoid the formation of a low-viscosity suspension of powder in the fused salt. In the electrolyte-starved regime, the magnesia is first mixed with the finely-ground electrolyte salt, and then cold compacted under high pressure into a plaque of the desired size and shape. On heating, the salt melts and is held in the pores of the magnesia plaque by surface tension. By adjusting the MgO-to-salt ratio, it is possible to fabricate a plaque in which the magnesia is bonded together to give integrity, while the ionic conductivity is still satisfactorily high. The electrolyte-starved cell has the advantage of no excess free liquid to corrode the case and seals, and to degrade the specific energy unnecessarily. Some workers prefer to mix magnesia powder into the constituents of the negative and positive electrodes as well, and then compact the entire cell assembly so that there is no discontinuity between the electrolyte and the electrodes.

9.6.5. Monopolar cells

Most laboratory cells have had a cylindrical configuration in which a single iron sulfide electrode is sandwiched between two Li–Al electrodes. Such an engineered cell, developed at the Argonne National Laboratory, is shown schematically in Fig. 9.18. These cells, of diameter 130 mm and

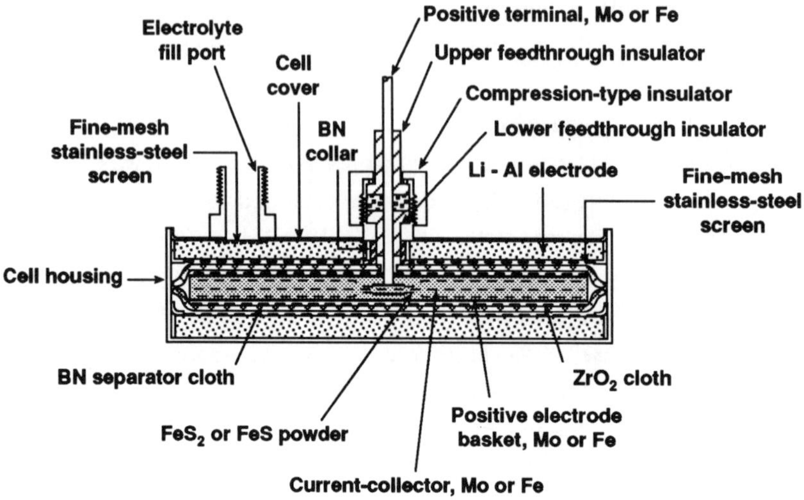

Fig. 9.18. Schematic diagram of cylindrical Li–Al/Fe$_x$ cell. (Taken from [66] with permission of Academic Press Limited, UK.)

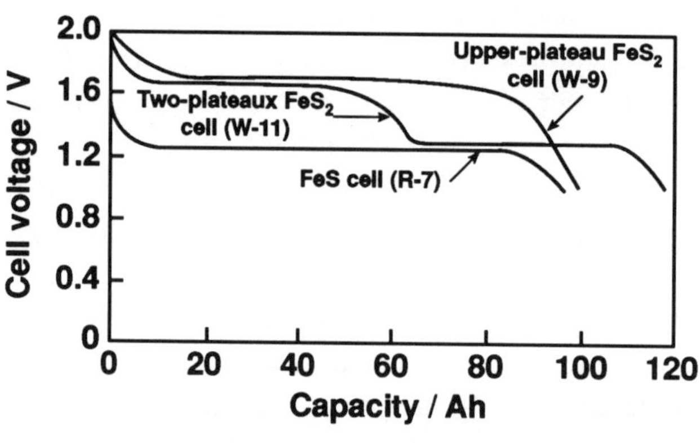

Fig. 9.19. Discharge curves for various Li–Al/FeS$_x$ cells. (Taken from [66] with permission of Academic Press Limited, UK.)

height 30–50 mm, weighed 1.0–1.5 kg. The capacity was ~100–120 Ah. Discharge curves for the UP FeS_2 cell, the two-plateaux FeS_2 cell and the FeS cell are shown in Fig. 9.19. These are somewhat idealized curves in that they relate to low discharge rates and early cycles. Under such conditions, the FeS_2 and FeS cells gave specific energies of up to 150 and ~100 Wh kg^{-1}, respectively.

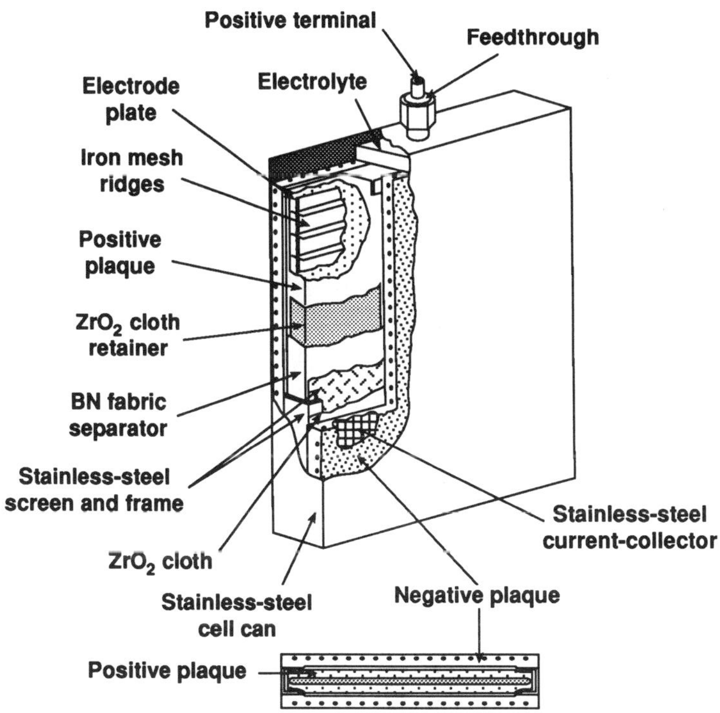

Fig. 9.20. Schematic diagram of prismatic Li–Al/FeS_x cell. (Taken from [66] with permission of Academic Press Limited, UK.)

In the mid-1970s, the Argonne National Laboratory designed and engineered prismatic cells of the type illustrated in Fig. 9.20. The dimensions were 130 mm (length) × 130 mm (height) × 20–40 mm (width). The metal cell case was in contact with the negative electrode and an insulated positive feed-through was welded into the top of the cell. Cells with FeS electrodes had a capacity of ~90 Ah and a specific energy of ~100 Wh kg^{-1}. On the other hand, the peak power declined dramatically as discharge proceeded to unacceptably low values. In Germany, VARTA fabricated a series of similar prismatic cells with a nominal capacity of 140 Ah [67]. These also had a specific energy of 100 Wh kg^{-1}, as well as a peak power of 80 W kg^{-1}. Although a life of several hundred cycles was obtained, it is clear that the early FeS cells did not have the power capability required for most EV applications.

Fig. 9.21. 36-V, 20-kWh battery module of monopolar Li–Al/FeS cells. (Taken from [65] with permission of Elsevier Science Limited, UK.)

As experience and confidence grew in the US programme, multi-plate cells were developed in which electrodes were parallel-connected with internal overhead straps to meet the power requirements of the urban EV. In the 1980s, full-size cells of this type were designed for electric van applications and assembled into 36-V, 20-kWh battery packs. A battery comprising 27 monopolar prismatic cells, each of nominal capacity 550 Ah and connected in series (i.e., a matrix of 3×9 cells), is shown in Fig. 9.21. The battery module was housed in a thermal enclosure which consisted of a double-walled steel container. The annulus of the container was filled with alternate layers of aluminium foil and glass-fibre paper, compressed together and evacuated before being hermetically sealed. A number of such batteries were built by different industrial contractors as part of the US government programme.

In parallel with the above battery engineering programme, research continued at the Argonne National Laboratory on the Li–Al/FeS_2 cell that previously had displayed very limited cycle life. Several advances were made, namely: (i) replacement of the LiCl–KCl electrolyte by the lithium-rich LiCl–LiBr–KBr eutectic; (ii) lowering of the cell operating temperature to ~425 °C; (iii) use of a dense FeS_2 electrode cycled over the upper plateau only. With these advances [63], a cycle life of >1000 was achieved, in 1986, with a flooded-electrolyte prismatic cell which employed a dense UP FeS_2 positive. In summary, monopolar prismatic batteries have the potential to meet the energy and power requirements of electric urban delivery vans, but not to meet the power requirements of either electric or hybrid cars.

9.6.6. Bipolar cells

A schematic of a four-cell bipolar Li–Al/FeS_x stack, with the upper cell 'exploded' to show detailed construction, is presented in Fig. 9.22. The attraction of the bipolar design is the short path length for the current. This results in a low internal resistance as well as a large surface area, and thus gives a high power output for a given current density. The following design criteria must be met before a bipolar design is practicable.

(i) *The provision of a mechanism whereby the state-of-charge of series-connected cells may be equalized.* This requires an ability to overcharge cells, e.g., by arranging for a lithium shuttle mechanism through the addition of a few mol% Al_5Fe_2 to the negative electrode mix, as discussed in Section 9.6.2.

(ii) *The use of a starved-electrolyte regime with a powdered MgO separator.* This is made possible by adopting the lithium-rich LiCl–LiBr–KBr electrolyte with a conductivity in the starved-electrolyte regime equivalent to that of the eutectic in the flooded regime. The absence of excess fused salt reduces the corrosion problems in the cell.

(iii) *The development of effective and reliable seals.* Bipolar cells are more difficult to seal than monopolar cells as the circumferential metal–ceramic seal is much longer than the feed-through seal of a monopolar cell. Moreover, each cell requires two metal–ceramic seals. This technical problem was finally solved in 1990 by the

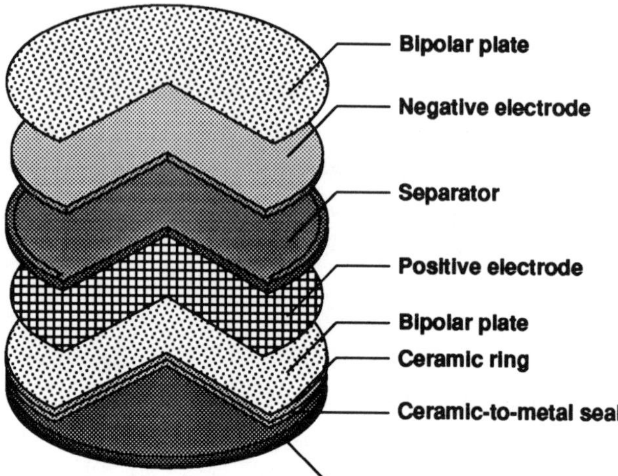

Fig. 9.22. Exploded view of a four-cell bipolar Li–Al/FeS_x stack. (Taken from [63] with permission of Elsevier Science Limited, UK.)

development and introduction of new chalcogenide-based sealing materials. A steel–ceramic–steel peripheral seal was found to be satisfactory for bipolar Li–Al/FeS cells with a chalcogenide ceramic ring as the intermediary (Fig. 9.22). For Li–Al/FeS_2 cells, it was necessary to fit a steel–ceramic–molybdenum seal; this is more technically challenging and costly. Large diameter (130 mm) seals between metals of differing expansion coefficient (e.g., molybdenum–steel) have been fabricated. Possible alternatives to molybdenum for the positive current-collector are coated steels or Grafoil. For the bipolar plate, a promising candidate is a coating of titanium nitride (TiN) on steel.

The scene was now set for the design and manufacture of prototype, bipolar Li–Al/FeS_x batteries. Several dozen such cells (diameter: 130 mm; thickness: 7 mm; weight: 0.25 kg) were constructed and tested in the early 1990s. These were cells of both the UP FeS_2 and the FeS type, and the two varieties were found to deliver 25 Ah at the $C_5/5$ rate (Fig. 9.23). The specific energy and specific power of the bipolar cells are summarized in Table 9.8. The specific energy values for both cell types comfortably meet

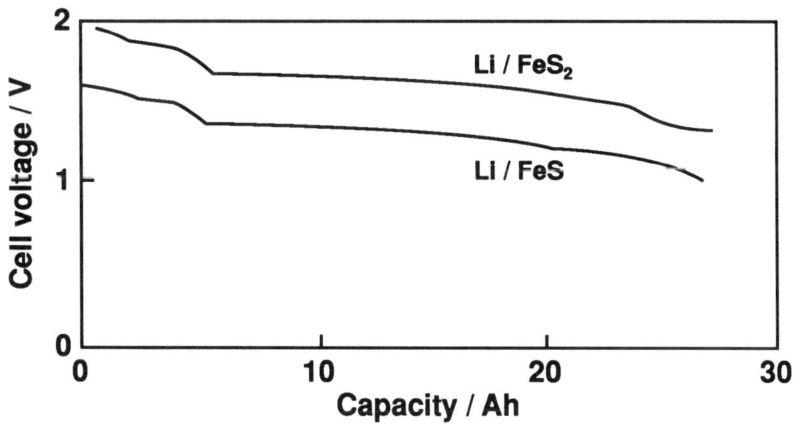

Fig. 9.23. Performance of prototype bipolar Li–Al/FeS_2 and Li–Al/FeS cells at 425 °C. (Taken from [63] with permission of Elsevier Science Limited, UK.)

Table 9.8. Performance characteristics of 130-mm diameter bipolar Li–Al/FeS_x cells [65].*

Cell type	Specific energy (Wh kg^{-1})	Specific power at 80% DoD (W kg^{-1})
Li–AlFeS_2	180 at 30 W kg^{-1}	400
Li–Al/FeS	130 at 25 W kg^{-1}	240

* Note, these data pertain to cells, not batteries, and exclude thermal enclosure.

the USABC target for the mid-term (80–100 Wh kg^{-1}), and the FeS_2 cell comes close to the long-term target of 200 Wh kg^{-1} (see Table 3.4, Chapter 3, but note that the targets relate to engineered batteries and not to cells alone). The most striking performance is that relating to specific power where the mid-term target at 80% DoD is 150–200 W kg^{-1} and the long-term target 400 W kg^{-1}; the FeS_2 cells approach the latter. Finally, cells of both the UP Li–Al/FeS_2 and the Li–Al/FeS design have demonstrated cycle lives in excess of 1000 cycles. This performance is a consequence of the many improvements which have been made during the past decade. The cells are also tolerant to overcharge, overdischarge, and freeze–thaw cycling.

Prismatic bipolar Li–Al/FeS cells have also been under investigation in the UK for many years by the Defence Research Establishment, and the technology is now at an advanced stage of development for prototype units. Sealed modules with a nominal voltage of 25 V and a capacity of 60 Ah have been constructed and tested. These modules have a specific energy of 75 Wh kg^{-1} and an energy density of over 200 Wh l^{-1}. Many hundreds of deep-discharge cycles have been obtained. By reducing the DoD to 50% of the practical capacity, >3000 cycles have been achieved with no reduction in overall capacity. All cell failures occur via a short-circuit condition and this precludes connecting single cells in parallel. Rather, it has been found that connecting series stacks in parallel provides the optimum solution to making large-capacity, high-voltage batteries suitable for EVs.

9.6.7. Safety

Although Li–Al/Fe_x cells and batteries do not seem to have been subjected to as rigorous safety and abuse testing as sodium batteries, there is now extensive practical knowledge of their laboratory operation without serious incident. The US safety experience is summed up by Henriksen and Vissers [65] as follows:

> 'Argonne National Laboratory has a 20-year history, working with a variety of industrial partners, in the development of these technologies without experiencing any safety problems. One of the most dramatic tests was an unplanned event that occurred with a 20-kWh Li–Al/FeS prismatic battery. This battery developed an internal short-circuit and subsequently discharged all of its stored energy in a 20-min period. Although significant internal heating occurred, the battery skin temperature only reached 130 °C, and the case was not breached. In planned safety tests, unprotected individual prismatic Li–Al/FeS cells were held at operating temperature while being subjected to a drop test that simulated a 30-mph (49-km h^{-1}) collision. Although the cell cases breached and their hot contents spewed on the ground, no explosion, fire or release of noxious gas occurred.'

In the UK, cells have been overheated (to >900 °C) and also quenched from normal operating temperature to room temperature without any untoward effects.

The favourable safety experience with Li–Al/FeS_x batteries may be attributed principally to the fact that the electroactive materials are solids and are coated with fused-salt electrolyte which provides a protective film. Reaction with either air or water vapour does not appear to pose a serious safety risk.

9.6.8. Conclusions

Research on high-temperature lithium batteries has, like that on high-temperature sodium batteries, been in progress for over thirty years. The Li–Al/FeS_x couple has been under investigation for twenty-five years. There have been many false dawns, but it does seem that, at last, progress is being made. Cells and small battery modules have been produced that combine acceptable values for specific energy and power, for at least some EV applications, with reasonable cycling performance. Nevertheless, it is perhaps unfortunate that the best performance is exhibited by the UP LiAl/FeS_2 cell in bipolar configuration as this is the design which is likely to have most manufacturing problems and highest cost. As has been found with sodium batteries, the fabrication of a satisfactory demonstration battery in the laboratory is achievable, but the setting up of a pilot production line which will produce quality batteries at realistic prices is a serious technological challenge.

Bipolar Li–Al/FeS_x batteries are of recent origin (<10 years). It appears that the basic requirements for a bipolar design have now been met, and the initial performance trials of four-cell stacks at the Argonne National Laboratory are most encouraging. The same is true for 18-cell stacks in the UK. Nevertheless, EV-size batteries have not yet progressed beyond the conceptual modelling of 40-kWh units [63,65] and much remains to be done. The major advantage of these lithium batteries over sodium/sulfur counterparts lies in the improved safety, and the major advantage over

$Na/NiCl_2$ lies in the higher specific power. Despite this, an EV traction battery has to meet many other goals (see Table 3.4, Chapter 3), and it is not yet clear whether, overall, the Li–Al/FeS_x battery will be able to compete effectively with the more highly developed sodium batteries. In summary, the advantages claimed for the high-temperature lithium system are:

- simple modular design and construction
- Li–Al/FeS designs have at least twice the energy storage of a lead/acid battery of similar volume and mass; even better performance is obtained from UP FeS_2 units
- good power output
- long cycle life
- straightforward heat management, with no auxiliary heating or cooling required during use
- safe, even when subjected to severe abuse.

It remains now to be seen whether all these claims can be substantiated in practice, and at an acceptable cost. Key tasks remaining are the conversion of the laboratory assembly (under high-purity argon in a glove box) to a commercial operation, the elimination of molybdenum current-collectors from FeS_2 cells, and the thermal management of EV-size batteries.

References

1. VINCENT, C.A., BONINO, F., LAZZARI, M., and SCROSATI, B. *Modern Batteries. An Introduction to Electrochemical Power Sources* (Edward Arnold, London, UK, 1984).

2. GABANO, J.-P. (Ed.) *Lithium Batteries* (Academic Press, London, UK, 1983).

3. WHITTINGHAM, M.S. *Science*, 1976, **192**, pp. 1126–1127.

4. WHITTINGHAM, M.S. *Prog. Solid State Chem.*, 1978, **12**, pp. 41–99.

5. WHITTINGHAM, M.S. *J. Electrochem. Soc.*, 1976, **123**, pp. 315–320.

6. JACOBSON, A.J., CHIANELLI, R.R., RICH, S.M., and WHITTINGHAM, M.S. *Mat. Res. Bull.*, 1979, **14**, pp. 1437–1448.

7. JOHNSON, J.W., and WHITTINGHAM, M.S. *J. Electrochem. Soc.*, 1980, **123**, pp. 1653–1654.

8. GAINES, L.H., FRANCIS, R.W., NEWMAN, G.H., and RAO, B.M.L. *Proc. 11th Intersoc. Energy Conver. Eng. Conf.*, 12–17 September 1976, State Line, Nevada, USA (American Institute of Chemical Engineers, New York, USA, 1976), pp. 418–423.

9. SUBBARAO, S., SHEN, D.H., DAWSON, S., DELIGIANNIS, F., TARASZKIEWICZ, J., and HALPERT, G. *J. Power Sources,* 1988, **22**, pp. 269–275.

10. BROADHEAD, J. *Power Sources 4, Research and Development in Non-Mechanical Electrical Power Sources* (Oriel Press, Newcastle upon Tyne, England, 1973), pp. 469–482.

11. BRANDT, K., and STILES, J.A.R. *The 1984 Goddard Space Flight Center Workshop,* NASA Conference Publn. 2382 (dated 1985).

12. DAHN, J.R., and McKINNON, W.R. *Physics in Canada,* July 1988, pp. 93–99.

13. BRANDT, K. *J. Power Sources*, 1986, **18**, pp. 117–125.

14. DEY, A.N., KUO, H.C., PILIERO, P., and KALLIANIDIS, M. *J. Electrochem. Soc.,* 1988, **135**, pp. 2115–2120.

15. FRANK, H., HALPERT, G., LAWSON, D.D., BARNES, J.A., and BIS, R.F. *J. Power Sources,* 1986, **18**, pp. 89–99.

16. IRISH, D.E., DENG, Z., and ODZIEMKOWSKI, M. *J. Power Sources,* 1995, **54**, pp. 28–33.

17. GUYOMARD, D., and TARASCON, J.M. *J. Power Sources,* 1995, **54**, pp. 92–98.

18. KOCH, V.R. *J. Power Sources,* 1981, **6**, pp. 357–370.

19. MATSUDA, Y. *J. Power Sources,* 1993, **43–44**, pp. 1–7.

20. AURBACH, D., ZABAN, A., GOFER, Y., ELY, Y.E., WEISSMAN, I., CHUSID, O., and ABRAMSON, O. *J. Power Sources*, 1995, **54**, pp. 76–84.

21. ARAKAWA, M., TOBISHIMA, S., NEMOTO, Y., ICHIMURA, M., and YAMAKI, J. *J. Power Sources*, 1993, **43–44**, pp. 27–35.

22. SELMAN, J.R., DeNUCCIO, D.K., SY, C.J., and STEUNENBERG, R.K. *J. Electrochem. Soc.,* 1977, **124**, pp. 1160–1164.

23. MEGAHED, S., and SCROSATI, B. *J. Power Sources,* 1994, **51**, pp. 79–104.

24. DAHN, J.R., SLEIGH, A.K., SHI, H., REIMERS, J.N., ZHONG, Q., and WAY, B.M. *Electrochim. Acta,* 1993, **38**, pp. 1179–1191.

25. CHUSID, O., ELY, Y.E., AURBACH, D., BABAI, M., and CARMELI, Y. *J. Power Sources,* 1993, **43–44**, pp. 47–64.

26. DAHN, J.R. *13th International Seminar on Primary and Secondary Battery Technology and Applications,* 4–7 March 1996, Boca Raton, Florida, USA.

27. MIZUSHIMA, K., JONES, P.C., WISEMAN, P.J., and GOODENOUGH, J.B. *Mat. Res. Bull.,* 1980, **15**, pp. 783–789.

28. GOODENOUGH, J.B., MANTHIRAM, A., and WNETRZEWSKI, B. *J. Power Sources,* 1993, **43–44**, pp. 269–275.

29. BROUSSELY, M., PERTON, F., LABAT, J., STANIEWICZ, R.J., and ROMERO, A. *J. Power Sources,* 1993, **43–44**, pp. 209–216.

30. YAMADA, S., FUJIWARA, M., and KANDA, M. *J. Power Sources,* 1995, **54**, pp. 209–213.

31. EBNER, W., FOUCHARD, D., and XIE, L. *Solid State Ionics,* 1994, **69**, pp. 238–256.

32. THACKERAY, M.M., DAVID, W.I.F., BRUCE, P.G. and GOODENOUGH, J.B. *Mat. Res. Bull.,* 1983, **18**, pp. 461–472.

33. GOODENOUGH, J.B., THACKERAY, M.M., DAVID, W.I.F., and BRUCE, P.G. *Rev. de Chim. Minér.,* 1984, **21**, pp. 435–455.

34. THACKERAY, M.M., JOHNSON, P.J., de PICCIOTTO, L.A., BRUCE, P.G., and GOODENOUGH, J.B. *Mat. Res. Bull.*, 1984, **19**, pp. 179–187.

35. THACKERAY, M.M., de KOCK, A., ROSSOUW, M.H., LILES, D., BITTIHN, R., and HOGE, D. *J. Electrochem. Soc.,* 1992, **139**, pp. 363–366.

36. THACKERAY, M.M., de PICCIOTTO, L.A., de KOCK, A., JOHNSON, P.J., NICHOLAS, V.A., and ADENDORFF, K.T. *J. Power Sources,* 1987, **21**, pp. 1–8.

37. HOSSAIN, S. *13th International Seminar on Primary and Secondary Battery Technology and Applications,* 4–7 March 1996, Boca Raton, Florida, USA.

38. TARASCON, J.M., and GUYOMARD, D. *Electrochim. Acta,* 1993, **38**, pp. 1221–1231.

39. OZAWA, K. *Solid State Ionics,* 1994, **69**, pp. 212–221.

40. DAHN, J.R., von SACKEN, U., JUZKOW, M.W., and AL-JANABY, H. *J. Electrochem. Soc.,* 1991, **138**, pp. 2207–2211.

41. NISHI, Y. *13th International Seminar on Primary and Secondary Battery Technology and Applications,* 4–7 March 1996, Boca Raton, Florida, USA.

42. HUGGINS, R.A. *In* MURPHY, D.W., BROADHEAD, J., and STEELE, B.C.H. (Eds.) *Materials for Advanced Batteries,* NATO Conference Series. Series: VI Materials Science (Plenum Press, London, UK, 1980), pp. 91–110.

43. ARMAND, M.B., CHABAGNO, J.M., and DUCLOT, M. *2nd International Meeting on Solid Electrolytes*, Conference held under the auspices of The Electrochemical Society (USA), The Electrochemistry Group, Royal Society of Chemistry (UK) and The Society of Chemical Industry (UK), St. Andrews, Scotland, 20–22 September 1978, Ext. Abstr. 6.5.

44. ARMAND, M.B. *Solid State Ionics,* 1983, **9–10**, pp. 745–754.

45. NEAT, R.J. *In* TUCK, C.D.S. (Ed.) *Modern Battery Technology,* (Ellis Horwood, Chichester, UK, 1991), pp. 528–543.

46. NEAT, R.J., HOOPER, A., GLASSE, M.D., and LINFORD, R.G. *In* POULSON, F.W. (Ed.) *6th International Symposium on Materials Science,* 1985, Riso, Denmark (Riso National Laboratory, Roskilde, Denmark), pp. 341–346.

47. TOFIELD, B.C., DELL, R.M., and JENSEN, J. *Rechargeable Solid State Batteries,* AERE R-11261 (April 1984), Harwell Laboratory, UK.

48. HOOPER, A., and NORTH, J.M. *Solid State Ionics,* 1983, **9–10**, p. 1161.

49. MACKLIN, W.J., NEAT, R.J., and SANDHU, S.S. *3rd International Symposium on Polymer Electrolytes,* June 1991, Annecy, France; subsequently published in *Electrochim. Acta,* 1992, **37**, pp. 1715–1720.

50. WALK, C.R., and MARGALIT, N. *13th International Seminar on Primary and Secondary Battery Technology and Applications,* 4–7 March 1996, Boca Raton, Florida, USA.

51. MACKLIN, W.J., and NEAT, R.J. *Solid State Ionics,* 1992, **52–56**, pp. 694–700.

52. MACKLIN, W.J., NEAT, R.J., and POWELL, R.J. *J. Power Sources,* 1991, **34**, pp. 39–49.

53. SCROSATI, B. *13th International Seminar on Primary and Secondary Battery Technology and Applications,* 4–7 March 1996, Boca Raton, Florida, USA.

54. STEVENS, T. *Industry Week,* December 19, 1994.

55. *Introducing a New Twist in Battery Technology — The Plastic Lithium Ion Cell* (Bellcore Strategies and Energy Storage Research, Redbank, New Jersey, USA); see also: SCHMUTZ, C., TARASCON, J.M., GOZDZ, A.S., WARREN, P.C., and SHOKOOHI, F.K. *Proc. Symposium on Rechargeable Lithium and Lithium-Ion Batteries* (The Electrochemical Society, Pennington, New Jersey, USA, 1994), Proc. Vol. 94-28, pp. 330–335.

56. WEAVER, R.D. *Proc. 19th Annual Power Sources Conf.*, 18–20 May 1965, Atlantic City, New Jersey, USA (PSC Publications Committee, Red Bank, New Jersey, USA, 1965), pp. 113–115.

57. WILCOX, H.A. *Proc. 21st Annual Power Sources Conf.*, 16–18 May 1967 (PSC Publications Committee, Red Bank, New Jersey, USA, 1965), pp. 39–42.

58. RIGHTMIRE, R.A., and JONES, A.L. *Proc. 21st Annual Power Sources Conf.*, 16–18 May 1967 (PSC Publications Committee, Red Bank, New Jersey, USA, 1967), pp. 42–45.

59. CAIRNS, E.J., and SHIMOTAKE, H. *Science*, 1969, **164**, pp. 1347–1355.

60. LAI, S.-C. *J. Electrochem. Soc.*, 1976, **123**, pp. 1196–1197.

61. WEN, C.J., BOUKAMP, B.A., HUGGINS, R.A., and WEPPNER, W. *J. Electrochem. Soc.*, 1979, **126**, pp. 2258–2266.

62. GAY, E.C., VISSERS, D.R., YAO, N.P., MARTINO, F.J., KAUN, T.D., and TOMCZUK, Z. *In* COLLINS, D.H. (Ed.) *Power Sources 6, Research and Development in Non-Mechanical Electrical Power Sources* (Academic Press, London, UK, 1977), pp. 735–749.

63. KAUN, T.D., NELSON, P.A., REDEY, L., VISSERS, D.R., and HENDRIKSEN, G.L. *Electrochim. Acta*, 1993, **38**, pp. 1269–1287.

64. SHARMA, R.A. *J. Electrochem. Soc.*, 1976, **123**, pp. 448–453.

65. HENDRIKSEN, G.L., and VISSERS, D.R. *J. Power Sources*, 1994, **51**, pp. 115–128.

66. WALSH, W.J., and SHIMOTAKE, H. *In* COLLINS, D.H. (Ed.) *Power Sources 6, Research and Development in Non-Mechanical Electrical Power Sources* (Academic Press, London, UK, 1977), pp. 725–733.

67. BORGER, W., KUNZE, D., and PANESAR, H.S. *In* EHRINGER, H., ZEGERS, P., and HOYAUX, G. (Eds.) *Energy Conservation by Energy Storage: Advanced Batteries and Fuel Cells*, Talence, France, 25–26 April 1983, European Union Publication EUR 8660EN, 1983, pp. 18–31.

CHAPTER 10

The Future for Electric Vehicles

10.1. Setting the Scene

Almost all electric vehicles manufactured to date have employed lead/acid batteries. The majority of these vehicles have been designed for off-road duties in factories, warehouses, hospitals, airports and mines, on golf courses, and indeed in any location where the need for silent, pollution-free operation is paramount. A common characteristic of such applications is that the required vehicle range is strictly limited, a feature which can be well accommodated by available lead/acid batteries. The various markets for off-road vehicles are mature. While most of these markets are stable, some may be expected to show steady growth as the advantages of using electric traction are more widely appreciated.

By contrast, most on-road EVs have been overwhelmingly confined to experimental vehicles and prototypes, many of which have finished their brief lives in motor museums. This situation is gradually changing, mainly due to universal concern over air pollution in cities. A few vehicles (vans, buses and cars) have been taken to the stage of series-production for fleet trials. Generally, these have been electric versions of production ICEVs. One road vehicle that has been truly commercialized is the electric milk-delivery truck ('milk float', see Fig 1.11, Chapter 1) which has found service in the UK for many years. Milk floats have been a commercial success on account of their required duty cycle (limited range, frequent stops at most houses, early morning delivery where silent operation is appreciated by the residents). At one stage, about 40 000 milk floats were operating in the UK. Nowadays, the number is declining rapidly because daily milk deliveries are being replaced by supermarket purchases. Small invalid carriages are a second important market for EVs. These battery-powered, 'quasi-road vehicles' are often seen in pedestrian areas, as well as on the roadway. To date, there has been no other large-scale commercialization of on-road battery EVs, either for general urban delivery duties or for passenger transport. Nevertheless, numerous companies are now offering EVs for fleet and customer acceptability trials. Most of the vehicles are fitted with lead/acid, but some are powered by advanced batteries.

In this book, the emphasis has been on reviewing the state-of-the-art and the prospects of advanced traction batteries with different chemistries that are designed for use in EVs (Chapters 5–9). In this final Chapter, the comparative advantages and limitations of the rival battery systems are summarized, and then consideration is given to the broader aspects of replacing ICEVs by EVs. The expectations for vehicles with fuel cells and with hybrid power sources are reviewed briefly, followed by a discussion of the infrastructure requirements which are associated with operating a fleet of urban EVs. Next, the subject is examined from the perspective of various parties interested in the 'EV economy' (viz., state governments and local authorities, manufacturing industry, and owners and drivers of EVs), and an attempt is made to show how the often conflicting viewpoints might be reconciled. Finally, and hesitantly, we present a realizable scenario of how events may develop into the twenty-first century.

10.2. Comparison of Advanced Traction Batteries

When evaluating the prospects for the various candidate batteries, it is necessary to take account of both technical factors (performance, life, safety, etc.) and commercial factors (development and production costs, market readiness, product liability, etc.). While it is a truism that if the product does not meet the technical specifications of the market then there is no product, it holds equally that a technically viable product will fail if it cannot be brought to market at the right time and at an acceptable price. This situation is especially complex for EVs because the anticipated market is being driven not by consumer demand but principally by environmental pollution concerns, and by the fact that the time-scale of advanced battery development and production is long and the associated cost is high.

The principal technical criteria to be met by an EV battery have been listed by the USABC (see Table 3.4, Chapter 3). Sixteen individual criteria are defined, with both mid-term and long-term goals. Many of these are interactive and need to be specified in greater detail than a simple list allows. For example, specific energy is dependent not only on discharge rate, but also on temperature and cut-off voltage. Also, the battery mass requires precise definition with respect to what elements are included. Further, the cycle life depends upon the entire cycling history of the battery, i.e., charge–discharge rates, temperature profile, extent of charge–discharge. Often, it is found that the cycle life in an EV under urban driving conditions is less than that in laboratory simulations. These interactive effects make it difficult to be quantitative when assigning performance levels to the individual battery parameters. Similarly, it is often difficult to assess the wider validity of published laboratory values for these parameters.

Despite the above complications, a semi-quantitative view of five key performance criteria for each of the main candidate batteries has been assembled in Table 3.6, Chapter 3. The values presented have been examined in detail in the discussion of the individual battery systems in Chapters 5–9. The principal attributes, both positive and negative, of the most promising EV battery types are now summarized in Table 10.1. In the medium term, it is concluded that the advanced batteries with the best technical prospects for demonstrating a range of at least 160 km (100 miles) in an electric car or van, and being available in pilot-plant quantities for fleet trials, are lithium-ion, nickel/metal hydride, sodium/nickel chloride, and zinc/air.

Aside from the technical performance parameters of the battery, a key commercial consideration is, of course, its cost. From the outset, it is very difficult for any battery to compete with the comparatively low cost of lead/acid. For example, nickel/cadmium batteries, although an established industry, are about four times the price of lead/acid counterparts, while

Table 10.1. Principal attributes of candidate EV traction batteries.

Battery	Principal attractions	Limiting factors
Lead/acid *(Chapter 5)*	• Established industry • Lowest cost battery • Sealed (maintenance-free)	• Low specific energy • Moderate cycle life (~500) • Poor low-temperature performance
Nickel/iron *(Section 6.2, Chapter 6)*	• Established, robust battery • Long cycle life	• Moderate specific energy • Gassing (high maintenance) • High cost
Nickel/cadmium *(Section 6.3, Chapter 6)*	• Established industry • Long cycle life	• Moderate specific energy • Poor performance above 35 °C • Highly toxic component • High cost
Nickel/zinc *(Section 6.4, Chapter 6)*	• Reasonable specific energy • Good all-round candidate battery	• Short cycle life (~300)
Nickel/metal hydride *(Section 6.6, Chapter 6)*	• Reasonable specific energy • High power • Long cycle life • Tolerance to overcharge and overdischarge • Sealed (maintenance-free)	• High self-discharge • Poor charge-acceptance at high ambient temperature • High cost
Metal/air *(Sections 6.7–6.9, Chapter 6)*	• Long EV range before re-plating or mechanical recharging	• Complexity • High volume (low energy density) • Low overall electrical efficiency • Re-plating technology not yet established
Zinc/bromine *(Section 7.3, Chapter 7)*	• Reasonable specific energy • Demonstrated in EVs • Low-cost components	• Low power output • Shunt-current problems • Safety concerns (Br_2 release)
Sodium/sulfur *(Section 8.7, Chapter 8)*	• High specific energy • Low-cost electrode materials • No self-discharge	• High operating temperature • Corrosion of cell components • No overcharge capability • Safety concerns (fire)

(continued)

Table 10.1. Principal attributes of candidate EV traction batteries *(continued)*.

Battery	Principal attractions	Limiting factors
Sodium/nickel chloride *(Section 8.8, Chapter 8)*	• High specific energy • Discharged-state assembly • Long cycle life • No self-discharge • Successful fleet trials	• High operating temperature • Low specific power at 80% DoD • Cost?
Lithium-ion *(Section 9.4, Chapter 9)*	• High specific energy and power • Long cycle life	• Early stage of development • Accurate control of charge voltage necessary • Safety? • High cost
Lithium/iron sulfide *(Section 9.6, Chapter 9)*	• Reasonable specific energy • Good power output • Long cycle life	• High operating temperature • Only a bipolar battery will meet specifications for EVs • No pilot production despite 20+ years of research

nickel/metal hydride batteries are currently even more expensive, although the higher costs of these batteries are offset to some degree by their longer cycle life. Both the initial capital cost and the running cost per km are important. The latter takes account of the possible daily range, the cycle life, and the electricity tariff. It is a reasonable assumption that all serious battery developers have calculated these costs, have extrapolated them to large-scale production rates, and have become convinced that it will be possible to achieve an acceptably low value. Others may not agree with this conclusion; so much depends upon the assumptions made. Some of the extrapolations are very significant, e.g., for lithium-ion batteries from US$ 1000–3000 per kWh at the pilot scale to less than US$ 200 per kWh for production scale [1]. Some authors have made attempts to gather together comparative production costs for different advanced batteries, but there is a tendency for them all to arrive at values around US$ 100–200 per kWh, no doubt to meet the USABC target. The problem of the high expense of purchasing advanced batteries may be resolved to some degree by battery leasing (see Section 10.7).

The data in Table 10.1 show that no battery system has all the required desirable attributes. It is likely that, progressively, development work will cease on the least promising candidates and only one or two battery types will be commercialized. A further problem, common to all new batteries, is that a developer will be unlikely to set up a full-scale production plant until firm orders have been received from one or more EV manufacturers, while potential EV manufacturers will be reluctant to place an order for batteries until they are certain that they can sell their vehicles. In addition, EV manufacturers will require firm prices, delivery dates and performance warranties, which the batterymaker will be hesitant to provide when based solely on the experience gained in operating a pilot plant. These are generic problems which have yet to be resolved.

10.3. Fuel Cells

In principle, EVs may be powered by either batteries or fuel cells and, therefore, it is appropriate to consider briefly the prospects for fuel cells in the transport scene. A full discussion of fuel cells lies outside the scope of this book and would merit a volume of its own.

Much of the interest in fuel cells has been directed towards stationary applications, e.g., the localized generation of electricity — particularly in connection with combined power and heat schemes ('co-generation'). For transport applications, i.e., powering EVs, the principal attractions are considered to be: (i) a vehicle range which is limited only by the amount of fuel carried; (ii) a conventional fuelling procedure (when using liquid fuels) with no requirement for a radically new infrastructure (see Section 10.5); (iii) high efficiency of fuel conversion, especially at low power. These attractions are, of course, in addition to the usual advantages of silent operation and no pollution that are common to all EVs. The added benefits of fuel cells are sufficiently compelling to cause a sharp divergence of view over the relative prospects for EVs powered by batteries and by fuel cells. At the present time, it appears that batteries have the edge for cars and light vans, while fuel cells are more suitable for buses and heavy-freight vehicles — but this situation could well change.

There are six generic types of fuel cell in various stages of development, namely:

- phosphoric acid fuel cells (PAFCs)
- alkaline fuel cells (AFCs)
- proton exchange membrane fuel cells (PEMFCs)
- molten carbonate fuel cells (MCFCs)
- solid oxide fuel cells (SOFCs)
- direct methanol fuel cells (DMFCs).

Of these, there is a broad consensus that only three (AFC, PEMFC, DMFC) have potential application in road transport. The DMFC is the most attractive option as it consumes directly a liquid fuel (methanol) which may be manufactured both readily and cheaply. Unfortunately, it is the least developed option, with the most technical problems, and can be discounted for the present. The remaining two candidate fuel cells both require hydrogen as a fuel, and this has either to be carried as a compressed gas, thus imposing major weight and volume constraints, or as liquid (methanol, gasoline) with an on-board fuel reformer. Originally, the development of the fuel reformer to provide hydrogen of a suitable purity has been seen as a task of comparable or greater magnitude than the development of the fuel cell itself. In the last few years, however, a major advance in this technology has been made by Johnson Matthey in the UK through the design of a compact, lightweight, methanol reformer which is regarded to be well suited for use in EVs.

The AFC has been promoted for transport applications, principally by the Belgian-Dutch consortium, Elenco NV. Full-scale cell stacks have been demonstrated in both commercial vehicles and buses. Initially, the cells used compressed hydrogen storage, but this is not truly viable on account of the mass and volume of the cylinders. If hydrogen is produced by on-board reforming, it must be thoroughly purged of the accompanying carbon dioxide as this gas reacts with the KOH electrolyte and reduces its conductivity. The same problem occurs when using air rather than pure oxygen and, therefore, it is necessary to pre-absorb the carbon dioxide content in an alkaline scrubber before introducing the air into the cell stack. Development work on the alkaline fuel cell continues in both Europe and the USA.

The PEMFC is now the preferred option for EV applications. In this system, a proton-conducting, solid polymer membrane serves as the electrolyte and is sandwiched between graphite-plate electrodes. The required electrocatalyst is finely dispersed platinum or platinum alloy and is coated on the two surfaces of the membrane. Research on this fuel cell has been carried out mainly in the USA (Texas A&M University, Los Alamos Scientific Laboratories) and in the UK (Johnson Matthey), while a company in Canada (Ballard Power Systems) has taken the lead in the engineering of commercial cell stacks. Recently, Ballard and Johnson Matthey, working together, have made major advances in improving the electrode kinetics, as well as in reducing the platinum inventory to the point where it is no longer seen to be a cost obstacle. The polymer electrolyte, a sulfonated polyfluorocarbon, is a high-cost item at present, but it is expected that this problem can be overcome. The most significant advance is that the latest fuel-cell stacks from this partnership have a power density of ~1 kW l^{-1}, i.e., considerably in excess of any traction battery [2]. When coupled to the new Johnson Matthey methanol reformer, there is an excellent prospect for a compact system. 200-kW cell stacks have now been assembled to power electric buses and these fit into the vehicle's standard engine compartment. Orders for a number of buses powered by PEMFCs have been placed by the Chicago Transit Authority and by the Province of British Columbia. It appears that the technology of these fuel cells is now in an advanced state, although there remains the need to reduce the cost to an acceptable level. A further development has been the announcement that Chrysler, together with Delphi (a subsidiary of General Motors), is to build by 1999 a 'proof of concept' EV based on the Ballard fuel cell with petroleum (gasoline) as the fuel and an on-board reformer to convert it to hydrogen. In September 1997, Mercedes-Benz revealed their New Electric Car, 'NECAR III', this is based on the Company's A-class sedan. The car is powered by a Ballard PEMFC using reformed methanol. This unit is fitted between the two layers of the sandwich-floor construction of the vehicle. It is claimed that 90% of the fuel cell's power is available within 2 s of pressing the accelerator, a claim which demonstrates the need for an auxiliary power unit such as that employed in a hybrid electric vehicle.

10.4. Hybrid Drive Systems

Hybrid vehicles are defined as vehicles which have two separate power sources — generally, one to provide range and one to supply peak power. Most conventional (ICEV) passenger cars are grossly overpowered for urban or highway driving, in order to ensure that sufficient power is available to meet the demands of acceleration and hill-climbing. As discussed in Chapter 4, the steady power requirement is only 10–20% of the peak power demand. It follows that the main propulsion unit could be sized for steady urban and highway driving if the peak power could be provided by an auxiliary source. This would permit considerable savings in the volume, mass and cost of the main propulsion unit, as well as in fuel consumption. Some hybrid vehicles are purely mechanical in nature. One example is a small ICE or gas turbine for steady driving, with a large flywheel to provide peak power when required. The flywheel is energized by the ICE, either while idling or when driving at low power. The discussion here, however, is concerned solely with hybrid vehicles which operate partly, or entirely, on electricity. In this respect, hybrid electric vehicles (HEVs) can use a variety of power-source configurations, namely, all-electric, electric–mechanical, or heat-engine–battery.

10.4.1. All-electric HEVs

These HEVs may be mains–battery, battery–battery, fuel-cell–battery, or battery–ultracapacitor. The advantage of an all-electric hybrid is that it is emission-free, but the disadvantage is the mass and cost associated with the storage of electrochemical energy. The mains–battery hybrid is a special case to allow trams, trolley buses or trains a limited 'off the rail' capability, which is both exceedingly useful for operational reasons and cost effective (see Section 1.9 and Fig. 1.15, Chapter 1). Many fuel cells and batteries are limited in power output and, in the case of batteries, rate of charge-acceptance. By hybridizing either of these power sources with a battery of high specific power, or an ultracapacitor, it is possible to improve both vehicle performance and the acceptance of regenerative-braking energy. To date, few vehicles of this type have been built.

Ultracapacitors, a relatively new concept, have the potential to provide the peak-power requirements of HEVs. Unlike a conventional dielectric capacitor, which stores energy as electronic charges that are separated by a dielectric medium, ultracapacitors utilize ions which are adsorbed from an electrolyte solution on to an electrode of high surface area (generally made from carbon black). Energy is stored as separated ionic charges and by redox reactions which involve the ions. Ultracapacitors store considerably more energy than conventional electronic capacitors, although the amount is still small compared with batteries — typically, in the range 1 to 10 Wh kg^{-1}. They have high power output (500–1000 W kg^{-1}) for relatively short periods (10–60 s). Even higher power outputs, up to 5000 W kg^{-1}, have been projected. Because the charge–discharge processes are all surface phenomena, ultracapacitors are capable of a very large number of cycles (>100 000). They are still in an early stage of development, but are foreseen to have many different applications as their performance improves [3,4]. In 1997, Maxwell Technologies, Inc. announced [5] that its ultracapacitors are to be evaluated by a major international car manufacturer for use in both EVs and HEVs. Whilst not an all-electric system, it is also interesting to note here that the Honda J-VX concept sports car uses an ultracapacitor to boost the power of a three-cylinder petrol engine in a parallel hybrid arrangement.

10.4.2. Electric–mechanical HEVs

Electric–mechanical HEVs possess a main traction battery with an auxiliary flywheel to provide peak power and to store the energy obtained from regenerative braking. Much work was done on this concept in the USA during the 1970s. For example, Fig. 10.1 shows an artist's impression of a battery–flywheel hybrid concept car designed at that time by the Garrett Corporation under contract to the US Department of Energy. The

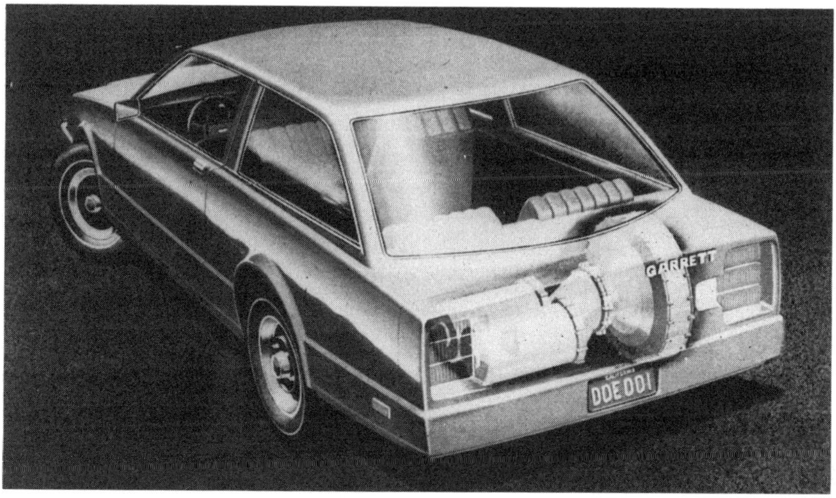

Fig. 10.1. Battery–flywheel hybrid concept car designed by Garrett Corporation in the USA.

flywheel was bulky, as well as heavy, and had limited energy-storage capacity. Since then, development work has been undertaken on high-speed flywheels which use lightweight composite materials for the rotor. Consequently, the energy storage per unit mass and volume have both improved markedly. Lawrence Livermore National Laboratory in the USA has proposed [6] coupling its zinc/air battery (see Section 6.9, Chapter 6) to a super-flywheel to produce a concept HEV with a range of 380 km. This vehicle would use a 70-kWh, 40-kW zinc/air battery in combination with a 0.5-kWh, 102-kW flywheel for acceleration and hill-climbing. Super-flywheels resemble ultracapacitors in having modest energy-storage capability, very high power output, long cycle lives, and high overall energy efficiencies.

10.4.3. Heat-engine–battery HEVs

The common objectives of heat-engine–battery HEVs are to reduce the urban pollution and the petroleum consumption that is associated with conventional ICEVs without restricting the driving range to that provided by battery-powered EVs. Within this HEV concept, there is scope for numerous designs and operating procedures to maximize range, provide a

Table 10.2. Power source, design and operational options for heat-engine–battery HEVs.

Fuel	petrol; diesel; mains electricity
Main traction unit	ICE; gas turbine; stirling engine; battery; fuel cell
Power boost unit	flywheel; battery; ultracapacitor
Design	series hybrid; parallel hybrid
Vehicle performance	conventional ICEV performance; restricted speed and range
Urban emissions	low; zero (i.e., for zero-emission vehicles, ZEVs)

high-performance vehicle, maximize overall energy efficiency, and minimize capital and running costs [7]. Before the vehicle design exercise can begin, it is necessary to decide which of these desirable goals has the highest priority. This approach is essential because meeting any one target can place constraints on the others.

Some of the options for heat-engine–battery HEVs are presented in Table 10.2. Hybrid vehicles of all types are still in an early stage of development, and a full technical, environmental and economic evaluation of the various options has yet to be made. With respect to the design of heat-engine–battery HEVs, there are two basic types: series hybrids and parallel hybrids, as discussed briefly in Chapter 1 and shown in Fig. 1.1. In the series configuration, all the traction power is delivered by an electric motor and a heat engine is used intermittently to generate electrical energy which is fed into the battery–controller system. This HEV is the simpler concept and has more scope for both lower fuel consumption and reduced emissions, but it requires more powerful and more efficient electric motors, a smaller ICE designed for running at constant speed, and a battery with a higher power rating. The parallel HEV, in which the wheels can be driven mechanically by a heat engine, or electrically by a battery–electric-motor, or by both, offers the easier option to manufacture. For example, the vehicle can make use of existing technology for engines, gear boxes, induction motors, and lead/acid batteries [8]. On the other hand, complex control systems are required, especially if a flywheel is incorporated as well.

In one typical version, the series HEV would have a battery which is sufficiently large to meet the daily range and peak-power requirements for city driving, and a small ICE (or gas turbine) which is used to generate electricity purely as a 'range extender' for out-of-town driving. This is essentially an EV, with an EV-sized battery and a small auxiliary engine. For urban driving, most of the energy consumed would be electricity from overnight charging. An alternative series concept is to have a larger engine and a smaller battery such that the power is provided almost entirely by the engine which is only switched off for urban operation. Volvo is following this latter route with its 'Environmental Concept Car, Urban Bus and City Truck'. In each of these designs, a gas turbine drives an alternator to power the EV and also to charge a nickel/metal hydride battery which supplies the traction power in environmentally sensitive areas.

The parallel HEV has two distinct drive-trains which are independent of each other. The engine is likely to be quite small and would be sized for steady highway driving. The battery provides auxiliary power for both acceleration and hill-climbing, and also accepts regenerative-braking energy. Limited urban driving in the battery electric mode would also be possible. Most, if not all, of the energy consumed in this version would be from petroleum. Renault has adopted such a configuration in its new, small hybrid car ('NEXT'). This vehicle has a 750-cc engine which provides 35 kW of power to the front axle, and two auxiliary 7-kW electric motors which drive the rear wheels. In July 1997, Toyota unveiled plans to mass-produce a parallel HEV (the 'Prius') powered by a 1500-cc engine and a nickel/metal hydride battery (see Section 6.6., Chapter 6).

The designs of heat-engine–battery HEVs become even more complex when flywheels or ultracapacitors are introduced as well. The flywheel requires the development of a continuously variable transmission together with a series of clutches. In addition, the power-control electronics are more complicated. The ultracapacitor resembles the flywheel in that it is a high power, low specific-energy device which is ideal for meeting short-duration, peak-power demands and for recovering regenerative-braking energy. Being an electrical device, it is readily integrated into an EV's control circuitry.

Opinions concerning the future prospects for heat-engine–battery HEVs are divided. Enthusiasts for the parallel hybrid point to the fact that it is essentially a conventional car, as accepted by the public, but with higher fuel efficiency and a larger battery. Moreover, such a vehicle could be built today with a standard lead/acid battery pack, which has suitably high power density. This is a very attractive proposition for the automotive companies. Indeed, Fiat in Italy is offering its well-known Panda car in both an electric version (the 'Panda Elettra') and as a parallel hybrid [9]. Since parallel HEVs possess essentially two independent power-trains, the failure of one or other should not leave the motorist stranded — a positive sales feature. Detractors point to the fact that such hybrids suffer from the complexity and cost of two propulsion systems, together with the difficulty of fitting them in the space available. On the other hand, it is contended that the series hybrid has all the advantages of electric traction, usually with a smaller battery than a pure-battery EV, as well as with greater range. Furthermore, the system has the merit of being practical with state-of-the-art lead/acid batteries. At the end of the day, the prospects for heat-engine–battery HEVs, with their more complex engineering, may well hinge upon whether or not advanced batteries will become available with suitable performance at a price which will make the pure-battery EV commercially viable.

10.4.4. The hypercar

In recent years, the abatement of urban-air pollution has assumed greater political importance than the conservation of energy. This priority, however, could easily change again — it depends upon the availability and price of petroleum, as determined by international politics or resource availability. The conventional ICEV with an Otto engine has difficulty in achieving high efficiency and low emissions over a wide operating range of speed and torque. A major advantage of HEVs lies in their improved energy efficiency when they employ a small engine which is running at constant speed. Also, with mains charging, there is scope to adjust the ratio of petroleum to electricity consumed, according to the design and the mode of operation.

Since the forces required to overcome rolling resistance, to accelerate and to move uphill are each proportional to the vehicle mass (eqns. (3.2), (3.5), (3.6) in Section 3.1, Chapter 3), it follows that a major saving in mass will lead to a corresponding significant improvement in energy efficiency. At present, the limitation on mass reduction of an ICEV is set by the extensive use of steel in the bodywork, but there is scope to lighten vehicles through the greater employment of aluminium, magnesium, and/or composites. Such design changes are equally applicable to HEVs. Pure-battery EVs, on the other hand, are inevitably heavy because of the weight of the traction battery; this is true even with advanced batteries since extending the driving range is generally considered more important than reducing the mass.

The concept of a 'hypercar' has recently been advanced [10] as a means to capitalize on the advantages of electric traction itself without incurring the disadvantages of pure-battery traction. This vehicle involves a synergy between an ultra-lightweight body, streamlined to have a low aerodynamic drag coefficient, and a series-hybrid drive-train. The projected curb weight is 480 kg for a 4–5 seater car. In the hypercar concept, the wheels are driven by one electric motor with a differential gear, or by a pair of motors for the front wheels, or by four small integral wheel motors each with an associated electronic controller. All the electricity is generated, as needed, by an on-board power plant which converts fuel into electricity by a number of different means which range from an ICE–electric-generator to a fuel cell. Thus, it is intended for the hypercar to operate without mains charging. A 'buffer' device (e.g., a battery) is used to store electrical energy produced in excess from the power plant or recovered from braking. The energy stored in the buffer augments the power plant for acceleration and hill-climbing. This arrangement permits the use of a smaller engine than in an ICEV as well as a relatively small battery which, in turn, compounds reduction in the structural mass of the vehicle. If the hypercar is fuelled by petroleum, then the engine needs to produce typically only ~10 kW of electrical power for steady driving on the flat. With respect to battery development, the hypercar requires a system with the capability for high power output–input rather than for high energy-storage.

Given the many technical options for HEVs, it is difficult to predict the ultimate outcome. What is quite clear is the massive advantage, from the standpoint of both energy conservation and pollution minimization, of sizing an engine for steady cruising and providing boost power from a separate unit. In view of these advantages, as well as the addiction of vehicle manufacturers to ICE technology and the in-place petroleum-refuelling infrastructure, it is very likely that a hybrid electric vehicle, most probably a hypercar, will be favoured for long-distance travel.

10.5. Infrastructure Considerations

The introduction of sizeable fleets of EVs will have serious infrastructure implications. Apart from the electrical-supply industry, interested parties include city planners (e.g., the provision of charging stations), the vehicle repair trade (garages), the fire service, and those concerned with electrical safety regulations and technician education. The establishment of a complete EV infrastructure, to supplement that which already exists for ICEVs, is a major undertaking — it will take many years, will incur substantial costs, and will have a significant impact on society as a whole. The key issues involved are outlined here.

10.5.1. Battery charging

By far the most important infrastructure consideration is the building of an extensive, energy-demanding and power-hungry, battery-charging network, together with the supply of the required electricity. In association, there is the need for procedures and facilities to connect EVs both safely and conveniently to the electricity grid via d.c. chargers.

For small vans and cars, the most common mode of charging will be overnight in a garage. By way of example, the normal domestic supply outlet in the UK (240 V, 13 A) is capable of providing 3 kW for short periods and, say, 2.5 kW continuously. As these small vehicles are likely to have a battery of 20–25 kWh capacity, and would not normally be discharged more than 80%, they could be comfortably recharged in 8–10 h (allowing for some inefficiency in the charger). In many instances, a lower current and a longer time might be employed. Thus, a standard domestic socket can be used to supply the charger, which may be located either permanently in the garage or on board the vehicle. Although the standard supply voltage varies from country to country (in the USA, for example, it is 120 V), most residences in developed countries are wired to supply multi-kW for cookers, water heaters, etc. and, consequently, overnight charging should present no supply problems, at least for those vehicles which are garaged.

Fleet vehicles (e.g., commercial delivery vans and, especially, buses) present a more difficult challenge. Even supposing that they are garaged, or have access to recharging points at their parking places, the power requirements are likely to be substantial. Taking a fleet of fifty urban buses, as a typical case, each equipped with a 100-kWh battery which has to be recharged between midnight and 5 a.m., a power supply of 1 MW will be required. This electrical power has to be distributed around the garage or parking area to fifty, separate, 20-kW charging stations. Such a requirement is by no means insignificant, in terms of both engineering and cost. Moreover, this infrastructure will occupy considerable space.

The only traction battery that is certain to be available at an acceptable cost is the lead/acid battery. The range limitation of EVs equipped with this system implies that it will be necessary to have either battery-exchange facilities or fast-charging stations for those occasions when it is required to drive longer distances. An EV which is limited to 70–80 km per day is unlikely to have wide commercial acceptability, even if the majority of journeys fall within this range. Designs have therefore been advanced for rapid battery-exchange facilities. For a delivery van with an underfloor battery pack, this may involve simply jacking up the vehicle and dropping the entire pack from below (Fig. 10.2), an exchange that takes just minutes. For cars, it has been proposed that the batteries should be positioned longitudinally along the centre line of the vehicle (in place of a transmission tunnel), and should be removed from the rear by means of a roller track. Whichever proposal is adopted, it is clear that the requirement for rapid battery-exchange will impose some limitations on vehicle design. Another

Fig. 10.2. Fitting of a replacement battery to a Bedford electric van.

feature of battery exchange is that it would inevitably involve the leasing of batteries (rather than outright purchase), as few would be prepared to exchange their new battery for an old one. Finally, it should be noted that the concept of rapid battery-exchange necessitates having more than one battery per vehicle and this would add to the overall costs of ownership.

The alternative approach to extending the range is that of day-time recharging. This is already widely practised with industrial battery-powered vehicles and is known as 'opportunity charging'. For example, a fork-lift truck in a factory or a baggage-handling truck at an airport may be used over two shifts rather than just one with judicious use of opportunity charging. This saves the labour which would be required for battery exchange after one shift and also the cost of having two batteries per truck. A similar regime might be applied to EVs, provided that recharging stations were made available at, say, car parks. For commuters or shoppers, who generally leave their cars for several hours at a time, normal recharging rates could be employed. In the USA, steps have been taken by Edison EV to support the General Motors EV–1 and other electric cars by installing a network of charging stations at workplaces and vehicle fleet locations in southern California. Charging stations are also planned for public use at shopping malls, hotels, entertainment centres, and public transit stations. Other States, namely, Arizona, New Jersey and Massachusetts, have also installed a number of public charging stations.

The more difficult charging situation arises with single journeys which are longer than the range provided by the battery. Here, it is desirable to recharge the battery in as short a time as possible, preferably 15 min or less. Recently, work on the rapid recharging of lead/acid batteries has been carried out by contractors to the Advanced Lead-Acid Battery Consortium (ALABC) [11]. The goal set by the ALABC is to return 50, 80 and 100% of the depleted charge in 5, 15 and 240 min, respectively. This goal has been achieved for monoblocs without any performance degradation and the study is to be extended to instrumented EV battery packs. Initial results from the Arizona Public Service Company show that the lead/acid batteries in its EV fleet can be routinely charged from 30 to 100% state-of-charge in just 13 min [11,12]. These charges enable the EVs to achieve daily ranges of well over 160 km; the vehicles have provided over 22 000 km of duty with in excess of 200 fast charges [13]. In a 24-h endurance trial, an EV equipped with lead/acid batteries covered 1630 km, with fast recharging (taking 12 min) every 70 km. Evidence has also been cited [14] that fast charging, far from being detrimental, can actually be beneficial to battery performance, although the effect on battery life has yet to be fully determined. Certainly, fast charging leads intrinsically to heating of the battery and, unless proper procedures are implemented to limit temperature rise, battery life may be impaired. Accordingly, research is presently being performed within the ALABC programme to develop appropriate fast-charging algorithms (see Section 5.5.4, Chapter 5).

There is another difficulty with fast charging. Although the technique may be feasible from a battery point of view, the infrastructure implications are considerable, not least from the utility's perspective (see Section 10.5.2). Consider a 25-kWh traction battery which is to be charged from 20 to 80% state-of-charge in 15 min. Assuming no electrical losses and a uniform rate of charging, this requires a 60-kW supply. Such a charger will be too heavy

Fig. 10.3. A US 'Electricar' being charged via a 150-kW fast charger at Phoenix, Arizona. (Taken from [11] with permission of UK & International Press.)

to carry on-board and is likely to be too costly to be dedicated to one vehicle. The charger must therefore be land-based with a high-power d.c. coupling to the EV. If the charger is 80% efficient, then it will require a 75-kW a.c. supply. At normal mains voltage (e.g., 240 V), the input current will be in excess of 300 A. For an 'electrical service station' (an analogue of a petrol station) to have a capability of servicing 20 EVs at a time (maximum 80 per hour), it will require a 1.5-MW supply. Such requirements will again be costly, as it will be necessary to install an electrical sub-station to provide this power. A US 'Electricar' being charged from a 150-kW fast charger at the Arizona Public Service facility is shown in Fig. 10.3 [11]. The substantial size of the electrical installation is evident.

There are two generic types of charger, namely, conductive and inductive. With conductive charging, direct wires are connected between the electricity supply, the charger, and the EV. The facility is also likely to include an earth (ground) wire and connections for battery monitoring, feedback control, and safety functions. It is considered vital to develop a standard supply socket for use on all EVs and the ALABC has a Standing Committee which is looking into the matter of connectors and connecting stations,

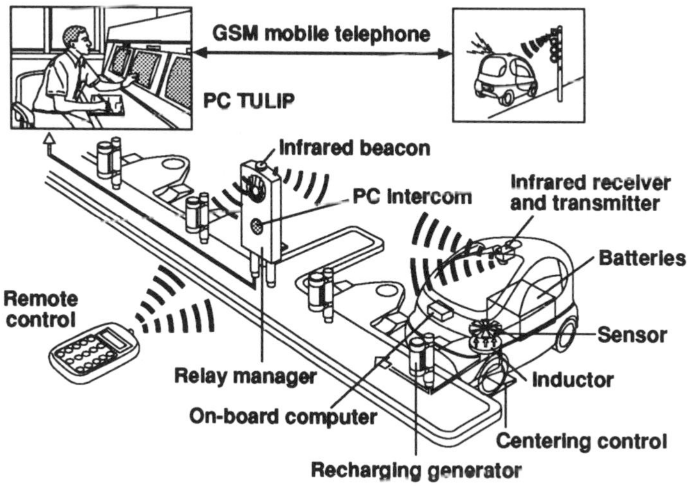

Fig. 10.4. Inductive charging as foreseen in a rental scheme for a PSA Peugeot-Citroën small urban car in Europe (TULIP). (Taken from [15] with permission of UK & International Press.) See Section 1.10, Chapter 1.

while a sub-committee is addressing fast charging. Inductive charging differs from conductive charging in that there is no direct physical connection between the supply and the EV. Rather, the a.c. supply activates the primary coil of a transformer, the secondary of which forms part of the EV. The EV 'docks' on to the charging station and the current is induced in the secondary coil, Fig. 10.4 [15]. For the transfer of high power, the inductive charger may incorporate a frequency booster (to ~10^5 Hz) so as to reduce the physical size of the coupling There is also likely to be a RF-coupled link across the interface to facilitate communications. When using an inductive charger, an on-board rectifier is required. As there is no conductive contact between the EV and the supply, there is no earthing and, therefore, safety relies on isolation via double insulation.

The Hughes Corporation in California has developed a high-frequency inductive charger for use with the General Motors 'Impact' car [16]. This is rated at 6 kW, although other models at much higher power levels have been projected. The transformer primary, in the form of an inductive paddle (somewhat like a table tennis bat), is connected via a cable to the off-board

Fig. 10.5. Inductive charging of the General Motors 'Impact'.

control cabinet and converts mains power at 60 Hz into high-voltage, low-current a.c. at 80–350 kHz (Fig. 10.5). Instead of docking the vehicle, the paddle is inserted manually into the charge port on the EV and, when locked in position, is in intimate contact with the transformer secondary. The charging needs of the battery are signalled to the control cabinet via a low-power RF signal which is generated on the EV and transmitted via a pick-up coil in the paddle. By operating at high frequency, the on-board part of the 6-kW inductive charger weighs only 9 kg. The downside of inductive charging is that it is an expensive technique compared with conductive charging.

Inevitably, most work on charger design has been for lead/acid batteries. Other battery systems will most likely have their own specific requirements as regards battery monitoring and charge control, which may make the design of universal connectors difficult. It should be remembered that charging is one half of the electrochemical cycle — often, the neglected or under-researched half — and control of the conditions prevailing during charging is just as important in determining battery life as control of those during discharging.

10.5.2. Electric utilities' perspective

The primary interest of electric utilities in EVs is to take advantage of night-time charging to smooth the demand load over the diurnal cycle. Generally, the demand for electricity varies with the season, according to the weather pattern, and also with the time of day. In Northern Europe, the peak demand is during the winter as electricity is still used for heating purposes, while there is comparatively little air-conditioning. In many parts of the USA, however, the situation is reversed — the peak demand is in summer as air-conditioning is employed extensively, whereas little electrical heating is used in winter. Thus, seasonal fluctuations in demand can differ from country to country and, indeed, from region to region.

The diurnal demand cycle is quite predictable. Usually, there is a broad maximum in demand during the course of the working day (8 a.m. to 6 p.m.) and a broad trough during the night (10 p.m. to 6 a.m.); an example is given in Fig. 10.6 [17]. Superimposed upon this sine-type curve are short-term peaks dictated by social factors, e.g., the times at which meals are cooked and popular television programmes are broadcast. At intervals in television programmes, the electricity demand can rise dramatically as viewers boil electric kettles to make beverages. These short-term peaks in demand are met by pumped-hydro schemes (if available) or by 'spinning reserve' generators which are already operational and poised to satisfy sudden demands.

Electricity supply is not just a matter of generation, but also of transmission across the country and subsequent local distribution. The networks are designed to meet the anticipated peak demands with a generous margin for security of supply. Utilities frequently buy and sell electricity from each other as their individual peak demands do not always coincide. In extreme circumstances, utilities may reduce the voltage ('brown-outs') or even shed load selectively. Major electricity consumers may negotiate a very favourable tariff if they agree to be disconnected at the utility's discretion.

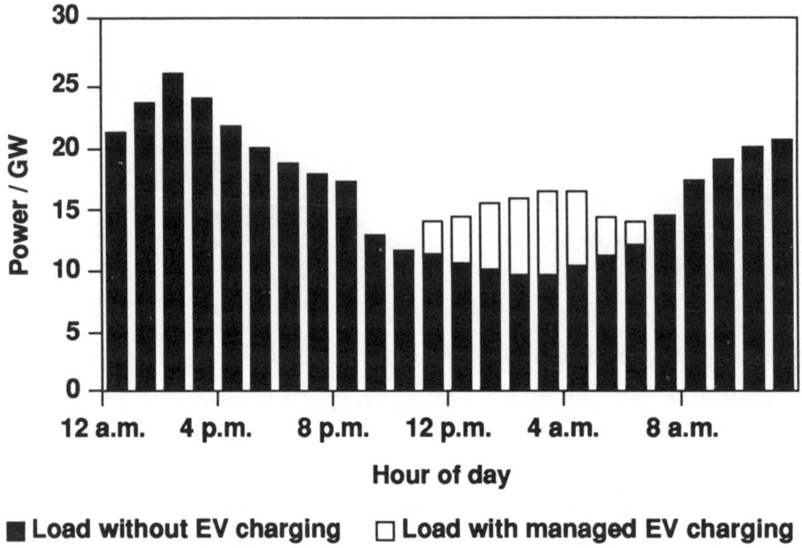

Fig. 10.6. A projected utility demand profile for year 2010: ■ without an EV charging demand; □ with a managed EV charging demand. (Taken from [17] with permission of Butterworths & Co. Publishers (Limited).)

The addition of a large number of battery EVs to the road transport fleet will create a substantial new market for electricity. Studies in several countries have shown that, with plausible assumptions about the rate of EV introduction, there will be no problems with meeting the demand for overnight charging, either in terms of generation or distribution. This is illustrated in Fig. 10.6 by the projected demand profile (year 2010) for a particular utility with and without a large, night-time charging load for two-million EVs. Indeed, as stated above, the utilities would welcome such an extra overnight load, which could be supplied at favourable tariffs. The same would certainly not be true for day-time charging during peak hours. A major national fleet of EVs which engage in opportunity charging at high rates would necessitate the progressive introduction of fresh generating, transmission and distribution systems. This added day-time demand would inevitably lead to an increase in the differential between day and night tariffs, either in terms of kWh consumed or kW supplied. This is a cost factor which has to be included when assessing the relative economics of short-range batteries with opportunity charging (lead/acid, nickel/cadmium), and of long-range advanced batteries which may only require overnight charging. One suggestion that has been advanced to reduce peak demand by EVs is to install land-based flywheels at the charging stations as energy-storage units. These would be recharged during off-peak periods. Work along these lines has begun in California.

10.5.3. EV servicing

The present road transport scene includes not only a network of petrol stations, but also a network of vehicle repair garages. The latter, in turn, are backed up by certified training programmes for mechanics and technicians. Clearly, the service and repair of EVs require rather different skills to those for ICEVs, i.e., knowledge of electrical as well as mechanical engineering, expertise in traction battery control and maintenance, and experience in dealing with high-voltage systems. This change is not expected to pose untoward difficulties, especially as it will be gradual. In recent years, the vehicle maintenance and repair business has adapted successfully to service the needs of ICEVs of increasing sophistication. The emphasis has been on electrical complexity because more and more electric motors and electronic circuitry have been introduced into cars to enhance comfort, safety and fuel economy. These developments are likely to continue, especially in the areas of security systems and automated vehicle control, and will translate directly to EVs. It may be envisaged that only technicians with a diploma in high-voltage maintenance will be permitted to work on the high-voltage side of the EV (traction battery, battery controller, and traction motor). Overall, there is confidence that the vehicle-maintenance industry will continue to evolve and will readily meet the servicing demands of EVs.

10.5.4. EV road accidents

Paradoxically, one of the greatest dangers posed by EVs may lie in one of their real virtues for urban use, namely, silent operation. An unanswered question which needs to be addressed is how many potential road accidents involving pedestrians are avoided as a result of the audible approach of ICEVs? Any increase in pedestrian accidents would be totally unacceptable and it may be necessary to provide some audible warning for EVs.

More attention has been directed towards road accidents in which there is major structural damage to the EV. In such situations, there are two types of potential hazard, namely, electrical and chemical. The electrical hazard associated with high-voltage operation is no different from that with mains EVs (i.e., trams, trolley buses, trains), although the latter are more sturdily built than cars and vans and are less frequently involved in accidents. It is felt that the electrical hazard in an EV accident situation may be readily controlled, possibly by some form of isolator switch which is impact initiated.

Rupture of the battery will present a chemical hazard. Its severity will depend entirely upon the battery chemistry which is involved. With lead/acid and alkaline batteries, there would be the dangerous release of strong sulfuric acid and strong potassium hydroxide, respectively. (Note, however, this possibility has been reduced substantially for lead/acid batteries by the trend towards valve-regulated technology in which the acid is immobilized in either a gel or an absorptive glass-mat, see Section 5.2.7, Chapter 5.) Experience with both lead/acid and nickel/cadmium batteries has generally been favourable, however, and there seems little likelihood that anyone other than the vehicle occupants will be exposed to any risk of skin burns. Added protection can be provided by placing absorptive material both in and around the battery. The hazard to the vehicle occupants is certainly less than that posed by a tankful of petrol.

Advanced batteries which are based upon more reactive chemicals (sodium, lithium, bromine) introduce a new class of potential danger. It may be taken as axiomatic that such chemicals will not be employed until proved safe in normal use and, therefore, only the vehicle accident situation needs to be addressed. The toxicity of chlorine, released as a gas, was one of several reasons why development of the zinc/chlorine battery was abandoned. Bromine is similarly toxic, but it has the slight advantage of being a volatile liquid at ambient temperature and, moreover, is contained as a complex in present zinc/bromine battery technology. There has been one serious accident in which bromine vapour leaked from a racing car powered by a zinc/bromine battery whilst engaged in a race in Arizona on a day when the ambient temperature was 44 °C [18]. It was reported that several of the spectators were overcome by bromine fumes and were admitted to hospital, although none seems to have been seriously injured. As a result of this accident, further safely assessments have been conducted and a consensus is developing that, in principle, zinc/bromine batteries can be made safe (see Section 7.3.8, Chapter 7).

The safety aspects of sodium-based traction batteries (Na/S, $Na/NiCl_2$) have attracted special attention, both because of the known hazards of sodium and because these batteries operate at high temperatures. Such problems for these two batteries in normal operation are discussed in Chapter 8. In the case of battery fracture, there is concern over the possibility of a major fire as well as an explosive reaction between sodium and water. Although there have been several fires involving sodium/sulfur traction batteries, in which the vehicle has been either damaged or destroyed, these have not been the result of vehicle accidents. Severe accident simulation tests of both sodium/sulfur and sodium/nickel chloride batteries have been conducted (Chapter 8), and have provided strong evidence that the batteries can be designed to be safe in the event of mechanical damage. Further work on crash simulation will be needed to demonstrate high levels of safety and build up public confidence as final designs of EVs powered by sodium batteries emerge. Ultimately, if these vehicles come into widespread use, it will be necessary to train the emergency services how to respond to accidents which involve crushed or ruptured sodium batteries.

10.5.5. Battery supply, recycling and disposal

As part of the battery–EV infrastructure, it will be necessary to have in place a scheme for the supply and fitting of traction batteries (both new and replacement) to vehicles, the removal and making safe of spent batteries, and the recycling or disposal of battery components. For both safety and environmental reasons, it is highly unlikely that statutory approval for large-scale manufacture and use of advanced traction batteries would be given until such a recycling scheme was in place. Disposal and recycling issues have been addressed by the Department of Energy in the USA through the setting up of a joint government–industry group (designated the 'EV Battery Readiness Group'). In addition, the International Seminar on Battery Waste Management is held annually in Florida.

Two of the heavy metals that are most commonly employed today in rechargeable batteries, namely, lead and cadmium, are toxic materials unsuited to waste disposal. Thus, for environmental as well as economic reasons, lead/acid and nickel/cadmium batteries are mostly dismantled and recycled. The technology for the recovery of metals from these batteries is well established. In the USA, for example, 95% of all lead/acid batteries used for the starting, lighting and ignition of ICEVs are recycled. Unfortunately, the recycling rate of small nickel/cadmium cells used in consumer applications is not so high because the economics are less favourable and the recycling infrastructure is not always so well established. The materials present in advanced secondary batteries (e.g., Zn/Br_2, Zn/air, Ni/MH, Na/S, $Na/NiCl_2$, LiAl/FeS, lithium-ion) are intrinsically less toxic than heavy metals and, on environmental grounds, the need for recycling is not so pressing. Nevertheless, with large-scale implementation of EVs, spent batteries would constitute a valuable resource of the component metals. Thus, any battery system employing an expensive metal (e.g., nickel) as a major component is almost certain to be recycled.

The handling of batteries which operate at high temperature (Na/S, $Na/NiCl_2$, LiAl/FeS) will require particular attention. In so far as these batteries will be conveyed from the factory to the point of use after cooling to ambient temperature, they do not present an intrinsic hazard in transit, although the possibility of an external fire may still cause the regulatory authorities to favour a discharged battery. (Note, this is principally a problem with sodium/sulfur since the other two systems can be assembled in the discharged state.) The battery re-heating stage is slow (many hours) and provision in the infrastructure will have to be made for this, as well as for many hours charging of a discharged battery. The entire infrastructure issue for hot sodium batteries is under scrutiny in the USA; working parties have been set up to consider transport to point of use, safety in normal operation and under accident conditions, and the dismantling and recycling/ disposal of used batteries.

Rechargeable lithium-metal batteries which are designed to operate at ambient temperature are known to present a special hazard. At present, such batteries are only available commercially in small sizes for consumer applications and, for transport on aircraft, the quantity of lithium metal that may be contained in a cell is subject to strict regulations. Greater assurance of safety would be needed before traction-size lithium batteries could become an item of international commerce. From this standpoint, as well as from a technical standpoint, lithium-ion batteries have much better prospects than lithium-metal batteries (see Section 9.4, Chapter 9).

With any large-scale use of traction batteries in EVs there will be a need for an internationally agreed code of practice to take account of all aspects of battery supply, fitting, removal, dismantling, and recycling/disposal. This code of practice would be specific to the given battery chemistry.

10.6. National / Local Perspectives

The interest of national governments and local authorities (state or regional governments, city councils) in EVs is based on one or more of the following three considerations: concern over supplies of liquid petroleum fuels; urban environmental pollution; transport strategy within cities. Each of these merits brief discussion.

10.6.1. Petroleum fuels

Concern over the long-term future of petroleum supplies first arose in 1973 as a consequence of the 'oil crisis' in that year. This crisis stemmed from the activities of the oil-producing countries who argued, reasonably enough, that the price of oil was far too low since it reflected the production cost rather than the value to the consumer. Oil from the Middle East and parts of the USA was very cheap to produce; all that was necessary was to sink a pipe and extract the oil from underground reservoirs. Thus, the producing nations received little income (not more than a dollar or two a barrel).

Around 1973, the world demand for petroleum approached the supply which was immediately available. Thus, the oil-producing nations were in a position to establish a cartel (the Organization of Petroleum Exporting Countries, OPEC) in order to force up prices. The number of oil-producing nations which became members of OPEC was sufficient to cause sharp rises (about four-fold) in the price of crude oil. This had the immediate effect of making many Middle Eastern states wealthy and resulted in significant increases in the pump price of petroleum fuels, and even some fuel shortages. In turn, oil-importing nations focused on the influence of the price rises on their international balance-of-payments. In addition, major oil-consuming nations began to question the sustainability of world oil supplies, given the ever-increasing demand for liquid petroleum fuels.

An additional effect of the steep price rises was to encourage the oil exploration companies to capture new sources of petroleum in sites which, formerly, were not regarded as economically viable. For example, the start of the off-shore oil industry in the North Sea and on other continental shelves may be dated to the 1970s. Soon, the supply and the demand for oil were back in balance, at the new prices, although there was a second, lesser, 'oil crisis' in 1978–9 that led to further price increases and again attracted the attention of national strategists.

A further outcome of these events was that many governments set up departments or ministries to develop national energy policies. Broadly, the role of these bodies was to investigate measures for energy conservation and petroleum substitution, and to encourage their adoption where economically viable. Interest in these twin topics was by no means confined to governments — industry also took a keen interest in reducing its energy costs through conservation measures. During the 1980s, the European Commission and the US Department of Energy ran several research and development programmes which were directed towards both energy conservation and the exploitation of new energy sources (including renewables). The development of EVs, and advanced batteries and fuel cells, fell within this broad energy framework.

In parallel with the above activities, events on the broader energy front were progressing throughout the 1980s and 1990s. Some of the principal developments were:

- the democratization of countries in Eastern Europe and, thereby, the opening up of huge new sources of natural gas in Siberia for supply to Europe, as well as the prospect of an enlarged market for petroleum products
- world-wide new discoveries of petroleum resources and, especially, of natural gas
- advances in the technology of extracting petroleum economically from deep-sea reservoirs
- the eclipse of nuclear power in many countries (although, interestingly, France, Japan and some Asian countries are still fully committed to this energy source)
- the development of combined-cycle gas-turbine technology and the use of natural gas for electricity generation
- major energy conservation measures in industry and improvements in the thermal insulation of buildings
- the introduction of smaller ICE cars with much higher energy efficiency
- increasing use of diesel and liquid petroleum gas to fuel road vehicles (in some countries).

Most of these developments were only glimpsed, or not foreseen at all, in the 1970s. The result, together with the impact of other advances in energy supply and conservation, is that now (i.e., the late 1990s) there is a more than adequate supply of both oil and natural gas to meet current world demand. Therefore, the interest in EVs from an energy substitution standpoint has largely evaporated. Nevertheless, this is certain to be a short-term situation. Although some of the energy supply and demand forecasts of the late 1970s have proved to be unduly pessimistic, it is only a matter of time, with rising world population and expectations, before petroleum once again will become in short supply and more expensive. Although it is difficult to predict when this will happen, and at what price level oil from new fossil fuels (e.g., tar sands, oil shales) will become competitive, the long-term prognosis must be for increasing prices. Clearly, this will again exacerbate balance-of-payments worries for oil-importing countries. Thus, the interest in EVs from the standpoint of petroleum substitution is expected to return.

10.6.2. Environmental factors

Atmospheric pollution from road transport gives rise to concern at three levels: global, regional and local. The relative contributions to global warming, regional 'smog', and local pollution by a petrol engine (with/without catalyst), a diesel engine, and an EV are shown in Fig. 10.7 [19]. The data are normalized to a modern petrol engine with a catalyst.

Global concern centres around the possibility of climate change through build-up of the so-called 'greenhouse gases', e.g., carbon dioxide, nitrous oxide and methane. The substitution of ICEVs by EVs does little to ameliorate this problem — the pollution is merely transferred from the tailpipe of the car to the electricity-generating station. Naturally, this is not the case if the electricity is of nuclear origin, or derived from renewable sources (wind, solar, tidal, hydro), and EVs would then make a positive contribution towards reducing the 'greenhouse effect'. In the more general case of electricity produced from fossil fuel, the precise comparison of EVs with ICEVs would depend on factors such as their overall relative fuel efficiency (traction-energy/thermal-energy of the fuel), the efficiency of clean-up of exhaust gases at the power station compared with that at the tailpipe of the ICEV, and so on.

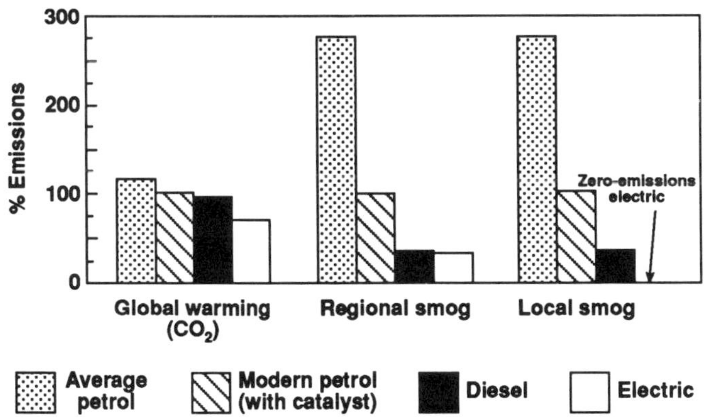

Fig. 10.7. The impact of different road vehicles on global, regional and local air-quality. (Taken from [19] with permission of UK & International Press.)

Regional atmospheric pollution ('smog'), as experienced in the Los Angeles basin and other large conurbations, results from photochemical reactions between the gaseous emissions and oxygen in the air, coupled with an atmospheric inversion layer which prevents the dispersion of the gases. Photochemical smog is now the subject of much concern in many major cities — particularly those which have sunny climates and are surrounded by hills — and results in serious health problems.

Local pollution is that experienced by drivers, cyclists and pedestrians when caught up in heavy traffic in cities, or even in traffic jams outside of cities. In these situations, the hazard arises from direct exposure to tailpipe emissions, i.e., carbon monoxide, unburnt hydrocarbons and NO_x from vehicles with inefficient catalytic converters, and particulates from diesels. This local pollution can be very unpleasant, if not positively injurious to health. Even today, a significant proportion of the petroleum sold is of the leaded variety, despite the evidence of the harm that it may do, particularly to children. It is also important to realize that, in addition to endangering the health of the community, pollution from the exhaust of ICEVs causes considerable damage to buildings and cultural heritage.

Throughout the world, governments and civic authorities have come to recognize the urgent need to develop strategies to combat both regional and local atmospheric pollution. A leading role in addressing these concerns has been taken by the USA, and in particular by the State of California which has enacted legislation in an attempt to address the air-quality issue (see Section 1.1, Chapter 1). Under this legislation, major manufacturers of ICEVs will be required to sell a set number of zero emission vehicles (ZEVs) in order to be permitted to market conventional vehicles in the State. For all practical purposes, ZEVs equate to EVs. Initially, the ZEV requirement was for 2% of all new vehicles offered for sale by each major manufacturer by 1998 and 10% by 2003, but the former figure has since been relaxed while the latter has been retained.

There is no doubt that the Californian law is focusing sharply the minds of those vehicle manufacturers who do not wish to be excluded from an important market. This may well be largely responsible for many of the recent EV and battery development programmes. Latterly, therefore, the impetus for EVs has swung markedly from petroleum substitution to abatement of atmospheric pollution. Nevertheless, a conflict remains between citizens and politicians on the one hand, who feel that the only way to tackle the pollution problem is by way of legislation (as was largely successful with catalytic converters for ICEVs), and many industrialists on the other hand, who consider that mandates to buy EVs (which, at the present stage of development, they see as inferior and expensive transport options) is not the correct approach. Certainly, if the Californian legislation holds and is copied elsewhere, the required EVs will be expensive to manufacture in the early years, at least until such time as the level of production increases markedly and the resultant economies-of-scale take effect.

10.6.3. Urban transport

In attempts to alleviate atmospheric pollution and road congestion, a growing number of city councils are adopting restrictive practices towards the entry of ICEVs to city centres. These may take the form of prohibitions (e.g., pedestrian precincts, or roads restricted to public service vehicles only) or disincentives (e.g., limited availability of parking at high prices). Often, the limited parking is associated with 'park and ride' schemes, whereby cars are left near the city boundary in large public parking facilities from where there are rapid and frequent bus services into the city centre.

These measures have not provided, however, a complete solution to the urban-transport problem. For instance, city workers who have private parking facilities at their place of work still drive their ICEVs into town; delivery vehicles need access to shops and offices, as do disabled people who are unable to use buses; streets have to be cleaned and refuse collected. Many cities, therefore, have a real interest in EVs as a means of further improving air quality. Among European cities that have been active in trials of electric buses are Esslingen, Dusseldorf and Mönchengladbach in Germany, and Oxford, Manchester and Sheffield in the UK. Delivery vans were extensively explored in London (UK) in the early 1980s in the 'London Goes Electric' programme. Many cities in France, Italy, the Netherlands and elsewhere have been engaged in trials of private electric cars, with one of the foremost exponents being La Rochelle in France. In Germany, there has been a major trial of EVs of various types, equipped with both traditional and advanced traction batteries, on the island of Rügen in the Baltic Sea. The latest European venture is the CITELEC scheme, whereby cities in different countries are co-operating in EV trials and exchanging experiences. All such major projects must involve co-operation between EV manufacturers and local authorities, often with grants-in-aid being provided by national governments.

The European activities are paralleled by EV demonstrations in many other countries, notably Japan and the USA. Australia also has a growing interest in EVs for use in large conurbations and, incidentally, hosts the well-known international solar electric car race across the 'Outback' from Darwin to Adelaide (see Section 1.4, Chapter 1). In the 1996 race, five different types of battery system were used, namely: lead/acid (25 cars), silver/zinc (13 cars), nickel/zinc (5 cars), nickel/cadmium (2 cars), and lithium-ion (1 car) [20].

In order to encourage the use of EVs in the city environment, various national and regional governments are setting targets and providing incentives. The Californian initiative has already been mentioned. In France, the government has aimed for 100 000 EVs by the year 2000 and, as part of a package of support measures, is offering a rebate of FF 5000 to purchasers of EVs [15]. The Swiss government has a target of 200 000 EVs by 2010, these would represent 8% of the total vehicle fleet [21]. The UK has not proposed a specific goal, but encourages the use of EVs by waiving the annual licence fee for road vehicles. At present, this amounts to a saving of £ 140 per annum for cars and more for larger vehicles. Clearly, these and similar politically motivated initiatives will hasten the introduction and widespread urban use of EVs (see Section 10.9).

10.7. Manufacturers' Perspectives

There has been uncertainty as to the level of commitment of the major ICEV manufacturers to the building of EVs. Have their activities to date been motivated by a desire to demonstrate environmental concern (e.g., to meet the Californian legislation), or just to show the public the limited performance and high cost of state-of-the-art EVs in the expectation that no large competitive markets will then emerge? Such cynicism, while perhaps arguable ten years ago, is hardly in keeping with the major investments which are now being made in EV technology by such large motor corporations as Daimler-Benz, Fiat, Ford, General Motors, Honda, Peugeot, Renault, Toyota, and Volkswagen. Electric cars and vans are presently available in small numbers for either purchase or hire from several of these manufacturers. The urgent need to address the environmental concerns which emanate from ICEV operation has now been accepted by the motor corporations who, accordingly, are taking all possible steps to resolve the problem. The most important of these steps is to improve the fuel efficiency and, thereby, to reduce the emissions from conventional ICEVs but, realizing that this does not completely solve the urban air-pollution issue, a strong interest has now developed in electric traction, either EVs or HEVs.

One matter on which all the manufacturers are agreed is that the car is a consumer product and that sales will be determined by market forces and not by government mandates. The most that governments and local authorities may hope to do is to provide a political framework, involving incentives for EVs and restrictions on the use of ICEVs, within which a choice may be made.

There are two basic approaches to the manufacture of an EV, namely, the conversion of a standard-production ICEV to electric propulsion, and the custom design of an EV. The adaptation of ICEVs is seen to offer the following advantages:

- there is no need for a dedicated production line
- many of the vehicle parts are common (e.g., in the Volkswagen 'CitySTROMer', an EV adaptation of the Golf ICEV, 90% of the parts are common [22])
- there is flexibility in the number of vehicles of each type produced
- the build may comply with the same standards of quality, reliability and materials recycleability
- the production cost is held to a minimum.

Most series runs of EVs have taken this approach. Notable examples are (see Chapter 1): the Lucas Chloride Bedford CF van; the General Motors G-Van; the Chrysler TEVan; the Renault Clio car; the Fiat Panda car; the Mercedes 190E.

The second approach, i.e., custom designing the EV from 'the-wheels-up', has two important advantages:

- the design can be optimized both for electric drive and for battery accommodation, as well as for a particular market sector (e.g., the commuting, family run-around vehicle)
- there is the opportunity to use lightweight materials and new construction techniques in initially short production runs which would not impact unduly on world commodity prices.

Against these benefits must be set the higher cost of a custom-designed vehicle. Examples of this approach are the General Motors 'Impact' car and the Peugeot-Citroën 'Tulip' city car.

The EV manufacturer has many other component decisions to make, namely:

- whether to use d.c. brushless motors, or a.c. motors in conjunction with an inverter
- whether to have a single motor in the drive-train or individual wheel motors
- whether or not to have a gearbox
- what type of battery controller and power-conditioning equipment to employ
- where to locate the battery in the vehicle chassis
- whether the battery and/or electric motor need(s) to be cooled
- whether or not to have an on-board charger
- how to heat and air-condition the cabin.

These and similar engineering issues are beyond the scope of this book. Suffice it to say that the prospects for EVs are of interest to a wide range of manufacturing industries besides the motor companies. Potentially, there is much new business for companies engaged in electrical engineering, electronics engineering, and the supply of advanced materials (plastics, composites, aluminium extrusions, etc.), as well as in the manufacture of traction batteries.

The impact of a commercially successful EV on the battery industry would be enormous, purely on account of the large unit size of traction batteries. Even a modest market penetration by EVs would create a whole new secondary battery industry, probably based eventually upon a new battery chemistry and, therefore, opening up new business opportunities for suppliers of raw materials such as lithium, nickel, rare earth metals, etc. There seems to be a consensus that the target cost of an EV before fitting its battery should approximate to that of a complete ICEV ready for the road.

This leaves the added cost of the traction battery to be found, which is in the range of US$ 2000–5000 given the type of vehicle and its battery size. Moreover, battery replacements will be necessary during the life of the EV itself. The usual response to the intrinsic problem of the high cost of battery packs is to point out that petrol is considerably more expensive than electricity, and that EVs may be expected to be rather more energy efficient than ICEVs. Thus, the increased capital outlay for the vehicle–battery package will be offset by reduced running costs, both in terms of fuel and, hopefully, maintenance. In general, though, people prefer to pay on-going charges rather than to find up-front capital. Also, it is probable that customers initially would be reluctant to make a major investment in a new battery technology which is untried in the market-place. It follows that the most likely scenario is for traction batteries to be leased rather than sold outright, with the responsibility for lifetime and maintenance remaining with the manufacturer. The aim is for the leasing charge to equate to the cost savings in using electricity rather than petroleum as fuel so that running an EV would be no more expensive than running an ICEV. Whether this target can be achieved remains to be seen, although it should be borne in mind that financial goals, unlike technical goals, tend to change with time in a manner which depends upon the prevailing economic and political climate.

10.8. Consumers' / Users' Perspectives

The purchaser of an EV, or indeed any other vehicle, is primarily interested in fitness for purpose and value for money. The purchaser may also be the driver of the vehicle, or may be a fleet owner. In the latter case, the priorities of the purchaser and the employed driver will not necessarily coincide. For example, the driver is more concerned with comfort and convenience, but less with cost, than the purchaser. Nevertheless, both are likely to be more influenced by consumer considerations than by the broader fuel-supply and environmental issues outlined in Section 10.6. The latter issues will impact the buying decision only indirectly, in so far as they affect the economics of vehicle operation or relate to restrictions placed upon the operation of ICEVs in cities.

General Motors has conducted 'voice of the customer' research and has identified the following as the top seven explicit purchase factors demanded for EVs [22]:

- daily range
- recharge time
- battery cost
- battery life
- availability of remote charging
- home charging
- charging method employed.

Presumably, the customer has taken for granted that other factors such as vehicle performance and traffic compatibility, vehicle purchase price, and cabin heating and air-conditioning will be comparable with those of ICEVs.

With respect to the above seven purchase factors for EVs, both range and refuelling time are acceptable for most urban duties although, clearly, these two performance characteristics cannot match those of ICEVs. The cost of the batteries is also a particular concern of battery developers. At the prototype production stage, the advanced batteries are especially expensive. For example, early nickel/metal hydride traction batteries cost more than US$ 2000 per kWh, although forecasts of meeting the USABC target of $150 per kWh for full-scale production have been made [23]. Lithium-ion batteries are even more expensive, although again optimistic projections have been put forward [1]. Sodium/nickel chloride (ZEBRA) batteries are currently being sold for fleet testing at a price similar to that of nickel/cadmium, although estimates for series-production indicate a lower price level that will be comparable with lead/acid [24]. Obviously, for all types of battery, a long cycle life is necessary in order to spread the unit cost over as long a period as possible. The remaining three purchase factors relate to battery charging. In particular, the availability of conveniently located recharging points throughout the city will play a vital role in meeting customer demands. A characteristic favouring the lead/acid battery is that fast recharge has been demonstrated to be quite feasible (see Section 5.5.4,

Chapter 5), whereas it is not so clear whether this is equally possible for the advanced batteries without the introduction of adverse side effects.

The EV has a number of special features which have been recognized by most people who have driven such vehicles. These features enhance the quality of the ride through smooth and silent operation, as well as the absence of fumes when the vehicle is stationary in traffic. Moreover, EVs display fast acceleration from a standing start, a common characteristic of electric traction (cf., electric trains).

10.9. Reconciliation of Perspectives

As seen earlier in this Chapter, many of the advantages of EVs accrue to society as a whole, while most of the disadvantages are borne by the owner or driver. These advantages/disadvantages are summarized in Table 10.3. Not all of the disadvantages apply equally to all batteries. For instance, the high-temperature batteries are independent of ambient temperature, but are more difficult to access should they require servicing. On the other hand, aqueous electrolyte batteries have the opposite characteristics.

In general, if society is to gain from the perceived advantages of EVs, it will have to develop a framework which will encourage the customer to purchase EVs. It is not thought that government mandates alone are likely to achieve this goal. Rather, the framework is likely to consist of a package of incentives for the EV user and restrictions on the ownership and use of ICEVs, particularly in an urban context. What might such a framework look like?

Incentives for EVs:

- Government subsidies for the purchase of EVs (such as those presently being provided in France and Switzerland), reduction of value-added tax or sales tax on EVs, and/or zero rating the vehicle road tax (as in the UK).

Table 10.3. The impact of EVs.

Advantages	Disadvantages
• Reduce urban air pollution • Silent in operation • Conserve oil supplies • Diversify the energy base of transport • Help national balance-of-payments by reducing imports • Help to load-level the electricity supply	• Limited daily range • Higher capital cost • Cost of domestic chargers • Extended recharge times • Availability of recharge points • Battery care and maintenance • Restricted ambient-temperature range of operation • Need for auxiliary heating and air-conditioning units

- The provision of an extensive infrastructure of charging points at popular locations and assistance with the installation of chargers in the home.
- Cheaper, night-time, electricity tariffs in which the financial benefits accruing to the utilities through increased load-levelling are passed on to EV owners.
- Free parking and/or the provision of parking places restricted to EVs.
- Permission for EVs to use car-pool and/or bus lanes; this would reduce journey times in heavy commuter traffic.

Restrictions on ICEVs:

- The banning of ICEVs from city centres on the grounds of pollution would certainly stimulate the purchase of viable EVs. An alternative to outright banning would be severe limitations on the number of parking places or very high parking tariffs.
- Increasing the tax on petroleum fuels would be a further strategy. A possible approach would be for governments to give advance warning of annual increases in petroleum tax over a 5–10 year period as the production capability for EVs is stepped up and the associated infrastructure is developed. This would facilitate a smooth and orderly transition to an urban EV society.

Clearly, in a democracy, none of these measures will be possible unless they are supported by a majority of the citizens through their elected representatives. It follows that the large-scale introduction of EVs into cities will be a matter of establishing an acceptable economic framework to stimulate vehicle ownership. This, in turn, will be predominantly a socio-political matter. By no means are all of the above measures needed to ensure a level playing field between EVs and ICEVs; it will be up to each national government and local administration to determine its own policy. In essence, the role of government is to aid industry in building customer awareness and acceptance of EVs as a new form of personal transport for the twenty-first century.

10.10. A Realizable Scenario

Forecasting the future is notoriously difficult, whether it be a forecast of technological, social or political developments. As the future of the EV in urban transport is dependent upon all three, to a greater or lesser degree, it is even more difficult than usual to make reliable predictions. The best that can be done is to engage in 'crystal-ball gazing' with a view to projecting a realizable scenario.

In 1979, a forecast was made of the future prospects for EVs in the European Community [25,26]. It was predicted that the combined sales of electric vans and cars, as a proportion of total sales, might be 5% by the year 2000 and 15% by 2025. It was further suggested that the most optimistic and pessimistic variants would be a factor of three higher and lower, respectively. These forecasts were made predominantly in the light of the 1973 and 1978 oil crises and the likely international response in the energy field; little regard was paid to urban pollution considerations. Now that the world is approaching the year 2000, it is unlikely that even the lower limit will be reached; the European programme has slipped by a decade or so and the justification for it has changed. With hindsight, this is hardly surprising in the light of the extraneous developments in oil supply–demand outlined in Section 10.6.1, together with the protracted time-scale for advanced battery development. Instead, urban pollution has become the driving force for EV development. A forecast, based on the Californian zero emission mandate and its adoption by the 12 North Eastern States, is that by 2003 there will be upwards of 460 000 new EVs registered annually in the USA with an integrated total of 1.4 million. By 2010, there could be as many as 14 million EVs in the USA. Although this would represent a very high growth rate, it would be no greater than that already achieved by many other new technologies (cellular telephones, video recorders, personal computers, car stereo systems, etc.). In the yet longer term, it is inevitable that oil prices will escalate as easily exploitable resources are exhausted and demand increases. Precisely when such events will happen is still a matter of conjecture.

The growing world consciousness for environmental protection has now led undeniably to an international growth of interest in EVs — both by industry and by government — and many motor corporations are presently building and leasing/selling delivery vans and small cars. Furthermore, this activity is increasing rapidly. In France, for example, several hundred Renault 'Clio electrics' have been manufactured and sold, in addition to Peugeot 106 electrics. Meanwhile, at least 20 companies in Japan are engaged in EV development programmes. Clearly, the EV bandwagon is beginning to roll and developments world-wide are likely to accelerate.

It is notable that before the advent of the California mandate, most EV developers were focusing on delivery vans rather than cars. The operation of ICEVs is poorly suited to door-to-door delivery duties. When used in this mode, fuel consumption is high, engine life is short, and maintenance is required frequently. Also, stationary vehicles left with engines running are irritating to residents and passers-by. These are precisely the reasons why, many years ago, dairies in the UK adopted basic EVs for house-to-house milk deliveries. For similar reasons, the post offices of the world, who also offer a door-to-door delivery service, are seriously interested in EVs. Trials have been carried out by postal authorities in many countries and it seems likely that, as EVs continue to improve in range and cost-competitiveness, postal delivery will be an early and significant market for electric vans. Utility meter readers (electricity, gas, water) are other door-to-door callers who would appreciate the convenience of small EVs, although it is possible that, with developments in data communications, such readings will become automated.

Next in terms of prospects for on-the-road EVs are those vehicles whose stops are more widely separated, but still frequent, e.g., urban/suburban parcel delivery services, catering, linen/laundry delivery services, florists, service engineers, tradesmen. Two classes of commercial vehicle have been developed to meet this market, namely, light vans (1-t payload) and medium vans (2-t payload). For these services, a traffic-compatible vehicle is needed with a longer daily range than is necessary for the strictly house-to-house caller. The anticipated new generation of lead/acid traction batteries will suffice for many of these duties. Thus, the choice between EVs and ICEVs will be dictated by considerations of cost, convenience, and legislative incentives/restrictions.

Urban buses present a further promising market for electric traction. Many buses are in use for up to 18 h per day and cover correspondingly long ranges. This market sector would require an advanced battery, such as zinc/air, or a fuel cell to be viable. Both in the USA and in Europe, fuel cells are seen as being most appropriate for buses and other long-range vehicles. Specialized buses, such as airport and hotel courtesy shuttles, cover much shorter daily ranges, and lead/acid batteries readily meet the range and power requirements. In general, buses for both urban and rural use may be ideal candidates for hybrid traction. In Japan, for example, there are about 100 diesel/battery–electric hybrid buses on city bus-routes in Osaka and Tokyo. An interesting hybrid bus is being developed by a government–industry consortium in Australia for service in the environs of Brisbane; the bus is equipped with an innovative, combined mechanical-drive and regenerative-braking system to provide power for acceleration and most hill-climbing duties, and a gelled-electrolyte, valve-regulated, lead/acid battery to supply power for cruising.

There are several occupations with duties that primarily involve house-to-house calls in the urban/suburban area. These include doctors, nurses, social workers, ministers of religion, insurance salespersons, and financial advisors. The limited mileage travelled daily would make the electrically-propelled car an attractive option for such users. A cautionary note should be added, however, that often these professional cars are used also for private purposes which involve much longer distances.

The near-term future for privately owned electric cars is considerably more problematical. It is frequently stated that the electric car would be ideal for commuting or as a second car in a family. Whilst it is true that the electric car could perform these duties, what is overlooked is that the second car often doubles as the first car when the latter is unavailable and, therefore, may be called upon occasionally for longer journeys. Moreover, economic circumstances often dictate that the second car is a comparatively old and, therefore, cheap model; this is not a market sector well suited to the EV. A better prospect may lie in the middle–upper income family where capital is available to purchase an electric car purely for shopping/commuting/school-trip purposes, with one or more ICEVs available for longer journeys. In point of fact, the private electric car may have greater appeal as the third car in a family rather than the second and this may represent a viable market sector, particularly in the USA where the ownership of three cars is commonplace.

An interesting niche market could be the Neighbourhood Electric Vehicle (NEV). This is a vehicle which is designed to provide low-speed, personal transportation in restricted areas. Examples would be university campuses, hospitals, theme parks, industrial parks, holiday resorts, residential communities and, of course, city centres. The NEVs would be restricted to speeds of 40–50 km h^{-1}. Various small electric cars with lead/acid batteries are being designed to meet this niche market. Already, the State of Arizona has enacted legislation specifically to cater for NEVs.

The above discussion relates to EVs propelled by a new generation of lead/acid batteries which will have higher specific energy, longer life and lower cost than hitherto. In this respect, lead/acid will serve as the trailblazer for advanced technologies which are waiting in the wings. The latter batteries will provide the long range which is required for non-urban use and will promote the widescale use of EVs. Nevertheless, the jury is still out regarding which battery technology will prove to be the best for pure-battery vehicles, or for hybrids. Everything depends on the levels of technical performance and cost that can be achieved by each of the competing candidate systems.

Overall, there will be a move towards smart batteries with computer-controlled management systems in which sensors will continuously monitor the battery state-of-health and maintain the discharge of each cell in harmony with others in the vehicle pack. There will also be intelligent charging systems which will provide the maximum amount of charge at the best rate so that the battery pack can deliver a dependable supply of energy. Indeed, the revolutionary progress being made in both electronics and information technology is reinforcing research into advanced traction batteries and alleviating many of their traditional problems. Looking still further ahead, it seems possible that, with improvements in the efficiency of solar cells and falling costs of their manufacture, the ultra-lightweight hypercar of the future will derive some of its propulsion energy from solar-generated electricity. Thus, with all these developments in the pipeline, we see a bright future for battery electric vehicles as a major component of the transport scene.

References

1. KALHAMMER, F.R., KOZAWA, A., MOYER, C.B., and OWENS, B.B. *Performance and Availability of Batteries for Electric Vehicles: A Report of the Battery Technical Advisory Panel,* paper prepared for the California Air Resources Board, El Monte, California, USA, 11 December 1995.

2. PRATER, K.B. *J. Power Sources,* 1996, **61**, pp. 105–109.

3. NARANG, S., VENTURA, S., and SMEDLEY, S. *Electric & Hybrid Vehicle Technology '95* (UK & International Press, Dorking, UK, 1995), pp. 131–135.

4. BLANK, E., FARAHMANDI, C.J., and TRIPPE, A.P. *Electric & Hybrid Vehicle Technology '95* (UK & International Press, Dorking, UK, 1995), pp. 135–137.

5. *Batteries International,* 1997, **30**, p. 11.

6. TOKARZ, F., SMITH, J.R., COOPER, J., BENDER, D., and ACEVES, S. *Electric & Hybrid Vehicle Technology '95* (UK & International Press, Dorking, UK, 1995), pp. 126–130.

7. WOUK, V. *Scientific American*, October 1997, pp. 70–74.

8. HOLMSTEAD, H.D. *Electric & Hybrid Vehicle Technology '95* (UK & International Press, Dorking, UK, 1995), pp. 175–177.

9. CAPPELLI, P.G., and BONINO, M. *Electric & Hybrid Vehicle Technology '95* (UK & International Press, Dorking, UK, 1995), pp. 52–55.

10. LOVINS, A.B. *Electric & Hybrid Vehicle Technology '95* (UK & International Press, Dorking, UK, 1995), pp. 43–46.

11. Advanced Lead-Acid Battery Consortium, *Electric & Hybrid Vehicle Technology '95* (UK & International Press, Dorking, UK, 1995), pp. 81–84.

12. MOSELEY, P.T. *J. Power Sources,* 1996, **59**, pp. 81–86.

13. MOSELEY, P.T. *J. Power Sources,* 1997, **67**, pp. 115–119.

14. LAM, L.T., OZGUN, H., LIM, O.V., HAMILTON, J.A., VU, L.H., VELLA, D.G., and RAND, D.A.J. *J. Power Sources,* 1995, **53**, pp. 215–228.

15. BUREAU, N. *Electric & Hybrid Vehicle Technology '95* (UK & International Press, Dorking, UK, 1995), pp. 20–22.

16. WOUK, V. *Batteries International,* 1994, **19**, pp. 52–53, 57.

17. FORD, A. *Energy Policy*, 1994, **22**, pp. 550–570.

18. GRIBBLE, D. *Batteries International,* 1992, **12**, p. 80.

19. ADAIR, R., and OWEN, D. *Electric & Hybrid Vehicle Technology '95* (UK & International Press, Dorking, UK, 1995), pp. 243–245.

20. MAYER, M.G., and RAND, D.A.J., *The Battery Man,* 1997, **39**, pp. 16, 18–22, 24, 29–31.

21. PIFFARETTI, M. *Electric & Hybrid Vehicle Technology '95* (UK & International Press, Dorking, UK, 1995), pp. 61–63.

22. BAKER, K. *Electric & Hybrid Vehicle Technology '95* (UK & International Press, Dorking, UK, 1995), pp. 38–40.

23. OVSHINSKY, S.R., FETCENKO, M.A., VENKATESAN, S., DHAR, S.K., HOLLAND, A., YOUNG, R., GIFFORD, P.R., CORRIGAN, D.A., NG, A., and TSANG, R. *13th International Seminar on Primary and Secondary Battery Technology,* 4–7 March 1996, Boca Raton, Florida, USA.

24. HAMMERLING, H., BOEHM, H., DUSTMANN, C.-H., TILLEY, A.R., and VAN ZYL, A. *13th International Seminar on Primary and Secondary Battery Technology,* 4–7 March 1996, Boca Raton, Florida, USA.

25. JENSEN, J., McGEEHIN, P., and DELL, R.M. *Electric Batteries for Energy Storage and Conservation,* Odense University Press, Odense, Denmark, 1979.

26. DELL, R.M., and JENSEN, J. *Drive Electric '80 Conference,* 14–17 October 1980; see also UK Atomic Energy Authority Report: AERE–R9958 (October 1980).

Abbreviations, Symbols and Units Used in Text

Abbreviations

a.c.	alternating current
AEVA	Australian Electric Vehicle Association
AFC	alkaline fuel cell
AGM	absorptive glass mat
AGV	automatic guided vehicle
ALABC	Advanced Lead-Acid Battery Consortium (USA)
ANL	Argonne National Laboratory (USA)
ASI	area-specific impedance
BAB	barium alumino-borate
BL	γ-butyrolactone
1BS	monobasic lead sulfate
3BS	tribasic lead sulfate
4BS	tetrabasic lead sulfate
CARB	California Air Resources Board
CSM	copper-stretch-metal
CVT	continuously-variable transmission
d.c.	direct current
DEE	diethoxyethane
DMC	dimethyl carbonate
DMFC	direct methanol fuel cell
DoD	depth-of-discharge
DOE	Department of Energy (USA)
DSEP	Dual-Shaft Electric Propulsion
DST	Dynamic Stress Test
EC	ethylene carbonate
EMD	electrolytic manganese dioxide
EV	electric vehicle

FUDS	Federal Urban Driving Schedule (USA)
FUDS79	Federal Urban Driving Schedule limited to 79 W kg^{-1} battery power (USA)
GSFUDS	Generic Simplified Federal Urban Driving Schedule (USA)
HC	hydrocerussite, $Pb(OH)_2 \cdot 2PbCO_3$
HEV	hybrid electric vehicle
ICE	internal-combustion engine
ICEV	internal-combustion-engined vehicle
IDSEP	'Improved' Dual-Shaft Electric Propulsion
IUPAC	International Union of Pure and Applied Chemistry
LBL	Lawrence Berkeley Laboratory (USA)
LEV	low-emission vehicle
LLNL	Lawrence Livermore National Laboratory (USA)
JRC	Joint Research Committee (Europe)
MA	methyl acetate
MCFC	molten carbonate fuel cell
MEMBr	N-methyl N-ethyl morpholinium bromide
MEPBr	N-methyl N-ethyl pyrrolidium bromide
MH	metal hydride
Mm	mischmetal (natural mixture of rare earth metals)
NBTL	National Battery Test Laboratory (USA)
NEV	Neighbourhood Electric Vehicle
OPEC	Organization of Petroleum Exporting Countries
PAFC	phosphoric acid fuel cell
PAN	polyacrylonitrile
PC	propylene carbonate
PCL	premature capacity loss
PEMFC	proton exchange membrane fuel cell
PEO	polyethylene oxide
PMMA	polymethyl methacrylate
PN	propionitrile
PTFE	polytetrafluoroethene
PVC	polyvinyl chloride
PVDF	polyvinylidene difluoride
RBSM	recombinant-battery separator mat (same as AGM)

RIMU	relaxable insufficient mass utilization
SAE	Society of Automotive Engineers
SoA	state-of-the-art
SoC	state-of-charge
SOFC	sold oxide fuel cell
SEI	solid electrolyte interface
SFUDS	Simplified Federal Urban Driving Schedule (USA)
SFUDS79	Simplified Federal Urban Driving Schedule limited to 79 W kg^{-1} battery power (USA)
SHE	standard hydrogen electrode
sp. gr.	specific gravity (relative density)
TCB	thermo-compression bond
TLEV	transitional low-emission vehicle
TULIP	Transport Urbain, Libre, Individuel et Public
UP	upper voltage plateau
ULEV	ultra-low-emission vehicle
UNSW	University of New South Wales (Australia)
USABC	United States Advanced Battery Consortium
VRLA	valve-regulated lead/acid battery
ZEV	zero-emission vehicle

Symbols and units — Roman letters

a	activity of species which is usually specified as a subscript (e.g., O, R, H^+, HSO_4^-)
atm	atmosphere
A	ampere
A_f	frontal area of vehicle
Ah	ampere hour (= 3600 coulombs)
b	gradient of a hill
°C	degree Celsius
C_A	charge acceptance
C_d	air-drag coefficient
C_T	capacity at temperature T (T may be quantified)

C_X/t	capacity of a battery at the X hour rate (X is usually quantified)
cc	cubic centimetre
cm	centimetre
cP	centipoise (= 10^{-3} pascals per second)
dm	decimetre
ΔG	Gibbs free energy of reaction
ΔG^o	standard Gibbs free energy of reaction
$\Delta G^{\ddagger}$	Gibbs free energy of activation
$\Delta G_a^{\ddagger}$	Gibbs free energy of activiation of an oxidation (anodic) reaction
$\Delta G_c^{\ddagger}$	Gibbs free energy of activation of a reduction (i.e., cathodic) reaction
e^-	electron
E^o	standard electrode potential
E_i	potential of electrode under a current load
E_r	reversible electrode potential
F	faraday (96 485 coulombs per mole)
F_a	force applied to accelerate a vehicle
F_h	force applied to climb hills
F_r	force applied in overcoming rolling resistance
F_w	force applied in counteracting wind resistance
g	gram
G	gravitational constant
h	hour
ΔH	enthalpy of reaction
Hz	hertz (= 1 cycle per second)
i	current density
i_0	exchange current density
i_a	anodic current density
i_c	cathodic current density
i_L	limiting current density under mass transport control
I	current
I_{ch}	current on charge
I_{corr}	grid corrosion current
I_d	current on discharge

I_g	current associated with decomposition of water
I_{H_2}	hydrogen evolution current
I_{O_2}	oxygen evolution current
$I_{O_2\text{-}red}$	oxygen reduction current
kg	kilogram
kJ	kilojoule (= 238.85 calories)
km	kilometre
kPa	kilopascal (= 9.87 x 10^{-3} atmospheres)
kW	kilowatt
kWh	kilowatt hour
l	litre
m	metre
m	molality (moles per kilogram)
$m_{\pm}$	mean ionic molality
M	molar (moles per cubic decimetre)
M	mass of a vehicle
M_i	molecular weight of an individual species
mA	milliampere
mAh	milliampere hour
min	minute
mm	millimetre
MN	meganewton (= 10^{11} dynes)
mol	mole
mV	millivolt
μm	micrometre
μV	microvolt
n	number of electrons involved in an electrode process
N	newton (= 10^5 dyne)
nm	nanometre
O	oxidized species
p	exponent in Peukert equation
Pa	pascal (= 9.87 x 10^{-6} atmospheres)
P_B	power demand on a battery
P_V	power required to drive a vehicle
R	reduced species

R	gas constant (8.3145 joules per degree per mole)
$\mathcal{R}$	resistance
$\mathcal{R}'$	sum of the internal resistance of a cell plus the equivalent resistances of the activation and concentration overpotentials
$\mathcal{R}_A$	resistance equivalent to activation overpotential
$\mathcal{R}_C$	resistance equivalent to concentration overpotential
rpm	rotations per minute
s	second
S	siemens (= reciprocal ohms)
ΔS	entropy of reaction
t	tonne
t	time
T	temperature
t_a	acceleration time
t_b	braking time
t_{ch}	charge time
t_{co}	coasting time
t_{cr}	cruising time
t_d	discharge time
t_d^*	discharge time to defined capacity
t_i	idle time
v	speed of a vehicle
V	volt
V	voltage
ΔV	shift in voltage
V^o	standard cell voltage
V_{ch}	instantaneous cell voltage on charge
$\overline{V}_{ch}$	average cell voltage on charge
V_d	instantaneous cell voltage on discharge
$\overline{V}_d$	average cell voltage on discharge
V_r	reversible cell voltage
W	watt
Wh	watt hour
wt.%	percentage by weight
x	variable in stoichiometry

X^O	Oxidized organic species
X^R	Organic species
z	charge on ion; the positive or negative sign is specified as a superscript

Symbols and units — Greek letters

α	transfer coefficient
γ	activity coefficient of species which is usually specified as a subscript (H^+, HSO_4^-, etc.)
$\gamma_\pm$	mean activity coefficient
ε_E	electrical efficiency of a drive-train
ε_M	mechanical efficiency of a drive-train
η	overpotential
η_A	activation overpotential
η_C	concentration overpotential
η_+	overpotential at positive electrode
η_-	overpotential at negative electrode
θ	angle of incline
ν_+	number of positive ions formed on dissociation of a compound
ν_-	number of negative ions formed on dissociation of a compound
ρ	air density
Ω	ohm

Glossary of Electric-Vehicle Battery Terms Used in Text

Activation overpotential. The overpotential which results from the restrictions imposed by the kinetics of charge transfer at an electrode.

Active material. The material in the electrodes that takes part in the electrochemical reactions which store–deliver the electrical energy.

Active-material utilization. The fraction, usually expressed as a percentage, of the active material present in a positive or a negative electrode that reacts during discharge before the battery can no longer deliver the required current at a useful voltage.

Ageing. Permanent loss of capacity as a result of battery use and/or the passage of time.

Alkaline battery. A battery which has a strong aqueous alkaline electrolyte.

Ampere-hour efficiency. The ratio, usually expressed as a percentage, of the ampere-hours removed from a battery during a discharge to the ampere-hours required to restore the initial capacity.

Anode. An electrode at which an anodic process is occurring.

Anodic current. The current flow associated with an anodic process.

Anodic process. An electrochemical reaction in which a chemical species in the active material is oxidized and electrons are transferred to the electrode.

Aprotic solvent. A non-aqueous solvent that does not provide hydrogen ions on dissociation.

Aqueous battery. A battery based on an electrolyte dissolved in water.

Automotive battery. A battery designed to provide electrical power for the starting, lighting and ignition of an ICEV.

Auxiliaries. Additional components required to operate and sustain the battery, e.g., heat exchangers, fans, pumps.

Average power. The total energy delivered by a battery divided by the discharge time.

Average voltage. The energy delivered by (or to) a battery divided by the charge withdrawn (or supplied).

Battery. A multiple of electrochemical cells of the same chemistry.

Battery conditioning. The initial application of charge–discharge cycling to establish full battery capacity.

Battery management. Regulation of both charging and discharging conditions, maintenance of battery components, and control of the operating temperature within the appropriate range.

Battery module. A battery unit manufactured as the basic component of a battery pack.

Battery pack. A number of batteries connected together to provide the required power and energy to operate an EV.

Binder. A substance added to the active material of an electrode to enhance mechanical strength.

Bipolar. Battery design in which the component cells are connected through plates that each, in turn, act as the current-collector for the positive electrode in one cell and for the negative in the adjacent cell.

Butler-Volmer equation. The relationship between the current flowing through an electrode and the potential across the electrode/electrolyte interface.

Capacity. The amount of charge, usually expressed in ampere hours, that can be withdrawn from a fully-charged battery under specified conditions; see also **Rated capacity**.

Cathode. An electrode at which a cathodic process is occurring.

Cathodic current. The current flow associated with a cathodic process.

Cathodic process. An electrochemical reaction in which a chemical species in the active material is reduced and electrons are taken from the electrode.

Cell. See **Electrolytic cell**.

Cell reversal. Inversion of the polarity of the electrodes of the weakest cells in a battery, usually as a result of overdischarge.

Charge. The supply of electrical energy to a battery for storage as chemical energy.

Charge-acceptance. The ability of a battery to convert active material during charge into a form that can be subsequently discharged; it is quantified as the ratio, expressed as a percentage, of the charge usefully accepted during a small increment of time to the total charge supplied during that time.

Charge profile. The sequence of current and voltage used to charge a battery.

Charge rate. The current applied to charge a battery to restore its available capacity. For a lead/acid battery, this current is usually expressed in terms of the rated capacity of the battery.

Charge retention. The ability of a battery to retain its charge under zero-current conditions.

Charge-transfer overpotential. Same as **Activation overpotential**.

Concentration overpotential. The overpotential which results from the kinetics of mass transfer of active materials to the electrode surface during the passage of current.

Coulombic efficiency. Same as **Ampere-hour efficiency**.

Coup de fouet. A term used to describe the transient dip in the potential of the positive electrode in a lead/acid battery at the beginning of discharge.

Couple. Combination of the positive and negative electrode materials which engage in electrochemical processes in a battery.

C-rate. The discharge rate or charge rate expressed relative to the rated capacity.

Current-collector. A material included in a battery to conduct electrons to or from the electrodes.

Current density. The current flowing per unit electrode area.

Cut-off voltage. The selected voltage at which charge or discharge is terminated.

Cycle. A single charge–discharge of a battery.

Cycle life. The number of cycles that can be obtained from a battery before it fails to meet selected performance criteria.

Deep discharge. A qualitative term that indicates discharge of a large proportion of the available capacity of a battery.

Depth-of-discharge. The ratio, usually expressed as percentage, of the ampere-hours discharged from a battery at a given rate to the available capacity under the same specified conditions.

Discharge profile. The current–time sequence used in the discharge of the battery.

Discharge rate. The current at which the battery is discharged. The current can be expressed in ampere-hours, but for lead/acid batteries, it is usually normalized to the rated capacity (C), and expressed as C_X/t, where X is the hour rate and t is the specified discharge time, usually in hours.

Double layer. The separation of charges in the electrode and adjacent electrolyte as a result of the potential across the interface.

Double-layer capacity. The charge required to establish the double layer at the electrode/electrolyte interface.

Drive-train. The elements of the propulsion system that produce and transmit mechanical energy to drive the wheels of an EV.

Driving profile. The sequence of acceleration, coasting, braking and stationary operations in driving an EV.

Driving schedule. A simulated driving profile used to test battery and/or vehicle characteristics.

Duty cycle. The driving profile required from an EV in a particular application.

Efficiency. The fraction, usually expressed as a percentage, of the available output from a battery that is achieved in practice.

Electrode. An electronic conductor that acts as a source or sink of electrons which are involved in electrochemical reactions.

Electrolysis cell. An electrolytic cell in which the electrochemical reactions are caused by supplying electrical energy.

Electrolyte. The medium that provides ionic conductivity between the two electrode polarities of a cell.

Electrolytic cell. An electrochemical cell which consists of a positive and a negative electrode and an electrolyte, and which is used to store or generate electricity.

End-of-life. The stage at which a battery can no longer achieve the required performance criteria.

Energy density. The energy output from a battery per unit volume, usually expressed in Wh l^{-1}.

Energy efficiency. The fraction, usually expressed as a percentage, of the energy used in charging the battery that is delivered on discharge.

Equalization. Same as **Equalizing charge**.

Equalizing charge. A charge regime designed to restore undercharged cells in a battery to the fully-charged state.

Equilibrium potential. See **Reversible potential**.

Equilibrium voltage. See **Reversible voltage**.

Exchange-current density. The current per unit area that flows equally in the forward and back directions when an electrode reaction is in equilibrium.

Expander. A compound added to the negative paste of a lead/acid battery to maintain plate porosity during charge–discharge service.

Failure mode. A process that results in a battery failing to meet the required performance criteria.

Float charging. A constant-voltage charge regime applied over extended periods to maintain a battery in the fully-charged state.

Flow battery. A battery system in which the active materials of one or both electrode polarities are stored externally and pumped to the battery during operation.

Formation. The initial charging process, during manufacture or installation, that converts the active materials into the required species for proper electrochemical operation.

Galvanic cell. An electrolytic cell in which chemical energy is converted into electrical energy on demand.

Gassing. The evolution of a gaseous product at one or both electrode polarities.

Gradeability. The maximum percent grade that a vehicle can traverse at a specific speed for a specified time.

Gravimetric energy density. Same as **Specific energy**.

Gravimetric power density. Same as **Specific power**.

Grid. The framework of a battery plate that supports the active material and also serves as the current-collector.

Hour rate. See **Discharge rate**.

Hybrid vehicle. A vehicle that operates on more than one power source.

Internal resistance. The opposition to current flow that results from the various electronic and ionic resistances within the battery.

$I\mathcal{R}$ drop. The decrease in battery voltage that results from current flow through the internal resistance.

Life. The duration over which a battery continues to meet the required performance criteria.

Load. When referred to battery operations, this term relates to the energy-consuming devices to which electric power is delivered.

Maintenance-free. A battery that does not require additions of water during its normal service life.

Mass-transfer overpotential. Same as **Concentration overpotential**.

Mechanical recharging. The recharging of a battery by the physical replacement of discharged active material.

Monobloc. Same as **Battery module**.

Monopolar. The conventional method of battery construction in which the component cells are discrete and are externally connected to each other.

Negative electrode. The electrode in an electrolytic cell that has the lower potential.

Nernst equation. The relationship between the reversible voltage of a cell and the activities (i.e., thermodynamic concentrations) of the electrochemically reacting species.

Nominal capacity. Same as **Rated capacity**.

Ohmic loss. The decrease in the output from a battery that results from the $I\mathcal{R}$ drop.

Open-circuit voltage. The voltage of a battery when there is no net current flow.

Overcharge. The supply of charge to a battery in excess of that required to return all the active materials to the fully-charged state.

Overdischarge. The discharge of a battery beyond the level specified for correct operation.

Overpotential. The shift in the potential of an electrode from its equilibrium value as a result of current flow.

Overvoltage. The shift in the voltage of a cell or battery from its equilibrium value as a result of current flow.

Parallel connection. The connection of like terminals of cells or batteries to form a system of greater capacity, but with the same voltage.

Passivation. The formation of a surface layer that impedes the electrochemical reactions at an electrode.

Paste. An intimate mixture of compounds that is applied to the grid and converted into the electrode active material.

Peak power. The sustained pulsed power that is obtainable from a battery under specified conditions, usually measured in W over 30 s.

Peukert equation. The relationship between the capacity available from a lead/acid battery and the discharge current.

Plate. A common term for battery electrode.

Polarity. Denotes which electrode is positive and which is negative.

Polarization. The shift in the potential of an electrode from its reversible value, or the lowering of the voltage of a cell, as a result of current flow.

Positive electrode. The electrode in an electrolytic cell that has the higher potential.

Post. Same as **Terminal**.

Power density. The power output of a battery per unit volume, usually expressed in W l^{-1} and quoted at 80% depth-of-discharge.

Power-train. The elements of a vehicle propulsion system that include all drive-train components plus an electrical power inverter and/or controller, but not the battery system.

Primary battery. A battery designed to deliver a single discharge.

Ragone plot. A graphical illustration of the specific energy of a battery as a function of its specific power.

Range. The distance that an EV can travel on a single battery discharge over a specified driving profile.

Rated capacity. The capacity of a battery as specified by the particular manufacturer.

Rated power. The power capability of a battery as specified by the particular manufacturer.

Recombination. For lead/acid batteries, this term refers to the reaction of oxygen evolved on charge with the active material of the negative electrode or with co-evolved hydrogen in the presence of a catalyst.

Rectifier. A device for converting alternating current (a.c.) to direct current (d.c.).

Redox battery. A battery in which the chemical energy is stored as dissolved redox reagents.

Reference electrode. An electrode with a reproducible, well-established potential, against which potentials of other electrodes can be measured.

Regenerative braking. The recovery of some fraction of the energy normally dissipated in braking and its return to a battery or some other energy-storage device.

Reversible potential. The potential of an electrode when there is no net current flowing through the cell.

Reversible voltage. The difference in the reversible potentials of the two electrodes that comprise the cell.

Round-trip efficiency. Same as **Energy efficiency**.

Secondary battery. A battery that is capable of repeated charging and discharging.

Self-discharge. The loss of capacity of a battery under open-circuit conditions as a result of internal chemical reactions and/or short-circuits.

Separator. An electronically non-conductive, but ion-permeable, material which prevents electrodes of opposite polarity from making contact.

Series connection. The combination of unlike terminals of cells or batteries to form a battery of greater voltage, but with the same capacity.

Service life. The duration over which a battery continues to meet the required performance criteria for a particular EV service.

Shedding. The loss of active material from battery grids.

Shelf-life. The period over which a battery can be stored and still meet specified performance criteria.

Short-circuit. The direct connection of positive and negative electrodes either internal or external to the battery.

Shunt current. A self-discharge current that occurs through the electrolyte manifold in bipolar batteries because electrodes of the same polarity are at different potentials.

Specific capacity. The capacity output of a battery per unit weight, usually expressed in Ah kg^{-1}.

Specific energy. The energy output of a battery per unit weight, usually expressed in Wh kg^{-1}.

Specific power. The power output of a battery per unit weight, usually expressed in W kg^{-1}.

Standard cell voltage. The reversible voltage of an electrochemical cell with all active materials in their standard states.

Standard electrode potential. The reversible potential of an electrode with all the active materials in their standard states. Note, usual standard states specify unit activity for elements, solids, 1 M solutions, and gases with 1 fugacity (~1 atmosphere).

State-of-charge. The fraction, usually expressed as a percentage, of the full capacity of a battery that is still available for further discharge, i.e., (100 – (% depth-of-discharge)).

String. A number of batteries connected in series.

Tafel equation. The relationship between the current flowing and the overpotential of an electrode.

Terminal. The external connection to a battery electrode.

Theoretical capacity. The charge output of a battery assuming 100% utilization of the active materials.

Theoretical specific energy. The energy output of a battery referred to the weight of only the active materials and a 100% utilization of these materials.

Thermal management. The means by which a battery is maintained within its operating temperature range during charging and discharging.

Thermal runaway. A condition under which an uncontrolled increase in temperature occurs and destroys the battery.

Throughput. The total energy output provided by a battery during its life, i.e., the sum of the energy output during each discharge.

Traction battery. A battery designed to provide motive power.

Transport number. The fraction of the total current flowing in an electrolyte that is carried by a particular ion.

Trickle charge. A low-rate charge sufficient to compensate for self-discharge and maintain a battery in the fully-charged state.

Utilization. Same as **Active-material utilization**.

Vent. A valve that allows the release of gases from a battery but prevents spillage of electrolyte.

Voltage. The difference in potential between the two electrodes of a cell, or the two terminals of a battery.

Voltage delay. Time interval at the start of discharge during which the operating voltage of a cell falls and then recovers (see also **Coup de fouet**). The transient voltage drop is due to the influence of passivating films on the electrode surface.

Voltaic efficiency. The ratio, usually expressed as a percentage, of the average voltage during discharge to the average voltage during charge.

Volumetric energy density. Same as **Energy density.**

Volumetric power density. Same as **Power density**.

Watt-hour efficiency. Same as **Energy efficiency.**

CITATION INDEX

The number in **bold** is the Chapter in which the reference appears.
The following number is the reference itself.
The number in (parenthesis) is the page on which the complete reference is listed.
The number in *italics* is the page on which the reference is first cited.
Note, Editors of journals, books, etc., are not included.

SUBJECT INDEX